ATOMIC COLLISIONS

ATOMIC COLLISIONS

Heavy Particle Projectiles

Earl W. McDaniel

J. B. A. Mitchell

M. Eugene Rudd

A WILEY-INTERSCIENCE PUBLICATION

JOHN WILEY & SONS, INC.

New York / Chichester / Brisbane / Toronto / Singapore

This text is printed on acid-free paper.

Library of Congress Cataloging in Publication Data:

McDaniel, Earl Wadsworth, 1926–
    Atomic collisions : heavy particle projectiles / Earl W. McDaniel,
J. B. A. Mitchell, M. Eugene Rudd.
        p.   cm.
    Includes bibliographical references and index.
    ISBN 0-471-85308-9
    1. Collisions (Nuclear physics) 2. Heavy particles (Nuclear
physics) I. Mitchell, J. B. A. (J. Brian A.) II. Rudd, M. Eugene.
QC794.6.C6M44   1993
539.7′57—dc20   92-26121

Printed in the United States of America

10 9 8 7 6 5 4 3 2 1

To my grandson, Johnny, granddaughter, Faith, and wife, Frances
−E.W.M.

To my wife, Elizabeth, and daughter, Polly
−J.B.A.M.

To my wife, Eileen, my sons, Eric and Leif, and my daughter, Nancy
−M.E.R.

PREFACE

This book deals mainly with elastic, inelastic, and reactive collisions between *heavy particles*, by which we mean atoms, molecules, and ions. Also included are discussions of negative ions, ion and electron swarms, and ion–ion and electron–ion recombination. One mechanism of the latter process, dielectronic electron–ion recombination, was introduced in Chapter 7 of our companion volume, *Atomic Collisions: Electron and Photon Projectiles*, by Earl W. McDaniel (1989), Wiley, New York. The preface to that volume is reproduced in this book because the two volumes are intended to form a unified whole and because much of the preface to the earlier volume also applies here.

Here the energy range of the collisions considered extends from a few meV to a few tens of MeV in a few cases; accordingly, the phenomena of interest are atomic, molecular, or chemical. Not included are high-energy collisions in which nuclear forces are important. Also, as in the companion volume, we place emphasis on collision mechanisms. We do not display large quantities of data, and we rely heavily on review articles and books, and less on individual papers than is usually done. The rapid proliferation of the literature of atomic collisions has necessitated this approach, which is consistent with the intended use of this book as a graduate text as well as a research reference.

In Chapter 3 of *Electron and Photon Projectiles*, a systematic treatment of collision theory was attempted at the intermediate level of sophistication, and practically all of the approximations used in atomic collisions were sketched. This arrangement accounts for the relatively small amount of theory appearing in this volume. However, some of the basic concepts and theory appearing in the companion volume are reproduced here in Chapter 1 to lessen the necessity to refer to the earlier volume.

The two volumes under discussion here were designed to replace *Collision Phenomena in Ionized Gases*, by Earl W. McDaniel (1964), Wiley, New York, and little of the 1964 volume appears here. New or expanded material presented here deals with scattering resonances, nozzle beam expansion sources, beam monochromators, particle detectors, coincidence measurements, merged beam techniques, alignment and orientation parameters, mass spectrometric sampling of ions in ion drift and recombination experiments, laser photodetachment and probing techniques.

The reader is assumed to be familiar with classical mechanics, electromagnetic theory, quantum mechanics, and atomic structure at about the graduate school entrance level. Some of the essential information on atomic structure is collected in Appendix I.

In the present volume, problems are provided at the end of each chapter (166 problems in all). In most cases, one or more sources are cited for a solution to the problem considered.

General areas of application of the phenomena described in this volume are listed in the preface to the 1989 companion volume. Many specific applications are discussed in H. S. W. Massey, E. W. McDaniel, and B. Bederson (Series Editors), *Applied Atomic Collision Physics*, 5 vols. (1982–1984), Academic, New York.

Finally, we wish to thank the experts in atomic collisions who have helped us in the writing of this book. It is a pleasure to acknowledge the contributions of D. L. Albritton, D. R. Bates, R. W. Crompton, K. T. Dolder, M. T. Elford, E. E. Ferguson, A. Fontijn, M. R. Flannery, J. W. Gallagher, W. R. Gentry, J. L. Gole, D. H. Jaecks, Q. C. Kessel, E. J. Mansky, T. M. Miller, R. A. Phaneuf, A. V. Phelps, R. F. Stebbings, L. H. Toburen, J. P. Toennies, A. T. Uzer, B. Van Zyl, and L. A. Viehland.

EARL W. McDANIEL

Atlanta, Georgia

J. B. A. MITCHELL

London, Ontario

M. EUGENE RUDD

Lincoln, Nebraska

PREFACE TO COMPANION VOLUME

In 1964 my book *Collision Phenomena in Ionized Gases* was published by John Wiley & Sons. It purported to cover, in 775 pages, the entire field of atomic collisions on the experimental side, as well as relevant topics in gaseous electronics. Some elementary scattering kinematics and collision theory were included, along with a few more advanced theoretical discussions. *Collision Phenomena* was designed to serve both as a text and as a reference, and it appears to have been widely used as both. However, in the quarter-century since its publication, the field of atomic collisions has matured very rapidly; *Collision Phenomena* is now being replaced by this volume (*Atomic Collisions: Electron and Photon Projectiles*) and another volume (*Atomic Collisions: Heavy Particle Projectiles*) that is in preparation for publication by Wiley several years hence. (By "heavy particles" we mean atoms, molecules, and ions.)

The present volume deals mainly with collisions of electrons and photons with heavy particles, with some discussion of electron-electron and photon-electron collisions. The energy range covered in *Electron and Photon Projectiles* extends from a few meV up to a few MeV (in rare cases). The upper cutoff excludes collisions in which nuclear forces are important, so the discussion here is limited to atomic, molecular, and chemical phenomena. The forthcoming book, *Heavy Particle Projectiles*, will include collisions between heavy particles, negative ions, electron and ion swarm phenomena, recombination, collisions in strong fields, and surface impact. The objectives of these two books taken together are much the same as those of *Collision Phenomena*. However, the field of atomic collisions has now progressed to the point that the phenomena are understood at a much deeper level; hence the heavy emphasis here on the *mechanisms* by which the various kinds of collisions take place, and the inclusion of much more theory than appeared in *Collision Phenomena*.

One part of the additional theoretical discussion that may be of special utility is the enumeration and brief description (in Sections 3-23 and 3-24) of almost all of the scattering approximations used in atomic collision theory. The author gratefully acknowledges the guidance and constructive comments of Professors A. Dalgarno, M. R. Flannery, and T. Uzer in the preparation of this new material. These experts are responsible for whatever success I have had in avoiding the rocks and shoals of this vast and difficult subject.

Another difference between the present book and *Collision Phenomena* relates to the inclusion of data. In the earlier book it was possible to present a large fraction of the available experimental data considered reliable at the time, but this is no longer possible. One must now rely on the many extensive data indices and compilations to access the enormous quantity of cross sections and reaction rate coefficients that are on hand. Here, I include, for the most part, only samples of data that are representative, or that illustrate trends, or that illuminate points of special interest. Similarly, the present availability of many review articles and books on various detailed aspects of atomic collisions has led me to rely heavily on these sources and not to attempt a more exhaustive coverage of the primary literature.

An effort has been made to stay abreast of current developments during the preparation of this book. Topics included here, but not in *Collision Phenomena*, include scattering resonances, coincidence measurements, merged-beam experiments, positron collisions, collisions between spin-polarized particles, GaAs polarized electron sources, alignment and orientation parameters, "complete" experiments, position-sensitive detection, synchrotron radiation sources, free electron lasers, cyclotron resonance masers, laser cooling and trapping, electron beam cooling, and multiphoton processes.

This book contains more than 200 problems, at the ends of the chapters, that deal with scattering kinematics, collision theory, radiation theory, charged particle optics, spin dynamics, optics, and experimental techniques. In most cases, one or more sources are cited for a solution to the problem at hand. Also included at the end of each chapter is a set of Notes designed to amplify points in the text or to cover some peripheral topics.

I have attempted to make this book as nearly self-contained as possible. I believe that the reader who knows quantum mechanics and special relativity at about the level of graduate school entry can effectively use this book without having to resort to standard physics texts to an excessive degree.

Actually, I hope that *Electron and Photon Projectiles* may prove as useful to chemists and engineers as *Collision Phenomena* apparently did. Atomic collisions impinges on many other fields, such as astrophysics, planetary atmospheres, controlled fusion, laser development, isotope separation, magnetogasdynamics, high-velocity projectiles and vehicles, surface physics, chemical kinetics, laser spectroscopy, plasma dynamics, electrical discharges, gaseous electronics, and gas dynamics. Most of this book is accessible to chemists and engineers working in these fields even if they have not had the full theoretical background described above.

I am indebted to many people for their help during the planning and writing of this book. Sir David Bates and the late Sir Harrie Massey played a special role in encouraging me to undertake this task. Helpful criticism of portions of the manuscript were given by C. Bottcher, R. N. Compton, S. Datz, H. Ehrhardt, D.

C. Gregory, R. Johnsen, J. Kessler, W. C. Lineberger, M. S. Lubell, E. J. Mansky, S. T. Manson, J. T. Moseley, W. R. Newell, W. Raith, F. H. Read, A. C. H. Smith, B. Van Zyl, R. N. Zare, and J. C. Weisheit. It is also a pleasure to acknowledge productive discussions with other experts: J. W. Gallagher, T. C. Griffith, D. W. O. Heddle, J. W. Humberston, W. E. Kauppila, M. H. Kelley, J. W. McConkey, M. A. Morrison, D. T. Pierce, R. Roy, and T. S. Stein. I wish to thank Ms. Audrey Ralston for expertly typing the manuscript, and Prof. E. W. Thomas and Dean Les Karlovitz for their support. Finally, sincere thanks to my family, especially my wife, Frances, for their help, which was so important during the years occupied by the writing of this book.

EARL W. McDANIEL

Atlanta, Georgia

CONTENTS

1

ELASTIC SCATTERING OF HEAVY PARTICLES

1-1. INTRODUCTION

As defined in Section 1-3 of the companion volume, (McDaniel, 1989), an *elastic collision* is one in which there is no permanent change in the internal excitation energy of either of the collision partners. The total kinetic energy of the system of colliding particles is the same before and after the collision, although it is different "during" the impact. Linear and angular momentum are conserved at all times throughout the event. By contrast, in an *inelastic collision*, linear and angular momentum are still conserved, but the total kinetic energy of the system decreases (or increases) as the result of the excitation (or deexcitation) of one or both of the particles. *Reactive collisions* involve a transfer, association, or rearrangement of atoms in the colliding structures. Elastic collisions of electrons with heavy particles are treated in Chapters 4 and 7 of the companion volume, and examples of elastic collisions of photons are provided in Chapter 8. The present chapter is concerned with elastic collisions between heavy particles, or to use alternative terminology, with the elastic scattering of heavy particles.

There are several incentives for studying the elastic scattering of atoms, molecules, and ions by other heavy particles. Perhaps the strongest comes from the possibility of deducing information on the interaction potential, which determines the mutual scattering behavior of the collision partners at a given impact energy (Levine and Bernstein, 1987; Scoles, 1988). It is also important that differential cross sections for elastic scattering be available to experimentalists so that they may account for the effects of scattering in various types of experiments involving beams of heavy particles. As we shall see in Chapters 7 and 9, the mobility of gaseous ions and the diffusion of both neutrals and ions through gases are determined mainly by elastic collisions (Mason and McDaniel, 1988), which are also important in the process of ion–ion recombination (Flannery and Mansky, 1988). Finally, there are applications in nature and in the laboratory (Massey et al., 1982–1984). For example, in hot plasmas,

1

fast ions lose energy and are scattered by Coulomb collisions with thermal ions and electrons (Cordey, 1984). Also, heavy particles that precipitate into the earth's atmosphere during geomagnetic storms undergo elastic scattering and charge transfer, and produce local heating, optical emission, ionization, and loss of atmospheric ions (Massey, 1982; Newman et al., 1985; Schafer et al., 1987). Elastic collisions of ions in the earth's atmosphere play a key role in the propagation of ELF (extremely low frequency; 3 Hz to 3 kHz) and VLF (very low frequency; 3 to 30 kHz) electromagnetic radiation. This application, and others, are discussed by Junker (1982) and McDaniel and Viehland (1984).

We must now reiterate the basic differences between the elastic scattering of electrons and heavy particles that were discussed at the beginning of Chapter 4 in the companion volume. A simple calculation based on the crude elastic sphere scattering model indicates that the mean fractional amount of energy lost by projectiles in elastic scattering with molecules initially at rest is*

$$\Delta = \frac{2mM}{(m + M)^2}, \qquad [1\text{-}5\text{-}8]$$

where m and M are the projectile and target masses, respectively. For electron projectiles, [1-5-8] becomes

$$\Delta \approx \frac{2m}{M} \leqslant 1.09 \times 10^{-3}, \qquad [1\text{-}5\text{-}21]$$

and accordingly, the energy loss experienced by an elastically scattered electron is usually not important. The elastic sphere calculation also predicts that the electrons are scattered isotropically, to a very high approximation, in the laboratory frame. Indeed, it is usually the wide-angle deflections in the electron collisions that are important. In elastic collisions of heavy particles, the situation is reversed. Equation [1-5-8] indicates that large fractional energy losses may occur, and the analysis predicts that the scattering is predominantly at small angles about the forward direction in the Lab frame. More realistic calculations corroborate these results and show that heavy particle elastic scattering becomes more strongly concentrated in the forward direction as the projectile energy is increased (see Section 1-9).

The considerations discussed above were influential in the organization of the present chapter, which we now outline. Part $\mathscr{A}$ deals with general aspects of heavy particle elastic scattering. We then split the remainder of the discussion into two parts according to the kinetic energy of relative motion of the collision partners. Part $\mathscr{B}$ covers the energy region below a few tenths of an electron volt, and Part $\mathscr{C}$ the region above this boundary. Admittedly, this division is arbitrary

---

*Equation numbers in brackets indicate equations that appeared in the companion volume and are reproduced in the present volume.

and procrustean, but it does make some sense and it is useful, provided that the authors are allowed to perform a little gerrymandering from time to time. Also, largely for convenience, ion–molecule collisions will generally be split off from neutral–neutral collisions; the interaction potentials are quite different, and many of the experimental techniques differ in important respects.

PART $\mathscr{A}$. GENERAL CONSIDERATIONS; THEORY

1-2. KINEMATICS OF ELASTIC SCATTERING; CLASSICAL CALCULATION OF CROSS SECTIONS

At this point it is appropriate to collect some results from our earlier discussion. These results, which are presented without proof here, will be useful in the present chapter and in later work. The reader is referred to Chapters 1 and 3 of the companion volume for derivations and full explanations.

A. Scattering Kinematics

Consider first the simple situation in which a projectile of mass m is elastically scattered by a target of mass M that is initially at rest in the Lab frame (the reference frame of the laboratory observer). Any structure possessed by the two particles is ignored here, and they are assumed to interact through a *central force* $\mathbf{f}(r)$ that is a function only of the instantaneous separation distance r (Problems 1-1 and 3-1 in the companion volume). Hence we can describe the interaction between m and M in terms of the potential energy function $V(r)$. The velocity of approach of the projectile is taken to be v_0 when it is at a great distance from the target, and we assume that $v_0 \ll c$, the speed of light in a vacuum, so that the collision will be nonrelativistic. Figure 1-2-1 shows asymptotic views of the event in the Lab frame, that is, the event as seen (*a*) long before, and then (*b*) long after, the period during which the particles interact. The impact parameter is b, and V_{CM} is the velocity of the center of mass in the Lab frame. The Lab scattering angle is ϑ, and the recoil angle of the target is θ. In the asymptotic region following the collision, the scattered projectile is receding from the collision site with fixed speed v, while the target recoils with fixed speed V.

Figure 1-2-2 shows the same event in the center-of-mass (CM) reference frame. We locate our coordinate system in this frame in such a way that its origin coincides with the CM of the colliding particles. The common angle of scattering in the CM system is Θ, and each particle has the same speed after the collision as it had initially. We shall find it convenient to use the *reduced mass* of m and M, defined as

$$M_r = \frac{mM}{m + M}.$$ [1-2-3]

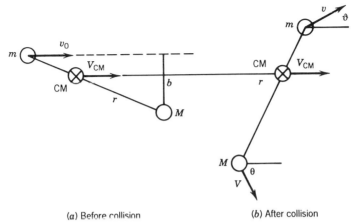

(a) Before collision   (b) After collision

Figure 1-2-1. Asymptotic views of an elastic collision in the Lab frame, in which the target M is initially at rest. In each view the distance between the particles r is much greater than the range of the interaction between the particles. The initial velocity of the projectile is v_0; it is scattered through the angle ϑ in the Lab system. The target M recoils through the Lab angle θ.

In Figs. 1-2-1 and 1-2-2 we are not concerned with the detailed nature of the interaction or with what takes place "during" the collision, but rather, with the scattering angles and the ultimate changes in the speeds. Application of energy and momentum conservation allows us to find the final velocities, v and V, in

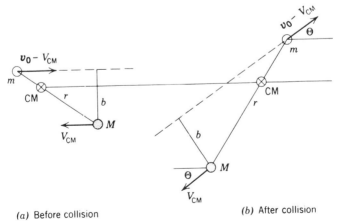

(a) Before collision   (b) After collision

Figure 1-2-2. Asymptotic views of the elastic collision of Fig. 1-2-1, now shown in the center-of-mass frame. Here both particles are scattered through the common CM angle Θ, so that the distinction between projectile and target vanishes.

terms of the initial velocity, v_0, and the scattering angles:

$$\left(\frac{v}{v_0}\right)^2 - 2\left(\frac{v}{v_0}\right)\left(\frac{m}{m+M}\right)\cos\vartheta - \left(\frac{M-m}{m+M}\right) = 0, \qquad [1\text{-}2\text{-}7]$$

$$v^2 = v_0^2\left(1 - \frac{4M_r^2}{mM}\cos^2\theta\right), \qquad [1\text{-}2\text{-}8]$$

$$V = 2v_0\frac{M_r}{M}\cos\theta. \qquad [1\text{-}2\text{-}9]$$

We may also express the total kinetic energy of the particles in the CM frame as

$$T_{CM} = \tfrac{1}{2}M_r v_0^2, \qquad [1\text{-}2\text{-}12]$$

and the total angular momentum of the pair about the center of mass as

$$J_{CM} = M_r v_0 b. \qquad [1\text{-}2\text{-}14]$$

In a similar fashion, we obtain relationships between the scattering angles:

$$\theta = \tfrac{1}{2}(\pi - \Theta), \qquad [1\text{-}2\text{-}18]$$

$$\tan\vartheta = \frac{\sin\Theta}{\gamma + \cos\Theta}, \qquad [1\text{-}2\text{-}20]$$

where

$$\gamma = \frac{m}{M} = \frac{V_{CM}}{v_0 - V_{CM}} \qquad [1\text{-}2\text{-}21]$$

and

$$v\sin\vartheta = v_0\frac{M_r}{m}\sin\Theta. \qquad [1\text{-}2\text{-}22]$$

Finally, we note three useful relationships between ϑ and Θ for three special values of the ratio m/M.

$$\text{If} \quad m \ll M, \quad \vartheta \approx \Theta. \qquad [1\text{-}2\text{-}23]$$

The Lab and CM systems are almost identical; ϑ increases monotonically from 0 to π as Θ increases from 0 to π.

$$\text{If} \quad m = M, \quad \vartheta = \frac{\Theta}{2}; \qquad [1\text{-}2\text{-}24]$$

ϑ varies from 0 to $\pi/2$ as Θ varies from 0 to π. No particles are backscattered in the Lab system. Head-on scattering corresponds to $\Theta = \pi$, $\vartheta = \pi/2$, $\theta = 0$.

$$\text{If} \quad m \gg M, \quad \vartheta \approx \frac{M}{m} \sin \Theta. \qquad \text{[1-2-25]}$$

When $m > M$, ϑ first increases from 0 to a maximum value $\vartheta_{max} = \sin^{-1}(M/m)$, which is less than $\pi/2$, as Θ increases from 0 to

$$\Theta = \cos^{-1}\left(-\frac{M}{m}\right);$$

ϑ then decreases to 0 as Θ increases further to π, so that Θ is a double-valued function of ϑ. No particles are scattered beyond ϑ_{max} in the Lab system. We may distinguish between the two values of Θ that give a particular value of ϑ between 0 and $\sin^{-1}(M/m)$ by comparing the energies of m after scattering in the two cases—the energy is greater for the smaller Θ.

The geometrical relationships we have presented apply in a quantum mechanical collision as well as in the classical case. The reason for this is that the equations are essentially relationships of momentum vectors and are applied only in the asymptotic region, remote from the collision site, where the particle positions do not have to be precisely known and where consequently their momenta can be precisely determined.

The results displayed above apply only to nonrelativistic elastic collisions, in connection with which it may be useful to consult Problems 1-3 and 1-18 of the companion volume. (All subsequent cross-references in this paragraph are also to the companion volume.) References to relativistic kinematics are provided in Section 1-3, and Problems 1-15, 1-16, and 1-17 are recommended. Inelastic collisions are treated in Section 1-3 and in Problems 1-5 and 1-8. The kinematics of crossed, merged, inclined, and colliding beams are discussed in Note 1-1, Section 7-1-C, and Problems 1-4, 1-6, 1-7, 7-5, 7-6, and 7-7. Problems 1-11, 1-12, 1-13, 7-8, and 7-9 deal with the kinematics of spontaneous dissociation. Reactive collisions are the subject of Problems 1-2 and 1-9. We deal at much greater length with these subjects later in the present volume.

B. Classical Calculation of Scattering Angles and Cross Sections

Figure 1-2-3 shows the CM trajectories of particles m and M undergoing mutual elastic scattering under the influence of a central interaction potential $V(r)$ consisting of an attractive component and a shorter-range repulsion. The motion occurs in a plane perpendicular to the total angular momentum. As shown in Section 3-2 of the companion volume, this two-body problem may be reduced to a one-body problem, described by Fig. 1-2-4, wherein a single hypothetical particle of mass M_r is considered to be scattered by a center of force

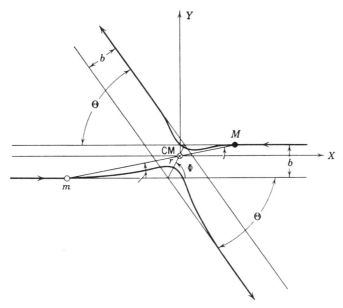

Figure 1-2-3. Mutual elastic scattering of particles m and M as viewed in the CM frame. The motion is two-dimensional and shown in the plane of the paper. An interaction potential $V(r)$ consisting of an attractive component and a shorter-range repulsion is assumed. The angle of closest approach is labeled Φ. The CM trajectories are symmetrical about the line of closest approach.

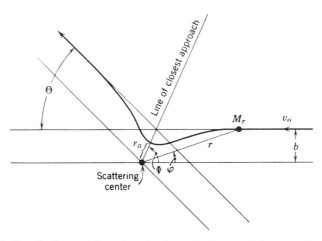

Figure 1-2-4. One-body problem that is dynamically equivalent to the two-body problem illustrated in Fig. 1-2-3. The distance of closest approach is r_a, and Φ is the corresponding angle of the line separating the particles, measured with respect to the X axis shown in Fig. 1-2-3.

fixed in the CM frame. The potential describing this force field is $V(r)$, where r is the separation distance between m and M. The CM velocity of M_r is considered to be equal to v_0, where v_0 is the initial relative velocity of m and M. The hypothetical particle M_r is moving toward the stationary scattering center with impact parameter b. The instantaneous distance of M_r from the scattering center is r, and M_r is scattered through the angle Θ. This angle is the scattering angle of both m and M in the CM frame. The azimuthal angle φ plays no role in the problem analyzed here, nor in any other central force problem.

At the angle of closest approach Φ we have

$$1 - \frac{V(r_a)}{M_r v_0^2/2} - \frac{b^2}{r_a^2} = 0. \qquad [3\text{-}4\text{-}3]$$

The largest real root of this equation is the distance of closest approach r_a.

We should note that a real solution of [3-4-3] does not always exist. If we assume an attractive potential of the form $V(r) \sim -r^{-n}$ and further assume that $n \geqslant 2$, there are values of the initial velocity and impact parameter for which no solution exists (see Section 3-6 of the companion volume). The particle spirals inward until it is stopped by some repulsive force and then spirals back out. A solution always exists for a repulsive potential, however.

If a solution exists,

$$\Phi = -\int_\infty^{r_a} \frac{d\varphi}{dr}\, dr = -\int_\infty^{r_a} \frac{(b/r^2)\, dr}{\left[1 - \dfrac{V(r)}{M_r v_0^2/2} - \dfrac{b^2}{r^2} \right]^{1/2}}. \qquad [3\text{-}4\text{-}4]$$

Equation (3-4-1) of the companion volume then shows Θ to be

$$\Theta(v_0, b) = \pi - 2b \int_{r_a}^\infty \frac{dr/r^2}{\left[1 - \dfrac{V(r)}{M_r v_0^2/2} - \dfrac{b^2}{r^2} \right]^{1/2}}. \qquad [3\text{-}4\text{-}5]$$

Let us now calculate the differential cross section $I_s(\Theta)\, d\Omega_{\mathrm{CM}}$ for scattering into the element $d\Theta$ between the angles Θ and $\Theta + d\Theta$. The necessary and sufficient condition for scattering particles of initial relative velocity v_0 into the element of solid angle $d\Omega_{\mathrm{CM}} = 2\pi \sin\Theta\, d\Theta$ is that the particles be incident in the annular impact ring of area $2\pi b\, db$ formed by circles of radius b and $b + db$, where b is related to Θ by [3-4-5]. Hence

$$|2\pi b\, db| = |I_s(\Theta)\, 2\pi \sin\Theta\, d\Theta|, \qquad [3\text{-}7\text{-}1]$$

since the differential cross section is the effective area presented by the scattering

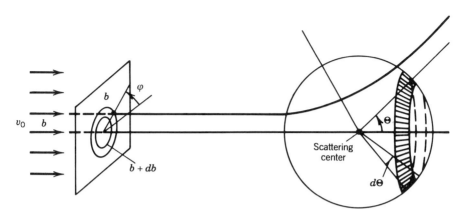

Figure 1-2-5. Relation between the impact parameter b and the CM scattering angle Θ for the one-particle equivalent problem of a purely repulsive Coulomb potential $V_{\text{coul}}(r)$. The outer circle in the plane at the left corresponds to the values b and Θ, the inner circle to the values $b - |db|$ and $\Theta + |d\Theta|$. [Adapted from Goldstein (1980).]

center for deflection of projectiles through angles between Θ and $\Theta + d\Theta$ (see Fig. 1-2-5). Finally, we have

$$I_s(\Theta)\,d\Omega_{\text{CM}} = \left| \frac{b}{\sin\Theta}\frac{db}{d\Theta} \right| d\Omega_{\text{CM}}. \qquad [3\text{-}7\text{-}2]$$

The differential cross sections in the Lab and CM systems are equal, that is,

$$I_s(\vartheta, \varphi)\,d\Omega_{\text{Lab}} = I_s(\Theta, \varphi)\,d\Omega_{\text{CM}}, \qquad [1\text{-}4\text{-}15]$$

because the same number of projectiles must be scattered into the Lab element of solid angle, $d\Omega_{\text{Lab}}$, as are scattered into the corresponding CM element of solid angle, $d\Omega_{\text{CM}}$. It then follows that

$$I_s(\vartheta, \varphi) = \frac{[(m/M)^2 + 2(m/M)\cos\Theta + 1]^{3/2}}{1 + (m/M)\cos\Theta} I_s(\Theta, \varphi) \qquad [1\text{-}4\text{-}16]$$

in the present case of a stationary target and scattering independent of the azimuthal angle φ.

The integral elastic scattering cross section is obtained by integrating the differential cross section over the complete solid angle:

$$q_s = \int_0^{2\pi}\int_0^{\pi} I_s(\Theta)\,d\Omega_{\text{CM}} = \int_0^{2\pi}\int_0^{\pi} I_s(\vartheta)\,d\Omega_{\text{Lab}}. \qquad [3\text{-}7\text{-}3]$$

These integrals diverge for potentials of infinite range, such as $V(r) \sim \pm r^{-n}$, because of a pole in the differential cross section in the forward direction, $\Theta = 0$. For potentials of this type, some deflection of the interacting particles occurs, no matter how large the impact parameter may be. Thus the effective cross-sectional area of the particles appears to be infinite. This paradox is resolved in Section 3-11 of the companion volume by quantum mechanical considerations.

It can be seen from [3-7-2] that $I_s(\Theta)$ will become infinite if $\Theta = n\pi \, (n = 0, 1, 2, \ldots)$ for $b \neq 0$ because of the $\sin \Theta$ term in the denominator; it will also become infinite if $d\Theta/db = 0$. Both are possible for potentials that are attractive at large distances and repulsive at small distances. Both classical scattering phenomena have optical analogs that become apparent in the quantum theory of scattering when the wave nature of the particles is explicitly introduced. The first effect is called a *glory* and the second, a *rainbow*. These effects are treated by Ford and Wheeler (1959a, b), Child (1974), Pauly (1979), Bernstein (1982), Bransden (1983), and Levine and Bernstein (1987); we return to them in Part ℬ and in Section 2-9-A. Another useful quantity is the *diffusion cross section*:

$$q_D = \int (1 - \cos \Theta) I_s(\Theta) \, d\Omega_{CM} = 2\pi \int_0^\pi I_s(\Theta)(1 - \cos \Theta) \sin \Theta \, d\Theta. \qquad [1\text{-}6\text{-}1]$$

It is apparent that q_D is a measure of the average forward momentum lost by the particles m in collisions with the molecules M, and q_D is therefore often called the *momentum transfer cross section*.

Scattering at small angles produces only a small contribution to q_D. The quantity $(1 - \cos \Theta)$ vanishes as Θ^2 for small deflections, and for this reason the diffusion cross section often has a finite value when the integral elastic cross section is (mathematically) infinite. The matter of theoretical predictions of infinitely large cross sections is discussed in Section 3-11 of the companion volume.

The diffusion cross section q_D differs appreciably from q_s, the integral elastic scattering cross section, only when the scattering is distinctly anisotropic. When the backward scattering dominates the forward scattering, $q_D > q_s$, whereas $q_D < q_s$ when the scattering is concentrated in the forward direction. If the differential elastic scattering cross section is independent of Θ, $q_D = q_s$.

The diffusion cross section arises naturally in the theory of diffusion. A similar weighted cross section

$$q_\eta = \int \sin^2 \Theta \, I_s(\Theta) \, d\Omega_{CM} = 2\pi \int_0^\pi I_s(\Theta) \sin^3 \Theta \, d\Theta \qquad [1\text{-}6\text{-}6]$$

appears in the theory of viscosity; it is appropriately named the *viscosity cross section*.

C. Notation

Several alternative sets of notation are used to represent cross sections, two of the most frequently employed being the following.

Angular differential elastic scattering cross section:

$$I_s(\vartheta, \varphi)\, d\Omega \equiv \frac{d\sigma(\vartheta, \varphi)}{d\Omega}\, d\Omega \quad (\text{see } [3\text{-}7\text{-}2]).$$

Integral elastic scattering cross section: $q_s \equiv \sigma_s$ (see [3-7-3]).

Total cross section: $q_t \equiv \sigma_t$. As pointed out in the introduction to Chapter 4 of the companion volume, q_t is the cross section for the totality of all possible kinds of collisions, that is, the sum of all (angle-integrated) cross sections for the various scattering channels, elastic and otherwise.

Momentum transfer (diffusion) cross section: $q_m \equiv \sigma_m \equiv q_D \equiv \sigma_D$ (see [1-6-1]).

We shall use the various notations interchangably and introduce minor variations as required to denote other cross sections that will appear later.

1-3. POTENTIAL ENERGY CURVES AND SURFACES

In Section 1-7 of the companion volume we discussed a variety of potential energy functions used to describe the interaction between particles. Emphasis was placed on spherically symmetric functions $V(r)$ that depend only on the separation distance r (Hirschfelder et al., 1964; Pauly, 1979; Maitland et al., 1987.) A typical *potential energy curve* for an *atom–atom* system is shown in Fig. 1-3-1, and under such an interaction, the scattering may be described by the analysis outlined in Section 1-2-B.

If, on the other hand, the interaction between the collision partners depends on the relative orientation of the particles, as indicated in Fig. 1-3-2, a *potential energy surface* of higher dimensionality is required in the calculation of elastic scattering cross sections (Stolte and Reuss, 1979). Such a surface, showing the anisotropy of the LiH–He interaction, is illustrated in Fig. 1-3-3. Scattering under a fixed potential that depends only on the relative coordinates of the collision partners is called *potential scattering*. Surfaces of still higher dimensionality are needed in the analysis of inelastic and reactive collisions, where internal coordinates of the colliding structures play a role (Schaefer, 1979; Kuntz, 1979, 1985; Dunning and Harding, 1985).

Interaction potentials are important not only in the calculation of elastic, inelastic, and reactive cross sections, but also in providing the basis of all theories of the equation of state for gases, liquids, and solids. In addition, they

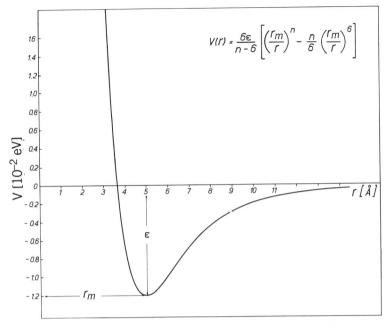

$$V(r) = \frac{6\varepsilon}{n-6}\left[\left(\frac{r_m}{r}\right)^n - \frac{n}{6}\left(\frac{r_m}{r}\right)^6\right]$$

Figure 1-3-1. Interaction potential for the Na–Xe system, which has been carefully studied by the molecular beam method. The curve shown here is based on the Lennard-Jones $(n, 6)$ model potential described in the inset, with the repulsive parameter n assigned the value 8. The values for the well depth ε and the equilibrium spacing r_m are derived from experiment. [From Pauly and Toennies (1968).]

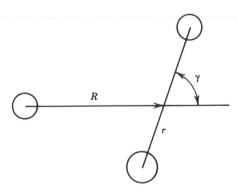

Figure 1-3-2. System whose interaction potential depends on the relative orientation of the collision partners. Here a spherically symmetric structure, say a rare-gas atom (*left*), interacts with a rigid diatomic molecule (*right*) whose orientation is specified by the angle γ. The bond distance in the molecule is denoted by r.

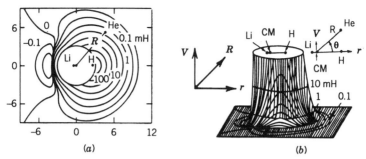

Figure 1-3-3. Potential energy surface for the LiH–He system, derived by *ab initio* calculation. (*a*) Equipotential contours for the interaction of a He atom with a rigid LiH molecule. Here the Li–H spacing is taken to be the equilibrium spacing for the $v = 0$ state. The unit of distance here is the bohr (1 bohr = 0.529 Å), and the unit of energy the millihartree (1 mh = 0.0272 eV). (*b*) Three-dimensional view of the same surface truncated at 0.5 eV, the contour levels being the same. [Adapted from Silver (1980).]

play a fundamental role in nonequilibrium statistical mechanics and underlie the calculation of transport coefficients and relaxation times for the deexcitation of molecules (Pauly and Toennies, 1968).

The main categories of methods used to determine interaction potentials are (1) theoretical calculations, (2) spectroscopic observations, (3) measurements of bulk properties, and (4) beam scattering (Mason, 1982). A few comments will now be made on each of these classes of methods.

A.   Theoretical Calculations

The recent explosive advances in computer technology and computational methods have led to accurate *ab initio* calculations of interaction potential curves and surfaces for many systems (Lawley, 1987). In less tractable cases, accurate experimental information is often introduced into the theoretical calculations to produce *semiempirical results* (Kuntz, 1979; Levine and Bernstein, 1987). In another kind of calculation, simplification is achieved by replacing the real problem by a *model problem* that possesses the main features of the original one. For discussion of the *electron-gas model*, see Waldman and Gordon (1979), and for the *damped-dispersion model*, see Tang and Toennies (1984, 1986, 1988); Ahlrichs et al. (1988), and Bowers et al. (1988). Scoles (1980) has provided a useful review of model calculations.

B.   Spectroscopic Observations

Noteworthy here is the determination of potential well shapes from observations of molecular bands of bound diatomics and the spacings of the rovibrational levels. Other techniques include inferences from pressure broadening, predissociation, continuum intensity distributions, and photodissociative processes. [See Section 8-15-B of the companion volume for a discussion of photofragment spectroscopy; see also Scoles (1991).]

C. Measurements of Bulk Properties

Here we refer mainly to the determination of interaction potentials from the temperature dependence of microscopic properties, especially second virial coefficients and low-density transport coefficients (Massey, 1971; Mason, 1982). Also, developments in the theory of gaseous ion transport that began around 1975 have led to techniques for extracting ion–neutral interaction potentials from measurements of the mobility of gaseous ions as a function of the ionic energy parameter E/N. Here E is the intensity of the electric field applied to the ions, and N is the gas number density (Mason and McDaniel, 1988). This application is discussed in Chapter 7.

D. Beam Measurements

The beam scattering technique is the most powerful and versatile of all the experimental techniques applied to the determination of interaction potentials, and in addition, it provides cross sections for various applications and for checks on collision theory (Mason and Vanderslice, 1962; Pauly and Toennies, 1965; Massey, 1971; Massey and Gilbody, 1974; Lawley, 1975; Scoles, 1988). Accurate measurements of angular scattering distributions and differential cross sections can be made at impact energies ranging from a few tenths of an electron volt up into the realm of nuclear phenomena. Hence information can be obtained on the interaction potential at very large values of the separation distance r and well into the repulsive region at very small r, as well as at intermediate separation distances. A high degree of energy selection is possible for the projectiles, so the incident beam can be made almost monoenergetic in many experiments. Good angular resolution is also possible for the scattered projectiles. One may work with pure beams of selected species of neutrals, positive ions, or negative ions, and techniques developed in the 1980s exist for generating homogeneous beams of positive ions that are highly charged (see Chapter 7 of the companion volume). In many cases it is also possible to work with projectiles in known excited states. Isotopic variation within either of the collision partners is also possible, and angular momentum state selection is now used routinely. The averaging over many variables that is required in the bulk measurements may be avoided in beam measurements, and much more detail can thus be observed. However, one must realize that the interaction potential determines the scattering behavior, and not vice versa, so extracting the potential from the scattering data involves an *inversion problem* and the concomitant concern about uniqueness of the solution (Wheeler, 1976). The same difficulty applies to the spectroscopic and bulk measurements, but substantial progress has also been made there (Mason, 1982; Mason and McDaniel, 1988).

Elastic beam scattering measurements are the main subject of this chapter; Part $\mathscr{B}$ will deal with techniques and data for low-energy collisions, and Part $\mathscr{C}$ will treat the high-energy regime. Additional material on interaction potentials appears in Section 1-7-D-3 and Note 1-1.

1-4. CENTRIFUGAL POTENTIAL; ONE-DIMENSIONAL RADIAL MOTION

A potential energy function that is completely different in origin from those discussed in Section 1-3 also proves useful in the analysis of collision problems. Here we refer to the *centrifugal potential energy*, which arises not from the mutual interaction of particles, but rather from a purely kinematic effect associated with the rotation of the collision partners about their center of mass. To see how this potential arises naturally in the discussion of the motion, we can start with the equations of motion in the CM system and obtain

$$\tfrac{1}{2}M_r v_0^2 = \tfrac{1}{2}M_r \dot{r}^2 + V_{\text{eff}}(r), \qquad\qquad [3\text{-}3\text{-}1]$$

where $V_{\text{eff}}(r)$ is the *effective potential*, defined by

$$V_{\text{eff}}(r) = V(r) + \frac{M_r v_0^2 b^2}{2r^2}. \qquad\qquad [3\text{-}3\text{-}2]$$

Equation [3-3-1] contains no terms in φ. It can be regarded as describing the one-dimensional motion (along the r axis) of a particle of mass M_r with total energy $M_r v_0^2/2$ in an effective potential field described by $V_{\text{eff}}(r)$. The second term on the right in the expression for $V_{\text{eff}}(r)$ is equal to $J^2/2M_r r^2$ and represents the rotational kinetic energy of the system. Note that this term is a positive monotonic decreasing function of r. It can therefore be considered the source of a fictitious outwardly directed force, the *centrifugal force*, and for this reason the second term on the right side of [3-3-2] is termed the *centrifugal potential*. We use the symbol V_c to denote this quantity.

1-5. CLASSIFICATION OF THE ORBITS IN THE CENTRAL FORCE PROBLEM

In Section 1-2-B it was stated that a solution of [3-4-3] always exists for a repulsive potential, but not always for an attractive potential of the form $V(r) \sim -r^{-n}$, if $n \geqslant 2$. This statement is easily understood in terms of the effective one-dimensional motion and the centrifugal potential.

First let us consider motion in a typical repulsive potential field such as that described by the potential function $V(r)$ in Fig. 1-5-1. It has already been pointed

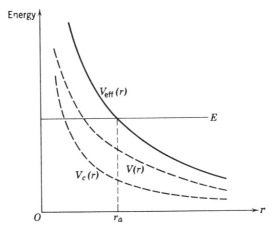

Figure 1-5-1. Potential functions used for analysis of elastic scattering in a typical repulsive potential field.

out that if we wish to analyze the scattering problem in the single dimension r we must add to the real potential $V(r)$ a fictitious centrifugal potential $V_c(r)$, given in terms of the angular momentum J of the system by the equation

$$V_c(r) = \frac{J^2}{2M_r r^2}.$$ [3-6-1]

This potential is plotted in Fig. 1-5-1 for some arbitrary but nonzero value of J. The total effective potential, $V_{\text{eff}}(r) = V(r) + V_c(r)$, is also shown in the figure. It is apparent that in this case the only significant effect of the centrifugal potential will be to supplement the real potential. If $V(r)$ decreases monotonically with increasing r, the effective potential will also, regardless of the value of the angular momentum. Hence in this one-dimensional formulation of the scattering problem a particle of total energy E will approach the scattering center and be reflected at the distance of closest approach, r_a, given by the intersection of the horizontal line at height E and the curve $V_{\text{eff}}(r)$. The radial velocity $\dot{r}$ at any point is given by the equation

$$E = \tfrac{1}{2}M_r \dot{r}^2 + V_{\text{eff}}(r),$$ [3-6-2]

while the angular motion must, of course, be such that the angular momentum is conserved throughout the scattering.

The shape of the effective potential curve is quite different if an attractive potential is assumed. However, if $V(r) \sim -r^{-n}$ with $n < 2$, the effective potential again decreases monotonically with increasing r, at sufficiently small r, and the one-dimensional motion is similar to that in the repulsive case in regard to reflection along the r axis at some finite distance of closest approach. (This

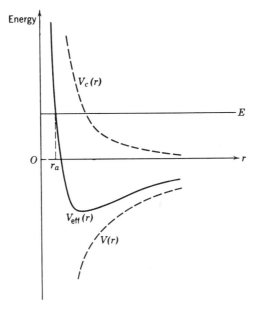

Figure 1-5-2. Potential functions used for analysis of elastic scattering in an attractive inverse-first-power potential field.

situation is illustrated in Fig. 1-5-2, in which an attractive inverse-first-power potential is assumed.) Stable orbits* are possible only for $n < 2$, and for values of E lying in the well.

If, on the other hand, the attractive potential falls off with increasing r faster than r^{-2}, the attractive potential must dominate the centrifugal potential at small r, and an effective potential of the form shown in Fig. 1-5-3 results.† (An inverse-third-power attractive potential is assumed in this illustration.) Let us consider the motion for several different values of the total energy. A particle of energy $E_1 < E_0$ starting at large r and moving toward the center of attraction will evidently be reflected at $r = r_a$ by the potential barrier shown. If, however, the total energy is $E_2 > E_0$, the particle will be able to pass over the potential barrier. It will experience a repulsion only for $r > r_p$ and will sense an attraction thereafter, as its radial distance from the scattering center decreases. When $E > E_0$, the particle actually passes through the center of attraction, which is assumed to be a mathematical point in the model under consideration here. A particularly interesting situation develops when the total energy just exceeds E_0, the value of $V_{\text{eff}}(r)$ at the peak of the potential curve. In this case the particle will spend a considerable time at a radial distance near the peak, where $\dot{r}$ is small by

---

*A particle is said to be in a *stable orbit* if the application of a small perturbation produces a small bounded excursion from the original orbit.

†The limiting case in which $n = 2$ must be treated separately. Here the shape of $V_{\text{eff}}(r)$ is determined by the relative magnitudes of J and the coefficient of r^{-2} in the expression for $V(r)$.

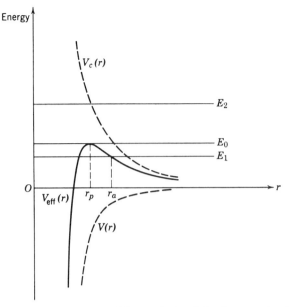

Figure 1-5-3. Potential functions used for analysis of elastic scattering in an attractive inverse-third-power potential field.

assumption, while spiraling inward toward the center. The particle is then said to "orbit" about the scattering center. The angular motion speeds up as r decreases in order to conserve angular momentum, and a large number of revolutions may be made. (This is an unstable orbit, unlike those for which $n < 2$.) Under certain conditions the scattering angle may approach $-\infty$. Figure 1-5-4 illustrates an orbiting collision for which Θ has a value between 6π and 7π. In the figure a long-range attractive potential and a short range, hard-core repulsive potential are assumed. A cusp appears in the trajectory when the particle is reflected from the hard core. We see that the orbits may thus be meaningfully classified as *spiraling* or *nonspiraling*, and this distinction is frequently encountered in the literature.

1-6. SEMICLASSICAL APPROXIMATIONS

A number of semiclassical approximations were discussed in general terms in Section 3-24 of the companion volume, and others were treated with reference to inelastic electron collisions in Section 6-13 of that volume. All of the former, that is, (1) the semiclassical impact parameter method, (2) the classical trajectory Monte Carlo method, (3) the classical impulse, binary encounter approximation, (4) the eikonal approximation, (5) the multichannel eikonal approximation, (6) the Glauber approximation, and (7) the semiclassical S-matrix method, are

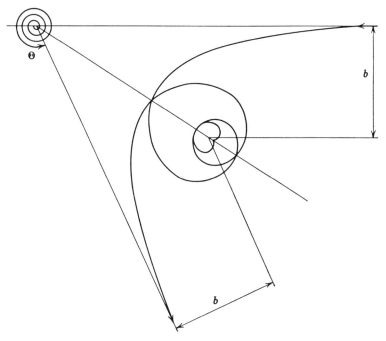

Figure 1-5-4. Orbiting collision for which the scattering angle Θ has a large value.

applicable to heavy particle collisions. (Of these, approximations 1, 2, and 7 are inappropriate for electron collisions; see Problem 3-23 of the companion volume.)

The term *semiclassical* implies, of course, a blend of classical and quantal methods, and in the present context, the salient feature of the semiclassical approach is the use of classical mechanics to calculate the trajectories of the colliding particles. We discussed the inadequacies of classical theory in Section 3-10 of the companion volume, but in Section 3-11 we used the standard argument (based on the uncertainty principle) to show that the concept of a well-defined classical trajectory is valid provided that

$$\vartheta \gg \frac{\hbar}{mv_0 b}. \tag{1-6-1}$$

Here a projectile of mass m is approaching a fixed target with impact parameter b and velocity v_0, and the scattering angle is ϑ. Two slightly different expressions for the condition of validity of the classical treatment are

$$\vartheta \gg \frac{\hbar}{mv_0 a} \tag{1-6-2}$$

(Mott and Massey, 1965, p. 111; Dalgarno, 1970, p. 184) and

$$\vartheta \gg \frac{\hbar}{L} \tag{1-6-3}$$

(McDowell and Coleman, 1970, p. 9). Here a is the range of the scattering potential and L is the classical orbital angular momentum. McDowell and Coleman (1970) point out that in electron collisions we may typically take L to be a few times $\hbar$ and ϑ to be about 1 rad, so we see that the validity of classical mechanics is severely limited in range. However, even for thermal velocity heavy particle collisions, L is on the order of $10^3\hbar$ and ϑ about 10^{-3} rad ≈ 0.1 deg. Hence classical theory should be valid throughout most of the angular range accessible in heavy particle scattering experiments, except in cases where quantum symmetry considerations arise.

Similarly, from (1-6-2), we see that classical trajectories are satisfactory over a wide range of scattering angles for fast heavy particles moving in a long-range potential, but that classical calculations must become invalid at small angles. As Dalgarno (1970) points out, however, large impact parameters leading to small-angle scattering are unimportant in heavy particle collisions involving reactions, which can occur only at small separation distances, so classical theory may be entirely satisfactory here.

There are several reasons for employing semiclassical methods in the study of heavy particle collisions (Dickinson and Richards, 1982). First, the routine application of pure quantal techniques is too expensive unless drastic approximations are made. Second, a method based on exact classical trajectories of heavy particles may provide better results than do many quantal approximations. Finally, a study of the classical trajectories may be more revealing of the collision dynamics and lead to new approximations or better insight into the old ones.

1-7.   QUANTUM THEORY

A.   Time-Independent Formulation of the Elastic Scattering Problem

Let us imagine an infinitely wide beam of structureless projectiles approaching along the $-Z$ axis a structureless scattering center fixed at the origin of coordinates. A spherically symmetric force field is assumed, so that the potential energy of interaction V is a function only of the separation distance r. The incident beam is to be homogeneous and monoenergetic, and the particles in the beam are to be dissimilar to the target particle representing the scattering center. (Identical projectiles and targets require special handling in quantum mechanics and are treated separately in Section 1-7-D-4.) Furthermore, the beam intensity is to be uniform across any plane normal to the Z axis and is to be constant in time. To each projectile is ascribed a fictitious mass $M_r = mM/(m + M)$, where

m and M are the true masses of the projectile and target, respectively. As is well known, the quantal representation of such a beam is a plane de Broglie wave of constant amplitude traveling in the $+Z$ direction. Let us suppose that the current density of the beam is A particles/cm^2-s, so that A particles pass through unit area normal to the beam direction per second.

For quantitative expression of the intensity of scattering in any direction (Θ, φ) let us refer to Fig. 1-7-1. In this figure dS represents a small surface at a large distance r, in the direction (Θ, φ), from the origin. It is assumed to be normal to the radius vector. The solid angle subtended at the origin by dS is $d\Omega_{CM}$. From our previous discussion of the significance of the differential cross section [see (1-4-1) of the companion volume] we see that $AI_s(\Theta)\, d\Omega_{CM}$ equals the number of projectiles scattered through dS per second.

The plane wave representing the incident beam of particles is described by a wave function $Ce^{i(\kappa z - \omega t)}$, where C is the amplitude, κ is the wave number, and ω is the angular frequency of the wave; κ is given in terms of the de Broglie wavelength λ, the particle velocity v_0, or the initial kinetic energy T_0 by the equation

$$\kappa = \frac{2\pi}{\lambda} = \frac{M_r v_0}{\hbar} = \frac{\sqrt{2M_r T_0}}{\hbar}, \qquad \text{[3-14-2]}$$

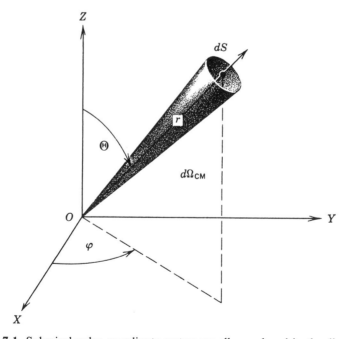

Figure 1-7-1. Spherical polar coordinate system usually employed in the discussion of collision problems. The beam of projectiles moves upward through the $X-Y$ plane and along the Z axis. The scattering center is located at the origin, O.

and ω is given in terms of the frequency v, or the total energy E, by the equation

$$\omega = 2\pi v = \frac{E}{\hbar}.$$ [3-14-3]

The scattering problem we have formulated is a steady-state problem, and the energy of the incident de Broglie wave is unchanged in the process of elastic scattering. Consequently, the term $e^{-i\omega t}$ is common to all the wave functions we shall write down, so that we may drop this common time factor and indicate only the spatial dependence of the wave functions. The *time-independent* wave function for the incoming plane wave is denoted by $\psi_{\text{inc}}(r, \Theta) = Ce^{i\kappa z}$. Note that dependence of ψ_{inc} on φ is not implied, since all aspects of the problem are independent of the azimuth angle.

We have taken the current density of the incoming beam to be A projectiles/cm²-s. The current density may also be expressed in terms of C, the amplitude of the wave function, and thus we may obtain the relationship between A and C:

$$C^2 = \frac{A}{v_0}.$$ [3-14-4]

The total wave function will consist of a component associated with the incoming plane wave and another representing that portion of the incident wave which experiences scattering. The scattered wave must have the form of an outgoing spherical wave whose amplitude decreases, for large r, as $1/r$, in order that the radial current density may fall off as the inverse square of the distance from the center of force and that the number of scattered particles may thus be conserved. The wave function for the scattered wave will then have the asymptotic form

$$\psi_{\text{scatt}} \approx \frac{C}{r} f(\Theta)e^{i\kappa r},$$ [3-14-5]

where $f(\Theta)$ is usually called the *scattered amplitude*. It then follows that

$$I_s(\Theta) = |f(\Theta)|^2;$$ [3-14-6]

hence it is not necessary to know $\psi(r, \Theta)$ completely to determine the differential or integral cross section—we need know only the asymptotic form of the scattered wave function. (In the foregoing discussion only the scattered wave function was used to calculate the number of particles passing through dS per second. The incident beam is assumed to be collimated in such a way that its unscattered component is prevented from reaching dS, which also is assumed to be at a great distance from the scattering center. Such collimation is always provided in experimental apparatus.)

To obtain $f(\Theta)$, and thus the cross sections, we must solve the time-independent wave equation for the motion of particles of mass M_r and positive total energy $(E > 0)$ in the potential field of the fixed scattering center. The solution is required to have the asymptotic form

$$\psi = \psi_{\text{inc}} + \psi_{\text{scatt}} \approx e^{i\kappa z} + \frac{e^{i\kappa r}}{r} f(\Theta), \qquad \text{[3-14-7]}$$

where we have set the amplitude of the incoming wave equal to unity for convenience. The wave number κ was expressed in terms of the initial kinetic energy by [3-14-2]. However, T_0 equals the total energy because the mutual potential energy of the particles is zero at very large r, and we may write

$$\kappa = \frac{\sqrt{2M_r E}}{\hbar}. \qquad \text{[3-14-8]}$$

The time-independent wave equation is

$$\nabla^2 \psi + \frac{2M_r}{\hbar^2} [E - V(r)]\psi = 0, \qquad \text{[3-15-1]}$$

where, in spherical coordinates, the Laplacian operator is

$$\nabla^2 = \frac{1}{r^2} \frac{\partial}{\partial r} \left(r^2 \frac{\partial}{\partial r} \right) + \frac{1}{r^2 \sin \Theta} \frac{\partial}{\partial \Theta} \left(\sin \Theta \frac{\partial}{\partial \Theta} \right)$$
$$+ \frac{1}{r^2 \sin^2 \Theta} \frac{\partial^2}{\partial \varphi^2}. \qquad \text{[3-15-2]}$$

The last term in this expression vanishes in the present problem because of the lack of dependence of any of our functions on φ. Let us put

$$U(r) = \frac{2M_r V(r)}{\hbar^2}, \qquad \text{[3-15-3]}$$

so that the wave equation may be written in the convenient form

$$\nabla^2 \psi + [\kappa^2 - U(r)]\psi = 0. \qquad \text{[3-15-4]}$$

We seek a solution of [3-15-4], which is everywhere bounded, continuous, and single valued and which has the asymptotic form indicated in [3-14-7].

B. Solution by the Method of Partial Waves

Since by assumption V depends only on r, the wave equation may be separated into two equations, one of which contains only the variable r and the other only Θ. (In the more general case in which there is dependence on the azimuthal angle we would have a third equation in φ.) To perform this separation, let us write

$$\psi(r, \Theta) = L(r)Y(\Theta), \tag{3-15-5}$$

and substitute this expression into [3-15-4]. Thus

$$\frac{1}{r^2}\frac{d}{dr}\left(r^2\frac{dL}{dr}\right) + \left[\kappa^2 - U(r) - \frac{l(l+1)}{r^2}\right]L = 0 \tag{3-15-7}$$

and

$$\frac{1}{\sin\Theta}\frac{d}{d\Theta}\left(\sin\Theta\frac{dY}{d\Theta}\right) + l(l+1)Y = 0. \tag{3-15-8}$$

Equation [3-15-8] is a special case of Legendre's equation. Because it is a second-order equation it has two linearly independent solutions, each of which may be expressed as a power series in $\cos\Theta$. Both solutions become infinite for $\Theta = 0$ unless l is zero or a positive integer, and therefore the only physically acceptable solutions of the wave equation are those corresponding to $l = 0, 1, 2, 3, \ldots$. As we shall soon see, l is a measure of the angular momentum of the projectile about the fixed scattering center. For this reason l is called the *angular momentum quantum number*. The first few acceptable solutions of [3-15-8] are

$$P_0(\cos\Theta) = 1$$
$$P_1(\cos\Theta) = \cos\Theta$$
$$P_2(\cos\Theta) = \tfrac{1}{2}(3\cos^2\Theta - 1)$$
$$P_3(\cos\Theta) = \tfrac{1}{2}(5\cos^3\Theta - 3\cos\Theta).$$

The functions $P_l(\cos\Theta)$ are the well-known Legendre polynomials.

We now see that the desired wave function may be written in the form

$$\psi(r, \Theta) = \sum_{l=0}^{\infty} A_l P_l(\cos\Theta)L_l(r), \tag{3-15-9}$$

where the A_l are arbitrary constants and the $L_l(r)$ are solutions of [3-15-7] for particular values of l. The terms in the foregoing sum are known as *partial waves*. Now let us assume that $U(r)$ falls off faster than $1/r$ at large r and that if $U(r)$ has a pole at the origin it is not of higher order than $1/r$. Then there are two independent solutions of [3-15-7], one of which is finite at the origin, the other being infinite there.

The solution of [3-15-7] that is finite at the origin will then have the asymptotic form

$$L_l(r) \approx \frac{1}{\kappa r} \sin\left(\kappa r - \frac{l\pi}{2} + \eta_l\right), \qquad \text{[3-15-15]}$$

where η_l is a constant for given κ and $U(r)$ called the *l*th-*order phase shift*; η_l is the phase shift of the *l*th partial wave due to the action of the scattering potential. The term $-l\pi/2$ is inserted into [3-15-15] so that η_l will be zero if $U(r)$ is zero. Evidently, it is then required that

$$A_l = (2l + 1)i^l e^{i\eta_l}, \qquad \text{[3-15-27]}$$

and the total wave function is seen to be

$$\psi(r, \Theta) = \sum_{l=0}^{\infty} (2l + 1)i^l e^{i\eta_l} L_l(r) P_l(\cos \Theta). \qquad \text{[3-15-28]}$$

C. Scattering Cross Sections

According to [3-14-7], the scattered wave function is $(e^{i\kappa r}/r)f(\Theta)$. The right side of (3-15-26) of the companion volume can have this form only if

$$f(\Theta) = \frac{1}{2i\kappa} \sum_{l=0}^{\infty} (2l + 1)(e^{2i\eta_l} - 1)P_l(\cos \Theta). \qquad \text{[3-15-29]}$$

Note that $f(\Theta)$ is complex, having the form

$$f(\Theta) = A + iB, \qquad \text{[3-15-30]}$$

where

$$A = \frac{1}{2\kappa} \sum_{l=0}^{\infty} (2l + 1) \sin 2\eta_l P_l(\cos \Theta) \qquad \text{[3-15-31]}$$

$$B = \frac{1}{2\kappa} \sum_{l=0}^{\infty} (2l + 1)(1 - \cos 2\eta_l) P_l(\cos \Theta) \qquad \text{[3-15-32]}$$

and i is the imaginary operator. Thus we see that

$$I_s(\Theta) = |f(\Theta)|^2 = A^2 + B^2$$

$$= \frac{1}{\kappa^2} \left| \sum_{l=0}^{\infty} (2l + 1)e^{i\eta_l} \sin \eta_l P_l(\cos \Theta) \right|^2. \qquad \text{[3-15-33]}$$

Integration of $I_s(\Theta)$ over the complete solid angle gives for the integral elastic scattering cross section

$$q_s = \frac{4\pi}{\kappa^2} \sum_{l=0}^{\infty} (2l + 1) \sin^2 \eta_l. \qquad [3\text{-}15\text{-}34]$$

In Note 1-1 we sketch the Massey–Mohr and Landau–Lifshitz derivations of the integral elastic cross section for an attractive potential of the form $V(r) \sim -Cr^{-n}$, where C is a positive constant and n is a positive integer. The results are useful in discussions of intermolecular potentials.

Methods of calculating the phase shifts are enumerated in Section 3-15-F of the companion volume and discussed by Bransden (1983), Burhop (1961), Joachain (1975), Mott and Massey (1965), Massey and Burhop (1969), and Massey (1969, 1971). Other approaches to the scattering problem (time independent and time dependent) are also described in the books cited here and in Sections 3-16, 3-19, and 3-23 of the companion volume. Problems 3-9 through 3-22 of the companion volume and Problems 1-4 and 1-5 at the end of this chapter may prove useful.

D. Extensions, Refinements, and Complications

Here we take note of some additional features of the elastic scattering problem.

1. *Multichannel Calculations*

First, recall that the phase-shift calculation in Section 1-7-B is a *single-channel* calculation that involves only the direct transition $i \to f$, where i and f denote the initial and final states of the system. By allowing the target to possess structure and by permitting transitions through intermediate states $(i \to m \to f)$, we can make a *multichannel* calculation that may provide distinctly better results. The intermediate states m may be discrete or lie in the continuum. (See the discussion of the close-coupling approximation in Section 3-23-I of the companion volume.)

2. *Modification of Elastic Scattering by Resonant Excitation Transfer and Resonant Charge Transfer*

Now consider a complication that arises when an excited atomic projectile A^*, or a singly charged ion A^+, collides with an atom A of the same species. Here, in addition to *elastic scattering*,

$$A^* + A \to A^* + A \qquad (1\text{-}7\text{-}1)$$

or

$$A^+ + A \to A^+ + A, \qquad (1\text{-}7\text{-}2)$$

we may have *resonant excitation transfer*,

$$A^* + A \rightarrow A + A^* \tag{1-7-3}$$

or *resonant charge transfer*,

$$A^+ + A \rightarrow A + A^+. \tag{1-7-4}$$

In the last two processes here, as in the first two, the total kinetic energy of the collision partners does not change, and the "energy defect", $\Delta E = T_f - T_i$, is zero. Experimentally, it is impossible to distinguish unambiguously between (1-7-1) and (1-7-3), and between (1-7-2) and (1-7-4), so the theoretical descriptions of the scattering of A^* and of A^+ must reflect the two physically different ways, in each case, of obtaining the same result. One must observe this general rule [see Feynman et al. (1965, Chaps. 3 and 4)]: The two alternatives in each pair of indistinguishable processes interfere with one another, and the scattering amplitude for an event is then the sum of the two interfering amplitudes. The scattering intensity is calculated as the modulus squared of the sum of the amplitudes. Note that sometimes the amplitudes are added with the same phase, sometimes with the opposite phases. We see that the elastic scattering processes (1-7-1) and (1-7-2) are modified by resonant excitation transfer and by resonant charge transfer, respectively. It is interesting to note that with both A^* and A^+ projectiles, there are two possible interactions, one arising when the electron distribution is symmetric with respect to the nuclei, and the other when the electron distribution is antisymmetric [see Problem 1-6; see also Massey (1971, pp. 1298 and 1301, Secs. 18-8 and 19-3-3)].

3. Elastic Scattering under Different Spin-Orbit Orientations

Consider first two atoms (say, two rare gas atoms) in their ground state with completely filled outer shells (Massey, 1971, Chap. 16). The atoms interact at a large separation distance mainly through the attractive van der Waals forces, the interaction potential having the form

$$V(r) \sim -\frac{C}{r^6}. \tag{1-7-5}$$

The mechanism at work here is the dynamic polarization of one atom by the other (see Note 1-1). At small r, strong repulsive forces come into play. These short-range forces increase exponentially with decreasing r; they are produced by averaged Coulomb interactions and by the exclusion principle, which has an increasing effect as the electron clouds of the atoms overlap more and more. The combination of these chemical (or intrinsic) short-range forces and the long-range van der Waals forces produces a potential curve of the general form shown in Fig. 1-3-1. Here there is a unique interaction, as in the case of a rare gas atom interacting with other atoms in doublet or triplet S states, provided that the atoms are not of the same kind. If one of the atoms is replaced by an ion, the main change that results is the addition of a term describing the polarization potential

$$V(r) \sim \frac{-\alpha e^2}{2r^4},$$ (1-7-6)

where α is the polarizability of the atom (see Section 1-8 of the companion volume).

With other combinations of atomic states, the interaction is not unique but depends on the relative orientation of the spin and orbital angular momentum. Accordingly, two atoms interact quite differently depending on whether the unpaired spins in each atom are parallel or antiparallel [see Massey (1969, Chaps. 12 and 13) and Massey (1971, Sec. 16-12)]. This behavior for two hydrogen atoms is illustrated in Fig. 1-7-2, where the separated atom limit potentials at large r split into two curves as r decreases and the united atom limit is approached. In collisions between unpolarized H atoms, one should take account of the existence of the two interactions and of the symmetry of the two protons as well [see Massey (1971, p. 1418) and the discussion in Section 1-7-D-4 below].

The elastic, diffusion, and viscosity cross sections contain contributions from scattering according to both potentials in the H + H case. The situation can get

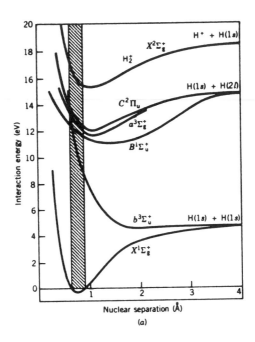

Nuclear separation (Å)

(a)

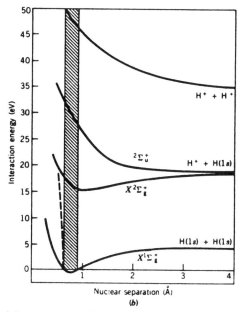

Figure 1-7-2. Potential energy curves for electronic states of H_2 and H_2^+: (a) within 20 eV of the ground state; (b) for higher-energy states. The united atom and separated atom limits of a few states of H_2 are given below:

United Atom Limit	Molecular State	Separated Atom Limit
$He(1^1S)(1\,s\sigma)^2$	$X^1\Sigma_g^+$	$(\sigma_g 1s)^2$
$He(2^1P)\,1s\sigma\,2p\sigma$	$B^1\Sigma_u^+$	$\sigma_g 1s\pi_u 2s$
$He(2^1P)\,1s\sigma\,2p\pi$	$C^1\Sigma_u$	$\sigma_g 1s\pi_u 2p$
$He(2^1S)\,1s\sigma\,2s\sigma$	$E^1\Sigma_g^+$	$\sigma_g 1s\sigma_g 2s$
$He(2^3S)\,1s\sigma\,2s\sigma$	$a^3\Sigma_g^+$	$\sigma_g 1s\sigma_g 2s$
$He(2^3P)\,1s\sigma\,2p\sigma$	$b^3\Sigma_u^+$	$\sigma_g 1s\sigma_u 1s$

The notation used here is explained by Sharp (1971). [This figure is identical to Fig. 5-5-1 of the companion volume and was adapted from Massey (1969).]

much more complicated in other cases; for example, two O atoms in their ground triplet P state can interact in 18 different ways.

4. Scattering of Identical Particles

Up to this point no mention has been made of the symmetry effects that appear when the colliding particles are identical, and we must now modify the equations for the scattering cross section to take these effects into account. Their physical basis is the fact that it is impossible to distinguish experimentally between identical projectile and target particles, and the wave function describing the system must therefore satisfy certain symmetry properties with respect to the interchange of the coordinates of the two particles. The wave function, in fact, must be symmetric or antisymmetric with respect to the interchange of the particles, depending on whether their total (combined) spin is even or odd, that is, an even or odd multiple of $\hbar$.

In order to justify the statement concerning the symmetry character of the wave function, we reason as follows. Because the collision partners are indistinguishable by assumption, the states of the system obtained from each other by simply interchanging the particles must be completely equivalent physically. This means that the wave function describing the system can change only by an unimportant phase factor as a result of a single interchange. Let $\psi(1, 2)$ represent the spatial part of the wave function of the system* before the interchange occurs. The numbers 1 and 2 refer to the two particles composing the system. The new wave function describing the system after the particles are interchanged must be given, according to what has been said, by the equation

$$\psi(2, 1) = e^{ia}\psi(1, 2),$$

where a is a real constant. If the interchange is repeated, we must recover the original wave function, that is,

$$\psi(1, 2) = e^{2ia}\psi(1, 2).$$

Thus $e^{2ia} = 1$ and $e^{ia} = \pm 1$, so that

$$\psi(2, 1) = \pm\psi(1, 2).$$

We see that the wave function must either be unchanged or simply undergo a change of sign if the two particles are interchanged. In the first case the wave function is said to be symmetric, whereas in the second it is antisymmetric.

An interchange of the positions of the particles is equivalent to a reversal of the direction of the radius vector joining them. In the center-of-mass coordinate system r remains unchanged in this reversal, whereas the angle Θ becomes $\pi - \Theta$ and z becomes $-z$, since $z = r \cos \Theta$. Therefore, if we assume that the interaction potential is a function only of r, the proper asymptotic expression for the spatial part of the wave function is now

$$\psi \approx e^{i\kappa z} \pm e^{-i\kappa z} + \frac{e^{i\kappa r}}{r} [f(\Theta) \pm f(\pi - \Theta)] \qquad [3\text{-}18\text{-}1]$$

*When the spin is considered, the total wave function is expressed as a product of two functions, one of which depends only on the spatial coordinates and the other on the spin. The statement above concerning the symmetry or antisymmetry of the wave function refers to the spatial part of the total wave function.

(with the positive sign to be used to obtain the symmetric wave function and the negative sign, the antisymmetric wave function) instead of

$$\psi \approx e^{i\kappa z} + \frac{e^{i\kappa r}}{r} f(\Theta), \qquad [3\text{-}14\text{-}7]$$

which is correct in the absence of symmetry. The wave function in [3-18-1] contains two incident plane waves of equal amplitude traveling in opposite directions, and the outgoing spherical wave takes into account the scattering of both waves. The differential cross section per unit solid angle is given by the square of the coefficient of the term $e^{i\kappa r}/r$ and is related to the probability that *either* of the particles will be scattered into the element of solid angle at Θ. Thus if the total spin of the particles is even, the differential cross section per unit solid angle is

$$(I_s)_{\text{sym}} = |f(\Theta) + f(\pi - \Theta)|^2, \qquad [3\text{-}18\text{-}2]$$

and for odd total spin

$$(I_s)_{\text{anti}} = |f(\Theta) - f(\pi - \Theta)|^2. \qquad [3\text{-}18\text{-}3]$$

The appearance of the interference term $[f(\Theta)f^*(\pi - \Theta) \pm f^*(\Theta)f(\pi - \Theta)]$ is characteristic of quantum mechanics. This term does not appear in the classical treatment of scattering, in which the cross section per unit solid angle would be given simply by

$$(I_s)_{\text{class}} = |f(\Theta)|^2 + |f(\pi - \Theta)|^2. \qquad [3\text{-}18\text{-}4]$$

In the foregoing discussion it was assumed that the total spin of the particles had some definite value, but we must as a rule consider a distribution among the possible spin states. To determine a cross section, we shall assume that all the spin states are equally probable. It can be shown that the total number of spin states is $(2s + 1)^2$ for two particles of spin s and that $s(2s + 1)$ of these states correspond to an even total spin and $(s + 1)(2s + 1)$ to an odd total spin, if s is half-integral. If s is integral, the weighting is reversed. Thus if s is half-integral, the probability that the system will have even total spin is $s(2s + 1)/(2s + 1)^2 = s/(2s + 1)$, and the probability of odd total spin is $(s + 1)/(2s + 1)$. The differential cross section per unit solid angle is therefore

$$(I_s)_{\text{FD}} = \frac{s}{2s + 1} (I_s)_{\text{sym}} + \frac{s + 1}{2s + 1} (I_s)_{\text{anti}} \qquad [3\text{-}18\text{-}5]$$

for half-integral s. By a similar analysis we can show that for integral s the appropriate expression is

$$(I_s)_{BE} = \frac{s}{2s+1}(I_s)_{anti} + \frac{s+1}{2s+1}(I_s)_{sym}. \qquad [3\text{-}18\text{-}6]$$

The subscripts FD and BE in [3-18-5] and [3-18-6] are abbreviations for *Fermi–Dirac* and *Bose–Einstein*. In relativistic quantum mechanics it is shown that particles with half-integral spin obey Fermi–Dirac statistics, hence [3-18-5] applies to the scattering of particles of this type, which are called *fermions*. Equation [3-18-6] applies to particles of integral spin, which obey Bose–Einstein statistics and are called *bosons*. Electrons, positrons, protons, neutrons, and nuclei of odd mass number are fermions, and nuclei of even mass number (such as deuterons and alpha particles) are bosons.

It is now of interest to write down the expressions for the cross section in terms of the phase shifts for particles obeying these types of statistics. For distinguishable particles we know that

$$I_s(\Theta) = \frac{1}{4\kappa^2}\left|\sum_{l=0}^{\infty}(2l+1)(e^{2i\eta_l}-1)P_l(\cos\Theta)\right|^2, \qquad [3\text{-}18\text{-}7]$$

which is just the square of the expression given in [3-15-29] for the scattered amplitude. Two changes are required, however, when we write the corresponding expressions for identical particles. First we must multiply I_s by a factor of 2, since we cannot distinguish between the scattered waves representing the two collision partners. Second, assuming for the moment that the total spin has a definite even value, terms of odd l must be excluded from the sum because $P_l(\cos\Theta)$ is odd for all odd l. Thus for even total spin

$$(I_s)_{sym} = \frac{1}{2\kappa^2}\left|\sum_{even\ l}(2l+1)(e^{2i\eta_l}-1)P_l(\cos\Theta)\right|^2, \qquad [3\text{-}18\text{-}8]$$

and similarly, for odd total spin

$$(I_s)_{anti} = \frac{1}{2\kappa^2}\left|\sum_{odd\ l}(2l+1)(e^{2i\eta_l}-1)P_l(\cos\Theta)\right|^2. \qquad [3\text{-}18\text{-}9]$$

Equations [3-18-5] and [3-18-6] are to be used with these equations to obtain the scattering distributions when there is a mixture of spin states.

The reader may find it useful to consult Section 2-15 and Problems 3-21 and 3-22 in the companion volume. Symmetry effects in the scattering of helium atoms and ions by helium atoms are discussed in Sections 1-12 and 1-13 of the present volume.

1-8. REFERENCES ON THEORY: CLASSICAL, SEMICLASSICAL, AND QUANTUM

Here we list books and reviews that are useful general references on the theory of elastic scattering of heavy particles. The symbols C, S, and Q are used to indicate whether the references are recommended for classical, semiclassical, or quantal treatments.

Bernstein (1966)—Q

Berry and Mount (1972)—S

Bransden (1983)—Q

Bransden and Joachain (1993)—Q

Burgess and Percival (1968)—C

Child (1974)—CSQ

Child (1980)—S

Dalgarno (1970)—SQ

Demkov (1983)—CS

Eu and Sink (1983)—S

Fano and Rau (1986)—Q

Ford et al. (1959)—S

Ford and Wheeler (1959a, b)—S

Joachain (1975)—Q

Landau and Lifshitz (1960)—C

Landau and Lifshitz (1965)—SQ

Levine and Bernstein (1987)—CQ

Massey (1969)—Q

Massey (1971)—CSQ

Massey and Burhop (1969)—Q

Massey and Gilbody (1974)—Q

McDowell and Coleman (1970)—CSQ

Messiah (1961, 1962)—Q

Miller (1974, 1975, 1976)—CSQ

Mott and Massey (1965)—SQ

Newton (1982)—Q

Pauly (1975)—CSQ

Pauly (1979)—CSQ

Pauly and Toennies (1965)—CSQ

Percival (1977)—S

Percival and Richards (1975)—C

Rodberg and Thaler (1967)—Q

Sakurai (1985)—Q

Stolte and Reuss (1979)—SQ

Su and Bowers (1979)—C

Taylor (1972)—Q

Wheeler (1976)—S

Wu and Ohmura (1962)—Q

PART ℬ. HEAVY PARTICLE COLLISIONS AT LOW ENERGIES

As indicated in Section 1-1, this part of our discussion of heavy particle elastic scattering is restricted mainly to CM energies below a few tenths of an electron volt. The most productive approach in this energy regime has involved the use of molecular beam techniques, the development of which is traced by Ramsey (1988) and by the various contributors to the book edited by Scoles (1988). A few historical comments will suffice here.

Following pioneering experiments by L. Dunoyer circa 1911, Otto Stern established a molecular beam laboratory at Frankfurt shortly after World War I. After he moved to Hamburg in 1922, his laboratory became the world's center for molecular beam research and remained so until he left Germany in 1933. From about that time until the mid-1950s, Isidor Rabi's laboratory at Columbia University occupied center stage. The emphasis there was not on collisions, but rather on molecular beam resonance spectroscopy. The research in this area at Columbia led to Nobel Prizes in Physics for Rabi, Kusch, Lamb, Townes, and Ramsey. (Stern had received the Nobel Prize in Physics in 1943.) Since the mid-1950s, molecular beam laboratories have been established at many universities and research institutes, and overall, the investigations have acquired a more chemical flavor. (Witness the award of Nobel Prizes in Chemistry to Dudley Herschbach, Y. T. Lee, and John Polanyi in 1986.) Elastic, inelastic, and reactive collisions at low energies have proved amenable to beam techniques, and we study the first type here (inelastic and reactive collisions are covered in subsequent chapters). For reasons that will soon become apparent, it is desirable to discuss some calculations of the angular distributions of elastically scattered heavy particles before treating the experimental studies.

1-9. ANGULAR DISTRIBUTIONS OF ELASTICALLY SCATTERED HEAVY PARTICLES

We have already pointed out in Section 1-1 that the scattering of atoms, molecules, and ions is very strongly peaked in the forward direction, in marked contrast to the behavior of electrons. This fact, which was deduced experimentally many years ago, is illustrated by a calculation made by Massey and Smith (1933). They computed the differential elastic cross section for 110-eV protons on helium and 72-eV protons on argon, with the results shown in Table 1-9-1. Calculations were made for the Hartree self-consistent field (Appendix I), and for the Coulomb field of the unshielded nuclei as well, to determine the screening effect of the orbital electrons. Polarization and electron exchange were not considered, but their neglect should not affect the basic conclusions. Massey and Smith showed that the classical theory of Section 1-2-B could be used legitimately for all angles except $\Theta \approx 0°$. The phase-shift treatment of Sections 1-7-B and 1-7-C was applied for the forward direction. The sharp peaking of the angular distribution about the forward direction is indeed dramatic. The forward concentration of the scattering becomes even more pronounced as the energy increases, and Massey and Smith state that for 1000-eV protons on argon, the intensity per unit solid angle at $0°$ is at least 10^5 times that at $10°$. Massey and Smith also calculated the integral elastic scattering cross sections for protons on helium and argon at several energies. Table 1-9-2 shows that their results do not differ markedly from the cross sections derived from kinetic theory.

The foregoing discussion suggests that it is fruitless to attempt to measure

Table 1-9-1. Differential Elastic Scattering Cross Sections per Unit Solid Angle Calculated for Protons by Massey and Smith (1933)

CM Scattering Angle (deg)	$I_s(\Theta)$ (units of a_0^2 per unit solid angle)			
	Argon (72-eV protons)		Helium (110-eV protons)	
	Hartree	Coulomb	Hartree	Coulomb
0	16×10^4	∞	9×10^3	∞
12	22.0	1.58×10^4	7.85	124.0
28	7.20	770	2.00	6.10
34	2.76	365	0.72	2.85
57	0.93	51	0.21	0.40
80	0.48	15	0.08	0.12
114	0.14	5.3	0.04	0.04
137	0.08	3.5	—	0.03
167	0.05	2.6	—	0.02

Table 1-9-2. Integral Elastic Scattering Cross Sections Calculated for Protons by Massey and Smith (1933)[a]

Gas	Proton Energy (eV)	Massey and Smith's Cross Sections	Gas Kinetic Cross Sections
He	90	3.75	2.6
	800	2.0	—
Ar	73	16.4	7.3
	650	10.7	—

[a]Cross sections are expressed in units of πa_0^2.

accurately the integral elastic cross section for heavy particles at high energies. The demands on angular resolving power are impossible to meet even at energies as low as a few electron volts, and the resolving power required increases roughly as the velocity of the projectile. A better course is to make careful determinations of the angular distribution of scattering, or to measure the scattering through angles greater than some minimum value that can be accurately determined. The analysis of these measurements can be conducted classically, since quantum effects become appreciable only at scattering angles comparable with the angle of resolution that would be required to measure the true integral elastic cross section. Fortunately, measurement of the true integral cross section is not required for interparticle potential determinations—measurements of the cross section for scattering through angles greater than some known minimum angle suffice (Massey, 1971; Massey and Gilbody, 1974; Mason, 1982). It is this *effective cross section* that is reported in the literature, although it is usually loosely referred to as the *integral elastic scattering cross section*.

The situation is more favorable at lower energies. Using the schematic angular distribution for elastic spheres derived by Massey and Mohr (1933), Massey (1971, p. 1350) has calculated the minimum angle of deflection ϑ_0 that must be counted as a collision in order that the measured cross section may be within 10% of the true value. If D represents the sum of the gas kinetic radii of the projectile and target particles measured in angstroms, m the molecular weight of the projectiles, and T the effective temperature of the projectiles in kelvin, ϑ_0 is given approximately as

$$\vartheta_0 \approx \frac{277}{D(mT)^{1/2}}, \qquad (1\text{-}9\text{-}1)$$

where ϑ_0 is measured in degrees in the Lab system [see (1-6-2)]. The target is assumed to be at rest in this coordinate system. If the resolution of the apparatus is such that a collision involving a deflection through the angle ϑ_0 can be observed, further increase in the resolving power will lead to only a slight increase in the measured cross section. Values of ϑ_0 for various projectile–target combinations and several energies are shown in Table 1-9-3. It is shown in the

Table 1-9-3. Angular Resolution Requirements for Accurate Measurement of the Integral Elastic Scattering Cross Section of Atomic Projectiles on Atomic Targets

Target	Lab Energy of Projectile (eV)	ϑ_0 (deg)[a] for Projectile:							
		He	Ne	Ar	Li	Na	K	Rb	Cs
He	0.0255	3.6	—	—	1.5	0.73	0.50	0.36	0.27
	0.0862	2.0	—	—	0.80	0.40	0.27	0.20	0.15
	1	0.59	—	—	0.23	0.12	0.08	0.06	0.04
Ne	0.0255	—	1.4	—	1.4	0.62	0.40	0.27	0.19
	0.0862	—	0.75	—	0.75	0.34	0.22	0.15	0.11
	1	—	0.22	—	0.22	0.10	0.06	0.04	0.03
Ar	0.0255	—	—	0.70	0.87	0.41	0.27	0.19	0.14
	0.0862	—	—	0.30	0.48	0.23	0.15	0.10	0.08
	1	—	—	0.11	0.14	0.07	0.04	0.03	0.02

Source. Massey (1971).
[a]ϑ_0 is the minimum angle of deflection that must be counted as a collision in order that the measured cross section may be within 10% of the true value.

next section that these demands on resolving power, although great, are not impossibly high.

1-10. APPARATUS AND TECHNIQUES

Here we describe representative apparatus and techniques used for studies of elastic scattering of heavy particles at low energies.

A. Bernstein's Neutral-Beam Integral-Cross-Section Apparatus

In the late 1950s, Bernstein and his coworkers at the University of Michigan conducted a series of experiments in which they measured the integral cross section for elastic collisions of K and Cs atoms (Rothe and Bernstein, 1959) and CsCl molecules (Schumacher et al., 1960) with a large number of targets of varied complexity and reactivity. Their single-beam apparatus is shown schematically in Fig. 1-10-1. Not shown is the vacuum envelope, which is divided by a slotted bulkhead at C into two separately pumped regions, the *oven chamber* and the *detector chamber*. Typical operating pressures in these chambers, which contain large liquid nitrogen traps and baffles, are 5×10^{-7} and 1×10^{-7} torr, respectively. The vacuum envelope has an 8-in. diameter and is 24 in. long.

The main components of the apparatus are a Monel over (A) and effusion slit (B) for production of the neutral beam (Note 1-2), a scattering chamber (F), and a Langmuir–Taylor surface ionization detector (I, J) for detection of the

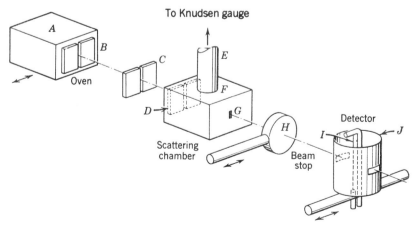

Figure 1-10-1. Bernstein's neutral beam–gas target apparatus for measurement of integral elastic scattering cross sections. The effusion oven and slit system produce a thermal beam of neutral atoms or molecules. The detector is of the surface ionization type. [Adapted from Rothe and Bernstein (1959).]

unscattered component of the beam (Note 1-3). Oven temperatures in the vicinity of 500 K are used to produce K and Cs beams. The oven is fitted with an *ideal slit** of width 0.0025 cm, and a similar slit is located at the entrance to the scattering chamber for beam collimation. The effective scattering path inside this chamber is 4.44 cm. Pressures in the range 1×10^{-6} to 2×10^{-4} torr are used in the scattering chamber. These pressures are measured with a Knudsen gauge.

A tungsten filament (I) about 5 cm long and 0.0025 cm in diameter is used in the surface ionization detector. The filament is heated to about 1500 K by 75 mA dc and is biased 90 V positive with respect to the ion collector (J). A preamplifier and inverse-feedback dc amplifier are used to measure the ion currents, which are typically about 10^{-10} A, corresponding to a beam intensity of 6×10^8 atoms/s. The distance from slit B to slit D is 11.12 cm, and slit D is located 19.68 cm from I. The calculated half-width of the unscattered beam at the detector is 0.007 cm; the observed value is about twice that size. The angle subtended at the midpoint of the scattering path is about 30 s, and the overall resolution of the apparatus is calculated to be 2 min of arc.

For each projectile–target combination the beam intensity is measured for 10 to 20 values of scattering gas pressure, corresponding to 5 to 95% attenuation of the beam. The 100% transmittance level I_0 is recorded before and after each series of measurements during evacuation of the scattering chamber. Plots of the logarithm of the attenuation ratio I/I_0 versus pressure are usually linear for $I/I_0 > 0.1$. From the slope of these plots the integral cross section q_s may be calculated by means of the *Rosin–Rabi equation:*†

$$q_s = 2\sqrt{\pi} \, J(z)nd \ln \frac{I}{I_0}, \qquad (1\text{-}10\text{-}1)$$

where n is the number density of the scattering gas, d the scattering path length, and $J(z)$ an integral whose value is a function of the masses of the projectile and target and of the temperatures of the beam and scattering gas (Rosin and Rabi, 1935). The experimental apparatus and technique used in the CsCl experiments were essentially the same as those we have described, but with some degradation in angular resolution.

The Bernstein apparatus clearly illustrates the basis for measurements of the integral cross section, and it produced valuable data. However, many important experimental advances have been made since that apparatus was put into operation, and great improvements in the flexibility and power of the technique have resulted. We now illustrate some of these advances by describing apparatus developed at Leiden University around 1980.

---

*The design of this type of slit is discussed by Miller and Kusch (1955).

†The Rosin–Rabi equation accounts for the thermal velocity spreads of the particles. See Pauly and Toennies (1968, pp. 283–287) for a general discussion of this subject.

B.   Leiden Neutral-Beams Integral-Cross-Section Apparatus

We first describe the general layout of the Leiden instrument and then explain the rationale behind the choice of various specific features. The integral cross section is determined here by measuring the attenuation of a neutral projectile beam in its passage through a neutral target beam intersecting it at 90° (Fig. 1-10-2). Many examples of this crossed-beam geometry appeared in Chapters 4 through 7 of the companion volume in the context of elastic and inelastic electron collisions.

The Leiden apparatus (Linse et al., 1979; van den Biesen et al., 1982) utilizes nozzle expansion in each of the two source chambers to produce the beams (Note 1-2). A third "main" chamber contains the collision volume, and a fourth (maintained at ultrahigh vacuum) houses the detection apparatus. The "primary" (i.e., projectile beam) enters the main chamber through a skimmer that defines the beam's transverse dimensions; it then passes through a cryopump and a slotted-disk velocity selector (Note 1-4). The projectiles next proceed through a shutter, a 30-Hz modulation chopper, a collimator, and the scattering region. When the beam enters the detection chamber, some of the projectiles are ionized by electron bombardment, and the resulting ions are subjected to mass selection by a quadrupole filter. Finally, the ions selected are detected by a Channeltron particle multiplier, and the signal is measured with a lock-in amplifier that selectively amplifies the 30-Hz component. Both the projectile and target beams can be interrupted by shutters, so that the background signal can

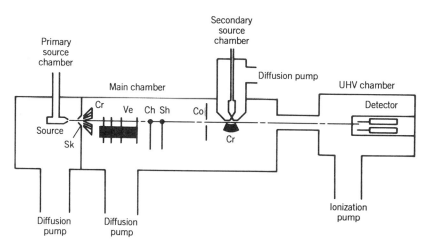

Figure 1-10-2. Leiden crossed-neutral-beam apparatus for integral cross-section measurement. Nozzle expansion is used to produce both the "primary" (projectile) beam and the "secondary" (target) beam. Slotted rotating disks (Ve) provide velocity selection on the primary beam. Sk, skimmer; Ch, 30-Hz chopper; Sh, shutter; Co, collimator. [From Linse et al. (1979).]

be measured as well as the signals for the attenuated and unattenuated projectile beam. A computer provides for control of the shutters and for on-line data handling. Normally, the projectile source is maintained at room temperature. Low projectile velocities can be achieved by cooling the source; higher velocities, by heating the nozzle or by using the technique of seeded beams (Note 1-2). The target nozzle is kept at room temperature. No absolute calibration of the density integral along the scattering path of the projectile beam is made, so only relative integral cross sections can be measured.

The consideration that determined the basic design of the Leiden apparatus was the desire for flexibility in the choice of collision partners. The surface ionization detector, used in the Bernstein apparatus of Fig. 1-10-1, played an extremely important role in the early days of molecular beam research because it was the only detector available for practical use at thermal energies. However, its use for a given projectile depends on that particle having an ionization energy that is lower than the work function of the metal used in construction of the detector. This requirement can be satisfied for the alkali atoms as thermal energy projectiles and for a relatively small number of other atomic and molecular species. The heavier alkalis are especially well suited for thermal energy detection and can be detected on a hot tungsten wire with nearly 100% efficiency (see Note 1-3). The electron bombardment ionizer used by the Leiden group can be used to detect any atomic or molecular species (although with efficiencies only on the order of 10^{-3} %), and it is now used almost universally. (Ultrahigh vacuum should be maintained in the detector chamber to minimize the spurious contribution from background gas. The mass spectrometer cannot distinguish between beam particles and residual gas particles of the same mass.) To achieve the greatest flexibility in the choice of collision partners, it is necessary to use an intersecting-beam technique rather than the single beam-static gas arrangement that was standard before about 1960. The crossed-beam approach permits the controlled use of unstable particles (such as atomic hydrogen, nitrogen, and oxygen) and state-selected species (such as atoms in known angular momentum states, and molecules in vibrationally excited states).

The use of a beam for the target and the inefficient electron bombardment ionizer for the detector necessitate the choice of beam sources that provide much greater particle densities than can be obtained with a simple effusion source. Nozzle expansion sources, such as used by the Leiden group, permit gains of about a factor of 100 for each beam. They also provide cold molecules in their lowest rotational states. Nozzle sources are discussed in Note 1-2 together with other types of beam sources. The mechanical velocity selector used here permits measurements on projectiles of well-defined energy and accordingly, yields much more precise information than would otherwise be possible. The rotating slotted-disk velocity selector is discussed in Note 1-4; time-of-flight techniques can be used as an alternative (Note 1-5).

Van den Biesen (1988) has emphasized the need for great care in aligning the beams in integral cross-section measurements and for determining the inter-section angle of the beams accurately. He also stresses the fact that a common source of error in experiments utilizing supersonic beams is the presence of dimers in the beams.

C. Göttingen Crossed-Neutral-Beams Differential Scattering Apparatus

The Göttingen apparatus, illustrated in Fig. 1-10-3, was designed for the study of both elastic and inelastic collisions at high angular and energy resolution (Faubel et al., 1982; Toennies, 1985). We describe it as it is used for measurement on collisions of He atoms with other species.

Here a high-intensity, nearly monoenergetic beam of He atoms with a velocity resolution of $\Delta v/v \leqslant 0.5\%$ is generated in a high-pressure nozzle expansion (Note 1-2). A 5- to 30-μm-diameter nozzle is employed; it operates at a backing pressure of 100 to 200 atm and expands into a 10^{-3} mbar $(7.5 \times 10^{-4}$ torr) vacuum, which is maintained by a 12,000-L/s diffusion pump.

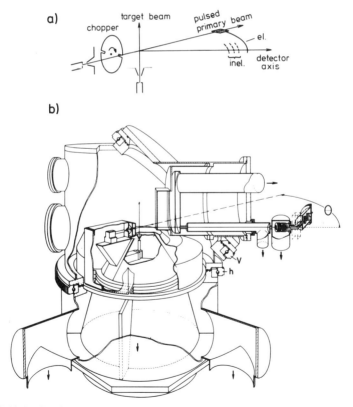

Figure 1-10-3. Göttingen crossed-neutral-beam differential scattering apparatus. This instrument is used for both elastic and inelastic studies at high angular and energy resolution. [From Toennies (1985).]

In the expansion, the gas is cooled adiabatically to an extremely low translational temperature with respect to an average velocity that is about 50% greater than the most probable velocity in the source. Almost all of the energy of the stagnation gas is converted into translational energy of the gas stream, and the internal gas temperature is reduced to less than 20 mK. If gases other than helium are used in a nozzle beam, dimerization sets in when internal gas temperatures below 1 to 5 K are achieved in the expansion. Temperatures of about 1 to 10 K and velocity spreads of 5% are typical. For the target gases N_2, O_2, CO, and CH_4, a nozzle of 100 μm diameter is used, and it is operated at a maximum pressure of 6 atm. Another fortunate feature of nozzle beam sources is that the temperature of the internal states is also cooled to a few kelvin, and the beam molecules are prepared in their rotational ground states.

Both the helium and the molecular beam are collimated to angular spreads less than 2° as they pass through slits in several stages of differential pumping, and the beams intersect at right angles in the center of the apparatus. Time-of-flight energy-loss spectroscopy can be performed by chopping one of the beams with a high-speed rotating slotted disk. One such chopper contained four slots in a 10-cm-diameter wheel and provided bursts of 10-μs duration. The neutral particle detector consists of a mass spectrometer and a particle counter located 165 cm from the scattering center in a housing maintained at 10^{-11} torr. The scattered particles can be detected over most of the upper hemisphere by rotating the detector flange about either of the two axes set at 45° with respect to each other. The overall energy resolution of this apparatus is about 0.5 meV, which is much better than can be achieved in similar ion-scattering apparatus.

D.   Linder's Ion-Neutral-Crossed-Beams Differential Scattering Apparatus

The powerful Linder instrument (Hermann et al., 1978) is shown schematically in Fig. 1-10-4a, and in more detail in Fig. 1-10-4b. Despite the well-known difficulties associated with handling beams of charged particles at low velocities,* this apparatus has provided ion beam energies E as low as 0.1 eV in the Lab frame, with energy spreads ΔE of 20 meV and an overall angular resolution of about $\pm 2.5°$ (Konrad and Linder, 1982).

A 90° crossed-beam geometry is employed, the scattering products being detected in a plane perpendicular to the neutral gas beam. Ions are produced in a sidearm, mass-selected at an energy of 200 eV (typically), decelerated to 3 eV, and injected into an electrostatic energy selector, which consists of two 127° cylindrical condensers in tandem. The final shape and energy of the projectile

---

*These difficulties arise mainly from space-charge repulsion in the ion beam, spurious electric fields produced by charging of low-conductivity surfaces, and stray electric and magnetic fields. These topics and others that relate to heavy particle collision apparatus are discussed in Sections 4-1-A and 4-3-B and Chapters 5 through 7 of the companion volume.

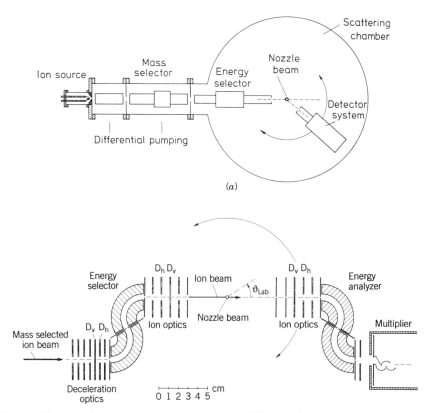

Figure 1-10-4. Linder's ion–neutral-crossed-beam differential scattering apparatus. Ion beam energies as low as 0.1 eV can be reached in the Lab frame. [From Hermann et al. (1978).]

ion beam is determined by zoom optics. The ion beam current in the collision region is typically about 2×10^{-10} A.

The neutral target beam is produced by a supersonic nozzle with an angular divergence of approximately $\pm 1.5°$ and a Mach number of about 26. The optical system for the ion beam emerging from the collision region is similar to that for the projectile ion beam, and it can be rotated in a plane perpendicular to the neutral beam. Digital techniques are used for detection of the scattered ions and processing the data. Ultrahigh-vacuum techniques are used in the main scattering chamber.

This apparatus was developed by F. Linder and his colleagues at the University of Kaiserslautern. It has been used for a variety of studies of elastic and inelastic collisions of ions, at energies on the order of 10 eV and down to well below 1 eV.

1-11. EXPERIMENTAL DATA;
THEIR INTERPRETATION AND APPLICATION

There now exists a substantial body of reliable cross sections for elastic collisions of heavy particles in the low-energy region (CM energies below a few tenths of an electron volt). Most of the data relate to neutral–neutral scattering, and a few to ion–neutral collisions, which, however, have been studied extensively in the high-energy regime. Integral cross sections as a function of collision energy or impact velocity generally show a great deal of structure, as do the differential cross sections as a function of scattering angle. Hence it is necessary to discuss the sources of this structure (Section 1-11-A) before we turn to the experimental data (Section 1-11-B).

We shall have occasion to use the *classical deflection function* $\Theta_{+}(b)$, which is related to the CM scattering angle $\Theta(b)$ by

$$\Theta = |\Theta_{\pm}|, \qquad 0 < \Theta < \pi. \tag{1-11-1}$$

Thus Θ is the magnitude of $\Theta_{\pm}$, modulo π. The positive or negative sign applies according as the collision has a net repulsive or attractive effect.

A. Origin of Structure in Elastic Cross-Section Curves

1. *Glory Undulations and Rainbows*

In our discussion of (3-7-2) in Section 1-2-B, we pointed out that the differential elastic cross section per unit solid angle, $I_{s}(\Theta)$, becomes infinite if $\Theta = n\pi (n = 0, 1, 2, \ldots)$ for $b \neq 0$, because of the $\sin \Theta$ term in the denominator, in which case we have a *glory*.* (Here, as usual, b is the impact parameter.) $I_{s}(\Theta)$ also becomes infinite if $d\Theta/db = 0$, with the appearance of a *rainbow*. Figure 1-11-1 shows in a striking fashion how these two effects arise. Here β is a reduced impact parameter defined as

$$\beta \equiv \frac{b}{r_{m}}, \tag{1-11-2}$$

r_{m} being the separation distance at the potential well minimum. The case of a head-on collision ($\beta = 0$) corresponds to $\Theta_{+} = \pi$. Increasing values of β are associated with smaller scattering angles, and for grazing collisions ($\beta \gg 2$), $\Theta_{+} \approx 0$. For the values of impact parameter covered in Fig. 1-11-1, negative values of Θ_{+} correspond to "net attractive" trajectories, that is, trajectories showing an overall attraction by the scattering center even though there may be a repulsive component in the interaction potential. For the *glory impact*

---

*The condition $\Theta = 2n\pi (n = 0, 1, 2, \ldots)$ produces *forward glories*, that is, singularities in the forward direction, whereas $\Theta = (2n + 1)\pi$ $(n = 0, 1, 2, \ldots)$ yields *backward glories*, singularities in the backward direction.

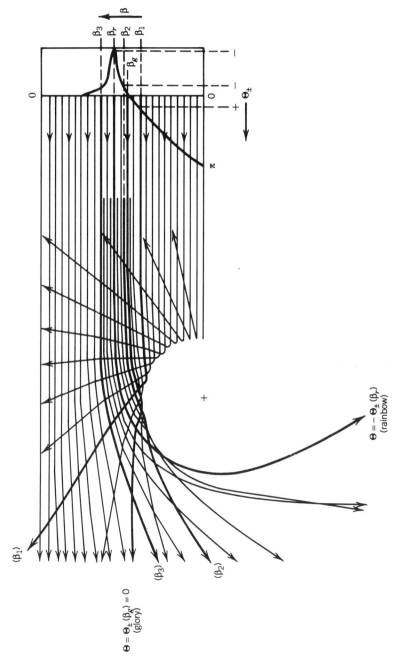

Figure 1-11-1. Classical trajectories plotted as a function of the reduced impact parameter β (main part of drawing). The small graph at the right of the figure shows the corresponding classical deflection function $\Theta_{\pm}(\beta)$. [From Pauli (1979, p. 127).]

45

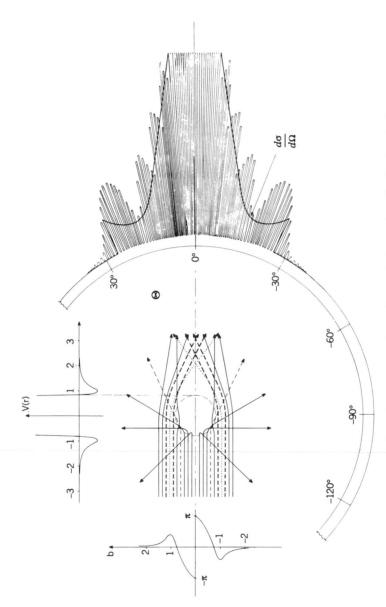

Figure 1-11-2. Schematic representation of a typical angular distribution in heavy particle elastic scattering (see Fig. 1-11-1). [From Faubel and Toennies (1978, p. 235).]

parameter β_g, the net deflection is zero, although the projectile is first attracted and then repelled by the scattering center. The classical deflection function has its greatest negative value for the *rainbow impact parameter* β_r, corresponding to the "most attractive" trajectory. Of course, both β_g and β_r depend on the impact energy and on the detailed shape of the potential function. Figure 1-11-2 displays some of the information contained in Fig. 1-11-1 but provides a useful indication of the angular scattering distribution as well.

The effect of glories on the integral elastic cross section is illustrated schematically in Fig. 1-11-3, which shows a plot of the reduced cross section $q_s^* = q_s/2\pi r_m^2$ as a function of the reduced impact velocity $v_0^* = \hbar v_0/\varepsilon r_m$. (Here again, r_m is the equilibrium separation distance and ε is the well depth.) The multiplicity of the glories leads to their description as *glory undulations*. The glories are numbered by the index N, in order of appearance, starting at the highest velocity. Dashed lines at the right of Fig. 1-11-3 indicate the behavior of the cross section for the attractive and repulsive components of the Lennard-Jones (12, 6) potential. The variation of q_s^* with v_0^* for the attractive component can be calculated from the Landau–Lifshitz equation (see Note 1-1). The indicated effects of orbiting resonances and symmetry oscillations on the curve in Fig. 1-11-3 are discussed below.

Effects of rainbows appear in a plot of the differential elastic cross section versus scattering angle, as in Fig. 1-11-4. For a potential of the usual shape such as shown here, the cross section is dominated by peaks at $\Theta_\pm \approx 0°$ that come (1) from scattering off the long-range attractive branch ($b \approx b_1$; see the "small

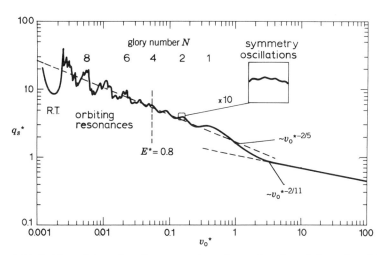

Figure 1-11-3. Plot of the reduced cross section q_s^* as a function of the reduced impact velocity v_0^* showing glories, orbiting resonances, symmetry oscillations, and the Ramsauer–Townsend effect. The dashed lines at the right show the behavior for the attractive and repulsive components of the Lennard-Jones (12, 6) potential. [Adapted from van den Biesen (1988 p. 473).]

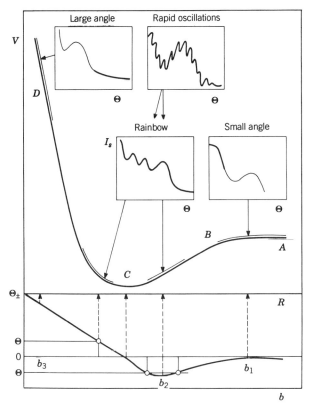

Figure 1-11-4. Differential elastic cross section versus scattering angle shown in the four insets at the top for scattering from four regions of an interaction potential of the usual shape (center of drawing). The variation of the classical deflection function $\Theta_{\pm}(b)$ with impact parameter is shown at the bottom. [Adapted from Buck (1988b, p. 500).]

angle" inset) and (2) from a rainbow associated with the minimum of the classical deflection function ($b = b_2$). As Buck (1988b) points out, the rainbow angle is given by

$$\Theta_r = C\frac{\varepsilon}{E},\qquad(1\text{-}11\text{-}3)$$

where C is a constant that depends on the form of the potential, having approximately the value 2 for the Lennard-Jones potential. The rainbow maximum in $I_s(\Theta)$ is important because its observation indicates the angular range ($\Theta < \Theta_r$) that is strongly affected by the attractive component of the potential, and the range ($\Theta > \Theta_r$) that is sensitive to the repulsive part (see Fig. 1-11-2). In Fig. 1-11-4 the inset labeled "large angle" refers to scattering of a projectile with a small impact parameter, such as b_3.

2. *Interference Oscillations*

These oscillations appear superimposed on the cross-section curve when trajectories with different impact parameters correspond to the same scattering angle (Buck, 1988b, pp. 500–501). If two trajectories (i and j) are involved, the angular spacing between adjacent extrema is

$$\Delta\Theta = \frac{2\pi}{\kappa(b_i - b_j)},\tag{1-11-4}$$

where κ is the wave number of the relative motion. The widely spaced oscillations in the inset of Fig. 1-11-4 marked "rainbow" are produced by interference of projectiles with different impact parameters near the rainbow position that are scattered through the same angle. Finally, the inset "rapid oscillations" in Fig. 1-11-4 with the small angular spacings is produced by the interference of projectiles scattered by the attractive and repulsive parts of the potential. This situation is illustrated by Fig. 1-11-5.

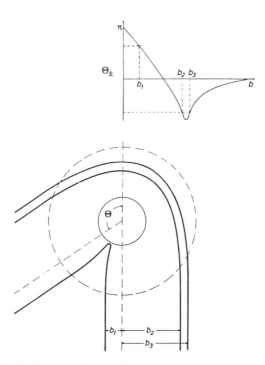

Figure 1-11-5. Classical trajectories produced by a Lennard-Jones potential (*bottom*), and the classical deflection function $\Theta_\pm(b)$ as a function of the impact parameter (*top*). The impact parameters b_1, b_2, and b_3 all produce scattering through an angle Θ. [From Pauly and Toennies (1968, p. 313).]

3. *Symmetry Oscillations in the Case of Identical Collision Partners*

In Section 1-7-D-4 we discussed the scattering of identical particles and demonstrated some of the symmetry effects that arise in this case. If the collision partners are in the same spin state, the total spin of the system has a definite value. Then [3-18-8] and [3-18-9] give the differential elastic cross section for bosons and fermions, respectively. The integral cross section [3-15-34] becomes

$$q_s = \frac{8\pi}{\kappa^2} \sum (2l + 1) \sin^2 \eta_l, \tag{1-11-5}$$

where the sum is over even values of l for bosons and over odd values of l for fermions (Pauly, 1979, pp. 157–159). Figure 1-11-3 illustrates schematically oscillations in the cross section that arise from the symmetry effects described here. For more detail, see van den Biesen (1988, pp. 472–474, 492–494), Buck (1988b, pp. 511–512), and Pauly (1975).

4. *Orbiting Resonances*

In Section 1-5 we described conditions under which a projectile can orbit about a scattering center. For the Lennard-Jones (12, 6) potential, orbiting is possible classically provided that the CM collision energy is less than 0.8ε (van den Biesen, 1988, p. 474). A semiclassical treatment (Ford et al., 1959) shows that resonances between the impact energy and the quasibound levels of the collision complex produce rapid changes in the phase shift of the orbiting particle. Only a few phase shifts are important at the low energies considered here, so a rapid change in one of them can result in a sharp resonance peak, as in Fig. 1-11-3.

5. *Ramsauer–Townsend Effect*

At extremely low impact energies, only the $l = 0$ partial wave is significant in [3-15-34] for the integral cross section (see Section 1-7-C). Under these conditions, the undulations in $\sin^2 \eta_0$ as a function of impact velocity translate into the Ramsauer–Townsend (RT) undulations shown in Fig. 1-11-3 (van den Biesen 1988 pp. 473, 475). The reader may wish to recall our discussion of the Ramsauer–Townsend effect in Sections 4-1-A and 4-8-A of the companion volume.

B.   Experimental Cross Sections; Potential Determinations

In this final section on elastic collisions of heavy particles at low energies we present some representative modern data on cross sections that were obtained in crossed-beam experiments. In such experiments the projectile and target particles are kept essentially free of perturbing factors until they enter the beam intersection region, where the particles interact in a controlled fashion. State selection of either the projectile or the target beam (or both) may be feasible, and is essential for certain types of studies. Only the crossed-beam configuration provides the extremely high angular and velocity resolution required to resolve the structure in the differential cross section (Buck, 1988b pp. 499, 502–503).

As stated earlier, our main interest in heavy particle elastic scattering relates to the determination of potentials. The attractive potential at large r (region A in Fig. 1-11-4) is probed by the average integral and small-angle differential cross sections. Near the inflection point (marked B) on the attractive branch, the rainbow structure of $I_s(\Theta)$ dominates. The region near the potential minimum (C) is connected with the glory effect in q_s. The shape of the repulsive part of the potential (in region D) can be determined by measurements of q_s at high energies and of $I_s(\Theta)$ at large angles. The number densities of the beams are difficult to measure, so indirect methods are generally used to determine absolute values of $I_s(\Theta)$. In many cases $I_s(\Theta)$ at small angles can be calculated accurately from theory and used for calibration of the corresponding experimental values. The absolute scale of the potential is determined by absolute measurements of q_s or by oscillations in $I_s(\Theta)$. If all the structure in the cross-section curves is fully resolved, the interaction is completely determined. In some cases it is possible to determine the potential by direct inversion (Buck, 1975, 1988b).

Before about 1965, neutral–neutral beam studies were restricted almost entirely to the scattering of alkali atoms by various target gases. By 1985, following the experimental improvements cited in Sections 1-10-B and 1-10-C, reliable data were available for a variety of projectiles that include the alkalis, the noble gas atoms, hydrogen atoms, mercury, and some simple molecules (such as H_2, N_2, and O_2). Results have also been obtained with excited and state-selected beams. The reduction of the raw scattering data to a form suitable for potential determination is not a simple matter. One must take into account the angular and velocity spreads of the beams, the finite size and shape of the apertures in the apparatus, and the nonhomogeneous density of the beams. Finally, there comes the problem of obtaining the interaction potential, which along with the other topics mentioned here, is treated by van den Biesen (1988), Buck (1988b), Pauly (1979), and others.

A set of integral cross sections for the elastic scattering of H atoms by each of the noble gases is shown in Fig. 1-11-6. The CM impact energy range extends from 0.01 to 1.00 eV. The hydrogen atoms were produced by thermal dissociation in a tungsten oven heated by electron bombardment to 2500 to 2600 K, and then velocity selected in a rotating disk selector. The curves are shifted vertically for clarity.

Integral scattering data for noble gas pairs in the glory region are displayed in Fig. 1-11-7. In each case, q_s is multiplied by the velocity factor $v_0^{2/5}$ to make the glory undulations appear on a horizontal line.

Figure 1-11-8 shows high-resolution angular distributions for Ar–Ar scattering at a CM impact energy of about 80 meV. A rainbow maximum is prominent at about 8°, and fast oscillations and symmetry interferences are in evidence, as

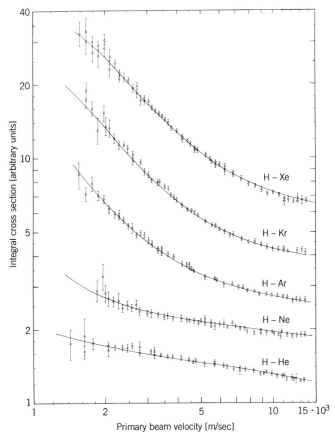

Figure 1-11-6. Experimental data on collisions of H atoms with rare-gas atoms, showing the typical variation of the "total" effective scattering cross section with velocity in the high-energy regime. [Adapted from Bickes et al. (1973).]

indicated. Rainbows and rapid oscillations are also clearly visible, at lower resolution, in earlier measurements by Buck and Pauly (1971), who studied differential scattering for Na–Hg at energies ranging from 0.179 to 0.251 eV.

Feltgen et al. (1982) have made definitive measurements of integral cross sections as a function of velocity that show the expected symmetry effects for ^{4}He–^{4}He and ^{3}He–^{3}He scattering and none for ^{3}He–^{4}He (Fig. 1-11-9). In their experiment, helium beams passed through a scattering cell that contained helium target gas at 1.6 K. Symmetry oscillations occur with identical collision partners because the detector cannot distinguish between the two different scattering events that produce the same result and therefore events wherein the projectile is detected at angle Θ interfere with those for which the target is detected at Θ. Finally, we should note the Ramsauer–Townsend effect manifested in the low-velocity region.

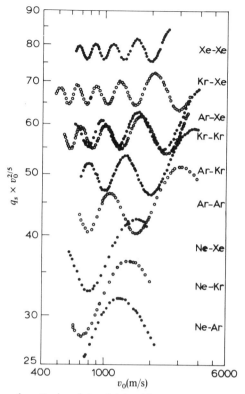

Figure 1-11-7. Integral scattering data obtained by measurements on noble-gas pairs in the glory region. The cross section q_s is multiplied by the velocity factor $v_0^{2/5}$ in order to present the glory undulations on a horizontal line. [From van den Biesen et al. (1982).]

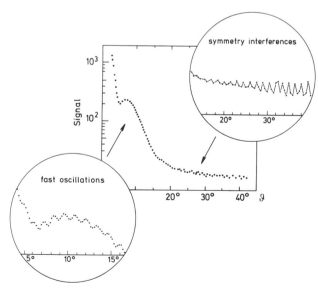

Figure 1-11-8. Angular distributions measured for Ar–Ar scattering at a CM energy of about 80 meV. [From Toennies (1985, p. 34).]

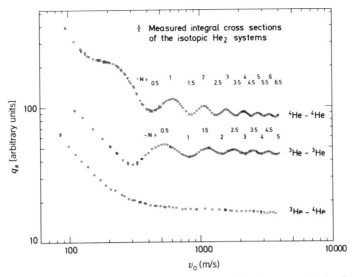

Figure 1-11-9. Integral cross sections as a function of the impact velocity for various isotopic helium collision systems. Symmetry oscillations are in evidence for ^{4}He–^{4}He and ^{3}He–^{3}He, but not ^{3}He–^{4}He, scattering. [From Feltgen et al. (1982).]

Figure 1-11-10 shows orbiting resonances in the integral cross section for collisions of H atoms with Ne, Ar, Kr, and Xe. The numbers attached to the curves are the rotational quantum numbers l of the resonating quasi-bound states. Here the vibrational quantum number equals zero.

Interesting studies have also been made with electronically excited states. Alkali atoms can be excited with lasers; electron impact excitation is applicable to other species. Beyer and Haberland (1984) have observed *g-u oscillations* in collisions of Ne* metastables with ground-state Ne atoms. Such oscillations arise when two potential curves of gerade (g) and ungerade (u) symmetry, respectively, are present (Morrison et al., 1976) because the two different potentials contribute coherently to the scattering amplitude.

A series of studies have been made at Göttingen on vibrationally excited molecules colliding with various species. In one case, Rubahn and Toennies (1988) generated a beam of Li_2 molecules that were excited to high vibrational states ($v = 0, 20$) by Franck–Condon pumping with a continuous-wave ring dye laser. They compared the scattering behavior of the $v = 0$ and $v = 20$ molecules in collisions with Kr atoms and observed a large increase in the anisotropy with vibrational excitation. In general, when molecules are scattered, interference effects are damped by the anisotropy of the potential, which also produces inelastic transitions. Hence studies of such systems provide information concerning not only the spherically averaged potential but also the anisotropy through the observed quenching of the structure.

A more powerful technique for investigating the anisotropy of the interaction

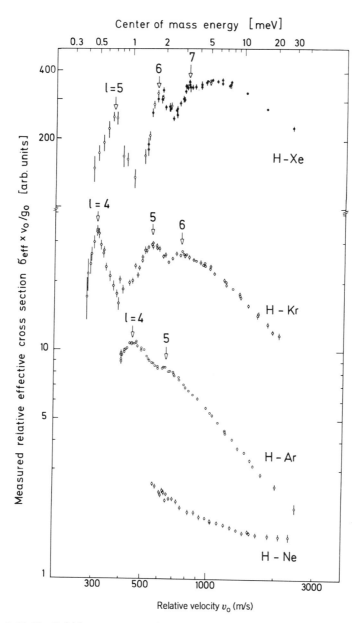

Figure 1-11-10. Orbiting resonances in the integral cross section for scattering of H atoms by noble-gas atoms. The numerical labels are values of the rotational quantum number l of the resonating quasi-bound states. The quantity v_0/g is the H projectile velocity divided by the most probable relative velocity. [From Toennies et al. (1979).]

potential is to prepare beams of molecular projectiles in specific orientation-dependent m_j states (Note 1-6) and compare the cross sections for parallel and perpendicular molecular axes with respect to a guiding field.

Finally, we should mention that measurements on inelastic processes not only provide information concerning the mechanisms of energy transfer between the various degrees of freedom, but also illuminate the portions of the interaction potential that control the energy transfer (Buck, 1988a). As explained in Sections 5-3, 5-4, and 5-7 of the companion volume, the potential must be angle dependent to induce rotational transitions, it must depend on the internuclear coordinate(s) if vibrational transitions are to be possible, and electronic transitions require a dependence on the electron coordinates. Inelastic collisions between heavy particles are treated at length in subsequent chapters.

PART $\mathscr{C}$ HEAVY PARTICLE COLLISIONS AT HIGH ENERGIES

1-12. NEUTRAL–NEUTRAL COLLISIONS AT ENERGIES ABOVE A FEW TENTHS OF AN ELECTRON VOLT

A. Experiments Designed for Determination of Interaction Potentials

Fast beam-scattering studies complement thermal and epithermal studies by providing information concerning particle interactions at closer distances of approach. We must iterate here that not all of the deflected particles can be detected, because of the strong concentration of the scattering in the forward direction. Indeed, what constitutes a scattering event at angles near $\Theta = 0$ is defined only through the uncertainty principle. As explained in van den Biesen (1988, pp. 477–486), q_s is determined by scattering through CM angles $\Theta > \Theta_C$, where Θ_C is a critical angle determined by the nature of the collision partners, the apparatus geometry, and the energy of the beam(s).* Van den Biesen derives equations for Θ_C, discusses the reduction of the scattering data, and develops equations for velocity and angular resolution corrections. He concludes that (1) only if the angular resolution is better than Θ_C can corrections for the finite angular resolution be applied successfully, and (2) only if the width of the velocity distribution is smaller than the characteristic width of the structural features in $q_s(v_0)$ is the deconvolution of the "effective cross section" to q_s feasible. If these two conditions are met, the angular and velocity resolution corrections can be calculated from any potential that describes the corrected data satisfactorily. If the conditions are not fulfilled or if no potential can be found that predicts the deconvoluted results, one must resort to a trial-and-error procedure and a multiparameter potential to obtain the interaction from measurements of q_s. Van den Biesen (1988) discusses all of these matters, as well as calibration procedures, in detail.

---

*This discussion is closely related to the discussions leading to (1-6-2) and (1-9-1).

Amdur and his colleagues at MIT made an important series of measurements of the elastic scattering of neutrals and ions by neutrals over a period of three decades starting in 1940. They used collision energies ranging from a few hundred to a few thousand eV and probed the repulsive potential energy range between about 0.1 and 20 eV. Much of their work is summarized by Smith (1971a). Van den Biesen (1988, p. 495) displays a graph showing data on He–He scattering obtained by the MIT group and by two other groups. The *apparent integral cross section*, defined as the cross section determined directly from the measured attenuation of the beam, is plotted as a function of the impact energy E. Each set of data shows a simple monotonic decrease as E increases, the differences in the magnitudes of the cross sections being ascribable to the differences in angular resolution in the various experiments.

We should also mention differential cross-section measurements made at SRI International for He* on He (Morgenstern et al., 1973) and for Ar* on Ar (Gillen et al., 1976) at CM energies between 5 and 10 eV. Scattering data on these systems were used to obtain information on their potential curves that is useful in the development of rare-gas excimer lasers.

B. Differential Cross-Section Measurements with H and He Projectiles in Stebbings' Laboratory

Johnson et al. (1988) have measured absolute differential cross sections for the small-angle scattering of H and He atoms by H_2 and N_2 molecules at 0.5, 1.5, and 5.0 keV over the Lab angle range 0.05 to 0.5°. Neither of the projectiles was spin polarized. Their apparatus is shown in Fig. 1-12-1. Ions are created in an electron-bombardment source, accelerated to the desired energy, and focused by an electrostatic lens. After momentum analysis by a pair of confocal 60° sector magnets, the beam then passes through a charge-transfer cell (CTC), about 10% of the particles emerging as neutrals. The emerging ions are swept aside by the electrostatic deflection plates (DP1). A high degree of collimation is obtained by using a 20-μm-diameter exit aperture in the CTC and a 30-μm entrance aperture in the target cell (TC). These apertures are separated by 49 cm, so the neutral beam is collimated to less than 0.003° divergence, and its intensity is reduced to about 10^3 per second in the target cell. In the CTC, helium atoms are produced by charge transfer of He^+ with He, a resonant process that produces essentially nothing but ground-state He atoms. Krypton is chosen to neutralize the H^+ ions. Here the charge transfer is near-resonant, so the electric field in DP1 quenches any $2s$ atoms by Stark mixing (Bransden and Joachain, 1989, pp. 527–533), and the H atoms are produced predominantly in the ground state. The TC is 0.36 cm thick with a 300-μm exit aperture. Gas pressures between 1 and 10 mtorr are used in the TC and are measured with a MKS Baratron capacitance manometer. Charged collision products are removed from the final beam by the deflection plates DP2. Both scattered and unscattered projectiles are detected by a position-sensitive detector [PSD; see Section 5-9 of the companion volume and Gao et al. (1988b)] located 109 cm from the TC on the

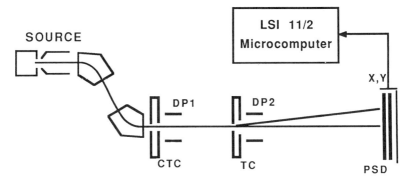

Figure 1-12-1. Apparatus used by Stebbings's group at Rice University for studies of small-angle elastic scattering and charge transfer. CTC, charge transfer cell; DP, deflection plates; TC, target cell; PSD, position-sensitive detector. [From Johnson et al. (1988).]

beam axis. The PSD has an active diameter of 2.5 cm, so the maximum observable scattering angle is about 0.7°.

Absolute differential cross sections for $H-N_2$ scattering are presented in Fig. 1-12-2, along with earlier results obtained in Stebbings's laboratory at Rice University (Newman et al., 1986). (The results for the other projectile–target combinations are similar.) Since no energy-loss measurements are made on the scattered projectiles, the cross sections contain inelastic contributions. *Diffrac-*

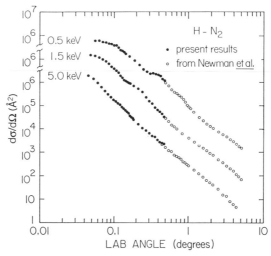

Figure 1-12-2. Absolute differential cross sections for $H-N_2$ scattering, measured by the Rice group. [From Johnson et al. (1988).]

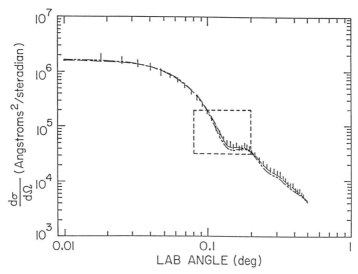

Figure 1-12-3. Absolute differential cross sections for ^{4}He–^{4}He small-angle scattering, measured by the Rice group. [From Nitz et al. (1987).]

*tion undulations** are observed in some of the plots of $d\sigma(\vartheta)/d\Omega$ versus ϑ for H and He projectiles at scattering angles near $0°$. The amplitudes of these undulations are smaller than those observed in previous experiments at Rice on the scattering of He projectiles by the noble-gas atoms (Nitz et al., 1987; Gao et al., 1987).

The absolute, small-angle differential cross section for ^{4}He–^{4}He elastic scattering at 0.5-keV impact energy is displayed as a function of the Lab scattering angle in Fig. 1-12-3. Classical theory predicts that $\sigma(\vartheta)$ should rise monotonically and diverge as $\vartheta \to 0$, but experiment shows an undulation superimposed on the classical curve and a flattening out that varies as $\exp(-c\vartheta^2)$ at small angles, in agreement with quantal calculations. The measurements by Nitz et al. (1987) were made at impact energies of 0.5, 1.5, and 5.0 keV and at very small scattering angles. Under these conditions, ^{4}He–^{4}He scattering does not involve excited states of the collision complex, and only one

*An undulation of this type is called a *forward diffraction peak* (Stebbings, 1988). It is produced by the interference of a large number of trajectories with a broad range of impact parameters associated with small-angle scattering from the tail of the potential. If observed as a function of impact energy, a given undulation feature resembles optical diffraction from a disk, moving to smaller angles as the de Broglie wavelength decreases (Beier, 1973). Diffraction undulations are unrelated to glory and rainbow scattering, which are associated with only a few impact parameters. The diffraction peak can be characteristic of either the attractive or repulsive part of the potential, depending on the impact energy and the nature of the potential. The forward diffraction peak is the same as the Fraunhofer peak observed at low energies. See Note 2-5 for a discussion of Fraunhofer diffraction.

potential produces scattering. Hence the differential cross section is given by (Feynman et al., 1965, pp. 30–32)

$$\frac{d\sigma}{d\Omega} = |f(\Theta) + f(\pi - \Theta)|^2, \qquad \text{[3-18-2]}$$

where the amplitude $f(\Theta)$ refers to scattering of projectiles at angle Θ and the amplitude $f(\pi - \Theta)$ relates to targets recoiling at angle Θ, [i.e., projectiles being scattered at angle $(\pi - \Theta)$]. For small-angle scattering, $f(\pi - \Theta)$ is negligible compared to $f(\Theta)$, so here we put

$$\frac{d\sigma}{d\Omega} = |f(\Theta)|^2. \qquad (1\text{-}12\text{-}1)$$

No interference effects are observed.

In Stebbings's laboratory, Chitnis et al. (1988) have also measured relative differential cross sections for He–He collisions in the keV region at *large* scattering angles (35 to 56°). They utilized two position-sensitive detectors to detect both collision partners in coincidence following the collision. Their data on ^{4}He–^{4}He are shown in Fig. 1-12-4. At the large scattering angles covered here, $f(\pi - \Theta)$ is not negligible in comparison with $f(\Theta)$, and interference oscillations are observed in accordance with [3-18-2], because the collision partners are indistinguishable, spin-zero bosons. Pronounced oscillations are also observed in the ^{3}He–^{3}He differential cross sections measured by Chitnis et

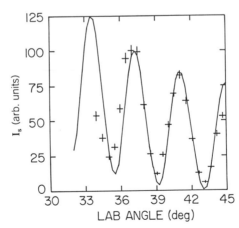

Figure 1-12-4. Relative differential cross sections (RDCS) for ^{4}He–^{4}He collisions measured at Rice for large scattering angles. The observed interference oscillations appear because the collision partners are indistinguishable, spin-zero bosons. [From Chitnis et al. (1988).]

al. Here, however, we are dealing with indistinguishable, unpolarized fermions having nuclear spin $\frac{1}{2}$, so the oscillation pattern is given not by [3-18-2], but by

$$\frac{d\sigma}{d\Omega} = \frac{|f(\Theta)|^2 + |f(\pi - \Theta)|^2}{2} + \frac{|f(\Theta) - f(\pi - \Theta)|^2}{2} \qquad (1\text{-}12\text{-}2)$$

[see Feynman et al. (1965, pp. 32–33)]. In measurements on the ^4He–^3He system, for which there is no nuclear symmetry, the differential cross section is given by (1-12-1), no oscillations being observed. For further discussion of the theory of symmetry effects in collisions, see Bransden and Joachain (1989, pp. 581–588).

1-13. ION–NEUTRAL COLLISIONS AT ENERGIES ABOVE A FEW TENTHS OF AN ELECTRON VOLT

A. Determination of Interaction Potentials

Here we do not attempt extensive coverage of this subject but provide only a few references to modern work of high quality. (A special case, that of the screened Coulomb potential, is treated in Section 1-13-D.) Gianturco et al. (1987) have determined ground-state potential energy curves for H^+ ions interacting with Ne and Ar from the interplay of theory and elastic differential cross-section measurements at 14.8 eV in Toennies's laboratory at Göttingen. Gislason and his associates have studied collisions of alkali ions with the noble gases and various molecules at Lab energies less than 1 keV. Representative references to their work are Polak-Dingels et al. (1982), Rajan and Gislason (1983), and Budenholzer et al. (1986). We should mention that in his recent measurements, Gislason has directly measured the angular resolution function $W(\vartheta)$ discussed by van den Biesen (1988, pp. 487–488, 495) and taken it into account properly in analyzing the data.

B. Interference Effects

In Section 1-13-C we concentrate on $^4\text{He}^+$–^4He collisions. First it is appropriate to discuss four kinds of interference effects that have been shown to be operative in them [see Marchi and Smith (1965) and Massey and Gilbody (1974, pp. 2546–2560); see also Section 1-11 and Problem 1-18].

1. *Interference between Waves Scattered by Even (g) and Odd (u) Potentials*

The underlying cause is electronic symmetry, illustrated in the double-minimum problem (Problem 1-6). One may refer to Section 1-11-B for our previous discussion of g-u oscillations in Ne*–Ne scattering.

2. Interference Arising because of Nuclear Symmetry

We illustrate the effects that come into play here by considering the scattering of He^+ ions by He atoms, for both indistinguishable and distinguishable nuclei. The differential cross section for $^4He^+$ ions to be scattered by ^4He atoms through the angle Θ into the solid angle $d\Omega$ is

$$\frac{d\sigma}{d\Omega} = \frac{1}{4} |f_g(\Theta) + f_u(\Theta) + f_g(\pi - \Theta) - f_u(\pi - \Theta)|^2, \qquad (1\text{-}13\text{-}1)$$

where f_g and f_u are the amplitudes for scattering by the potentials V_g and V_u, respectively. Here we see the interference occurring between (1) direct scattering of the projectile ion at angle Θ, and (2) resonant charge-transfer scattering of the projectile at angle $\pi - \Theta$ (see Section 1-7-D-2). In the latter, the target atom becomes singly ionized and recoils into the detector at angle Θ; it is indistinguishable from the projectile being directly scattered at Θ in the direct-scattering case.

On the other hand, for $^4He^+$ ions impinging on ^3He atoms, the nuclei of the collision partners are distinguishable. The differential cross section for scattering of the $^4He^+$ ions is

$$\frac{^4d\sigma}{d\Omega} = \frac{1}{4} |f_g(\Theta) + f_u(\Theta)|^2, \qquad (1\text{-}13\text{-}2)$$

whereas that for the charge-transfer reaction, with $^3He^+$ ions detected at angle Θ, is

$$\frac{^3d\sigma}{d\Omega} = \frac{1}{4} |f_g(\pi - \Theta) - f_u(\pi - \Theta)|^2, \qquad (1\text{-}13\text{-}3)$$

with no interference between direct elastic scattering and charge transfer (Massey and Gilbody, 1974, p. 2550).

3. Interference Produced by Curve Crossing

For details, see Massey, 1971, pp. 1914–1918; Massey and Gilbody, 1974, pp. 2388–2392; Child, 1974, pp. 161–174, 263–277; Marchi and Smith, 1965.) Figure 1-13-1a shows the potential energy curves for two electronic states of a molecule AB that dissociates into separated atomic states $(A_1 + B_1)$ and $(A_2 + B_2)$, respectively, calculated to zero-order approximation. (In the present problem, let A be the projectile ion and B the neutral target.) Suppose that these curves intersect at point S in the absence of interaction. Then if the properties of the states are such that they can interact, this interaction will modify the potential energy drawing to that shown in Fig. 1-13-1b, in which the curves no longer cross. Thus curve Ib has the character of an $(A_1 + B_1)$ combination at

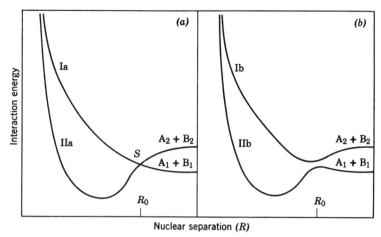

Figure 1-13-1. Molecular potential energy curves used in discussion of curve crossing in Section 1-13-B-3. [From McDaniel (1964, p. 585).]

small nuclear separations but the character of an $(A_2 + B_2)$ combination at large separations. The situation is reversed for curve IIb. This means that if particles A and B in states A_1 and B_1, respectively, are allowed to come together adiabatically (i.e., with infinite slowness), the interaction between them will follow curve IIb. On the other hand, if the particles in these states come together with a finite relative velocity (v_r), there is a finite probability that a transition will occur near S in which the system jumps from IIb to Ib. If this probability is denoted by $P(v_r)$, the probability that the system will continue along IIb is $1 - P$. When the particles reach their distance of closest approach, their relative motion will reverse, and there is again a probability P that a transition will occur at S. Thus if we imagine the particles to be brought together from infinite separation with finite relative velocity v_r and then allowed to separate again, the probability of finding the separated particles in the initial states A_1 and B_1 is consequently $1 - 2P(1 - P)$, whereas the *Landau–Zener probability* of finding them in the A_2 and B_2 states is $Q \equiv 2P(1 - P)$. The net probability, $2P(1 - P)$, that the particles will finally withdraw from one another in their original configuration will be small if P is small or nearly unity. The former case is the nearly adiabatic one of a very slow approach; the latter corresponds to the interaction between two molecular states being very weak. (In the limit of vanishing interaction the curves actually cross at S, as shown in Fig. 1-13-1a, and the identifications of the curves inside R_0 are interchanged.) The maximum value of the net transition probability is $\frac{1}{2}$.

If curve crossing occurs, the elastic differential cross section contains an oscillatory term that arises from interference between the two amplitudes for scattering from the two potentials through the same angle. This fact requires that the interference occur between waves corresponding to different impact parameters.

4. *Rainbow Scattering*

This phenomenon was discussed in Sections 1-2-B, 1-11-A-1, and 1-11-B; additional information is presented in Section 2-9-D.

C. Experimental Results

We first look at some celebrated results obtained by Lorents and Aberth (1965) on the scattering of ^{4}He$^+$ ions by ^{4}He atoms. They used the apparatus shown in Fig. 1-13-2 to measure relative differential cross sections for elastic scattering, and then normalized them to absolute cross sections obtained at small scattering angles by using a Faraday cup instead of the electron multiplier as a detector. Their data, shown in Fig. 1-13-3, were analyzed in an important paper by Marchi and Smith (1965), who were able to explain the structure of the curves. The regular oscillations below about 30° in most of the curves are due to the electronic symmetry described in Section 1-13-B-1. The modulation of these oscillations at large angles is produced by nuclear symmetry (Section 1-13-B-2). Curve crossing (Section 1-13-B-3) of V_g with the potential curve for the $(1\sigma_g)^2 2\sigma_g$ state is responsible for the slight perturbation visible for the 100 eV data for $\Theta = 34$ *to* 50° (Smith et al., 1965). The rapid rise in the curves for the lower energies as $\Theta \to 0°$ is due to rainbow scattering (Section 1-13-B-4). Measurements at SRI on other systems are discussed by Smith et al. (1967), Jones et al. (1974), and Lorents and Conklin (1972). More recently, high-quality elastic differential cross sections have been measured for H$^+$ on Kr and Xe in Toennies's laboratory at CM energies of 30.6 and 51.7 eV (Friedrich et al., 1987; Baer et al., 1987).

Linder and his colleagues have studied orbiting structures in the angular

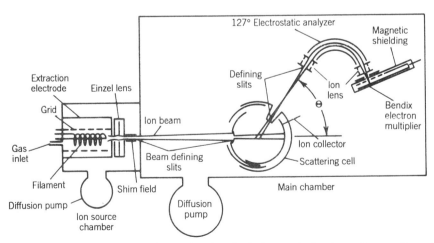

Figure 1-13-2. Apparatus used at the Stanford Research Institute for studies of differential scattering of ions by neutrals. [From Lorents and Aberth (1965).]

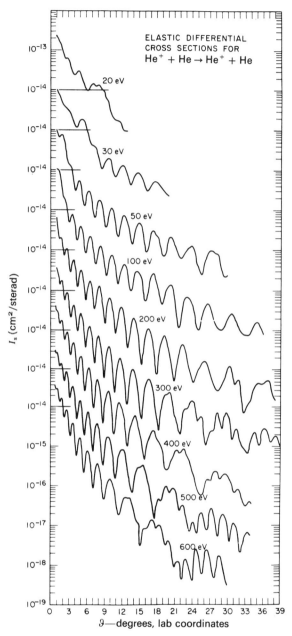

Figure 1-13-3. Measured differential elastic cross sections per unit solid angle for $^4\text{He}^+$–^4He collisions at impact energies ranging from 20 to 600 eV. The scales for the various energies are shifted from one another, the proper scale for each energy being identified at $10^{-14}\,\text{cm}^2/\text{sr}$ by intersection of the horizontal line with each curve. [From Lorents and Aberth (1965).]

scattering of H^+ ions by He and Ar (Konrad and Linder, 1982, 1983) and of He^+ ions by He, Kr, and Xe (Reinig et al., 1983) at energies up to a few electron volts. Related theoretical work has been published by Thylwe (1983) and Korsch and Thylwe (1983).

Park and his colleagues have made extensive measurements of the differential cross section for the elastic scattering of ions. Some of their recent work dealt with 25- to 60-keV protons on atomic hydrogen (Rille et al., 1984), and with Na^+ and Li^+ on H, and Mg^+ and Be^+ on He at impact energies between 30 and 150 keV (Peacher et al., 1989).

We now return to the work of Stebbings's group and present their small-angle differential cross sections for $^4He^+$ scattering by ^4He atoms Fig. 1-13-4. They obtained these data with the apparatus shown in Fig. 1-12-1 but with no gas in the charge transfer cell (CTC). Because of the symmetry of the Hamiltonian here, two electronic states must be considered when the ground-state helium atom and ion approach one another adiabatically. One state (g) is symmetric in the nuclei, the other (u) is antisymmetric, and as we have seen in previous examples, scattering takes place from both the g and u potentials. The differential elastic cross section is now given by (1-13-1), but the amplitudes $f_g(\pi - \Theta)$ and $f_u(\pi - \Theta)$ may be neglected here because the scattering is confined to very small angles [see Gao (1988a) and Massey and Gilbody (1974, p. 2550)].

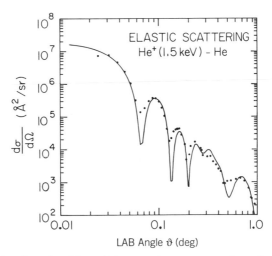

Figure 1-13-4. Small-angle differential scattering cross sections for $^4He^+ - ^4He$, as measured by the Rice group. [From Gao et al. (1988a).]

D. Investigations of the Screened Coulomb Potential by Fast Ion Beam Scattering

During the 1950s and 1960s, N. V. Fedorenko in Leningrad and E. Everhart at the University of Connecticut made many innovative and important contributions to the physics of high-energy heavy particle collisions. Their work is surveyed by Massey and Gilbody (1974), Hasted (1972), and McDaniel (1964). Here we discuss only one facet of their considerable body of work—study of the screened Coulomb potential in *violent collisions*, defined as collisions in which the impact parameter is small enough, for a given impact energy, that a significant change in the projectile's direction of motion results.

Long ago it was suggested that the interaction between two atomic structures during a violent collision may be describable to a good approximation by a *screened Coulomb potential energy function*:

$$V(r) = \frac{zZe^2}{r} \, e^{-r/a} \tag{1-13-4}$$

(See Sections 3-8 and 3-17 of the companion volume.) The first factor is the Coulomb function, which expresses the potential energy of two bare nuclei of charges ze and Ze separated by the distance r. The exponential factor accounts for the screening of the nuclei by the orbital electrons, the spatial extent of the screening being measured by the screening length a. Bohr (1948) has suggested use of the quantity

$$a = \frac{a_0}{(z^{2/3} + Z^{2/3})^{1/2}} \tag{1-13-5}$$

for the screening length. Here a_0 is the radius of the first Bohr orbit of the hydrogen atom, 0.53×10^{-8} cm. Firsov (1958) has used a different modification of the pure Coulomb potential to describe the shielding effect of the electrons, one calculated from the Thomas–Fermi statistical model of the atom (Morrison et al., 1976).

The studies made by Everhart and his students in this connection involved the scattering of ion beams at Lab energies in the range 25 to 100 keV (Fuls et al., 1957; Lane and Everhart, 1960). Strictly speaking, most of the scattering that occurs at these high energies is inelastic, but the fraction of the projectile energy lost in a collision is so small that it has a negligible effect on the projectile trajectory. The scattering behavior of the beam is almost exactly that corresponding to elastic scattering by the projectile–target potential, and observations of the scattering distributions thus permit accurate calculation of the potential energies of interaction at small separation distances. Here we treat these collisions as elastic.

Everhart and his colleagues measured differential scattering cross sections for

He^+ in He, Ne, and Ar; Ne^+ in Ne and Ar; and Ar^+ in Ar at beam energies of 25, 50, and 100 keV. The angular range covered (about 4 to 40° in the Lab system) was large enough to permit determination of $V(r)$ from the experimental data without any assumptions regarding the form of $V(r)$ except that it is monotonically decreasing. The calculations were made by classical scattering theory (see the discussion in the following paragraphs). The calculated potential curves were found to fit the Bohr form of the screened Coulomb potential very well; the fit to the Firsov form was even better. Normalization was not necessary. The potential energies corresponding to the small distances of approach (on the order of 10^{-2} Å) achieved in these high-energy experiments are in the range 1 to 60 keV.

A classical calculation of the differential scattering cross section will be valid if (1) the de Broglie wavelength λ of the projectile is negligible compared with any significant dimension of the scattering center, and (2) the collision is well defined within the limitations imposed by the uncertainty principle (see Section 3-11 of the companion volume). Condition 1 requires that

$$\lambda \ll a \qquad (1\text{-}13\text{-}6)$$

and

$$\lambda \ll d = \frac{zZe^2}{\frac{1}{2}M_r v_0^2}, \qquad (1\text{-}13\text{-}7)$$

where M_r is the reduced mass of the projectile–target combination and v_0 is the initial relative velocity. The length d is known as the collision diameter; it represents the distance of closest approach in a head-on collision in the absence of screening (see Problem 7-19 of the companion volume). The collision diameter is a good measure of the size of the scattering center when d/a is small. Condition 2 leads to a lower limit, Θ^*, on the scattering angle, above which the classical calculation is valid (Bohr, 1948):

$$\Theta^* \approx \frac{\lambda}{2\pi a}. \qquad (1\text{-}13\text{-}8)$$

For most collisions between atomic structures in the energy range from about 100 eV to several hundred keV the classical calculation of the differential scattering cross section is valid except at very small angles. For example, in 50-keV Ne^+–Ar measurements (Everhart et al., 1955), $a = 160 \times 10^{-11}$ cm, $d = 78 \times 10^{-11}$ cm, $\lambda = 0.28 \times 10^{-11}$ cm, and $\Theta^* = 2.08 \times 10^{-4}$ rad $= 0.016°$. Thus both conditions on the validity of a classical description are satisfied for angles greater than 0.016°.

It is interesting to note that in the regime specified by (1-13-6) and (1-13-7) the quantum mechanical Born approximation solution for the differential cross section is valid only for angles smaller than Θ^*. Thus in this regime

($\lambda \ll a$; $\lambda \ll d$) the classical and Born solutions are valid in mutually exclusive angular regions. In the regions in which $\lambda \gg d$, the Born approximation is valid for all angles and the classical solution is nowhere valid (Bohr, 1948; Mott and Massey, 1965; Williams, 1945).

1-14. NOTES

1-1. *Dispersion and van der Waals Forces; Integral Elastic Scattering Cross Section for an Attractive Inverse-Power-Law Potential.* Long-range attractive forces usually play the dominant role in determining the scattering behavior of atoms and molecules at low energies. For neutral nonpolar molecules, these forces arise in the following manner. At any instant the electrons in one molecule have a configuration that results in an instantaneous dipole moment. This moment induces a dipole moment in a neighboring nonpolar molecule, and these two moments interact to produce a force of attraction between the two molecules irrespective of the orientation of the instantaneous dipole in the first molecule. This type of interaction was first investigated quantum mechanically by F. London around 1930. The forces are called "dispersion" because they may be expressed in terms of oscillator strengths (Section 5-7-F of the companion volume) that arise in the theory of the dispersion of light. Both the induced dipole–induced dipole interaction and higher-order contributions to the dispersion energy are discussed in Section 1-3 of the companion volume and in Hirschfelder et al. (1964, Part III).

In many gases the long-range intermolecular attraction is identified with the attractive induced dipole–induced dipole *London dispersion force* only. In the event that one (or both) of the colliding molecules has a permanent dipole moment, dipole–induced dipole (and dipole–dipole) interactions may also be significant. To a first approximation, all three types of interaction are expressible in terms of an inverse-sixth-power dependence on the separation distance at large r. Long-range forces such as those described here are the familiar *van der Waals forces* (Section 1-7-D-3). They can be described rigorously in terms of the physical properties of the separated molecules. For short-range *valence* (or *chemical*) *forces*, on the other hand, rigorous treatment is not possible in terms of the properties of the separated molecules. For these forces it is necessary to consider each pair of interacting molecules as a special case. The valence forces (Section 1-7-D-3) do not concern us in this note, but they are important in high-energy scattering.

As pointed out in Section 3-9 of the companion volume, classical theory is not applicable to calculation of the differential elastic scattering cross section $I_s(\Theta)$ at small angles or to the integral cross section q_s. Here we derive approximate quantal equations for q_s for two particles interacting through an inverse-power-law potential. This discussion is based on Massey and Mohr (1934) and Massey (1971).

We consider a potential of the form

$$V(r) \sim -Cr^{-n}, \tag{1-14-1}$$

where C is a positive constant and n is a positive integer. According to [3-15-34], q_s may be expressed in terms of the phase shifts η_l as

$$q_s = \frac{4\pi}{\kappa^2} \sum_{l=0}^{\infty} (2l + 1) \sin^2\eta_l, \tag{1-14-2}$$

where

$$\kappa = \frac{M_r v_0}{\hbar} \tag{1-14-3}$$

is the wave number associated with the relative motion of the interacting particles. The reduced mass of the pair of particles is denoted by M_r, and v_0 is their relative velocity of approach at large r.

The phase shifts are given by Jeffreys's approximation (Section 3-15-F in the companion volume) as

$$\eta_l \approx \int_{r_0}^{\infty} \left[\kappa^2 - \frac{2M_r V(r)}{\hbar^2} - \frac{l(l + 1)}{r^2} \right]^{1/2} dr - \int_{r_0'}^{\infty} \left[\kappa^2 - \frac{l(l + 1)}{r^2} \right]^{1/2} dr, \tag{1-14-4}$$

where the lower limit of each integral is the zero of the corresponding integrand. This approximation is good in the present context except at very low temperatures. Since we are dealing with heavy particles, a large number of phase shifts is required. By putting $-M_r/\hbar^2$ equal to $\alpha/2$ and performing a binomial expansion of the square root, we have, for large l,

$$\eta_l \approx \int_{r_0'}^{\infty} \left[\kappa^2 - \frac{l(l + 1)}{r^2} \right]^{1/2} \left[1 + \frac{\alpha V/2}{\kappa^2 - l(l + 1)/r^2} + \cdots - 1 \right] dr$$

$$= \int_{r_0'}^{\infty} \frac{\alpha V/2}{[\kappa^2 - l(l + 1)/r^2]^{1/2}} dr. \tag{1-14-5}$$

The right-hand side of (1-14-5) is sometimes called the *Massey–Mohr approximation to the phase shifts*. This result is also given by the Born approximation, and (1-14-5) may be used for all values of l. If $\alpha V = -Cr^{-n}$, we have, for large l,

$$\eta_l \approx \frac{C}{2} \int_a^\infty \left\{ r^n \left[\kappa^2 - \frac{l(l+1)}{r^2} \right]^{1/2} \right\}^{-1} dr$$

$$= \frac{C}{2\kappa} \int_a^\infty [r^{n-1}(r^2 - a^2)^{1/2}]^{-1} dr, \qquad (1\text{-}14\text{-}6)$$

where $a = (l + \tfrac{1}{2})/\kappa$. [In replacing $l(l+1)$ by $(l + \tfrac{1}{2})^2$, we have made the *Langer modification*.] Integration gives

$$\eta_l \approx \frac{C}{2\kappa a^{n-1}} f(n) = \frac{C\kappa^{n-2}}{2(l + \tfrac{1}{2})^{n-1}} f(n), \qquad (1\text{-}14\text{-}7)$$

where

$$f(n) = \begin{cases} \dfrac{n-3}{n-2}\dfrac{n-5}{n-4}\cdots\dfrac{1}{2}\dfrac{\pi}{2} & (n \text{ even}) \\[2ex] \dfrac{n-3}{n-2}\dfrac{n-5}{n-4}\cdots\dfrac{2}{3} & (n \text{ odd}) \\[2ex] 1(n = 3); \quad \dfrac{\pi}{2}(n = 2). \end{cases} \qquad (1\text{-}14\text{-}8)$$

Massey and Mohr have shown that this equation holds accurately if η_l is less than 0.5, and then $\sin \eta_l \approx \eta_l$. Therefore, if m denotes that value of l for which $\eta_l = 0.5$, the contribution to q_s arising from the phases $l > m$ is

$$\frac{4\pi}{\kappa^2} \sum_{l=m}^\infty (2l + 1) \sin^2 \eta_l \approx \frac{8\pi}{\kappa^2} \sum_{l=m}^\infty \frac{C^2}{4} \frac{\kappa^{2n-4}}{(l + \tfrac{1}{2})^{2n-3}} f^2$$

$$\approx 2\pi C^2 \kappa^{2n-6} f^2 \int_m^\infty (l + \tfrac{1}{2})^{-2n+3} dl = \frac{2\pi C^2 \kappa^{2n-6}}{2n - 4}(m + \tfrac{1}{2})^{-2n+4} f^2$$

$$= \frac{4\pi}{n-2} \frac{(m + \tfrac{1}{2})^2}{\kappa^2} \eta_m^2 = \frac{\pi}{n-2} \frac{(m + \tfrac{1}{2})^2}{\kappa^2} \qquad (1\text{-}14\text{-}9)$$

since $\eta_m = 0.5$. The contribution to q_s from waves $l < m$ is

$$\frac{4\pi}{\kappa^2} \sum_{l=0}^{m-1} (2l + 1) \sin^2 \eta_l \approx 2\pi \frac{m^2}{\kappa^2}. \qquad (1\text{-}14\text{-}10)$$

Therefore, the integral cross section is approximately equal to

$$\pi\left(2 + \frac{1}{n-2}\right)\frac{(m+\frac{1}{2})^2}{\kappa^2}, \tag{1-14-11}$$

since m is large. If we now use (1-14-7) and the fact that $\eta_m = 0.5$, we obtain the *Massey–Mohr* result

$$q_s = \gamma_{MM}(n)\left(\frac{C}{\hbar v_0}\right)^{2/(n-1)}, \tag{1-14-12}$$

where

$$\gamma_{MM} = \pi\,\frac{2n-3}{n-2}\,[2f(n)]^{2/(n-1)}. \tag{1-14-13}$$

The summation (really an integration) over l can be carried out accurately—the Massey–Mohr device of splitting it into two pieces is not necessary. Landau and Lifshitz (1965, pp. 488–489) repeated the calculation without dividing the range of integration. They used the approximation

$$\eta_l \simeq A_n l^{-n+1} \quad\text{with}\quad A_n = \frac{M_r f(n)C\kappa^{n-2}}{\hbar^2} \tag{1-14-14}$$

in the equation

$$q_s = \frac{8\pi}{\kappa^2}\int_0^\infty l\,\sin^2\eta_l\,dl \tag{1-14-15}$$

and evaluated the integral directly. The *Landau–Lifshitz result* is

$$q_s = \gamma_{LL}(n)\left(\frac{C}{\hbar v_0}\right)^{2/(n-1)}, \tag{1-14-16}$$

with

$$\gamma_{LL} = \pi^2[2f(n)^{2/(n-1)}]\,\operatorname{cosec}\left(\frac{\pi}{n-1}\right)\Big/\Gamma\left(\frac{2}{n-1}\right). \tag{1-14-17}$$

The MM and LL expressions differ only in the numerical factors $\gamma(n)$, and each may be written in the form

$$q_s = B\left(\frac{C}{v_0}\right)^{2/(n-1)}, \tag{1-14-18}$$

where B is a constant whose value is determined by the value of *n*. This result holds for potentials that fall off faster than the inverse third power of the separation distance. The differential cross section $I_s(\Theta, v_0)\, d\Omega_{CM}$ cannot be expressed in simple terms. Exact numerical calculations made after the MM and LL results were published favor the LL approximation, which is now used almost exclusively. The MM approximation is accurate for $n = \infty$ but is 7% too low for $n = 6$ (see Massey, 1971, p. 1325).

1-2. Beam Sources

A. *Effusion (Oven) Sources.*
The first beams of neutral particles with thermal velocities and straight trajectories were obtained from an effusion source and reported by Dunoyer in 1911 (Pauly, 1988a). A source of this type is shown at the upper left of Fig. 1-10-1. A gas or vapor is permitted to effuse from a plenum (labeled "oven" in this figure) through a narrow slit into an evacuated region under conditions of *molecular flow*. These conditions require that the pressure inside the plenum be sufficiently low that the mean free path λ inside is much greater than the dimensions of the exit orifice; then the neutral particles move through the orifice and in the beam without undergoing collisions. Effusion sources have been operated at temperatures ranging from about 4 K to over 2800 K.

For a thin-walled circular orifice and axisymmetric geometry, the envelope of effusive flux vectors is a sphere tangent to the source exit orifice Fig. 1-14-1*a*), and the flux has a cosine distribution. The resulting beam intensity downstream a distance *x* on the axis of symmetry can be calculated from the kinetic theory of gases and is given by

$$I_0 = \frac{n_0 \bar{c}}{4}\, A\pi x^2 \quad \text{(thin-walled circular orifice),} \quad (1\text{-}14\text{-}19)$$

where n_0 is the number density of molecules in the source, $\bar{c}$ their average speed, and *A* the area of the source orifice. Several types of orifices (circular holes, slits, single channels, and multiple channels) offer certain advantages and have been used extensively.

Effusion sources dominated the scene from 1911 until the early 1950s, when they began to be replaced by nozzle expansion devices. Today the use of effusion sources is for the most part restricted to experiments in which nozzle sources are incompatible with the demands of the experiment, or in which the beam material is scarce or has a very low vapor pressure. Pauly (1988a) discusses effusion sources in considerable detail and describes other types of low-energy sources for the production of radicals and atoms in excited states. In addition, his discussion of the use of highly refractory materials in sources operated at temperatures above 2800 K is useful.

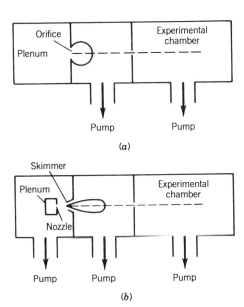

Figure 1-14-1. (*a*) Conventional effusion source; (*b*) supersonic nozzle beam source. [Adapted from Fenn (1982).]

B. *Nozzle Expansion (Free Jet) Sources* (see also Section 1-10-C). Figure 1-14-1*b* has already foreshadowed the discussion in this section by showing the envelope of flux vectors from a nozzle source and inviting comparison with the envelope for the effusion source illustrated in Fig. 1-14-1*a*. In a nozzle source, gas at high density ($\lambda_0 \ll d$, the nozzle diameter) expands through the nozzle (where the Mach number M reaches unity) and produces a supersonic jet, as shown in Fig. 1-14-2.* The conical *skimmer* in panel (*b*) of Fig. 1-14-1 replaces the circular orifice in panel (*a*). The "ellipsoidal" shape of the flux envelope in (*b*) is the result of the free-stream convective velocity being superimposed on the random thermal velocity, the ratio of the major axis to minor axis being given as the ratio of the two velocities, or roughly the Mach number. For $M > 4$, the intensity a distance x downstream from the skimmer tip is approximately given (Fenn, 1982) as

$$I_n = \frac{n_s u_s A_s (\gamma M_s^2)}{2\pi x^2},\tag{1-14-20}$$

where n_s is the number density of molecules, u_s the convective velocity, A_s the area, and M_s the Mach number. The subscript s refers to

*The *Mach number* is defined as $M = v/a$, where v is the local flow velocity and a is the local speed of sound. For an ideal gas, $a = (\gamma kT/m)^{1/2}$. A detailed discussion of Fig. 1-14-2 and the formation of the free jet is provided by Fenn (1982) and Miller (1988).

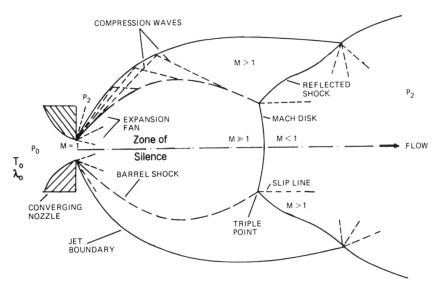

COMPRESSION WAVES

$M > 1$

P_2

REFLECTED SHOCK

P_2

P_2

EXPANSION FAN

MACH DISK

Zone of

$M \gg 1$ $M < 1$

P_0 $M = 1$

FLOW

T_0
λ_0

Silence

BARREL SHOCK

SLIP LINE

CONVERGING NOZZLE

$M > 1$

JET BOUNDARY

TRIPLE POINT

Figure 1-14-2. Schematic representation of the structure of an underexpanded free jet from a source at pressure p_0 into a region of ambient pressure p_2. [From Fenn (1982).]

conditions at the skimmer tip. Fenn (1982) calculates the ratio of the nozzle-beam to effusion-beam intensities for the case where $A_s = A_0$, $u_s = 2\bar{c}$, and $n_s = n_0$. For a monatomic gas and $M_s = 20$, Fenn argues that n_s in his analysis should be reduced by a factor of 20 to avoid perturbation of the flow by the skimmer. On this basis he obtains $I_n/I_0 \approx 250$, and he points out further that 90% of the molecules are within 5% of the most probable velocity! Higher Mach numbers are easily obtainable ($M = 400$ has been reached in the case of helium), and if one recognizes the imperative need for velocity selection when an effusion source is used, the great superiority of the nozzle source is apparent as far as intensity and velocity spread are concerned.

Another advantage that nozzle expansion sources have over effusion sources is that they can provide beams with much greater molecular translational energies. Because of the large mean free path of the molecules in the effusion source plenum, each molecule makes many collisions with the walls before it exits, and hence the gas temperature cannot exceed the wall temperature. The upper limit here is about 3000 K, corresponding to a thermal energy of 0.39 eV. In the nozzle source, on the other hand, the pressure of the gas in the plenum is maintained at a high value such that gas temperatures greatly in excess of the wall temperature can be used. A higher plenum temperature translates, of course, into higher beam particle energy. In addition, the *seeded beam technique* can provide aerodynamic acceleration to still higher energies. Here a "carrier gas" of light molecules containing a low concentration of heavy "seed gas" molecules undergoes free jet expansion through the nozzle, and the heavy molecules are swept along at a much higher velocity than they would otherwise achieve as a pure gas. In the limit of infinite dilution, the final translational energy of the heavy molecules is, in fact, enhanced by a factor approaching the ratio of the molecular weights of the heavy and light species. With source heating combined with the seeded beam technique, translational energies of about 40 eV per molecule have been obtained. Thus the most interesting part of the energy spectrum for chemistry (about 1 to 20 eV) has been covered, at least for many species.

All of the jet properties scale with downstream distance in units of the nozzle diameter. Figure 1-14-3 provides axial profiles of the gas temperature and number density in the zone of silence (Fig. 1-14-2). The gas velocity docs not depend on the nozzle diameter, so for very small nozzles, the rate of change of the state variables can be enormous. One should note in examining Fig. 1-14-3 that nozzle diameters as small as

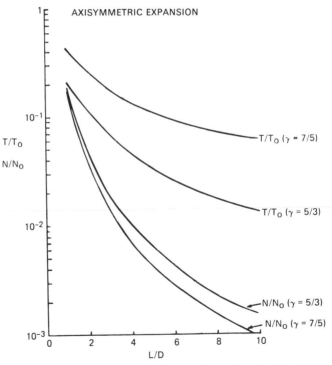

Figure 1-14-3. Axial profiles of temperature T and number density N in the zone of silence of a supersonic free jet. The abscissa scale is in terms of the axial distance L from the nozzle exit divided by the nozzle diameter D; T_0 and N_0 are the values of T and N in the source. [From Fenn (1982).]

$5 \mu m$ are often used. Flow velocities are usually about 10^5 cm/s, so cooling rates of 10^{10} K/s may be reached! The gas number density decreases extremely rapidly from the nozzle exit, and the collision frequency quickly becomes too low to maintain kinetic equilibrium (the relaxation processes "freeze"). A highly nonequilibrium steady-state condition is reached.

During the isentropic expansion of a molecular gas in the jet, the internal degrees of freedom are collisionally cooled, and both random translational energy and internal energy are converted to directed translational motion. The rotation–translation energy transfer process is especially efficient in intermolecular collisions because only 10 to 100 collisions are usually required for rotational deexcitation, and a molecule typically makes 100 to 1000 collisions during free jet expansions [see Levine and Bernstein (1987, pp. 243–247, 300–312)]. Hence the rotational states of molecules in the collision-free region of the jet have essentially a thermal Boltzmann distribution characterized by an extremely low temperature (frequently below 5 K), and only the lowest one or two rotational states in a given vibrational state are present. Simple diatomic molecules may require more then 10^4 collisions for vibrational deexcitation and hence will not be vibrationally relaxed in the expansion. On the other hand, large polyatomic molecules will be vibrationally deexcited because typically only 10 to 100 collisions are required. In this case, most of the molecules will end in their ground vibrational state.

As pointed out in Section 1-10-C, helium must be considered separately from all other gases because the helium dimer is not stable and hence helium does not condense under typical free jet conditions. Quantum effects in helium jets are discussed by Toennies and Winkelmann (1977).

C. *Low-Energy Pulsed Beam Sources.* We may define a *molecular beam* as a collimated, collision-free stream of molecules passing through a "vacuum." Then a *pulsed molecular beam* is a segment of such a beam that is truncated in space and time, with physical length Δx and duration Δt. For this definition to be meaningful, Δt must be small enough that Δx is much smaller than the relevant dimensions of the vacuum chamber. If a typical molecular speed of $v = 10^3$ m/s and a chamber dimension $x = 1$ m are assumed, the equation $\Delta t = \Delta x/v$ implies $\Delta t \approx 100 \mu s$ as a reasonable upper limit to the pulse duration. In his definitive review, Gentry (1988) suggests that beam segments of duration longer than the transit time across the chamber should be associated with the terms *beam gating* or *modulation* rather than with beam pulsing. Beam pulses may be produced by pulsing the source itself, or by *chopping* the beam with a shutter after it is formed.

At the University of Minnesota, Gentry and Giese (1978) have developed a pulsed supersonic nozzle source that sacrifices the large duty factor associated with a continuous source in order to achieve (1) a great enhancement of the instantaneous scattered signal, and (2) a great reduction of the instantaneous background signal. Advantage 1 stems from the fact that the beam intensity is not limited by differential pumping requirements because each gas pulse expands into a volume that has been evacuated since the previous gas pulse entered. Therefore, much larger nozzle diameters and much higher source pressures can be tolerated than in continuous sources, and instantaneous beam intensities can be larger by orders of magnitude. Advantage 2 arises from the fact that time-of-flight (TOF) discrimination eliminates most of the background signal (Gentry, 1979, 1985), as shown in the following discussion.

In "background limited" experiments where the background signals of concern are those produced by the molecular beam(s), an enormous enhancement of the signal-to-background ratio (S) may be realized by repetitively pulsing the beam(s). This is especially true if the interval between pulses (β) is comparable to or greater than the pump-out time constant (τ), because then most of the background gas from previous pulses will have been pumped out before a given pulse enters the chamber. Another important factor is that each primary beam pulse (only a few centimeters long) does not contribute to the background until it crosses the collision chamber and is reflected from the vacuum chamber wall. By contrast, molecules scattered by the target pass directly to the detector, so each background pulse is delayed significantly with respect to the corresponding primary beam pulse interaction with the target. Gentry (1988) shows that for a 10-Hz source delivering 100-μs pulses in a single-beam experiment, considerations of duty factor alone indicate an S enhancement factor of 10^3. If, however, the relevant background pressure is that which is present *during* the beam pulse expansion, not the average value, the pump-out time becomes important. For a 10-Hz, 100-μs source, the S enhancement factor with $\tau = 100$ ms increases only slightly, but it rises to 2×10^6 if τ is reduced to 10 ms. In a crossed-beam experiment with $\beta/\tau \geqslant 10$, the enhancement factor may be greater than 10^{12}, although factors other than the beam gas partial pressures may reduce its value markedly.

Gentry (1988) makes an important general comment concerning pulsed molecular beam experiments that implies that additional benefits can be derived: "Except in unusual circumstances, if one element of an experiment is pulsed, the optimum signal-to-background ratio will be obtained by pulsing all elements."

The *current-loop-actuated pulsed nozzle source* first described by Gentry and Giese (1978) has been used extensively. It employs a nozzle of up to 0.75 mm diameter that is located in a metal faceplate and surrounded by an O-ring on the high-pressure side. The center of a metal bar, 1.8 mm wide, presses against the O-ring and makes the seal. The ends of the bar are attached to the faceplate, and a current may be passed up through the plate and down through the bar in a loop. If a strong current pulse is applied, the antiparallel current elements cause the bar to be repelled from the O-ring for a predetermined period, and a pulse of gas passes through the nozzle. With the most recent version of this source, repetition rates of 20 Hz can be maintained. Pulse durations of 10 to 20 μs are typical, and source pressures are usually in the range 5 to 20 atm with nozzle diameters of about 0.5 mm. The beam intensity at the target region may be from two to four orders of magnitude greater than what is attainable in a continuous beam experiment, and the signal-to-background ratio is greatly increased.

A noteworthy achievement by the Minnesota group has been to obtain speed ratios $v/\Delta v$ greater than 1000 and temperatures less than 1 mK by expansion of pure He gas from a pulsed beam source into a vacuum (Wang et al., 1988). Gentry (1988) also discusses pulsed sources that utilize solenoid and piezoelectric mechanisms, as well as techniques used in pulsed photolysis and heating and in the control of cluster populations.

D. *Spin-Polarized Hydrogen-Atom Beam Source.* A relatively simple and inexpensive source that delivers spin-polarized H atoms has been developed by Chan et al. (1988) for use in studies of e + H collisions of various types at the City University of New York (see Sections 4-6 and 6-11 of the companion volume). Note that (1) the beam particles are free radicals, and (2) they have been subjected to state selection. Hydrogen gas is admitted into a radio-frequency discharge in a water-cooled Pyrex tube and dissociated to the extent of 80 to 85%. Atoms and molecules emerge through a 1-mm-diameter source nozzle and are formed into a beam by a stainless steel skimmer of 1.4 mm diameter. The beam then passes into a pair of state-selecting hexapole magnets, each with 152 mm length and a 6.4-mm pole gap. The magnets are separated axially by a 19-mm gap that facilitates pumping. Hydrogen atoms in the $m_s = +\frac{1}{2}$ state are transmitted, whereas those in the $m_s = -\frac{1}{2}$ state are deflected. The beam is mechanically chopped between the magnets at a frequency of less than 10 Hz, with open and closed times of 50 ms each. Finally, the beam passes into a spin-guiding solenoid that can adiabatically rotate the spins into the desired orientation. In the interaction region where the H-atom beam and electron beam cross, the averaged H-atom number density is $\rho \approx (1.7 \pm 0.7) \times 10^{10}$ atoms/cm^3. The low-field polarization is $\mathbf{P} \equiv \langle \sigma \rangle \approx 0.52$, and the high-field-state selection parameter is $s \equiv (N^+ - N^-)/(N^+ + N^-) \approx 0.99$. The discussion of experiments with spin-polarized beams in Iannotta (1988) and in Sections 4-6, 5-10, and 6-11 of the companion volume may prove useful here.

E. *High-Energy Beam Sources.* Pauly (1988b) reviews beam sources that provide molecules with energies ranging from the epithermal region up to about 1 MeV. Included are sources that utilize charge transfer, electron detachment, aerodynamic acceleration, sputtering, mechanical acceleration, electrical discharges, laser pulsing, and evaporation of accelerated micropellets. We shall not discuss these sources here, although we describe some of them at other places in the book.

1-3. Particle Detectors. A useful treatment of detection principles has been provided by Bassi (1988a), who has also discussed ion production in surface ionizers, electron impact ion sources, and field emission and field ionization sources (Bassi, 1988b). Finally, Bassi (1988c) treats mass selection (quadrupole mass spectrometers, magnetic sector mass spectrometers, and time-of-flight mass spectrometers) as well as ion detection (Faraday cups, electron multipliers, and scintillation detectors).

Hefter and Bergmann (1988) deal with spectroscopic detection methods, while Zen (1988) covers accommodation and accumulation techniques and methods of metastable detection. We give examples of various detection methods throughout the book but do not attempt systematic coverage of this large and important subject.

1-4. Velocity Selection by Mechanical Means. Van den Meijdenberg (1988) has prepared an excellent review of this technique. He points out that the earliest devices for mechanical velocity selection were based on the velocity-selecting properties of a rotating helix, such as that provided by a rotating cylinder with helical slots cut into its surface. The same principle underlies the slotted-disk velocity selector (SDVS), which consists of several disks mounted on a single rotating shaft, the disks being spaced axially and azimuthally so that the slots in the disk rims define a helical path. Van den Meijdenberg (1988) mentions several other kinds of velocity selectors but concentrates on the SDVS, which is the most commonly used. A SDVS that is often cited is the one constructed by Hostettler and Bernstein (1960). It is 10 cm long, 16 cm in diameter, and consists of six slotted aluminum alloy disks. The velocity resolution ($\Delta v/v$ for Δv at half-intensity) is 0.047, and no velocity sidebands are transmitted. The effective fractional time open to the incident beam is 0.35. At the highest rotor speed (17,000 rpm), the transmitted velocity is 1.05×10^5 cm/s. Other instruments that were constructed later and embody certain improvements are described by van den Meijdenberg (1988).

1-5. Velocity Measurements by Time-of-Flight Methods. This subject has been treated in the context of molecular collisions by Auerbach (1988). In Section 4-1-A of the companion volume we described time-of-flight (TOF) experiments on electron scattering at the University of Bielefeld, and later in this volume we discuss experiments in which TOF techniques are used. It is not necessary to cover them here.

1-6. State-Selection Methods. State selection by deflection and by focusing in electric and magnetic fields is discussed by Reuss (1988). Optical methods are treated by Bergmann (1988); these methods provide an especially versatile and general tool.

It is appropriate here to cite a few examples of the use of state-selected species. In the first successful experiment with state selectors, Berkling et al. (1962) measured total cross sections for open shell atoms on rare gases. A magnetic field of the Rabi type was used to select M_J states of the projectiles. Bennewitz et al. (1964) were able to measure the anisotropy of van der Walls potentials by preparing molecules in definite quantum states and studying their collisions with rare-gas atoms. An electrostatic quadrupole field was used to produce the molecules in specified rotational states, and the M_J dependence of the total cross section was measured (see Note 3-3). Toennies (1965) used electrostatic quadrupole fields for state selection before and after collisions and thereby was able to make the first successful measurements of cross sections for resolved state-to-state rotational transitions in molecular collisions. The pioneering experiments mentioned thus far were all performed in Paul's laboratory at the University of Bonn (Stolte, 1988). We have already mentioned experiments by Rubahn and Toennies (1988) on Li_2 molecules in high vibrational states, and the development of a spin-polarized H-atom beam by Chan et al. (1988). Laser-induced alignment of ground-state molecules has been employed by Mattheus et al. (1986) in studies of rotationally inelastic collisions of Na_2 molecules in the levels $M_J = 0$ and $|M_J| = J = 6$. On pages 312 and 320 of his review, Bergmann (1988) explains how vibrational state selection in molecular electronic ground states is more difficult than rotational state selection in the vibrational ground state or selection of an excited atomic state.

Scattering experiments with state-selected molecules provide a deep understanding of collision dynamics, and the subject is of great importance. We shall need to discuss a number of such studies at various points in this book.

In addition to the reviews of state-selection techniques already mentioned here, there are important reviews of scattering experiments with state selectors (Stolte, 1988), with laser-excited atomic beams (Düren, 1988), and with spin-polarized beams (Iannotta, 1988). Also, Bernstein (1982) provides a very useful survey of the entire subject.

1-15. PROBLEMS

Solutions to most of these problems appear in the references indicated.

1-1. Using the rigid-sphere model, make a few calculations to show that [1-5-21] is a reasonable result.

1-2. This problem has relevance to the thermal balance in the earth's atmosphere. [See Massey (1982, pp. 87–88); Massey's notation is used here.]

(a) Consider an *inelastic* collision between two particles, a and b, with masses m_a and m_b, respectively, in which an amount of kinetic energy ΔE is transferred from a to b. Let $Q_{in}(u)$ be the cross section for the process as a function of the relative velocity u. Now, if a Maxwellian distribution of n(a) particles of type a at temperature T_a interacts with a Maxwellian distribution of n(b) particles of type b at temperature T_b, the rate of energy loss by the ensemble a to ensemble b is

$$\frac{dE_{in}}{dt} = \Delta E\, n(a)n(b) \int\!\!\int f_a(v_a)f_b(v_b)Q_{in}(u)u\, dv_a\, dv_b, \qquad (1\text{-}15\text{-}1)$$

where f_a and f_b are the distribution functions of the ensembles. Justify this expression, substitute the explicit forms of f_a and f_b, and perform several integrations to obtain

$$\frac{dE_{in}}{dt} = 4\pi^{-1/2}n(a)n(b)\,\Delta E\alpha^{3/2} \int_0^\infty Q_{in}(u)u^3 e^{-\alpha u^2}\, du. \qquad (1\text{-}15\text{-}2)$$

Here

$$\alpha = \frac{m_a m_b}{2k(m_a T_b + m_b T_a)}. \qquad (1\text{-}15\text{-}3)$$

If the particles labeled a are electrons, $m_a \ll m_b$; then $\alpha = m_a/2kT_e$.

(b) The energy transferred in an *elastic* collision between a and b moving with velocities v_a and v_b, respectively, in which the direction of relative motion is turned through an angle θ is

$$\Delta E_{el}(\theta) = \frac{m}{M}\left[m_a v_a^2 - m_b v_b^2 + (m_b - m_a)v_a \cdot v_b\right](1 - \cos\theta), \qquad (1\text{-}15\text{-}4)$$

where m is the reduced mass $m_a m_b/(m_a + m_b)$ and $M = m_a + m_b$. Then if $I_{el}(\theta)\, d\omega$ is the differential elastic cross section for scattering into the solid angle element $d\omega$ about the direction θ, in relative coordinates,

$$\frac{dE_{el}}{dt} = \frac{m}{M}n(a)n(b)\int d\omega \left[\int\!\!\int f_a(v_a)f_b(v_b)I_{el}(\theta)\,\Delta E_{el}(\theta)u\, dv_a\, dv_b\right],$$

$$(1\text{-}15\text{-}5)$$

with $u = |\mathbf{v}_a - \mathbf{v}_b|$. Finally, we obtain

$$\frac{dE_{el}}{dt} = -8\pi^{-1/2} n(a)n(b) \frac{m}{M} k(T_a - T_b)\alpha^{5/2} \int_0^\infty Q_m(u)u^5 e^{-\alpha u^2} du, \quad (1\text{-}15\text{-}6)$$

where Q_m is the momentum transfer cross section

$$Q_m = 2\pi \int_0^\pi (1 - \cos\theta) I_{el}(\theta) \sin\theta \, d\theta. \quad [1\text{-}6\text{-}1]$$

1-3. Redraw Fig. 1-2-5, adding the trajectory for a projectile with initial velocity v_0 and impact parameter $b + |db|$ and showing it being scattered at the angle $\theta - |d\theta|$. A crude sketch will suffice. Now repeat the problem for a purely attractive Coulomb potential, showing the two trajectories and two scattering angles. Recall that if the two Coulomb potentials differ only in algebraic sign, the magnitudes of the scattering angles will be the same for given b and v_0. [See Evans (1955, pp. 836–843) and Goldstein (1980, pp. 105–114).]

1-4. From [3-15-7] with $l = 0$, we see that the S-wave ($l = 0$) phase shift corresponding to a potential $V(r)$ is defined by the second-order differential equation

$$\left[\frac{d^2}{dr^2} + \kappa^2 - U(r)\right] L_0(r) = 0 \qquad \left[U \equiv \frac{2m}{\hbar^2} V(r)\right] \quad (1\text{-}15\text{-}7)$$

with the boundary conditions that

$$L_0(r) \to 0 \qquad\qquad \text{as } r \to 0 \qquad\qquad (1\text{-}15\text{-}8)$$

$$L_0(r) \sim \sin(\kappa r + \eta_0) \qquad \text{as } r \to \infty. \qquad\qquad (1\text{-}15\text{-}9)$$

(See [3-15-15].) At the present time, these equations are usually solved numerically, to any precision required. The solution requires only that one write a short Fortran program, which will execute in a few seconds, even on the smallest personal computer.

(a) The Numerov algorithm represents a fast and simple way to integrate a second-order equation with no first derivative term. Imagine that the function L_0 is evaluated on a grid of uniformly spaced points $r_n = nh$, where h is the step size and $n = 0, 1, 2, \ldots$, so that $L_0(r_n) = y_n$. Then successive values of y_n are related by

$$\left(1 + \frac{h^2}{12} a_{n+1}\right) y_{n+1} - \left(2 + \frac{10h^2}{12} a_n\right) y_n + \left(1 + \frac{h^2}{12} a_{n-1}\right) y_{n-1} = 0,$$

$$(1\text{-}15\text{-}10)$$

where $a_n = \kappa^2 - U(r_n)$. Thus the solution at increasing values of r can be propagated out from the origin, starting with the estimates

$$y_0 = 0, \qquad y_1 = Cr_1, \qquad (1\text{-}15\text{-}11)$$

where C is any convenient normalization constant. At a sufficiently large value of r_n (one where the potential U is negligible compared with κ^2), the solution can be matched to the asymptotic condition at two points, for example,

$$y_{n-1} = A \sin \kappa r_{n-1} + B \cos \kappa y_{n-1}$$
$$y_n = A \sin \kappa r_n + B \cos \kappa r_n. \qquad (1\text{-}15\text{-}12)$$

Solving for A and B, we obtain the phase shift

$$\eta_0 = \tan^{-1} \frac{B}{A}. \qquad (1\text{-}15\text{-}13)$$

(b) Write a computer program to carry out this procedure, and test it on the static potential of a ground-state hydrogen atom (i.e., the potential seen by an electron close to an unperturbed H atom):

$$V(r) = \frac{e^2}{a_0} e^{-2r/a_0} \left(1 + \frac{a_0}{r} \right). \qquad (1\text{-}15\text{-}14)$$

[See Massey (1976, Chap. 1).] A sensible step size would be $h = 0.01a_0$, and one could integrate for 500 steps. Compare your answers with the values in Mott and Massey (1965, p. 128). A sample computer program is given in Koonin (1986, p. 50). Additional information on numerical algorithms is contained in Smith (1971b). See also Allison (1970, 1972).

(c) *Corollary* 1: At very small energies

$$\kappa \cot \eta_0 \approx -\frac{1}{a_s} + \frac{1}{2} r_s \kappa^2. \qquad (1\text{-}15\text{-}15)$$

This is called the *effective range expansion*: a_s and r_s are known as the *scattering length* and *effective range*, respectively. They play an important role in analyzing experimental data and in characterizing semiempirical potentials. The cross section at very low energy tends to the limit

$$\sigma_0 = 4\pi a_s^2. \qquad (1\text{-}15\text{-}16)$$

For the potential above, use your program to show that $a_s \approx 9.45a_0$.

(d) *Corollary 2:* By expanding V and L_0 in powers of r, show that a better estimate for starting the integration is given by

$$L_0(r) \approx Cr\left(1 - \frac{r}{a_0}\right). \tag{1-15-17}$$

Test this formula in your program. (This problem was provided by C. Bottcher.)

1-5. The phase shifts for higher values of l are obtained in a way similar to that explained in Problem 1-4. One has to solve the differential equation

$$\left[\frac{d^2}{dr^2} + \kappa^2 - U(r) + \frac{l(l+1)}{r^2}\right]L_l(r) = 0 \tag{1-15-18}$$

with the boundary conditions

$$L_l(r) \to 0 \qquad \text{as } r \to 0 \tag{1-15-19}$$

and

$$L_l(r) \sim \sin\left(\kappa r - \frac{l\pi}{2} + \eta_l\right) \qquad \text{as } r \to \infty. \tag{1-15-20}$$

(See [3-15-15].) As before, the equation can be integrated from the origin by the Numerov method. The asymptotic matching is best done with spherical Bessel functions, which have the properties that

$$j_l(x) \sim \sin\left(x - \frac{l\pi}{2}\right) \qquad \text{as } r \to \infty, \tag{1-15-21}$$

$$n_l(x) \sim \cos\left(x - \frac{l\pi}{2}\right) \qquad \text{as } r \to \infty. \tag{1-15-22}$$

The first few functions are explicitly given by

$$j_0(x) = \frac{\sin x}{x} \qquad\qquad n_0(x) = -\frac{\cos x}{x}$$

$$j_1(x) = \frac{\sin x}{x^2} - \frac{\cos x}{x} \qquad\qquad n_1(x) = -\frac{\cos x}{x^2} - \frac{\sin x}{x}$$

$$j_2(x) = \left(\frac{3}{x^3} - \frac{1}{x}\right)\sin x - \frac{3}{x^2}\cos x \qquad n_2(x) = -\left(\frac{3}{x^3} - \frac{1}{x}\right)\cos x - \frac{3}{x^2}\sin x.$$

$$\tag{1-15-23}$$

You will notice that n_l diverges as $x \to 0$, while j_l is well behaved. In the notation of Problem 1-4, one starts with the estimates

$$y_0 = 0, \qquad y_1 = Cr_1^{l+1} \tag{1-15-24}$$

and matches the asymptotic condition in the form

$$y_{n-1} = Aj_l(\kappa r_{n-1}) + Bn_l(\kappa r_{n-1})$$
$$y_n = Aj_l(\kappa r_n) + Bn_l(\kappa r_n). \tag{1-15-25}$$

The lth-order phase shift is given by

$$\eta_l = \tan^{-1} \frac{B}{A}. \tag{1-15-26}$$

(a) For the potential of Problem 1-4, calculate η_1 and η_2 for a range of κ between $\kappa = 0.01$ and $\kappa = 5.0$. Show that as $\kappa \to 0$,

$$\eta_l \sim a_l \kappa^{2l+1}. \tag{1-15-27}$$

(b) Repeat part (a) with a model polarization potential

$$V_{\text{Pol}} = -\frac{\alpha}{2r^4} [1 - e^{-(r/r_0)^6}] \tag{1-15-28}$$

added to the static field. For the hydrogen atom, $\alpha = 9/2$ (in atomic units), and r_0 can be taken as $2a_0$. Verify that the low-energy behavior changes to

$$\eta_l \sim \frac{\pi \alpha \kappa^2}{(2l-1)(2l+1)(2l+3)}. \tag{1-15-29}$$

(This problem was provided by C. Bottcher.) [See Bottcher (1983, 1985), Sandhya Devi and Garcia (1983, 1984), and Kulander (1987).]

1-6. *Double Minimum Problem*

(a) Consider a one-dimensional rectangular potential well $V(x)$ of width $2W$ and depth D, centered about the origin of the X axis. For each of several different ratios W/D, sketch the set of energy levels that you would expect for a particle trapped in the well. Then sketch the wave functions (energy eigenfunctions) for the two lowest stationary states (i.e., for the ground state and the first excited state).

(b) Now consider two rectangular potential wells of width W and depth D centered about $x = 0$, with their centers separated by a very large distance d. Because of the finite "strength" of the barrier between the wells, it is possible to introduce a particle into this potential system in such a way that it is *shared* by the two wells. The quantum states to which we refer here are stationary states with wave functions that are either symmetric (S) or antisymmetric (A) with respect to the origin of coordinates. If the particle is in any one of these S or A stationary states, the probability of finding it in one well at any instant is the same as finding it in the other well at that instant, and the probabilities are time independent. Sketch the wave functions for the lowest S state and for the lowest A state. In the limit $d \rightarrow \infty$, the energies of these two states approach equality, but for finite d, an energy splitting occurs, and the energies of the corresponding S and A states are different. Show that the A state has a higher energy than the corresponding S state for finite d.

(c) In part (b), if you decrease the spacing d until the two potential wells are in contact, the single well of part (a) is formed. Sketch the wave functions of the lowest-energy pair of S and A states for several intermediate values of d. Then sketch curves showing how the energy of these S and A states varies as a function of d, letting d go down to the value W.

(d) Now consider a nonstationary state: Show that if we put the particle into one of the potential wells so that we know that it is there at $t = 0$, the particle will oscillate from one well to the other, provided that it does not lose energy by radiation. [See Park (1974, pp. 116–123).]

(e) Explain how the considerations outlined in this problem provide a qualitative explanation of attractive and repulsive molecular states, covalent bonding, the splitting of the infrared lines of the NH_3 molecule, and the ammonia maser. [See Sproull and Phillips (1980), Feynman et al. (1965), Rapp (1971), and Morrison et al. (1976, Chap. 13).]

1-7. Justify the statement in Section 1-7-D-3 that two oxygen atoms in their triplet P state can interact in 18 different ways. [See Morrison et al. (1976, Chap. 14) for an excellent treatment of the separated atoms and united atoms limits of molecules, and of the correlation diagrams that connect them.]

1-8. Consider the scattering of alpha particles (Lab energy E) by copper nuclei ($Z = 29$, $M = 63$ amu, radius $R = 5 \times 10^{-15}$ m). Calculate the value of E at which deviations from Coulomb (Rutherford) scattering occur in head-on collisions. [See Bransden and Joachain (1983, p. 51).]

1-9. Carbon-12 nuclei are elastically scattered by oxygen-16 nuclei at a CM energy of 1 MeV.

(a) Show by a classical argument that only the Coulomb interaction need be considered, and calculate the classical CM differential cross section accordingly.

(b) Suppose that in (1-13-4) the screening distance a has the value 10^{-10} m, and use the first Born approximation (Chapter 3 of the companion volume) to calculate the differential cross section for scattering from the screened potential (1-13-4). Over what angular range does this cross section differ significantly from that calculated in part (a)? [See Bransden and Joachain (1989, p. 599).]

1-10. Projectiles are elastically scattered by a potential of range a. Estimate the uncertainty in the scattering angle for the following cases:

(a) protons with $E = 10$ MeV, $a = 2 \times 10^{-15}$ m;

(b) protons with $E = 10$ keV, $a = 10^{-10}$ m; electrons with $E = 5$ eV, $a = 10^{-10}$ m. [See Bransden and Joachain (1989, p. 597).]

1-11. Figure 1-15-1 shows a Stern–Gerlach apparatus (vintage 1922) consisting of an oven (O), a slit system (S), an inhomogeneous magnet (M), and a collecting plate (P) that is perpendicular to the axis of transmission (the Y axis). The magnet is symmetric about the Y–Z plane and has a field gradient $dB_z/dz = 10^3$ T/m in the Z direction. The length of the pole pieces in the Y direction is $L = 0.1$ m. The distance of the collecting plate from the nearest edge of the magnet is $l = 1$ m. The oven is maintained at 600 K, and the atomic beam is composed of ground-state silver atoms, whose magnetic moment given by

$$\mathscr{M} = -\frac{g\mu_B J}{\hbar}. \tag{1-15-30}$$

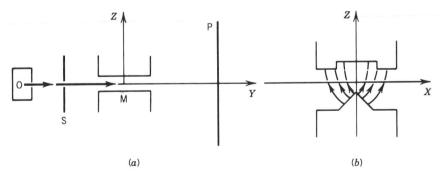

(a)   (b)

Figure 1-15-1. Schematic views of a Stern–Gerlach apparatus. (*a*) in the original experiment silver atoms from the oven (O) passed through the collimating aperture S, were deflected by the electromagnet (M), and condensed on the collecting plate (P). (*b*) Inhomogeneous field produced by the magnet M. [Adapted from Cohen-Tannoudji et al. (1977).]

g is the appropriate gyromagnetic ratio, which is a measure of the ratio of the magnetic moment to the angular momentum of the state in question (see Section 8-8, p. 550 of the companion volume). μB is the Bohr magneton, and **J** the total angular momentum of the ground-state silver atom. In the case considered here, $\mathcal{M} = \pm \mu_B$. The net force on each atom while it is in the inhomogeneous field is

$$\mathbf{F} = -\nabla W, \tag{1-15-31}$$

where

$$W = -\mathcal{M} \cdot \mathbf{B} \tag{1-15-32}$$

is the potential energy of the atom in the field. The torque on each atom is

$$\Gamma = \mathcal{M} \times \mathbf{B}. \tag{1-15-33}$$

(a) What is **J** for a ground-state silver atom, and what is the spectroscopic notation for the state? Sketch the trajectories of the initial and final atomic beams. Using the rms thermal value for the velocity of the silver atoms, show that the separation of the spots on the collecting plate at which the atoms condense is 0.078 m.

(b) Sketch the trajectories of the emerging beams for the cases where the atoms have $J = 1$, and $J = 3/2$. Discuss the way in which this problem illustrates the basis of Stern–Gerlach state selection.

(c) Can the Stern–Gerlach technique be used as a spin filter for free electrons? [See Section 4-6 of the companion volume and Baym (1973, Chap. 14), see also Bransden and Joachain (1983, pp. 40–46, 51), Cohen-Tannoudji et al. (1977, pp. 387–398), and Feynman et al. (1965, Chaps. 5–7).]

(d) Draw a large version of Fig. 1-15-1-b with a more realistic plot of the magnetic field lines and demonstrate that $\partial \mathbf{B}/\partial z$ has a significant value near the Z axis.

1-12. Consider a bunched beam of potassium atoms approaching a pair of Stern–Gerlach magnets (first A and then B) from the left. The magnets are spaced a distance X apart, and they transmit only $S_z = +\frac{1}{2}$ states. Larmor precession occurs between the magnets. The bunched beam is subjected to a spatially uniform magnetic field **B** in the region between A and B for 1 μs. What magnitude and direction should **B** have in order to prevent the atoms from reaching a tightly collimated detector D on the axis, off to the right of A and B? [See Chen (1974, p. 276) and Feynman et al. (1965, Secs. 7-5, 10-6, 10-7). An interesting set of problems on Stern–Gerlach filters appears in the Exercise Book that accompanies Feynman et al.]

1-13. Consider an electrostatic slit lens consisting of a long, narrow slit of height y_0 in a thin metal plate, with an electric field E_1 normal to the plate on the left and a field E_2 normal to the plate on the right. Both **E** vectors are directed toward the right. A beam of positive ions (energy eV_0) approaching the plate from the left is brought to a focus at a distance x_1 in front of the plate (on the left side). Show that the beam is refocused on the other side of the plate (the right side) at a distance x_2 from the plate given by

$$\frac{1}{x_2} + \frac{1}{x_1} \approx \frac{E_2 - E_1}{2V_0}. \tag{1-15-34}$$

Take $V_0 \gg E_1 x_1$, $V_0 \gg E_2 x_2$, $x_1 \gg y_0$, $x_2 \gg y_0$. The slit lens is the simplest electrostatic focusing device. [See Cronin et al. (1979, p. 19).]

1-14. Consider a particle of charge q and momentum p moving in a static magnetic field **B**. Show that it is possible to calibrate the orbit of the particle by determining the configuration of a perfectly flexible wire in the field if the current I in the wire and its tension T have values consistent with the equation $I = -Te/p$. (*Suggestion:* Write down the general differential force equation for the particle in terms of e, p, **B**, and $d\mathbf{r}/ds$, where ds is the element of path length, and compare it with the equation for $d^2\mathbf{r}/ds^2$ for a current-carrying wire in its equilibrium configuration. This technique has been extensively used in nuclear physics research.) [See Cronin et al. 1979, p. 20).]

1-15. (a) Calculate the phase shifts η_l for a reduced potential of the form $U(r) = A/r^2$, where A is positive and $U(r) = 2mV(r)/\hbar^2$, $V(r)$ being the true potential.

(b) Show that $n_l \approx -\pi A/2(2l + 1)$ when l is large.

(c) Is $I_s(\Theta)$ finite in the forward direction?

(d) Is q_s finite?

[See Bransden and Joachain (1983, p. 497).]

1-16. How must the treatment of Problem 1-15 be modified if A is negative? Show that the radial equation [3-15-11] has physically acceptable solutions only if $A > -\frac{1}{4}$. [See Bransden and Joachain (1983, p. 497).]

1-17. Consider elastic scattering under a central potential such that the classical deflection function $\Theta_{\pm}$ as a function of angular momentum has the shape shown in Fig. 1-11-1. Show that if Θ_r is the rainbow angle corresponding to the angular momentum $L = L_r$, then classically

$$\frac{d\sigma}{d\Omega} = \begin{cases} \dfrac{1}{2m^2v^2} \dfrac{L_r}{\sin\theta_r} \left| \dfrac{1}{B(\theta - \theta_r)} \right|^{1/2}, & \theta < \theta_r \\ 0, & \theta > \theta_r \end{cases} \tag{1-15-35}$$

where B is a constant. [*Suggestion*: Expand $\Theta_{\pm}$ about Θ_r to get $\Theta_{\pm} \approx -\Theta_r + B(L - L_r)^2$.] [See Bransden and Joachain (1983, pp. 538–539) and Bransden (1983, pp. 103–106, 108).]

1-18. Equations (1-12-1), (1-12-2), (1-13-1), (1-13-2), and (1-13-3) give the differential elastic cross section in terms of the scattering amplitudes for various collision situations. Justify fully the form of each equation. Note that collision partners that are identical if spin is ignored become distinguishable when their spin is considered unless each has the same component of spin along the quantization axis. [See Feynman et al. (1965, Chaps. 3 and 4), Baym (1973, pp. 400–408), Taylor (1972, pp. 441–450), Feynman and Hibbs (1965), Holstein (1992), and Landau and Lifshitz (1965, pp. 523–526).]

1-19. Deuterons are bosons with spin 1. Write down an expression analogous to (1-12-2) for elastic deuteron–deuteron scattering under the assumption that there is no change of the spins during the scattering. [A good set of more advanced problems of this type appears in the Exercise Book that accompanies Feynman et al. (1965).]

1-20. Derive (1-14-19) for the axial beam intensity at a distance x downstream from an effusion source. [See Kennard (1938).]

1-21. Show that the differential elastic cross section for small-angle scattering under a potential of the form $V(r) = -Cr^{-n}(C > 0)$ is given classically, for $\Theta \ll 1$ rad, by

$$\frac{d\sigma}{d\Omega} = g(n)\left(\frac{C}{E\theta}\right)^{2/n}\frac{1}{\theta\sin\theta}, \qquad (1\text{-}15\text{-}36)$$

where $g(4) = (3\pi)^{1/2}/8$ and $g(6) = (15\pi/2)^{1/3}/12$ for the important cases of $n = 4$ and $n = 6$, respectively. [See Bransden and Joachain (1983, pp. 536–537) and Kennard (1938, pp. 118–120).]

1-22. Integrate the small-angle classical approximation (1-15-36) in Problem 1-19 from $\Theta = \Theta_0$ to $\Theta = \pi$ for the case $n = 6$ to obtain an effective integral cross section. In practice, Θ_0 would be a small angle determined by the design of the scattering experiment. What happens for $\Theta_0 = 0$?

1-23. An approach to the scattering problem that is more rigorous than the approach used in this book involves the use of *wave packets*. Study the wave packet method as described in Taylor (1972).

REFERENCES

Ahlrichs, R., H. J. Bohm, S. Brode, K. T. Tang and J. P. Toennies (1988), *J. Chem. Phys.* **88**, 6290.

Allison, A. C. (1970), *J. Comp. Phys.* **6**, 378.

Allison, A. C. (1972), *Comp. Phys. Commun.* **3**, 173.

Auerbach, D. J. (1988), "Velocity Measurements by Time-of-Flight Methods," Chap. 14 in G. Scoles, Ed., *Atomic and Molecular Beam Methods*, Vol. 1, Oxford University Press, New York.

Baer, M. B., R. Düren, B. Friedrich, G. Niedner, M. Noll, and J. P. Toennies (1987), *Phys. Rev. A* **36**, 1063.

Bassi, D. (1988a), "Detection Principles," Chap. 6 in G. Scoles, Ed., *Atomic and Molecular Beam Methods*. Vol. 1, Oxford University Press, New York.

Bassi, D. (1988b), "Ionization Detectors I: Ion Production," Chap. 7 in G. Scoles, Ed., *Atomic and Molecular Beam Methods*. Vol. 1, Oxford University Press, New York.

Bassi, D. (1988c), "Ionization Detectors II: Mass Selection and Ion Detection," Chap. 8 in G. Scoles, Ed., *Atomic and Molecular Beam Methods*. Vol. 1, Oxford University Press, New York.

Baym, G. (1973), *Lectures on Quantum Mechanics*, W. A. Benjamin, Reading, Mass.

Beier, H. J. (1973), *J. Phys. B* **6**, 683.

Bennewitz, H. G., K. H. Kramer, W. Paul, and J. P. Toennies (1964), *Z. Phys.* **177**, 84.

Bergmann, K. (1988), "State Selection by Optical Methods," Chap 12 in G. Scoles, Ed., *Atomic and Molecular Beam Methods*, Vol. 1, Oxford University Press, New York.

Berkling, K., Ch. Schlier, and P. Toschek (1962), *I. Phys.* **168**, 81.

Bernstein, R. B. (1966), "Quantum Effects in Elastic Molecular Scattering," Chap. 3 in J. Ross, Ed., *Molecular Beams*, Interscience, New York.

Bernstein, R. B. (1982), *Chemical Dynamics via Molecular Beam and Laser Techniques*, Oxford University Press, New York.

Berry, M. V., and K. E. Mount (1972), "Semiclassical Approximation in Wave Mechanics," *Rep. Prog. Phys.* **35**, 315.

Beyer, W., and H. Haberland (1984), *Phys. Rev. A* **29**, 2280.

Bickes, R. W., B. Lantsch, J. P. Toennies, and K. Walaschewski (1973), *Faraday Discuss. Chem. Soc.* **55**, 167.

Bohr, N. (1948), *K. Dan. Vidensk. Selsk. Mat. Fys. Medd.* **18**, 8 (in English).

Bottcher, C. (1983), *ICPEAC XIII*. Berlin, p. 187.

Bottcher, C. (1985), *AAMP* **20**, 241.

Bowers, M. S., K. T. Tang, and J. P. Toennies (1988), *J. Chem. Phys.* **88**, 5465.

Bransden, B. H. (1983), *Atomic Collision Theory*, 2nd ed., Benjamin-Cummings, Menlo Park, Calif.

Bransden, B. H., and C. J. Joachain (1983), *Physics of Atoms and Molecules*, Longman, Harlow, Essex, England.

Bransden, B. H., and C. J. Joachain (1989), *Introduction to Quantum Mechanics*, Wiley, New York.

Bransden, B. H., and C. J. Joachain (1993), *The Theory of Electronic and Atomic Collisions*, Oxford University Press, New York.

Buck, U. (1975), *Adv. Phys. Chem.* **30**, 313.

Buck, U. (1988a), "General Principles and Methods," Chap. 18 in G. Scoles, Ed., *Atomic and Molecular Beam Methods*, Vol. 1, Oxford University Press, New York.

Buck, U. (1988b), "Elastic Scattering II: Differential Cross Sections," Chap. 20 in G. Scoles, Ed., *Atomic and Molecular Beam Methods*, Vol. 1, Oxford University Press, New York.

Buck, U., and H. Pauly (1971), *J. Chem. Phys.* **51**, 1929.

Budenholzer, F. E., E. A. Gislason, and A. D. Jorgensen (1986), *Chem. Phys.* **110**, 171.

Burgess, A., and I. C. Percival (1968), "Classical Theory of Atomic Scattering," AAMP, **4**, 109.

Burhop, E. H. S. (1961). "Theory of Collisions," in D. R. Bates, Ed., *Quantum Theory*, Vol. 1, Academic Press, New York.

Chan, N., D. M. Crowe, M. S. Lubell, F. C. Tang, A. Vasilakis, F. J. Mulligan, and J. Slevin (1988), *Z. Phys. D* **10**, 393.

Chen, M. (1974), *Berkeley Physics Problems with Solutions*, Prentice Hall, Englewood Cliffs, N.J.

Child, M. S. (1974), *Molecular Collision Theory*, Academic Press, New York.

Child, M. S., Ed. (1980), *Semiclassical Methods in Molecular Scattering and Spectroscopy*, D. Reidel, Boston.

Chitnis, C. E., R. S. Gao, J. Pinedo, K. A. Smith, and R. F. Stebbings (1988), *Phys. Rev. A* **37**, 687.

Cohen-Tannoudji, C., B. Diu, and F. Laloë (1977), *Quantum Mechanics*, 2 vols., Wiley, New York.

Cordey, J. G. (1984), "Trapping and Thermalization of Fast Ions," in Vol. 2, *Plasmas*, in H. S. W. Massey, E. W. McDaniel, and B. Bederson, Series Eds. (1982 1984), *Applied Atomic Collision Physics*, 5 vols., Academic Press, New York, p. 327.

Cronin, J. A., D. F. Greenberg, and V. L. Telegdi (1979), *University of Chicago Graduate Problems in Physics with Solutions*, Addison-Wesley, Reading, Mass.

Dalgarno, A. (1970), "Theory of Ion–Molecule Collisions," Chap. 3 in E. W. McDaniel, V. Čermák, A. Dalgarno, E. E. Ferguson, and L. Friedman, *Ion–Molecule Reactions*, Wiley, New York.

Demkov, Yu. N. (1983), "New Results in General Theory of Atomic Collisions: Small Angle Classical and Semiclassical Forward and Backward Scattering," *ICPEAC XIII*, Berlin, p. 171.

Dickinson, A. S., and D. Richards (1982), "Classical and Semiclassical Methods in Inelastic Heavy Particle Collisions," *AAMP* **18**, 166.

Dunning, T. H., and L. B. Harding (1985), "Ab Initio Determination of Potential Energy Surfaces for Chemical Reactions," Chap. 1 in M. Baer, Ed., *Theory of Chemical Reaction Dynamics*, Vol. 1, CRC Press, Boca Raton, Fla.

Düren, R. (1988), "Scattering Experiments with Laser-Excited Atomic Beams," Chap. 26 in G. Scoles, Ed., *Atomic and Molecular Beam Methods*, Vol. 1, Oxford University Press, New York.

Eu, B. C., and M. L. Sink (1983), *Semiclassical Theories of Molecular Scattering*, Springer-Verlag, Berlin.

Evans, R. D. (1955), *The Atomic Nucleus*, McGraw-Hill, New York.

Everhart, E., G. Stone, and R. J. Carbone (1955), *Phys. Rev.* **99**, 1287.

Fano, U., and A. R. P. Rau (1986), *Atomic Collisions and Spectra*, Academic Press, New York.

Faubel, M., and J. P. Toennies (1978), *AAMP* **13**, 229.

Faubel, M., K. H. Kohl, J. P. Toennies, K. T. Tang, and Y. Y. Yung (1982), *Faraday Discuss. Chem. Soc.* **73**, 205.

Feltgen, R., H. Kirst, K. A. Köhler, H. Pauly, and F. Torello (1982), *J. Chem. Phys.* **76**, 2360.

Fenn, J. B. (1982), "Collision Kinetics in Gas Dynamics," in Vol. 5, *Special Topics*, in H. S. W. Massey, E. W. McDaniel, and B. Bederson, Series Eds. (1982–1984), *Applied Atomic Collision Physics*, 5 vols., Academic Press, New York, p. 349.

Feynman, R. P., and A. R. Hibbs (1965), *Quantum Mechanics and Path Integrals*, McGraw-Hill, New York.

Feynman, R. P., R. B. Leighton, and M. Sands (1965), *The Feynman Lectures on Physics*, Vol. 3, *Quantum Mechanics*, Addison-Wesley, Reading, Mass.

Firsov, O. B. (1958), *Sov. Phys.-JETP* **6**, 534; **7** 308.

Flannery, M. R., and E. J. Mansky (1988), *J. Chem. Phys.* **88**, 4228.

Ford, K. W., and J. A. Wheeler (1959a), *Ann. Phys.* **7**, 259.

Ford, K. W., and J. A. Wheeler (1959b), *Ann. Phys.* **7**, 287.

Ford, K. W., D. L. Hill, M. Wakano, and J. A. Wheeler (1959), *Ann. Phys.* **7**, 239.

Friedrich, B., G. Niedner, M. Noll, and J. P. Toennies (1987), *Z. Phys. D* **6**, 49.

Fuls, E. N., P. R. Jones, F. P. Ziemba, and E. Everhart (1957), *Phys. Rev.* **107**, 704.

Gao, R. S., L. K. Johnson, D. E. Nitz, K. A. Smith, and R. F. Stebbings (1987), *Phys. Rev. A* **36**, 3077.

Gao, R. S., L. K. Johnson, D. A. Schafer, J. H. Newman, K. A. Smith, and R. F. Stebbings (1988a), *Phys. Rev. A* **38**, 2789.

Gao, R. S., W. E. Robert, G. J. Smith, K. A. Smith, and R. F. Stebbings (1988b), *Rev. Sci. Instrum.* **59**, 1954.

Gentry, W. R. (1979), *ICPEAC XI*, Kyoto, p. 807.

Gentry, W. R. (1985), *ICPEAC XIV*, Stanford, Calif., p. 13.

Gentry, W. R. (1988), "Low-Energy Pulsed Beam Sources," Chap. 3 in G. Scoles, Ed., *Atomic and Molecular Beam Methods*, Vol. 1, Oxford University Press, New York.

Gentry, W. R., and C. F. Giese (1978), *Rev. Sci. Instrum.* **49**, 595.

Gianturco, F. A., G. Niedner, M. Noll, E. Semprini, F. Stefani, and J. P. Toennies (1987), *Z. Phys. D* **7**, 281.

Gillen, K. T., R. P. Saxon, D. C. Lorents, G. E. Ice, and R. E. Olson (1976), *J. Chem. Phys.* **64**, 1925.

Goldstein, H. (1980), *Classical Mechanics*, 2nd ed., Addison-Wesley, Reading, Mass.

Hasted, J. B. (1972), *Physics of Atomic Collisions*, 2nd ed., Elsevier, New York.

Hefter, U., and K. Bergmann (1988), "Spectroscopic Detection Methods," Chap. 9 in G. Scoles, Ed., *Atomic and Molecular Beam Methods*, Vol. 1, Oxford University Press, New York, (1988).

Hermann, V., H. Schmidt, and F. Linder (1978), *J. Phys. B* **11**, 493.

Hirschfelder, J. O., C. F. Curtiss, and R. B. Bird (1964), *Molecular Theory of Gases and Liquids*, Wiley, New York.

Holstein, B. R. (1992), *Topics in Advanced Quantum Mechanics*, Chap. II, Addison-Wesley, Redwood City, Calif.

Hostettler, H. U., and R. B. Bernstein (1960), *Rev. Sci. Instrum.* **31**, 872.

Iannotta, S. (1988), "Experiments with Spin-Polarized Beams," Chap. 27 in G. Scoles, Ed., *Atomic and Molecular Beam Methods*, Vol. 1, Oxford University Press, New York.

Joachain, C. J. (1975), *Quantum Collision Theory*, North-Holland, Amsterdam.

Johnson, L. K., R. S. Gao, K. A. Smith, and R. F. Stebbings (1988), *Phys. Rev. A* **38**, 2794.

Jones, P. R., G. M. Conklin, D. C. Lorents, and R. E. Olson (1974), *Phys. Rev. A* **10**, 102.

Junker, B. R. (1982), "Military Applications of Atomic and Molecular Physics," in Vol. 5, *Special Topics*, in H. S. W. Massey, E. W. McDaniel, and B. Bederson, Series Eds. (1982–1984), *Applied Atomic Collision Physics*, 5 vols., Academic Press, New York, p. 379.

Kennard, E. H. (1938), *Kinetic Theory of Gases*, McGraw-Hill, New York.

Konrad, M., and F. Linder (1982), *J. Phys. B* **15**, L-405.

Konrad, M., and F. Linder (1983), *ICPEAC XIII*, Berlin, Contributed Papers, p. 351.

Koonin, S. E. (1986), *Computational Physics*, Addison-Wesley, Reading, Mass.

Korsch, H. J., and K. E. Thylwe (1983), *J. Phys. B* **16**, 793.

Kulander, K. C. (1987), *Phys. Rev.* **35**, 445; **36**, 2726.

Kuntz, P. J. (1979), "Interaction Potentials II: Semiempirical Atom–Molecule Potentials for Collision Theory," Chap. 3 in R. B. Bernstein, Ed., *Atom–Molecule Collision Theory: A Guide for the Experimentalist*, Plenum Press, New York.

Kuntz, P. J. (1985), "Semiempirical Potential Energy Surfaces," Chap. 2 in M. Baer, Ed., *Theory of Chemical Reaction Dynamics*, Vol. 1 CRC Press, Boca Raton, Fla.

Landau, L. D., and E. M. Lifshitz (1960), *Mechanics*, Addison-Wesley, Reading, Mass.

Landau, L. D., and E. M. Lifshitz (1965), *Quantum Mechanics, Nonrelativistic Theory*, 2nd ed., Addison-Wesley, Reading, Mass.

Lane, G. H., and E. Everhart (1960), *Phys. Rev.* **120**, 2064.

Lawley, K. P., Ed. (1975), "Molecular Beam Scattering," *Adv. Chem. Phys.* **30**.

Lawley, K. P., Ed. (1987), "Ab Initio Methods in Quantum Chemistry," Parts 1 and 2, *Adv. Chem. Phys.* **67** and **69**.

Levine, R. D., and R. B. Bernstein (1987), *Molecular Reaction Dynamics and Chemical Reactivity*, Oxford University Press, New York.

Linse, C. A., J. J. H. van den Biesen, E. H. van Veen, C. J. N. van den Meijdenberg, and J. J. M. Beenakker (1979), *Physica* **99-A**, 145.

Lorents, D. C., and W. Aberth (1965), *Phys. Rev. A* **139**, 1017.

Lorents, D. C., and G. M. Conklin (1972), *J. Phys. B* **5**, 950.

Maitland, G. C., M. Rigby, E. B. Smith, and W. A. Wakeham (1987), *Intermolecular Forces: Their Origin and Determination*, Oxford University Press, New York. Reprint of 1981 edition.

Marchi, R. P., and F. T. Smith (1965), *Phys. Rev. A* **139**, 1025.

Mason, E. A. (1982), "Determination of Intermolecular Potentials," in Vol. 5, *Special Topics*, in H. S. W. Massey, E. W. McDaniel, and B. Bederson, Series Eds. (1982–1984), *Applied Atomic Collision Physics*, 5 vols., Academic Press, New York, p. 255.

Mason, E. A., and E. W. McDaniel (1988), *Transport Properties of Ions in Gases*, Wiley, New York.

Mason, E. A., and J. T. Vanderslice (1962), in D. R. Bates, Ed., *Atomic and Molecular Processes*, Academic Press, New York.

Massey, H. S. W. (1969), *Electron Collisions with Molecules and Photoionization*, Vol. 2 of H. S. W. Massey, E. H. S. Burhop, and H. B. Gilbody, Eds., *Electronic and Ionic Impact Phenomena*, 2nd ed. in 5 vols., Clarendon Press, Oxford.

Massey, H. S. W. (1971), *Slow Collisions of Heavy Particles*, Vol. 3 of H. S. W. Massey, E. H. S. Burhop, and H. B. Gilbody, Eds., *Electronic and Ionic Impact Phenomena*, 2nd ed. in 5 vols., Clarendon Press, Oxford.

Massey, H. S. W. (1976), *Negative Ions*, 3rd ed., Cambridge University Press, Cambridge.

Massey, H. S. W. (1982), "The Thermal Balance in the Thermosphere at Middle Latitudes," in Vol. 1, *Atmospheric Physics and Chemistry*, in H. S. W. Massey, E. W. McDaniel, and B. Bederson, Series Eds. (1982–1984), *Applied Atomic Collision Physics*, 5 vols., Academic Press, New York, p. 77.

Massey, H. S. W., and E. H. S. Burhop (1969), *Collision of Electrons with Atoms*, Vol. 1 of H. S. W. Massey, E. H. S. Burhop, and H. B. Gilbody, Eds., *Electronic and Ionic Impact Phenomena*, 2nd ed. in 5 vols., Clarendon Press, Oxford.

Massey, H. S. W., and H. B. Gilbody (1974), *Recombination and Fast Collisions of Heavy Particles*, Vol. 4 of H. S. W. Massey, E. H. S. Burhop, and H. B. Gilbody, Eds., *Electronic and Ionic Impact Phenomena*, 2nd ed. in 5 vols., Clarendon Press, Oxford.

Massey, H. S. W., and C. B. O. Mohr (1933), *Proc. R. Soc. London* **A-141**, 434.

Massey, H. S. W., and C. B. O. Mohr (1934), *Proc. R. Soc. London* **A-144**, 188.

Massey, H. S. W., and R. A. Smith (1933), *Proc. R. Soc. London* **A-142**, 142.

Massey, H. S. W., E. W. McDaniel, and B. Bederson, Series Eds. (1982–1984), *Applied Atomic Collision Physics*, 5 vols., Academic Press, New York.

Mattheus, A., A. Fischer, G. Ziegler, E. Gottwald, and K. Bergmann (1986), *Phys. Rev. Lett.* **56**, 712.

McDaniel, E. W. (1964), *Collision Phenomena in Ionized Gases*, Wiley, New York.

McDaniel, E. W. (1989), *Atomic Collisions: Electron and Photon Projectiles*, Wiley, New York.

McDaniel, E. W., and L. A. Viehland (1984), *Phys. Rep.* **110**, 333.

McDowell, M. R. C., and J. P. Coleman (1970), *Introduction to the Theory of Ion–Atom Collisions*, North-Holland, Amsterdam.

Messiah, A. (1961, 1962), *Quantum Mechanics*, Vols. 1 and 2, Interscience, New York.

Miller, W. H. (1974), "Classical Limit Quantum Mechanics and the Theory of Molecular Collisions," *Adv. Chem. Phys.* **25**, 69.

Miller, W. H. (1975), "Classical S-Matrix in Molecular Collisions," *Adv. Chem. Phys.* **30**, 77.

Miller, W. H., Ed. (1976), *Dynamics of Molecular Collisions*, Vols. 1 and 2, Plenum Press, New York.

Miller, D. R. (1988), "Free Jet Sources," Chap. 2 in G. Scoles, Ed., *Atomic and Molecular Beam Methods*, Vol. 1, Oxford University Press, New York.

Miller, R. C., and P. Kusch (1955), *Phys. Rev.* **99**, 1314.

Morgenstern, R., D. C. Lorents, J. R. Peterson, and R. E. Olson (1973), *Phys. Rev. A* **8**, 2372.

Morrison, M. A., T. L. Estle, and N. F. Lane (1976), *Quantum States of Atoms, Molecules, and Solids*, Prentice Hall, Englewood Cliffs, N.J.

Mott, N. F., and H. S. W. Massey (1965), *The Theory of Atomic Collisions*, 3rd ed., Clarendon Press, Oxford.

Newman, J. H., K. A. Smith, R. F. Stebbings, and Y. S. Chen (1985), *J. Geophys. Res.* **90**, A-11, 11045.

Newman, J. H., Y. S. Chen, K. A. Smith, and R. F. Stebbings (1986), *J. Geophys. Res.* **91**, A-8, 8947.

Newton, R. G. (1982), *Scattering Theory of Waves and Particles*, 2nd ed., Springer-Verlag, New York.

Nitz, D. E., R. S. Gao, L. K. Johnson, K. A. Smith, and R. F. Stebbings (1987), *Phys. Rev. A* **35**, 4541.

Park, D. (1974), *Introduction to the Quantum Theory*, 2nd ed., McGraw-Hill, New York.

Pauly, H. (1975), "Collision Processes; Theory of Elastic Scattering," in H. Eyring, D. Henderson, and W. Jost, Eds., *Physical Chemistry*, Vol. 6-B, Academic Press, New York, pp. 553–628.

Pauly, H. (1979), "Elastic Scattering Cross Sections I: Spherical Potentials," Chap. 4 in R. B. Bernstein, Ed., *Atom-Molecule Collision Theory: A Guide for the Experimentalist*, Plenum Press, New York.

Pauly, H. (1988a), "Other Low-Energy Beam Sources," Chap. 4 in G. Scoles, Ed., *Atomic and Molecular Beam Methods*, Vol. 1, Oxford University Press, New York.

Pauly, H. (1988b), "High-Energy Beam Sources," Chap. 5 in G. Scoles, Ed., *Atomic and Molecular Beam Methods*, Vol. 1, Oxford University Press, New York.

Pauly, H., and J. P. Toennics (1965), "The Study of Intermolecular Potentials with Molecular Beams at Thermal Energies," *AAMP* **1**, 195.

Pauly, H., and J. P. Toennies (1968), "Neutral–Neutral Interactions: Beam Experiments at Thermal Energies," *M.E.P.* **7-A**, 227.

Peacher, J. L., E. Redd, D. G. Seely, T. J. Gay, D. M. Blankenship, and J. T. Park (1989), *Phys. Rev. A* **39**, 1760.

Percival, I. C. (1977), "Semiclassical Theory of Bound States," *Adv. Chem. Phys.* **36**, 1.

Percival, I. C., and D. Richards (1975), "The Theory of Collisions Between Charged Particles and Highly Excited Atoms," *AAMP* **11**, 1.

Polak-Dingels, P., M. S. Rajan, and E. A. Gislason (1982), *J. Chem. Phys.* **77**, 3983.

Rajan, M. S., and E. A. Gislason (1983), *J. Chem. Phys.* **78**, 2428.

Ramsey, N. F. (1988), *Z. Phys. D* **10**, 121.

Rapp, D. (1971), *Quantum Mechanics*, Holt, Rinehart and Winston, New York.

Reinig, P., G. Bischof, and F. Linder (1983), *ICPEAC XIII*, Berlin, Contributed Papers, p. 354.

Reuss, J. (1988), "State Selection by Nonoptical Methods," Chap. 11 in G. Scoles, Ed., *Atomic and Molecular Beam Methods*, Vol. 1, Oxford University Press, New York.

Rille, E., J. L. Peacher, E. Redd, T. J. Kvale, D. G. Seely, D. M. Blankenship, R. E. Olson, and J. T. Park (1984), *Phys. Rev. A* **29**, 521.

Rodberg, L. S., and R. M. Thaler (1967), *Introduction to the Quantum Theory of Scattering*, Academic Press, New York.

Rosin, S., and I. Rabi (1935), *Phys. Rev.* **48**, 373.

Rothe, E. W., and R. B. Bernstein (1959), *J. Chem. Phys.* **31**, 1619.

Rubahn, H. G., and J. P. Toennies (1988), *J. Chem. Phys.* **89**, 287.

Sakurai, J. J. (1985), *Modern Quantum Mechanics*, Benjamin-Cummings, Menlo Park, Calif.

Sandhya Devi, K., and J. D. Garcia (1983), *J. Phys. B* **16**, 2837.

Sandhya Devi, K., and J. D. Garcia (1984), *Phys. Rev.* **A-30**, 600.

Schaefer, H. F. (1979), "Interaction Potentials I: Atom–Molecule Potentials," Chap. 2 in R. B. Bernstein, Ed., *Atom–Molecule Collision Theory: A Guide for the Experimentalist*, Plenum Press, New York.

Schafer, D. A., J. H. Newman, K. A. Smith, and R. F. Stebbings (1987), *J. Geophys. Res.* **92**, 6107.

Schumacher, H., R. B. Bernstein, and E. W. Rothe (1960), *J. Chem. Phys.* **33**, 584.

Scoles, G. (1980), *Annu. Rev. Phys. Chem.* **31**, 81.

Scoles, G., Ed. (1988), *Atomic and Molecular Beam Methods*, Vol. 1, Oxford University Press, New York. Part I deals with basic techniques, Part II with molecular scattering.

Scoles, G., Ed. (1991), *Atomic and Molecular Beam Methods*, Vol. 2, Oxford University Press, New York. Part I deals with spectroscopy, Part II with surface collisions.

Sharp, T. E. (1971), *Atomic Data* **2**, 119.

Silver, D. M. (1980), *J. Chem. Phys.* **72**, 6445.

Smith, F. T. (1971a), *ICPEAC VII*, Amsterdam, p. 1.

Smith, K. M. (1971b), *The Calculation of Atomic Collision Processes*, Wiley, New York.

Smith, F. T., D. C. Lorents, W. Aberth, and R. P. Marchi (1965), *Phys. Rev. Lett.* **15**, 742.

Smith, F. T., R. P. Marchi, W. Aberth, D. C. Lorents, and O. Heinz (1967), *Phys. Rev.* **161**, 31.

Sproull, R. L., and W. A. Phillips (1980), *Modern Physics*, 3rd ed., Wiley, New York.

Stebbings, R. F. (1988), *AAMP* **25**, 83.

Stolte, S. (1988), "Scattering Experiments with State Selectors," Chap. 25 in G. Scoles, Ed., *Atomic and Molecular Beam Methods*, Vol. 1. Oxford University Press, New York.

Stolte, S., and J. Reuss (1979), "Elastic Scattering Cross Sections II: Noncentral Potentials," Chap. 5 in R. B. Bernstein, Ed., *Atom–Molecule Collision Theory: A Guide for the Experimentalist*, Plenum Press, New York.

Su, T., and M. T. Bowers (1979), "Classical Ion–Molecule Collision Theory," Chap. 3 in M. T. Bowers, Ed., *Gas Phase Ion Chemistry*, Vol. 1, Academic Press, New York.

Tang, K. T., and J. P. Toennies (1984), *J. Chem. Phys.* **80**, 3726.

Tang, K. T., and J. P. Toennies (1986), *Z. Phys. D* **1**, 91.

Tang, K. T., and J. P. Toennies (1988), *Chem. Phys. Lett.* **151**, 301.

Taylor, J. R. (1972), *Scattering Theory*, Wiley, New York.

Thylwe, K. E. (1983), *J. Phys. B* **16**, 1915.

Toennies, J. P. (1965), *Z. Phys.* **182**, 257.

Toennies, J. P. (1985), "Low-Energy Atomic and Molecular Collisions," in J. Bang and J. de Boer, Eds., *Semiclassical Descriptions of Atomic and Nuclear Collisions*, North-Holland, Amsterdam, p. 29.

Toennies, J. P., and K. Winkelmann (1977), *J. Chem. Phys.* **66**, 3965.

Toennies, J. P., W. Welz, and G. Wolf (1979), *J. Chem. Phys.* **71**, 614.

van den Biesen, J. J. H. (1988), "Elastic Scattering I: Integral Cross Sections," Chap. 19 in G. Scoles, Ed., *Atomic and Molecular Beam Methods*, Vol. 1. Oxford University Press, New York.

van den Biesen, J. J. H., R. M. Hermans, and C. J. N. van den Meijdenberg (1982), *Physica* **115-A**, 396.

van den Meijdenberg, C. J. N. (1988), "Velocity Selection by Mechanical Methods," Chap. 13 in G. Scoles, Ed., *Atomic and Molecular Beam Methods*, Vol. 1, Oxford University Press, New York, Scoles (1988).

Waldman, M., and R. G. Gordon (1979), *J. Chem. Phys.* **71**, 1325, 1340, 1353.

Wang, J., V. A. Shamamian, B. R. Thomas, J. M. Wilkinson, J. Riley, C. F. Giese, and W. R. Gentry (1988), *Phys. Rev. Lett.* **60**, 696.

Wheeler, J. A. (1976), "Semiclassical Analysis Illuminates the Connection Between Potential and Bound States and Scattering," in E. H. Lieb, B. Simon, and A. S. Wightman, Eds., *Studies in Mathematical Physics*, Princeton University Press, Princeton, N.J., pp. 351–422.

Williams, E. J. (1945), *Rev. Mod. Phys.* **17**, 217.

Wu, T. Y., and T. Ohmura (1962), *Quantum Theory of Scattering*, Prentice Hall, Englewood Cliffs, N.J.

Zen, M. (1988), "Accomodation, Accumulation, and Other Detection Methods," Chap. 10 in G. Scoles, Ed., *Atomic and Molecular Beam Methods*, Vol. 1, Oxford University Press, New York.

2

EXCITATION, DISSOCIATION, AND ENERGY TRANSFER IN HEAVY PARTICLE COLLISIONS

2-1. INTRODUCTION

This chapter deals with rotational, vibrational, and electronic excitation in binary heavy particle collisions (Part $\mathscr{A}$), with dissociation, which may be viewed as excitation into the vibrational continuum (Part $\mathscr{B}$), and with the collisional transfer of energy from one heavy particle to another (Part $\mathscr{C}$). Our treatment of excitation and dissociation here will parallel the discussion in Chapter 5 of the companion volume, which was concerned with the corresponding electron-impact processes. On the other hand, the present treatment of energy transfer shows little resemblance to the brief remarks on the subject in Chapter 5, for here we must deal with the full panoply of energy transfer mechanisms, in which energy is transferred among rotational (R), vibrational (V), electronic (E), and translational (T) degrees of freedom. The processes discussed in this chapter are of great practical importance. Some areas of application are listed below, the first (Roman) number given in parentheses in each case referring to the volume, and the second (Arabic) number to the chapter, in the books edited by Massey et al. (1982–1984):

Airglow and auroras (I–6)

Coronal ions (I-10)

Molecules in interstellar space (I-12)

Low-density plasmas (II-3)

Hot, dense plasmas (II-8)

Gas lasers (III-1, III-4, and III-7 through III-10, V-16)

Radiation detectors (V-6 and V-7)

Gas dynamics (V-15)

Modeling of disturbed atmospheres (V-16)

High-energy particle beams (V-16)

High-velocity vehicle plumes and reentry signatures (V-16)

Atomic magnetometers (V-16)

An especially well-known application is the near-resonant V–V' transfer of energy from the ($v = 1$) vibrational mode of N_2 to the (001) mode of CO_2 (see Note 2-1) in the CO_2 laser. Much more efficient vibrational excitation of the CO_2 molecule results than is possible with direct electron-impact excitation.

Studies of excitation and dissociation are important in their own right; they also contribute to a much deeper understanding of the mechanisms of energy transfer and to a more accurate description of the parts of interaction potentials that are responsible for the transfer. In Figs. 2-1-1, 2-1-2, and 2-1-3, Phelps (1990) plots as functions of Lab energy, cross sections for excitation and dissociation, momentum transfer, charge transfer, and ionization for H, H^+, and H_3^+ projectiles colliding with stationary H_2 molecules. (See Figs. 5-1-1 and 5-1-2 in the companion volume for similar data on collisions of electrons

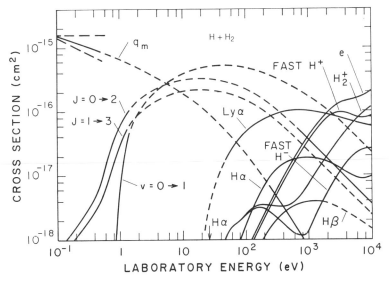

Figure 2-1-1. Cross sections for collisions of H atoms impinging on stationary H_2 molecules as a function of the Lab energy of the projectile. The solid curves are based on experiment or theory, whereas the short-dashed curves are extrapolations or interpolations. Momentum transfer, q_m; rotational excitation, $J = 0 \rightarrow 2$, $J = 1 \rightarrow 3$; vibrational excitation, $v = 0 \rightarrow 1$; ionization and ion-pair formation, product H_2^+, fast H^+, and/or H^-, electronic excitation, $Ly\alpha$, $H\alpha$, and $H\beta$ emission; electron production, e. The arrow indicates the threshold for $H\alpha$ excitation. [From Phelps (1990, p. 667).]

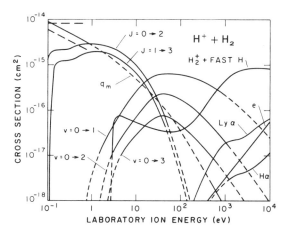

Figure 2-1-2. Cross sections for collisions of H^+ ions impinging on stationary H_2 molecules as a function of the Lab energy of the projectile. The solid curves are based on experiment or theory, whereas the short-dashed curves are extrapolations or interpolations. Momentum transfer, q_m; rotational excitation, $J = 0 \rightarrow 2$, $J = 1 \rightarrow 3$; vibrational excitation, $v = 0 \rightarrow 1$, $0 \rightarrow 2$, $0 \rightarrow 3$; charge transfer, H_2^+ and fast H products; electronic excitation, Lyα and Hα emission; electron production, e. The long-dashed lines are extrapolations to higher energies of fits of constant cross section and constant collision frequency (Langevin) models to 300 K mobility data. [From Phelps (1990, p. 655).]

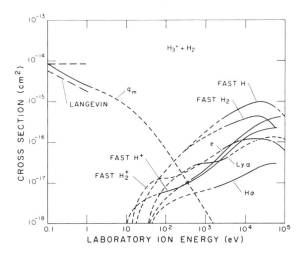

Figure 2-1-3. Cross sections for collisions of H_3^+ ions impinging on stationary H_2 molecules as a function of the Lab energy of the projectile. The solid curves are based on experiment or theory, whereas the short-dashed curves are extrapolations or interpolations. Momentum transfer, q_m; charge transfer, slow H_2^+ and fast H and slow H^+ and fast H_2 products; dissociation, fast H^+ and fast H_2^+ products; electronic excitation; Lyα and Hα emission; electron production, e. The long-dashed lines are extrapolations to higher energies of fits of constant cross section and constant collision frequency (Langevin) models to 300 K mobility data. [From Phelps (1990, p. 664).]

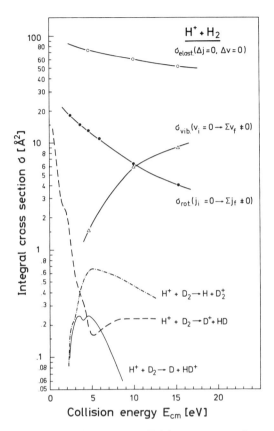

Figure 2-1-4. Integral cross sections versus CM impact energy for various channels of low-energy $H^+ - H_2(D_2)$ collisions. The nonreactive cross sections are derived from theory, the reactive data from experiment. [From Linder (1979, p. 543).]

with H_2 and N_2 molecules.) Figures 2-1-1 through 2-1-3 provide an immediate indication of the magnitudes of the cross sections for the various processes indicated and their relative importance over a wide range of impact energy.* Figure 2-1-4 displays integral cross sections for various kinds of H^+ collisions with H_2 and D_2 molecules at low CM energies ($\leqslant 17\,eV$).

Many of the most detailed and valuable experimental data on the phenomena covered here have come from complex and difficult *state-to-state* crossed molecular beam studies, which require both initial state selection and final state analysis. Measurements of this type were first made in the 1960s, and despite

*Phelps (1990) presents similar graphs for collisions of H_2, H_2^+, and H^- projectiles with H_2 molecules, as well as tabular data on all of the processes covered in his valuable paper. In addition, Phelps includes swarm coefficients where available. Cross sections and swarm coefficients are provided for ions and neutrals on N_2 and Ar by Phelps (1991) and for hydrogen and argon ions on Ar and H_2 (Phelps, 1992).

their complexity and difficulty, they have become highly refined. State-to-state techniques involving crossed neutral molecular beams are especially well suited for studies of rotational transitions at impact energies below 1 eV, because of the low-energy thresholds, small rotational-level spacings, and the strong coupling mechanism (see Sections 2-4 and 2-9). Vibrational excitation in neutral–neutral collisions even at a few eV is more difficult to study with state-to-state techniques, so *ionic projectiles* with energies on the order of 10 eV are frequently used to provide the excitation. However, the last decade has seen steady progress in vibrational measurements (Section 2-5). More severe difficulties arise in the application to electronic and dissociative transitions (Sections 2-6, 2-7, 2-10, and 2-16). Here conventional single-beam and crossed-beam techniques are dominant even in the low-eV impact energy range; at higher energies, their use is mandatory. A variety of absorption cell, acoustical, shock tube, gas dynamics, and spectroscopic techniques are employed in investigations of energy transfer, and a few of them are described in Part $\mathscr{C}$.

2-2. BACKGROUND INFORMATION

It will now be useful to recall some of our previous discussion. First, we have already shown, in Section 5-3 of the companion volume, how an isotropic interaction potential such as represented in Fig. 1-3-1 cannot produce rotational excitation. An angle-dependent potential is required, as in Fig. 1-3-3 (see Problems 2-1 and 2-2). Vibrational excitation can occur only if the projectile can couple to one of the vibrational coordinates of the target (see Section 5-4 of the companion volume). Electronic excitation involves coupling to one of the electronic position coordinates.

Second, Fig. 5-3-1 of the companion volume displays vibration–rotation energy diagrams for the o-H_2, p-H_2, o-N_2, p-N_2, HCl, LiF, and KI molecules. For ease in reference, it is resurrected here as Fig. 2-2-1.* Typically, for light

---

*It is interesting to compare the energy levels in Fig. 2-2-1 with those calculated from the simplest models in which vibration and rotation are treated as uncoupled (Problem 2-3). The vibrational levels for a diatomic molecule are given in zeroth approximation by the harmonic oscillator equation

$$E_v = \hbar\omega_0(v + \tfrac{1}{2}), \tag{2-2-1}$$

where v is the vibrational quantum number and ω_0 is the angular frequency of the oscillator calculated classically. The rotational energy of a molecule is given in the rigid rotor approximation by the equation

$$E_J = \frac{\hbar^2 J(J + 1)}{2I}, \tag{2-2-2}$$

J being the rotational quantum number, and I the moment of inertia of the molecule. See Problem 5-2 of the companion volume for the energy levels of a diatomic molecule when the coupling of the rotation and vibration is taken into account.

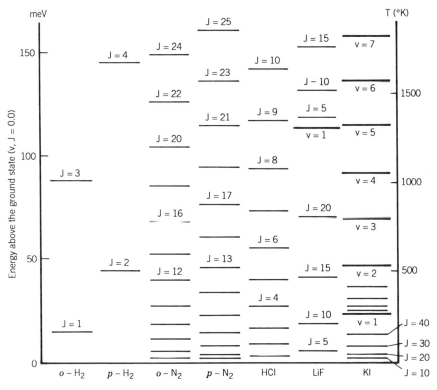

Figure 2-2-1. Vibrational–rotational levels (quantum numbers v and J) of a few diatomic molecules. The ($v = 1$, $J = 0$) level of H_2 lies 0.54 eV above the ground state ($v = 0$, $J = 0$). Rotational level spacings for H_2 are uniquely large, about $15J$ meV, where J is the quantum number for the upper level. For the ortho species of $H_2(o\text{-}H_2)$, the nuclear spins are parallel; for the para version ($p\text{-}H_2$), the nuclear spins are antiparallel. [From Shimamura (1984).]

molecules, adjacent rotational levels are spaced a few meV apart, corresponding to far-infrared and microwave radiation. (H_2 is singular, with much greater rotational spacings.) Vibrational levels are typically spaced about 0.1 eV apart, corresponding to infrared frequencies in radiative transitions. We may take typical outer-shell electronic level spacings to be several eV and note that radiative transitions between such levels usually involve visible and ultraviolet radiation. Inner-shell transitions correspond to the X-ray region and may involve energy changes of hundreds of keV. As indicated by Figs. 2-1-1 through 2-1-4, rotational, vibrational, and electronic excitation can be important up to impact energies that are much larger than the respective threshold energies. It is important to recall (pp. 267–269 of the companion volume) that the rotational, vibrational, and electronic motions are coupled, although pure transitions involving a change in only one quantum number are possible. Also, in events involving ionization, charge transfer, and/or chemical reactions, one or both of the reactants may be left in excited or dissociated states. Such processes are discussed in Chapters 3, 4, and 5.

Third, the cross section σ_{ij} for a deexcitation process $j \to i$ is related to the cross section σ_{ji} for the inverse excitation process by the principle of detailed balancing, derived in Section 8-2 of the companion volume. Specifically [(5-3-3) of the companion volume],

$$g_j E_j \sigma_{ij} = g_i E_i \sigma_{ji},$$

where i and j denote the quantum states of the target molecule, g_i and g_j the statistical weights of the states, and E_i and E_j the initial kinetic energies of the projectile at large distances from the target. We shall occasionally mention direct measurement of a deexcitation process.

Finally, at the end of Section 1-2-A, references are given to portions of the text and to problems that relate to the kinematics of inelastic collisions. These references, as well as Buck (1988b) and Dagdigian (1988), should be useful throughout this chapter.

2-3. EXPERIMENTAL CONSIDERATIONS: STATE-TO-STATE TECHNIQUES

In this chapter we are interested primarily in measurements based on particle beam techniques. Many examples of the use of *conventional single-beam* and *crossed-beam methods* were given in Chapters 4 through 7 of the companion volume (electron projectiles) and in Chapter 1 in this book (heavy particle elastic scattering). In this chapter we have a special interest in *state-to-state crossed-beam techniques*, and in this section we confine our discussion to them.

Buck (1988a) lists and critiques four methods for *initial state preparation:*

1. Supersonic nozzle beams (Section 1-10-C; Note 1-2-B; Bernstein, 1982; Miller, 1988; Gentry, 1988)
2. Separation and focusing in inhomogeneous electric and magnetic fields (Reuss, 1988; Bernstein, 1982)
3. Population depletion by optical pumping with lasers (Bergmann, 1988)
4. Excitation of the desired state with lasers (Bernstein, 1982; Bergmann, 1988)

In method 1, the lowest rotational–vibrational state of the molecule is present almost exclusively, except in the case of small rotational spacings or unless the gas plenum is maintained at high temperature. Method 2 suffers from the restriction to molecules with a permanent electric or magnetic moment. It also requires high field gradients and long trajectories in order to reject the unwanted states, so the beam intensity would thereby be reduced by a large factor if it were not for the focusing action of the beam selector. In method 3 the selected rotational state is depleted by continuous-wave (CW) laser excitation to an excited level, which radiates mainly to other vibrational levels in the ground electronic state. A hole results in the rotational state distribution of the ground vibrational level. By measuring the difference in the final-state populations with and without radiation from the pump laser, one can assess the scattering from the depleted rotational level. Method 4 involves the preparation of molecules in a given rotational level of an excited vibrational level, often by direct infrared laser excitation with a pulsed or CW laser. Thus far this method, as described here, has not been used in molecular beam scattering experiments, although excitation of the desired state with lasers has been utilized in atomic scattering measurements.

Buck (1988a) also describes five techniques for *final-state analysis*:

1. Velocity analysis of the scattered beam by mechanical means (van den Meijdenberg, 1988), or by time-of-flight techniques (Auerbach, 1988)
2. Separation of beam components in inhomogeneous electric or magnetic fields (Reuss, 1988; Bernstein, 1982)
3. Laser-induced fluorescence (LIF) detection (Bernstein, 1982; Hefter and Bergmann, 1988; Bergmann, 1988)
4. Emission of radiation (Bergmann, 1988)
5. Excitation by lasers followed by bolometric detection (Bergmann, 1988; Zen, 1988)

Method 1 has wide applicability but requires very high-velocity resolution, with a large concomitant intensity loss (a factor of 10 to 100), for the resolution of single rotational states. Buck (1988a) points out that the intersecting reactant beams must also be extremely well defined in speed and direction if method 1 is used, and the scattered beam intensity is further reduced by this requirement. Method 2 is very useful, but it has the same shortcomings here as when used for initial-state selection. The technique of laser-induced-fluorescence (LIF) detection, method 3, was introduced by R. N. Zare and his coworkers in the late 1960s.[*] Individual states of the scattered molecules are detected by tuning a pulsed or CW laser across a ro-vibrationally resolved electronic band system. The molecules that become electronically excited then fluoresce, and the relative ro-vibrational populations in the ground electronic state can be derived from the fluorescence intensities. The use of method 3 does not entail the increase of any beam path lengths and losses of intensity. LIF detection does require,

---

[*]For an early paper that deals with LIF spectroscopy, see Tango et al. (1968). LIF has been used by Yamasaki and Leone (1989) in the study of quenching and energy transfer of single rotational levels of Br_2, and by Dagdigian (1989) to measure state-resolved cross sections for rotationally inelastic collisions (see Section 2-16). For important applications of the LIF technique to the study of gas-phase chemical reactions, see Cruse et al. (1973), Zare (1979), and Chapter 3.

however, that the molecule have a bound–bound electronic transition that can be spanned with a tunable laser. Furthermore, the state that is excited by the laser must have a fairly short lifetime (say, less than 10^{-5} s) and an appreciable fluorescence quantum yield. Method 4 has seen little use in scattering experiments. Method 5 provides state analysis by CW infrared laser excitation of a specific (v, J) level of the beam molecules and bolometric measurement of the increase in the energy content of the beam. The vibrationally excited molecules must have a radiative lifetime long enough to permit the excited molecules to reach the bolometer before deexcitation occurs. This method has been applied mainly in spectroscopic studies.

In Note 1-6 we mentioned some of the pioneering experiments on state-selective collision experiments by W. Paul's group at the University of Bonn. More recent measurements are described in the discussion that follows.

PART $\mathscr{A}$. EXCITATION BY HEAVY PARTICLE IMPACT

2-4. ROTATIONAL EXCITATION

A. Introduction

The cross sections for rotational excitation in heavy particle collisions are large over a wide range of impact energies that extends downward to well below 1 eV, and it has been possible to acquire a wealth of reliable experimental data on this process. Also, the theory of rotational excitation is highly developed, so that the process is now well understood, largely because of the interplay between the theory and the state-to-state data. Experiment and theory have been discussed in a number of reviews: Pauly and Toennies (1965, 1968), Toennies (1976, 1985), Faubel and Toennies (1978), Dickinson and Richards (1982), Faubel (1983), Buck (1988a, b), Dagdigian (1988), Stolte (1988), and Düren (1988). Certain aspects of the theory are discussed in Section 2-9.

Most of the state-to-state cross-section measurements have been made by one of the following techniques (Buck, 1988a):

1. Preparation of a ground-state incident beam by a nozzle source, coupled with high-resolution time-of-flight (TOF) analysis of the scattered particles. Note that we have here an *energy-loss method*, the principle of which is the same as in the first reliable measurements of cross sections for rotational excitation by electron impact in 1968 (see pp. 270 and 275–279 of the companion volume). See Note 2-2 for a discussion of general aspects of the energy-loss method.

2. Initial-state preparation by laser population depletion (Jones et al., 1982), inhomogeneous electric fields [reviewed by Dagdigian and Bullman

(1985)], or supersonic beams (Andresen et al., 1984), followed by state-specific detection by LIF. (Most of the data obtained with this arrangement are integral cross sections.)

It should be noted that the only cross sections for inelastic collisions involving prepared excited states are for excited alkali *atoms* (Düren, 1988).

B.   Apparatus, Measurements, and Experimental Data

Here we discuss some of the experimental work that has been performed recently at several of the centers of research on heavy particle inelastic collisions.

1.   *Max Planck Institut für Strömungsforschung at Göttingen*

The crossed nozzle-beams apparatus of Fig. 1-10-3 (Faubel et al., 1982; Toennies, 1985), with TOF analysis of the scattered beam, has been used for many important studies of inelastic heavy particle collisions at Göttingen. Buck et al. (1985) measured total differential and rotationally inelastic cross sections for He + CH_4 in the CM angular range 3 to 70°. A typical time-of-flight spectrum for an impact energy $E_{CM} = 34.1$ meV and a perpendicular-plane laboratory scattering angle of 51° is shown in Fig. 2-4-1. At this energy, they obtained the state-resolved cross sections for the $0 \rightarrow 3$, $1 \rightarrow 2$, and $1 \rightarrow 3$ rotational transitions displayed in Fig. 2-4-2. [The TOF spectra were first published by Faubel et al. (1980).] See Faubel et al. (1982) for similar measurements on He + N_2. In TOF measurements on Ar + O_2, Faubel and

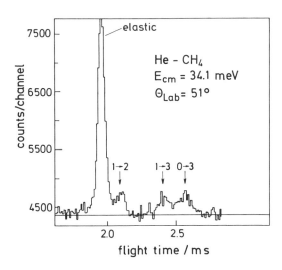

Figure 2-4-1. Time-of-flight spectrum measured for He + CH_4 at 34.1-meV CM energy and 51° Lab angle. [From Buck et al. (1985, p. 1260).]

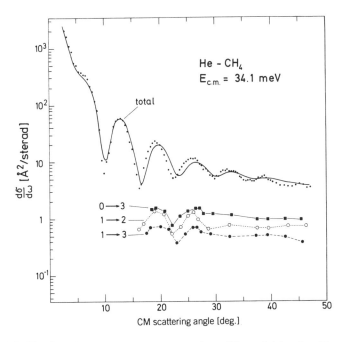

Figure 2-4-2. Total and state-to-state cross-section differential in the CM scattering angle, measured for He + CH$_4$ at 34.1-meV CM energy. The absolute values for the experimental data points are obtained by fitting the measurements to a theoretical curve. The solid curve representing the total differential cross section is based on a multichannel calculation in the CS (coupled states) approximation. [From Buck et al. (1985, p. 1263).]

Kraft (1986) observed individual rotational state-to-state transitions up to the value of $\Delta J = 18$. They found that at $E_{CM} = 97$ meV, a maximum of excitation occurs near the $1 \rightarrow 13$ rotational transition. In a different vein, Gundlach et al. (1988) have made experimental and theoretical studies of translational and rotational relaxation in Li/Li$_2$ and Na/Na$_2$ beams.

Gierz et al. (1984) have studied rotational excitation in the scattering of Li$^+$ ions from N$_2$ and CO at impact energies between 4 and 17 eV. Rotational rainbows (Section 2-9-D) appear in their energy-loss spectra shown in Fig. 2-4-3. This figure also displays rotational rainbows in TOF spectra obtained for Ar–O$_2$ and Ar–CO scattering. Friedrich et al. (1987a) have made a comparative study of rotational excitation in H$^+$ collisions with HF and CO$_2$ molecules. As shown in Fig. 2-4-4, the TOF spectrum for H$^+$ collisions with HF at $E_{CM} = 9.8$ eV indicates a broad range of rotationally excited states but practically no vibrational excitation, whereas the spectrum for CO$_2$ is dominated by vibrational excitation. As the CM energy is increased to 30 eV, there is a suggestion of vibrational excitation of HF into the $v = 1$ level only, but at 50 eV, vibrational excitation has become dominant, with at least the four lowest vibrational levels of HF being excited.

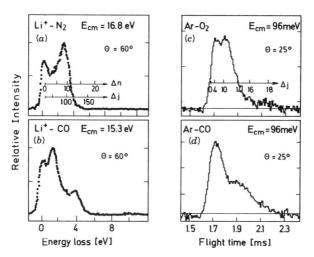

Figure 2-4-3. Rotational rainbow structures in (*a*) Li^+-N_2 and (*b*) Li^+-CO energy loss spectra at about 16-eV CM energy and in (*c*) $Ar-O_2$ and (*d*) $Ar-CO$ time-of-flight spectra at 0.096-eV CM energy. [From Faubel (1985, p. 509).]

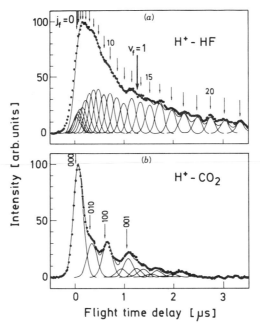

Figure 2-4-4. Typical time-of-flight spectra for (*a*) H^+-HF and (*b*) H^+-CO_2 at a CM energy of 9.8 eV and Lab scattering angle of 10°. The full curves at the bottom of each drawing represent the estimated relative contributions from the various excited states of the molecules. [From Friedrich et al. (1987a, p. 3728).]

The Göttingen group has also made extensive studies of anisotropic potentials for various molecules interacting with rare gas atoms (Meyer and Toennies, 1981, 1982; Faubel et al., 1982, 1983; Pack et al., 1984; Fuchs and Toennies, 1986; Beneventi et al., 1988; Rubahn and Toennies, 1988). These studies involved a combination of rotational state-to-state scattering data, transport data, and various types of theory. Pack (1984) has made a useful contribution to this work by developing equations for the CM-to-Lab transformation and the averages over beam velocity and spatial profiles, detector slits, and so on, that are necessary to compare theoretical differential cross sections with experimental data.

2.   *University of Minnesota*

The crossed, pulsed molecular beams instrument shown in Fig. 2-4-5 was developed in Gentry's laboratory (Hall et al., 1986; Gentry, 1985) for the measurement of state-to-state cross sections of ro-vibrational excitation. It utilizes a laser beam for LIF state-selective detection; this beam enters and leaves the vacuum system through Brewster's angle windows and long light baffles (1). The two pulsed molecular beam sources (2 and 3) are of the type described in Note 9-2-C. They are housed in differentially pumped chambers that can be rotated about the beam intersection site (4) by rotating the end plates

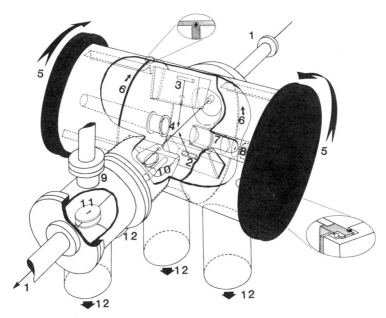

Figure 2-4-5. Crossed, pulsed molecular beam apparatus developed in Gentry's laboratory for the measurement of state-to-state cross sections. [From Hall et al. (1986, p. 1403).]

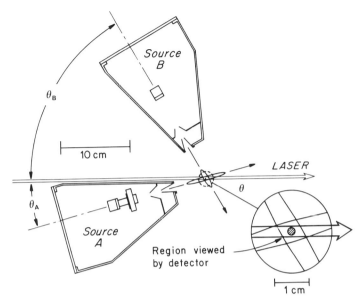

Figure 2-4-6. Double rotatable source arrangement in Gentry's crossed, pulsed molecular beams apparatus. The CM energy of the collisions between molecules from the two sources may be varied by changing either θ_A or θ_B, or both, to change the beam intersection angle, θ. [From Shamamian (1989).]

(5), to which they are rigidly attached (see Fig. 2-4-6). The internal partition disks (6) provide seals between the source differential pumping chambers and the collision chamber; these disks are attached to the end plates. Optical elements (7) image the center of the beam intersection site onto a photomultiplier tube (8). A second detector chamber (9) contains a second LIF detector (10) and ion extraction optics (11) for a multiphoton ionization detector, both of which are intended for differential cross section measurements. (Thus far, only integral measurements have been made with this apparatus.) Pumping is provided through valved ports for the four chambers of the apparatus by four 20-cm-diameter oil diffusion pumps (12). With both beam sources off, the quiescent pressure in the central chamber is about 2×10^{-7} torr; it rises by about a factor of 10 when both sources are running.

The CM energy of the collisions may be varied continuously without changing the source conditions by rotating the end plates to vary the intersection angle of the two beams. The impact energy is

$$E_{CM} = \frac{M_r(v_1^2 + v_2^2 - 2v_1v_2 \cos \theta)}{2}, \tag{2-4-1}$$

where M_r is the reduced mass of the system, v_1 and v_2 the speeds of the molecules

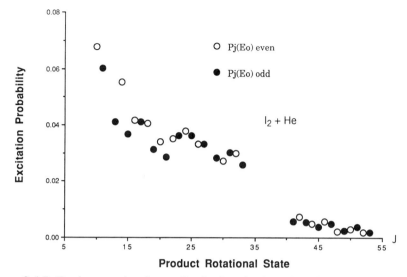

Figure 2-4-7. Product rotational state distributions for (I_2, He) collisions at 75-meV CM energy. The open circles represent even states, and the filled circles, odd states. [From Shamamian (1989).]

in the two beams, and θ the intersection angle. Shamamian (1989) has used this apparatus to study the state-resolved rotational excitation dynamics of iodine and *para*-difluorobenzene in collisions with helium. Figure 2-4-7 shows the product rotational state distribution of I_2 for an impact energy of 75 meV, and Fig. 2-4-8 displays the impact energy dependence of the excitation probability for the single product state $J = 30$.

Using other apparatus, Gentry and Giese (1977) resolved single-quantum rotational excitation in (HD, He) collisions. Hoffbauer et al. (1983) measured speed- and angle-resolved differential cross sections for scattering of Ar by Cl_2 at CM energies ranging from 0.09 to 0.16 eV and observed rotational rainbow scattering (see Section 2-9-D).

3.   *University of Kaiserslautern*

Linder and his colleagues have used both the apparatus of Fig. 1-10-4 and that of Fig. 2-4-9 to study rotational and vibrational excitation in collisions of H^+ ions with H_2 molecules. They covered the CM energy range 4 to 20 eV and obtained differential cross sections and transition probabilities for various ΔJ and Δv processes (Schmidt et al., 1978; Linder, 1979). At $E_{CM} = 10$ eV, rotational excitation in (H^+, H_2) collisions was found to be more probable in simultaneous vibrational transitions than in vibrationally elastic events (Hermann et al., 1978). At 4 eV, rotational excitation is the dominant inelastic process, but vibrational excitation is of comparable magnitude at 6 eV and becomes increasingly more important as the impact energy is increased.

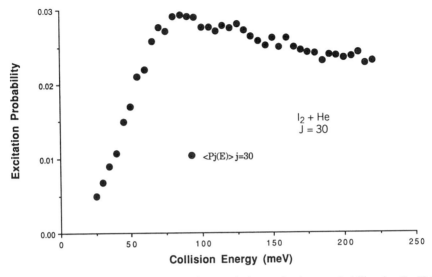

Figure 2-4-8. Collision energy dependence of the excitation probability in (I_2, He) collisions for the single product state $J = 30$. [From Shamamian (1989).]

Bergmann and his coworkers (1980) have introduced an interesting new technique, illustrated in Fig. 2-4-10, and applied it to the study of the rotational excitation of Na_2 molecules. A modulated pump laser beam intersects a supersonic Na/Na_2 beam and modulates the population of one specific Na_2 rotational state by excitation of that state, which then fluoresces to a broad distribution of lower states. (Here we have an example of rotational state selection by optical pumping depletion of the initial level J.) The Na_2 projectiles then intersect the supersonic beam of target particles 10 mm downstream. The radiative lifetimes of the excited molecules are so short that they decay within $10 \, \mu m$, and none reach the collision region. Molecules scattered at the Lab angle ϑ into the final states J' are detected by the fluorescence induced by a second modulated laser (the probe laser), and their state distribution is determined. Both the probe laser and LIF signals are transported by flexible fiber optics cables for processing at a remote site, thus permitting the scattering angle to be scanned easily. The differential cross sections are determined by a method discussed in the original papers, and in the review by Dagdigian (1988). Figure 2-4-11 shows differential cross sections for the rotational excitations

$$He + Na_2(v = 0, J) \rightarrow He + Na_2(v = 0, J' = 28). \qquad (2\text{-}4\text{-}2)$$

We put $\Delta J = J' - J$, and plot the differential cross section against both ΔJ and ϑ, the Lab scattering angle, in the figure. Small ΔJ transitions are seen to be strongly favored over large ΔJ transitions. Similar measurements have been made with the other rare gases (Bergmann et al., 1981; Hefter et al., 1981; Jones

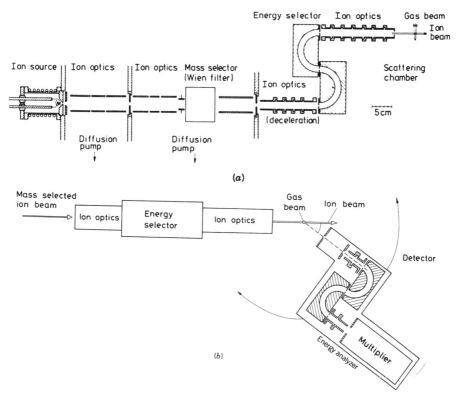

Figure 2-4-9. Linder's ion–neutral crossed-beam scattering apparatus. Both the energy selector (*a*) and energy analyzer (*b*) consist of two 180° hemispherical condensers in series. The scattering products in (*b*) are detected in a plane perpendicular to the gas beam. The detector system can be rotated in this plane, and the angular and energy distribution of the scattered ions can be measured. [From Krutein and Linder (1977a, p. 1365).]

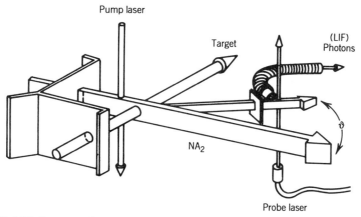

Figure 2-4-10. Bergmann's apparatus for the study of rotational excitation of Na_2 molecules. Signals from the pump-and-probe lasers are carried into the experimental region by single optical fibers. Similarly, LIF signals are transported away by flexible fiber optics cables for processing at a remote site. [From Bergmann et al. (1980, p. 4779).]

117

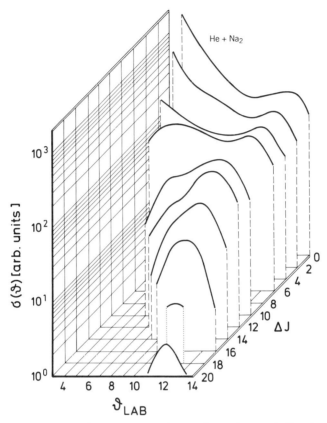

Figure 2-4-11. Differential cross sections for the rotational excitations $He + Na_2(v = 0, J) \rightarrow He + Na_2(v = 0, \; J' = 28)$. The Lab scattering angle is ϑ, and $\Delta J \equiv J' - J$. [From Bergmann et al. (1980, p. 4787).]

et al., 1983). Additional work in Bergmann's laboratory is described in publications by Hefter and Bergmann (1988), Bergmann (1988), Rubahn and Bergmann (1990), Ziegler et al. (1991), and Keller et al. (1992).

2-5. VIBRATIONAL EXCITATION

A. Introduction

At CM impact energies below 1 eV, the cross sections for vibrational excitation may be several orders of magnitude smaller than those for rotational excitation, and accordingly, the former tend to be more difficult to measure at these energies. Also, the vibrational spacings are much larger than the rotational (Fig. 2-2-1), so in neutral–neutral studies it is difficult to reach sufficiently high energies to make accurate vibrational measurements. Hence ionic projectiles are often used as an alternative, as we pointed out in Section 2-1. Despite the difficulties, in recent years many important state-resolved vibrational studies have been made, and it is these investigations that we emphasize. But first we refer the reader to some of the material presented in Section 2-4-A which is also applicable here, namely, the reviews of inelastic scattering cited there and the comments made on experimental techniques.* In addition, a look back at the discussion of electron-impact energy-loss measurements on pp. 282–289 of the companion volume may prove useful. General aspects of heavy particle energy-loss spectroscopy are discussed in Note 2-2.

B.   Apparatus, Measurements, and Experimental Data

1.   *Max Planck Institut für Strömungsforschung at Göttingen*

At the higher impact energies required to excite vibrations, the vibrational energy-loss distributions tend to be swamped by rotational excitation (Toennies, 1985). The Göttingen group has circumvented this difficulty by using protons and other light projectiles which produce very little energy transfer into rotational motion. In this way they have been able to study pure vibrational excitation in a number of diatomic and polyatomic molecules.

Polyatomic molecules generally have many vibrational modes (see Note 5-1 of the companion volume). Figure 2-5-1 illustrates the set of four normal modes for tetrahedral molecules such as CF_4 and the six for octahedral structures such as SF_6. The maximum level spacing is $\hbar\omega_0 = 0.159\,eV$ ($\tau_{vib} = 3 \times 10^{-14}\,s$) for the v_3 mode of CF_4. As the impact energy is increased, the energy-loss spectra become cluttered because of the increasing number of ways to form intercombination bands and because of level splitting resulting from anharmonic coupling. Toennies (1985) states that at an energy of 0.9 eV, the density of vibrational states is estimated to be much higher than $25\,meV^{-1}$. Noll (1988) and his colleagues have used crossed-beam TOF apparatus similar to that shown in Fig. 2-5-2 for energy-loss spectroscopy of high-overtone mode-selective vibrational excitation of CF_4 by H^+ and D^+ ions in the impact energy range 10 to 30 eV. Here a mass- and energy-selected ion beam is electrostatically chopped, and the resulting short ion pulses ($\Delta t = 10\,ns$) intersect the molecular beam directly beneath the nozzle source opening. Scattered ions are detected by an electron multiplier as a function of their arrival time after passage through a 3-m-long flight tube. For measurements at higher-energy resolution (10 to 15 meV), they employed apparatus similar to that shown in Fig. 2-4-9. Some of Noll's data are displayed in Figs. 2-5-3 and 2-5-4. Previous results on CF_4 and on SF_6 as well were reported by Gierz et al. (1985b).

Friedrich et al. (1987b) have made similar studies of vibrationally resolved inelastic scattering of H^+ by H_2O. Their differential cross sections for $E_{CM} = 27.0$ and 46.0 eV are shown in Fig. 2-5-5.

---

*Among recent reviews of *vibrational* excitation is the one by Sidis (1989).

Vibrational Modes and Levels of CF$_4$

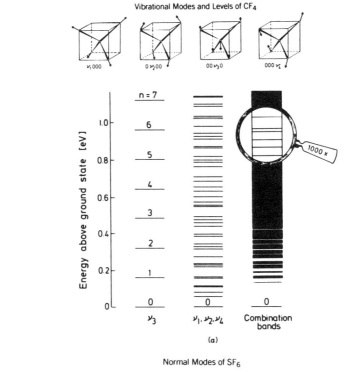

(a)

Normal Modes of SF$_6$

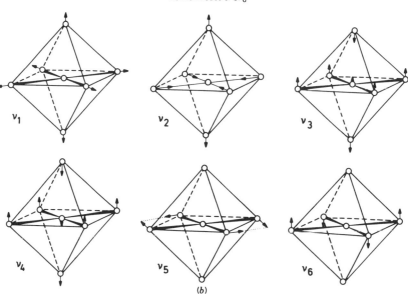

(b)

Figure 2-5-1. Normal vibrations of (a) tetrahedral and (b) octahedral molecules, exemplified by CF$_4$ and SF$_6$, respectively. The symmetries of the vibrational motions determine the active modes (see pp. 283–285 of the companion volume). In both molecules, only the triply degenerate ν_3 and ν_4 modes are infrared active. All modes are Raman active, with the exception of ν_3, ν_4, and ν_6 in the case of SF$_6$. [From Gierz et al. (1985b, p. 2264).]

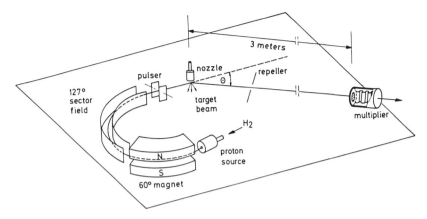

Figure 2-5-2. Crossed-beam apparatus used for studies of inelastic collisions of mass- and energy-selected H^+ ions. [From Friedrich et al. (1987b, p. 5257).]

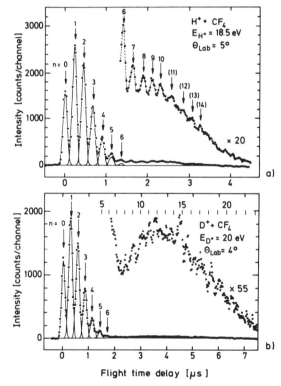

Figure 2-5-3. Mode-selective excitation to high-overtone vibrational states, where here n is the number of quanta in the v_3 mode of CF_4 illustrated in Fig. 2-5-1a. Time-of-flight spectra are shown for (a) H^+ and (b) D^+ scattered by CF_4 at Lab angles 5° and 4°, respectively. The Poisson distribution predicted by a forced oscillator theory for the vibrational transition probabilities is shown at $n = 0$ to 5 by solid-line fits. Additional contributions at larger n are attributed to small-impact-parameter trajectories that exert stronger forces on the molecule. [From Noll (1988, Fig. 2).]

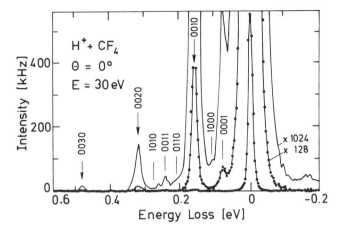

Figure 2-5-4. H^+ energy-loss spectrum determined with electrostatic energy analysis after scattering of 30-eV H^+ ions from CF_4. The v_3 mode excitation $(00n0)$ is dominant (see Fig. 2-5-1a), but weak fundamental and combination transitions of other modes can be recognized. [From Noll (1988, Fig. 4).]

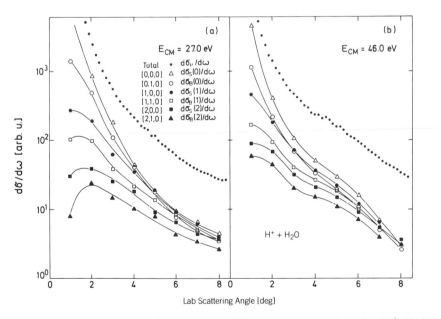

Figure 2-5-5. Total and state-selected relative differential cross sections for (H^+, H_2O) collisions at $E_{CM} = 27.0$ and 46.0 eV. [From Friedrich et al., 1987b, p. 5260).]

Niedner et al. (1987) have investigated selective vibrational excitation in collisions of H^+ ions with CO_2 and N_2O. Figure 2-5-6 shows some of their vibrational transition probabilities for CO_2. Vibrational energy diagrams for CO_2 and N_2O and their singly charged positive ions are displayed in Fig. 2-5-7. They are useful in analyzing the excitation experiments considered here as well as the charge-transfer measurements made by Niedner and his colleagues.

In addition, at Göttingen, Gierz et al. (1985a) have obtained TOF spectra for H^+ ions in state-resolved scattering from 11 small fluorohydrocarbon molecules at 9.8 eV. In a different kind of experiment, Knuth and his colleagues (1986) have studied vibrational relaxation of hydrogen fluoride by HF dimers in a laser-excited nozzle beam.

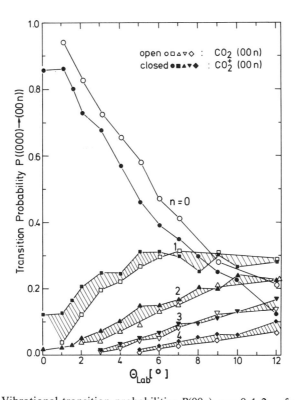

Figure 2-5-6. Vibrational transition probabilities $P(00n)$, $n = 0, 1, 2, \ldots$ for $v_3(H^+, CO_2)$ inelastic scattering (open symbols) compared to charge-transfer scattering (closed symbols) as a function of angle. The hatched areas for $n \geqslant 1$ denote the deviations between the transition probabilities for the same vibrational state in the two product channels (H^+, CO_2) and (H, CO_2^+). The notation is explained in Fig. 2-5-7. [From Niedner et al. (1987, p. 2073).]

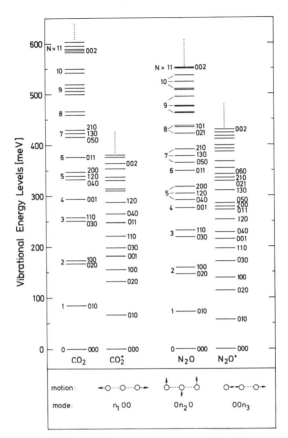

Figure 2-5-7. Vibrational energy-level diagram of the electronic ground state molecules CO_2, CO_2^+ (*left*), and N_2O, N_2O^+ (*right*). At the bottom are shown the three normal-mode vibrations of the linear triatomic molecules. The triads of numbers (n_1, n_2, n_3) specify the vibrational states in the upper part of the figure. The labels N on the left side of the CO_2 and N_2O columns refer to groups of states that cannot be individually resolved in the experiment mentioned in the text. [From Niedner et al. (1987, p. 2070).]

2.   *University of Minnesota*

Gentry and his associates have used the crossed, pulsed molecular beam apparatus of Fig. 2-4-5 for state-to-state studies of the vibrational excitation of I_2 in collisions with He (Hall et al., 1984) and with H_2 and D_2 (Hall et al., 1986). The Minnesota group has also studied state-to-state vibrational excitation in collisions of S_0 aniline with He (Liu et al., 1984) and *p*-difluorobenzene (Hall et al., 1985).

3. *University of Kaiserslautern*

Schinke et al. (1977) have measured differential cross sections for vibrational excitation in (H^+, H_2) collisions at $E_{CM} = 15.3$ and $20\,eV$, and at scattering angles that include the rainbow region. Their results are in good agreement with predictions of an impact parameter theory. Krutein et al. (1979) have studied vibrational excitation of H_2 by H^+ impact at CM energies covering the range 15 to $120\,eV$. They observed glory oscillations in the differential cross section for $v = 0 \rightarrow 1$ scattering at $0°$, and attributed them to contributions from small-impact-parameter trajectories interfering with outer trajectories.

Krutein and Linder (1977b, 1979) have also measured small-angle differential cross sections for vibrational excitation of CO, N_2, and NO by H^+ ions in the Lab energy range 5 to $90\,eV$. Krutein and Linder (1977a) have measured differential cross sections for vibrational excitation in H^+–CO_2 collisions at $E_{Lab} = 15$ to $50\,eV$, and Bischof et al. (1982) made similar studies at $E_{CM} = 15$ to $100\,eV$. The overall energy resolution in the two experiments ($50\,MeV$ and 20 to $40\,meV$ FWHM, respectively) was good enough for clear resolution of the vibrational modes of the CO_2 molecule. Hege and Linder (1985) have used the apparatus of Fig. 2-4-9 to measure forward-direction differential cross sections for vibrational excitation of H_2, N_2, O_2, and CO_2 by H^- ion impact at $E_{CM} = 20$ to $180\,eV$.

2-6. ELECTRONIC EXCITATION (OUTER SHELL)

The electrons in the outer shells of atoms and molecules are bound with energies ranging from a few eV to a few tens of eV, whereas inner-shell binding energies cover a range of the order of 10^2 to $10^6\,eV$ per electron. This difference in binding energy affects the basic physics of excitation and ionization processes, and is responsible for large differences in the experimental techniques by which they are studied. Hence as in Chapter 5 of the companion volume, a division of our discussion according to outer-shell and inner-shell excitation is warranted.

Two main methods are used here: (1) direct measurement of the loss of kinetic energy by one collision partner as a result of its exciting the other partner, and (2) detection of radiation emitted by one collision partner following excitation by the other. Energy-loss studies of electron-impact electronic excitation were described on pp. 306–307 of the companion volume and the radiation detection method on pp. 307–311.

In the case of electronic excitation, even higher projectile energies are required than in vibrational excitation, with the result that energy-loss studies can be complicated by many collision channels being open, and many degrees of freedom being excited (in the case of molecules). A few state-resolved measurements have been made at low energies, but in many cases only partially resolved spectra have been obtained, and their interpretation is difficult. In ion–neutral studies, there is no difficulty in producing CM energies ranging from threshold values to the maximum value desired. This is not the case in neutral–neutral studies. Seeded-beam techniques can be employed to reach energies up to only a few eV in most cases, so fast beams of neutral particles must be obtained by charge transfer to an ion beam. A new set of resolution problems arise in doing this. The relatively small amount of work done on electronic excitation at low impact energies is discussed in Buck (1988b, pp. 526, 534–535, 545–546) and Dagdigian (1988, p. 574). We now turn to studies made at higher energies.

A.   Energy-Loss Measurements

See Note 2-2.

1.   *Stanford Research Institute*

At SRI, Morgenstern et al. (1973) studied excitation in (He, He) collisions by a TOF technique over the Lab energy range 200 to 350 eV. They observed energy losses corresponding to both of the channels

$$He(1s^2) + He(1s^2) \rightarrow He(1s^2) + He(1s2l) \qquad (2\text{-}6\text{-}1)$$

$$He(1s^2) + He(1s^2) \rightarrow He(1s2l) + He(1s2l) \qquad (2\text{-}6\text{-}2)$$

and obtained angular differential cross sections for each. Sidis et al. (1977) employed electrostatic energy analysis of the scattered ions to measure differential cross sections for direct excitation in (He$^+$, Ar) collisions over the Lab energy range 100 to 2000 eV. (In the same experiments, they investigated charge transfer in this system by a TOF method.)* Hollstein et al. (1969) measured integral cross sections for the deexcitation of helium 2^1S and 2^3S atoms by ground-state He atoms at Lab energies ranging from 150 to 2200 eV. Observations were made of the radiation emitted when the metastable He atoms were deexcited in collisions, and the spatial decay rate, together with the collision chamber pressure, yielded the cross sections.

2.   *Groups at Paris, Aarhus, and Copenhagen*

Groups at Paris, Aarhus, and Copenhagen have made a multipronged study of collisions of Na$^+$ ions with stationary Ne atoms in the energy range 0.2 to 12 keV (Olsen et al., 1979). In addition to making a theoretical attack on the problem, they reported four different kinds of experimental results:

1. Cross sections doubly differential in projectile ion scattering angle and energy loss, obtained by electrostatic analysis

---

*Previously, Brenot et al. (1975b) and Barat et al. (1976) had studied (He$^+$, He) and (He$^+$, Ne) collisions, respectively, and Brenot et al. (1975a) had investigated He–He, Ne–Ne, Ar–Ar, and Kr–Kr processes in the range 0.5 to 4 keV.

2. Energy-loss spectra for the neutralized component of the scattered beam, obtained by TOF techniques

3. Energy spectra of electrons ejected in autoionization

4. Cross sections for some important Ne II transitions

By bringing all of their findings to bear on the problem, Olsen et al. (1979) were able to draw the following conclusions: For distances of closest approach, R_{min}, greater than about 1.63 au, the scattering is purely elastic. For smaller values of R_{min}, however, excitation is almost certain. If $1.39 < R_{min} < 1.63$, only one $2p$ electron is excited ($4f\sigma$ orbital promotion). The electron is predominantly excited to a Ne($3p$) level or transferred to the Na($3p$) level, in each case with photon emission showing *Rosenthal oscillations* (Rosenthal and Foley, 1969; Rosenthal, 1971; Bobashev, 1978; Kessel et al., 1978.) On the other hand, if $R_{min} < 1.39$ au, excitation of two $2p$ neon electrons usually occurs, with four main groups of exit channels:

1a. Transfer of one electron to the projectile with simultaneous excitation of a neon level, or

1b. Doubly excited autoionizing neon

or, following a core L-vacancy sharing process that transfers one of the Ne $2p$ vacancies to the Na $2p$ subshell

2a. Simultaneous excitation of one electron of both collision partners, or

2b. Excitation of autoionizing levels in neutral sodium plus a Ne$^+$ ion in the ground state

Channels 2a and 2b display Rosenthal interferences that produce oscillations in the cross sections for electron emission matching the oscillations in the photon spectrum. A large fraction of the *violent collisions* (Andersen and Nielsen, 1982, p. 268) lead to the excitation of electrons to continuum states.

This study indicates how complex collision phenomena can be even for a relatively simple system, but also how powerful the modern tools are for their study. Incidentally, we should point out that autoionizing states were discussed on pp. 236–238, 391–396, 463, 466, and 471 of the companion volume. We shall encounter them again in Chapter 4.

The Paris group (Dowek et al., 1981, 1982, 1983) has also performed collision spectroscopy on helium ions and atoms interacting with diatomic molecules. Dowek et al. (1981) made small-angle energy-loss measurements for direct-excitation and charge-transfer collisions of He$^+$ with N_2 and O_2 in a single-beam experiment at projectile ion energies of 0.2 to 4.0 keV. They found charge transfer to be strongly dominant, although the endothermic channels (He* + M_2^+) and (He + M_2^+*) were sometimes observed. Exothermic channels were only weakly excited. Dowek et al. (1982) made similar studies of He and

He$^+$ colliding with stationary H$_2$ molecules. They discovered that for violent collisions ($\tau \geqslant 3$ keV-deg, where $\tau \equiv E\vartheta$), the excitation processes are very similar for both the He–H$_2$ and He$^+$–H$_2$ systems, the same highly selective mechanisms governing the population of direct and exchange channels involving one and two $2p$ electrons. Here one- and two-electron transitions proceed through molecular orbital (MO) promotion (Note 2-3) and level crossings. For *soft collisions* (Andersen and Nielsen, 1982, p. 268) the He$^+$–H$_2$ system shows a nonselective excitation of various overlapping Rydberg series, but the neutral system does not. In studies of He$^+$ and He colliding with CO and NO, Dowek et al. (1983) were led to conclude that at the energies studied:

1. Direct transitions of the *Demkov type* (Demkov, 1964) populate the exothermic charge-exchange channels and produce diatomic ions in valence states. These states may then populate other exothermic channels far off resonance.

2. MO promotion is common to all of the neutral and ionic systems and leads to electronic transitions at the related MO crossings. [Here the *diabatic I (D-I) mechanism* is operative. See Brenot et al. (1975a). Diabatic and adiabatic processes are discussed in general terms in Section 2-9 and Note 2-3.]

3. In He$^+$ collisions with CO and NO, direct excitation and charge transfer are produced by correlated two-electron transitions [the *diabatic II (D-II) mechanism*; see Brenot et al. (1975a)].

3.   *University of Missouri at Rolla*

Park and his coworkers at Rolla have made an extensive set of measurements in the Lab energy range 15 to 150 keV with a large ion energy-loss spectrometer that provides high resolution in both energy loss and scattering angle (Kvale et al., 1985; Park, 1983). Their apparatus, shown in Fig. 2-6-1, is similar in concept to the spectrometers described in Sections 2-4 and 2-5 but is measured in meters rather than centimeters. An acceleration–deceleration scheme is employed to avoid degradation of the energy-loss resolution by voltage fluctuations of the 15- to 200-keV high-voltage power supply. A high-energy resolution ($E/\Delta E \geqslant 10^5$) is thereby achieved. The ion accelerator at the right of Fig. 2-6-1 can be rotated in the plane of the drawing about the center of the scattering chamber, which is stationary, as are the magnet and decelerator. The magnet transmits only projectiles that are scattered without undergoing charge changing and directs them into the decelerator column. After deceleration, the scattered projectiles are energy-analyzed by a 135° electrostatic hemispherical analyzer not shown in the drawing. The hemispheres in this device have a mean radius of 2.540 cm and are separated by 0.635 cm. In Problem 2-8 the reader is asked to fill in the details of the Park apparatus, data acquisition method, and deconvolution techniques used to obtain excitation cross sections from the raw energy-loss data.

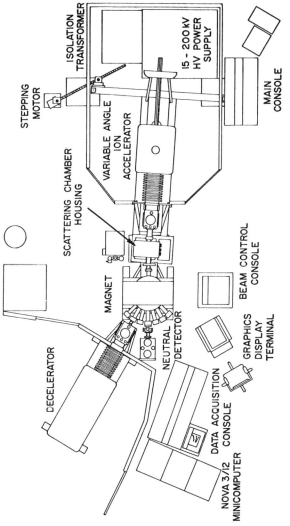

Figure 2-6-1. Apparatus used in Park's laboratory for measurements of energy loss in electronic excitation at high energies. [From Kvale et al. (1985, p. 1370).]

We now cite a few of the more recent studies made by Park's group. Park et al. (1980) reported angular differential cross sections for H^+ ions on H atoms, and Kvale and his colleagues (1985) provided similar data for H^+ on He. Peacher et al. (1984) obtained differential and integral cross sections for excitation of H atoms by H_2^+ ions. Redd and his coworkers (1987) measured differential cross sections for

$$Mg^+(3s) + He \rightarrow Mg^+(3p, \Theta) + He \qquad (2\text{-}6\text{-}3)$$

and

$$Na^+ + H(1s) \rightarrow Na^+(\Theta) + H(n = 2) \qquad (2\text{-}6\text{-}4)$$

(see Figs. 2-6-2 and 2-6-3). In both cases we have quasi-one-electron collision systems, but the $MgHe^+$ cross sections decrease monotonically, whereas the NaH^+ system exhibits structure. Further, the $MgHe^+$ differential cross sections are about an order of magnitude larger at equivalent angles and velocities. Analysis of the data leads to the conclusion that the (Mg^+, He) collisions are dominated by *direct excitation*, a one-electron process involving a glancing collision, wherein an impulsive, delocalized Coulomb interaction between the valence electron and the closed core is responsible for the excitation. On the other hand, the (Na^+, H) collisions exhibit significant *molecular excitation*,

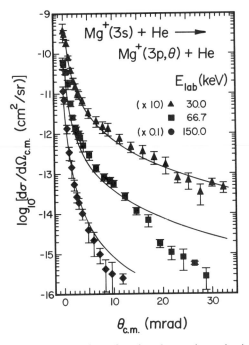

Figure 2-6-2. Differential cross sections for the electronic excitation of Mg^+ by He impact. [From Redd et al. (1987, p. 3476).]

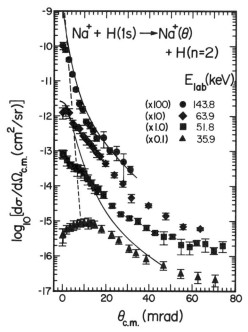

Figure 2-6-3. Differential cross sections for electronic excitation in collisions of Na$^+$ ions with H(1s) atoms. [From Redd et al. (1987, p. 3476).]

produced by violent encounters involving significant core penetration, in which excitation occurs at well-localized quasimolecular curve crossings.

4. University of Connecticut

As pointed out in Section 1-13-D, Everhart initiated a productive program of heavy particle collision research at the University of Connecticut in the 1950s that continues up into the 1990s. The early work is surveyed by McDaniel (1964), Kessel (1969), Hasted (1972), Kessel and Fastrup (1974), Massey and Gilbody (1974), and Kessel et al. (1978). Certain aspects of the Connecticut research are described in detail in Chapters 4 and 5. Here we mention briefly two experiments that deal with outer-shell excitation.

Andersen et al. (1980) measured energy losses in (Ne$^+$, H$_2$), (Ne$^+$, D$_2$), and (Ne, D$_2$) collisions for beam energies in the range $0.5 \leqslant E \leqslant 3.5$ keV and Lab scattering angles ϑ of less than 5°. They observed rotational and vibrational excitation of the target molecules, but very little direct electronic excitation. Martin et al. (1987) made similar measurements on (Ar, D$_2$) collisions in the Lab energy range 0.75 to 3.00 keV. They were concerned primarily with energy-loss scaling for collisions that are electronically elastic but may be ro-vibrationally inelastic. Let us define ΔE as the most probable energy loss in this "quasi-elastic channel," and ΔE_{elas} as the smaller energy loss that would result if the scattering were completely elastic. Martin et al. (1987) demonstrate that the *reduced energy loss*

$$f = \frac{\Delta E}{2\Delta E_{elas}} \tag{2-6-5}$$

scales with the *reduced scattering angle* $\tau = E\vartheta$, by which we mean that values of f for various beam energies lie close to a single plot of f versus τ.

The purely elastic energy loss ΔE_{elas} is determined by the collision kinematics alone. In their paper, Martin et al. use the expression

$$\Delta E_{elas} = E - \frac{[\cos \vartheta + (m^2/M^2 - \sin^2 \vartheta)^{1/2}]^2 EM^2}{(M + m)^2}, \tag{2-6-6}$$

where M is the projectile mass and m is the target mass. The only assumption made in the derivation of (2-6-6) is that $M > m$. This expression for ΔE_{elas} gives better results when applied in (2-6-5) than do previous expressions based on more stringent assumptions [see the review by Russek (1983)].

B.   Photon Detection Measurements

Let us first consider the determination of emission and excitation cross sections* by the observation of radiation emitted from a target cell in which gas is excited by a beam of heavy projectiles (Thomas, 1972). The problem is basically the same as that described on pp. 307–310 of the companion volume for the case of electron impact. Carried over directly are the difficulties associated with

1. Absolute radiometry (Van Zyl et al., 1980, 1985)
2. Target gas density measurements (Van Zyl, 1976; Van Zyl et al., 1976a)
3. Excitation transfer and radiation trapping (pp. 315–316 of the companion volume)
4. Cascading (pp. 311–313 of the companion volume; Van Zyl and Gealy, 1987)
5. Radiation polarization (pp. 313–315 of the companion volume)

Projectile beam production and flux measurement usually pose no additional problems if *ions*, rather than electrons, are used as projectiles, but they are much more difficult if excitation by *neutrals* is studied (Van Zyl and Gealy, 1986; Van

---

*As pointed out on p. 307 of the companion volume, an *emission cross section* is related to the totality of the radiation of a particular wavelength emitted by particles after they have been excited in collisions, regardless of the initial state of excitation. If it proves possible to subtract out the effects of cascading from higher states that were also collisionally excited, it may be possible to obtain the true *excitation cross section for a given level*, rather than the cross section for the emission of a particular wavelength of radiation. Problems 2-10 and 2-11 deal with the matter of accounting for cascading.

Zyl et al., 1976b; Ray et al., 1979). Successful attacks on many of these problems are described in the celebrated paper by Van Zyl et al. (1980a), which deals with benchmark cross-section measurements for electron-impact excitation of n^1S levels of He (pp. 308–309 of the companion volume).

If on the other hand, the measurements are based on detection of radiation emitted from *projectiles* excited in collisions with gas target particles, additional problems arise. The effects of cascading and projectile lifetime vary as the beam progresses into the target gas. Also, the branching ratios in the projectile decay are significantly influenced by small electric fields arising from contact potentials, space charge, and fringe fields, or from the $v \times B$ Lorentz force produced by the projectile motion through the earth's magnetic field. Van Zyl et al. (1988) show that the 1.5-V/cm electric field experienced by a 50-keV H atom moving across a 0.5-G magnetic field is strong enough to have serious effects. All things considered, it is difficult in heavy particle excitation measurements to approach the level of accuracy obtained in the best of the electron-impact work, but comparable accuracy can be achieved if great attention is paid to detail.

1.   *University of Denver*

Van Zyl's group in Denver has studied the emission of Balmer-α, Balmer-β, and Lyman-α radiation from H atoms excited in collisions with the rare-gas atoms, H_2, N_2, and O_2 (Van Zyl et al., 1980b, 1985, 1986b, 1989; Van Zyl and Gealy, 1987; Van Zyl and Neumann, 1988). The Lab energy of the H-atom projectiles was varied from about 40 to 2500 eV. The apparatus used in some of the most recent Denver measurements is shown in Fig. 2-6-4, which may be compared with Fig. 5-7-2 of the companion volume, constructed for the electron-impact benchmark experiments. Absolute emission cross sections were obtained in each case, and the polarization of the emitted radiation was studied. In some cases it was possible to evaluate the cascade contributions to the measured radiation signals and to deduce the cross sections for direct excitation to the $2p$ state of the H atoms (Van Zyl and Gealy, 1987; Van Zyl and Neumann, 1988). Measurements were also made on H atoms left in excited states which were formed by charge transfer of H^+ projectiles with rare-gas target atoms, N_2 and O_2 (Van Zyl et al., 1986a, 1987, 1989; Van Zyl and Neumann, 1988).

2.   *University of Nebraska*

Jaecks's group in Lincoln has also studied electronic excitation by making observations of the emitted radiation, but by different techniques and with different motivations. A beam of ions with a well-defined Lab energy (a few keV) passes through a thermal target beam; some of the ions are neutralized in charge-transfer collisions and left in electronically excited states. The neutrals left in the state to be studied decay with the emission of photons of characteristic wavelength. These photons are selected, polarization-analyzed, and detected in delayed coincidence with the deexcited neutrals from which they were emitted. (See pp. 324, 330–331, and 350–356 of the companion volume for discussions of

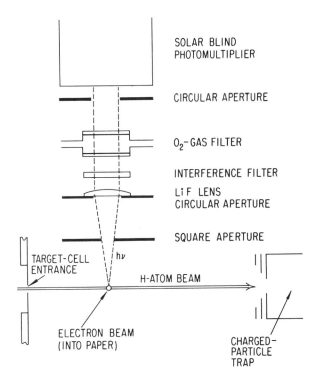

Figure 2-6-4. Apparatus used by Van Zyl's group for studies of electronic excitation. The photon detector shown here is designed for observations of Lα emission. The electron beam is turned on only during calibration of the photon detector. [From Van Zyl et al. (1985, p. 2923).]

electron–photon coincidence experiments.) Measurements are made as a function of the incident beam energy and of the Lab scattering angle. Eriksen et al. (1976) studied (He$^+$, He) collisions by this method. Eriksen and Jaecks (1978) investigated (He$^+$, H$_2$), and Goldberger et al. (1984) extended the studies to He$^+$ colliding with H$_2$ and D$_2$. In each of these experiments, the 3889 Å radiation from He(3^3P) was detected. Similar studies were made on the (H$^+$, He) system by Mueller and Jaecks (1985), with the observation of Lyman-α radiation from H($2p$) atoms.

These experiments have shed considerable light on the excitation mechanisms. For example, let us consider the investigation by Mueller and Jaecks (1985). If we take a two-state approach to the excitation (i.e., assume that the collision takes place by a series of single transitions from one molecular state to another), the rate of transitions between states k and n, defined in the rotating frame of the transitory molecule, is determined by the matrix element

$$\left\langle u_k \left| -\frac{\partial}{\partial t} \right| u_n \right\rangle = -\left\langle u_k \left| \dot{R}\frac{\partial}{\partial R} + i\omega \hat{L}_y \right| u_n \right\rangle. \qquad (2\text{-}6\text{-}7)$$

Here R is the time-dependent internuclear separation, ω the angular velocity of the internuclear axis, and $\hat{L}_y$ the Y component of the orbital angular momentum operator. The matrix element involving $\dot{R}$ on the right side of (2-6-7) is the *radial-* or *translational-coupling* matrix element; the term involving $\hat{L}_y$ gives the *rotational-coupling* matrix element (see Note 2-3). Mueller and Jaecks (1985) showed that the differential cross sections for population of the magnetic substates $m_l = 0$ and $m_l = \pm 1$ are of comparable magnitude, and hence the translational and rotational coupling are about equally important in the excitation here. These matters are discussed further by Jaecks et al. (1987) in their paper on isotope-dependent ion–molecule collisions.

Extensive excitation measurements based on the detection of radiation have also been made at the FOM Institute in Amsterdam in a long-term program spearheaded successively by J. Kistemaker, F. J. de Heer, and F. W. Saris. Much productive work has also been done by R. Geballe's group at the University of Washington, H. B. Gilbody and his colleagues at the Queen's University of Belfast, and E. W. Thomas's group at the Georgia Institute of Technology.

2-7. ELECTRONIC EXCITATION (INNER SHELL)

The interesting and important processes of inner-shell excitation and ionization by electron impact were discussed in Sections 5-7-H and 6-7 of the companion volume. Excitation by heavy particle impact in inner-shell collisions is also important. It is closely related to inner-shell ionization and charge transfer and for that reason is not treated here. Brief discussions of these processes appear in Chapters 4 and 5; a comprehensive source of information is the book edited by Carlson et al. (1991).

2-8. ALIGNMENT, ORIENTATION, AND STATE MULTIPOLES IN HEAVY PARTICLE COLLISIONS

In Section 5-10 of the companion volume we introduced the orientation vector, the alignment tensor, and state multipoles while discussing fine details of the electron-impact excitation process. One kind of measurement that is especially fruitful in this connection is that of electron–photon angular correlation, in which a subensemble of excited atoms is selected from the target, which can be characterized by the special properties of the target state. If the relationships between the scattered amplitudes (including their phases) and the target parameters (the orientation, alignment, and state multipoles) are established, the excitation process can be understood at a much deeper level than would otherwise be possible by measurements of only integral and differential cross sections.

Actually, the target parameters introduced here are of much greater generality than we have implied. For example, the alignment, orientation, and state multipoles have been the subject of many investigations of heavy particle collisions. However, as Blum and Kleinpoppen (1983) point out, there are some new features that must be considered when we turn to heavy particles. First, the heavy particle projectiles may have their quantum or charge states changed in collisions and thereby produce changes in the angular correlation. (Recall that in electron-impact collisions, only the spin state of the projectile may be affected.) Second, a heavy particle may be bound in a temporary quasi-molecular state during its interaction with the target. Hence molecular effects must be considered in an analysis of angular correlations with heavy particle projectiles.

There now exists a substantial literature on the determination of orientation, alignment, and state multipoles in heavy collisions. Among the important early theoretical papers are those of Macek and Jaecks (1971), Fano and Macek (1973), and Greene and Zare (1983). Valuable discussions of theory together with presentations of theoretical and experimental data appear in Blum and Kleinpoppen (1979, 1983) and Andersen et al. (1988, 1993a, b). We shall not treat this subject further at this point; additional discussion does appear, however, in Sections 2-8, 3-6, and 7-5, as well as in Note 3-4.

2-9. THEORY OF HEAVY PARTICLE EXCITATION PROCESSES

We begin by discussing several topics that relate to the theory of excitation in heavy particle collisions. Then we enumerate several simple model collision problems for which analytic solutions have been published. These solutions provide considerable insight into excitation processes without requiring an inordinate amount of computation. Finally, we present some important theoretical references to methods used in current research.

A. Adiabatic Hypothesis

In 1949, H. S. W. Massey put forward this hypothesis to provide a means of estimating the impact energy at which the integral cross section, σ_t, for a given nonresonant collision process should have its maximum value. He showed that σ_t might be expected to peak when

$$v \approx v^* \equiv \frac{a\Delta E}{h},\qquad(2\text{-}9\text{-}1)$$

and that σ_t should be very small if

$$v \ll \frac{a\Delta E}{h}\qquad\text{(the }adiabatic\ criterion\text{)}.\qquad(2\text{-}9\text{-}2)$$

Here v is the relative impact velocity, a the range of the interaction between the collision partners, and ΔE the *energy defect* [see (2-9-3)]. Massey proceeded by analogy with a classical system and used the Bohr correspondence principle, as follows.

First consider the target classically as a mechanical system in which vibrations may be excited by a disturbing force. [To be specific, suppose that this vibrational excitation corresponds to an internal transition of the target molecule from initial state i (energy E_i) to final state f (energy E_f). The transition may be rotational, vibrational, or electronic.] Let the natural frequency of the mechanical system be v. Express the disturbing force by a Fourier integral, in which only components of the expansion with frequencies near v will be effective in producing a forced oscillation. Thus the excitation will be maximized if the duration of the disturbance τ equals one period of the natural vibration, $1/v$. In the collision problem, $\tau \approx a/v$. In making the transition to quantum mechanics, we replace v by $\Delta E/h$, where the energy defect is

$$\Delta E = E_f - E_i \quad \text{(for R, V, or E transitions).} \quad (2\text{-}9\text{-}3a)$$

Thus we expect σ_t to be a maximum if $a/v \approx h/\Delta E$ (i.e., if $v \approx a\Delta E/h = v^*$). The collision is said to be *nonadiabatic* at velocities near v^* since then the collision time is comparable with the transition time $h/\Delta E$.

If $v \ll a\Delta E/h$, v is small compared to the characteristic velocity of the target (rotational, vibrational, or electronic, depending on the transition under discussion.) The target will have time to adjust to the perturbation imposed by the interaction without an electronic transition occurring, and the collision is said to be *adiabatic*.* Hence, at low impact velocities, σ_t will be small unless ΔE is very small. At very high impact velocities, σ_t must decrease with increasing impact energy since the interaction time ultimately becomes too short for the transition to be likely. Here we are in the *sudden* or *spectator regime*.

The adiabatic hypothesis has been tested with experimental cross sections for heavy particle excitation, ionization, and charge transfer collisions. A detailed assessment appears in Hasted (1972, pp. 621–625). To avoid uncertainty in the *a priori* assignment of values to the ill-defined interaction range a, Hasted worked backward from the cross-section data, assumed (2-9-1) to be an equality, and derived values for a from the observed positions of the cross-section maxima. For nonresonant transfer of a single electron from a neutral B to a singly charged ion A^+, the energy defect is

$$\Delta E_{ct} = E_{ion}(A) - E_{ion}(B) \quad \text{(single charge transfer),} \quad (2\text{-}9\text{-}3b)$$

the difference in the ionization energies, E_{ion}, of A and B. Hasted states that a large volume of experimental charge-transfer data can be fitted with reasonable

---

*Adiabatic denotes a change in a system without loss or gain of heat, or energy, and is the negative of *diabatic*. *Nonadiabatic* is the double negative of diabatic!

accuracy by taking $a_{ct} = 7 \text{ Å}/n$, where n is the number of electrons transferred in the collision. However, the adiabatic hypothesis is usually used only as a qualitative guide.

Massey (1971) has recast the inequality $a \Delta E/h \gg v$ into two forms that apply approximately to vibrational and rotational transitions, respectively, and shown them to be in accord with experiment (see Problems 2-14 and 2-15).

B. Setting Up the Basic Theory of Inelastic Collisions: Coupling Schemes

Here we follow Child (1974, pp. 86–91, 141–144) and use his notation. When we discuss scattering between collision partners with structure, we distinguish between their relative position vector $\mathbf{r}$ and the $3N - 3$ internal degrees of freedom $\boldsymbol{\rho}$. We wish to take into account a possible change in the internal state $\phi_i(\boldsymbol{\rho})$ of the system, where $\phi_i(\boldsymbol{\rho})$ is usually a product of rotational–vibrational–electronic functions for the collision partners, evaluated at $r = \infty$. Each of these states represents a different collision channel (see p. 151 in the companion volume). We begin by writing down the Hamiltonian operator

$$H = H_{int}(\boldsymbol{\rho}) - \frac{\hbar^2}{2m} \nabla_r^2 + V(\mathbf{r}, \boldsymbol{\rho}), \qquad (2\text{-}9\text{-}4)$$

where m is the reduced mass of the collision partners. We then set up the equations of relative motion based on a known set of internal states $\phi_i(\boldsymbol{\rho})$ and a known interaction potential $V(\mathbf{r}, \boldsymbol{\rho})$ that is defined so as to vanish at infinity. The time-independent Schrödinger equation

$$H\Phi(\mathbf{r}, \boldsymbol{\rho}) = E\Phi(\mathbf{r}, \boldsymbol{\rho}) \qquad (2\text{-}9\text{-}5)$$

can be expressed in various alternative forms, each of which offers advantages when one solves for the relative motion in particular physical situations. Several of these forms are discussed briefly here.

1. *Diabatic Formulation of the Collision Problem*

The first form of (2-9-5) that we consider is based on the *potential coupling scheme*, in which the full wave function is expanded in terms of the unperturbed (*diabatic*) orthonormal eigenstates $\phi_j(\boldsymbol{\rho})$ of $H_{int}(\boldsymbol{\rho})$:

$$H_{int}(\boldsymbol{\rho})\phi_j(\boldsymbol{\rho}) = E_j\phi_j(\boldsymbol{\rho}). \qquad (2\text{-}9\text{-}6)$$

It follows that

$$\Phi(\mathbf{r}, \boldsymbol{\rho}) = \sum_j \psi_j(\mathbf{r})\phi_j(\boldsymbol{\rho}), \qquad (2\text{-}9\text{-}7)$$

and if we substitute (2-9-7) into (2-9-5), multiply by $\phi_i(\rho)^*$, and integrate over the internal variables, we obtain

$$[\nabla_r^2 + k_i^2]\psi_j(\mathbf{r}) = \sum_j U_{ij}(\mathbf{r})\psi_j(\mathbf{r}), \tag{2-9-8}$$

where

$$k_i^2 = \frac{2m(E - E_i)}{\hbar^2} \tag{2-9-9}$$

and

$$U_{ij}(\mathbf{r}) = \frac{2m}{\hbar^2} \int \phi_i^*(\rho) V(\mathbf{r}, \rho) \phi_j(\rho)\, d\rho. \tag{2-9-10}$$

Child points out that the diagonal terms $U_{ii}(\mathbf{r})$ in (2-9-8) contribute directly to the elastic scattering, and they also affect the inelastic cross section by distorting the $\psi_i(\mathbf{r})$ from their plane-wave forms. The off-diagonal terms $U_{ij}(\mathbf{r})$ couple different channels (ϕ_i and ϕ_j) together and thereby may produce inelastic collisions, as well as polarization contributions to the elastic scattering.

2. Adiabatic Formulation

When the relative velocity is slow on the time scale of relaxation of the internal motion, it may be profitable to choose an adiabatic formulation in which (2-9-7) is replaced by an expansion in the adiabatic internal states $\chi_j(\mathbf{r}, \rho)$. These states are the orthonormal eigenfunctions of the internal Hamiltonian at a given $\mathbf{r}$:*

$$[H_{\text{int}}(\rho) + V(\mathbf{r}, \rho)\chi_j(\mathbf{r}, \rho) = W_j(\mathbf{r})\chi_j(\mathbf{r}, \rho). \tag{2-9-11}$$

Hence $\chi_j(\mathbf{r}, \rho)$ and $W_j(\mathbf{r})$ depend parametrically on r; as $r \to \infty$, they go over to their counterparts $\phi_j(\rho)$ and E_j defined by (2-9-6). If we now make the expansion

$$\Phi(\mathbf{r}, \rho) = \sum_j \tilde{\psi}_j(\mathbf{r})\chi_j(\mathbf{r}, \rho), \tag{2-9-12}$$

we obtain the following equation for $\tilde{\psi}_j(\mathbf{r})$:

$$[\nabla_r^2 + k_i^2(\mathbf{r})]\tilde{\psi}_i(\mathbf{r}) = \sum_j [\mathbf{X}_{ij}(\mathbf{r}) \cdot \nabla_r + Y_{ij}(\mathbf{r})]\psi_j(\mathbf{r}), \tag{2-9-13}$$

---

*The approach taken here is consistent with our discussion of the adiabatic hypothesis in Section 2-9-A. One set of basic functions (say, the χ) may be more meaningful physically than another set (say, the ϕ) for the solution of a given problem, and it may allow a much simpler expansion of the wave function than does any other set.

where

$$k_i^2(\mathbf{r}) = \frac{2m[E - W_i(\mathbf{r})]}{\hbar^2} \tag{2-9-14}$$

$$\mathbf{X}_{ij}(\mathbf{r}) = -2 \int \chi_i^*(\mathbf{r}, \boldsymbol{\rho}) \nabla_r \chi_j(\mathbf{r}, \boldsymbol{\rho}) \, d\boldsymbol{\rho} = -2 \langle \chi_i | \nabla_r | \chi_j \rangle \tag{2-9-15}$$

$$Y_{ij}(\mathbf{r}) = - \int \chi_i^*(\mathbf{r}, \boldsymbol{\rho}) \nabla_r^2 \chi_j(\mathbf{r}, \boldsymbol{\rho}) \, d\boldsymbol{\rho} = - \langle \chi_i | \nabla_r^2 | \chi_j \rangle. \tag{2-9-16}$$

We have described here the *adiabatic*, or *kinetic*, *coupling scheme*.

Child shows how as an alternative to (2-9-13), the coupling terms $\mathbf{X}_{ij}$ and Y_{ij} may be expressed as matrix elements of ∇V or $\nabla^2 V$. He also demonstrates how another form for the equations of motion may be obtained by transforming, in the adiabatic formulation, to a set of *rotating axes*.

3. Semiclassical Approximation

In Sections 3-24 of the companion volume and 1-6 of this volume, we discussed semiclassical methods used in collision theory. Here we return to the subject briefly to see how the general inelastic collision problem is formulated semiclassically. The basis of the semiclassical method is the assumption that there exists over the collision region a mean classical trajectory for the relative motion of the collision partners, and that it is determined by an average central potential $V(\mathbf{r})$. Here we replace the kinetic energy operator for the relative motion of the collision partners by a time-dependent interaction potential $V[\mathbf{r}(t), \boldsymbol{\rho}]$, so that the time-dependent Schrödinger equation for the relative motion becomes

$$i\hbar \frac{d\Phi}{dt} = [H_{\text{int}}(\boldsymbol{\rho}) + V(\mathbf{r}(t), \boldsymbol{\rho})]\Phi(\boldsymbol{\rho}, t). \tag{2-9-17}$$

The relative position vector is denoted by $\mathbf{r}(t)$, and the internal coordinates by $\boldsymbol{\rho}$. Now we introduce the functions $\phi_i(\boldsymbol{\rho})$, the eigenstates of $H_{\text{int}}(\boldsymbol{\rho})$, which are solutions of

$$H_{\text{int}}(\boldsymbol{\rho})\phi_n(\boldsymbol{\rho}) = E_n \phi_n(\boldsymbol{\rho}), \tag{2-9-18}$$

and expand the time-dependent internal state wave function as follows:

$$\Phi(\boldsymbol{\rho}, t) = \sum_n a_n(t)\phi_n(\boldsymbol{\rho})e^{-iE_n t/\hbar}. \tag{2-9-19}$$

The expansion coefficients a_i are determined by the equations

$$i\hbar \frac{da_n}{dt} = \sum_m V_{nm}(r)e^{i\omega_{nm}t}a_m(t) = \sum_m H'_{nm}(t)a_m(t), \qquad (2\text{-}9\text{-}20)$$

where

$$V_{nm}(\mathbf{r}) = \langle \phi_n | V(\boldsymbol{\rho}, \mathbf{r}) | \phi_m \rangle, \qquad (2\text{-}9\text{-}21)$$

$$\hbar\omega_{nm} = E_n - E_m, \qquad (2\text{-}9\text{-}22)$$

and

$$H'_{nm}(t) = V_{nm}(\mathbf{r})e^{i\omega_{nm}t}. \qquad (2\text{-}9\text{-}23)$$

The classical trajectory $\mathbf{r}(t)$ must be determined from the equations of relative motion before (2-9-17) can be integrated.

Now suppose that the equations (2-9-20) have been solved subject to the initial boundary conditions

$$a_m(-\infty) = \delta_{nm}, \qquad (2\text{-}9\text{-}24)$$

with the probability amplitude being unity for the entrance channel, n. The *transition probability* P_{nm} for the transition $n \to m$ is then given by

$$P_{nm} = |a_m(\infty)|^2. \qquad (2\text{-}9\text{-}25)$$

At any given impact energy, a particular trajectory is specified by an initial impact parameter, b, and azimuth angle, ϕ, so the integral cross section for the transition $n \to m$ is given by

$$\sigma_{nm} = \int_0^{2\pi} \int_0^{\infty} P_{nm}(b, \phi)b \, db \, d\phi. \qquad (2\text{-}9\text{-}26)$$

Child (1974, pp. 143–144) discusses the more complicated matter of expressing the differential cross section in terms of P_{nm}.

C.   Rate Constant for Inelastic Processes

Often experimental data are presented in terms of a *rate coefficient*, $k(T)$, which is related to a cross section σ_{ij} by the equation

$$k(T) = \langle k_{ij}(T) \rangle = \left\langle \int_0^{\infty} v\sigma_{ij}\left(\frac{1}{2}mv^2\right)P(v)v^2 \, dv \right\rangle. \qquad (2\text{-}9\text{-}27)$$

Here $P(v)$ is the distribution function for the relative velocity; it is usually Maxwellian in form (see p. 15 of the companion volume and Problem 2-16).

If the velocity distribution is Maxwellian at temperature T, and if σ_{ij} is zero for $E < E_{ij}$, then $k_{ij}(T)$ has the Arrhenius form

$$k_{ij}(T) = A_{ij}(T) \exp\left(\frac{-E_{ij}^0}{kT}\right), \tag{2-9-28}$$

where $A_{ij}(T) \sim (Tm)^{1/2}$, m being the reduced mass of the collision partners. If, on the other hand, σ_{ij} has the form

$$\sigma_{ij}\left(\frac{1}{2} mv^2\right) \simeq \sigma_{ij}^0 \exp\left(\frac{-v_0}{v}\right), \tag{2-9-29}$$

then $k_{ij}(T)$ is given by

$$k_{ij}(T) = B_{ij}(T) \exp\left[-\frac{3}{2}\left(\frac{mv_0^2}{kT}\right)^4\right], \tag{2-9-30}$$

where $B_{ij}(T)$ is a weak function of T. Child (1974, p. 8) points out that (2-9-30) is important in the theory of vibrational relaxation. Additional useful information on rate coefficients appears in Levine and Bernstein (1987, pp. 173–182).

D. Rotational Rainbows; Comparison with Elastic Rainbows

In Sections 1-2-B, 1-11-A-1, and 1-11-B, we discussed rainbows in angular distributions of heavy particle elastic scattering, restricting the theoretical part of the treatment to a classical description and a central interaction potential. Here we expand our former discussion and introduce rotationally inelastic rainbows. Our discussion is based on Faubel (1983) and on Chapter 3 of Levine and Bernstein (1987), to which the reader is referred for more detail.

1. Rainbows in Elastic Scattering

Equation (3-7-2) of the companion volume gives the CM differential elastic cross section $I_s(\Theta)$ in terms of the impact parameter b and the scattering angle Θ, on the assumption that there is a unique relationship between b and $\Theta = |\Theta_\pm|$. Here $\Theta_\pm$ is the classical deflection function (Section 1-11). For certain forms of interaction potential, however, there may be three or more values of b that correspond to a given value of Θ, as illustrated by Figs. 1-11-1, 1-11-4, and 1-11-5. In such cases we must sum over all of the values of b that contribute to the scattering at angle Θ, and we rewrite (3-7-2) as

$$I_s(\Theta) = \sum_{i=1} \frac{b_i}{\sin\Theta |d\Theta_\pm/db|_i} \qquad (\Theta < \Theta_r). \tag{2-9-31}$$

Only a single term contributes for scattering at angles greater than the rainbow angle, Θ_r, so there is a discontinuity in $I_s(\Theta)$ at Θ_r, which is an extremum angle. Since $|d\Theta_\pm/db| = 0$ at $b = b_r$, we have a range of impact parameters associated with scattering at the same angle, and a divergence in $I_s(\Theta)$ results. Our use of the classical scattering approximation has produced these two artifacts: the discontinuity and the divergence at Θ_r.

The interference terms that are present when quantal scattering theory is applied lead naturally to the elastic rainbow, and no divergence appears. There is an oscillatory pattern in the angular scattering distribution, with an envelope that exhibits bumps and depressions for $\Theta < \Theta_r$ and a pronounced maximum (the *primary rainbow*) near Θ_r. The other maxima are called *supernumerary rainbows*. $I_s(\Theta)$ then falls on the other side (the "dark side") of the primary rainbow. Figure 2-9-1 illustrates the comments made above with plots of classical and quantal data.

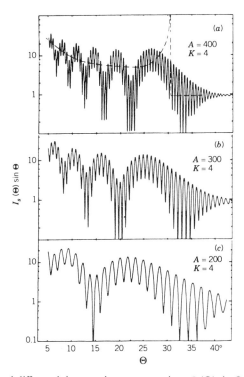

Figure 2-9-1. Quantal differential scattering cross sections $I_s(\Theta) \sin \Theta$ as a function of the CM deflection angle, Θ, calculated for a Lennard-Jones $(12, 6)$ potential (full curves). The reduced energy $(K \equiv E_{CM}/\varepsilon)$ is the same in all three panels, but the reduced wave number $(A \equiv kR_0)$ has the value 400 in (a), 300 in (b), and 200 in (c). The most nearly classical behavior occurs in (a), where the largest number of partial waves contribute; the classical cross section is shown there as a dashed line. Here ε is the depth of the potential well, and R_0 is the range parameter of the potential. [From Pauly (1975, p. 581).]

2. *Rotational Rainbows*

For simplicity we consider a collision between an atom and a nonrotating diatomic molecule (Fig. 1-3-2), but now we take the molecule to be rigid and represent it by an ellipsoid, as shown in Fig. 2-9-2. *If* the force exerted on the molecule by the impinging atom were a central force, there would be a unique scattering angle Θ for every impact parameter b (or initial orbital angular momentum $L = M_r v_0 b$). However, the reverse is not true—there can be more than one value of L for a given Θ; hence the interference pattern in elastic scattering. Now with the noncentral force that actually is applied in Figs. 1-3-2 and 2-9-2, the associated torque will set the molecule into rotational motion, with some "final" rotational angular momentum J. To each initial L and initial orientation angle γ, there corresponds a unique deflection function Θ_+ and final value of J. Unless the impact velocity is extremely low, we are dealing with a sudden collision, during which the molecule does not rotate appreciably, and the calculation of $\Theta(L, \gamma)$ and $J(L, \gamma)$ is thereby simplified.

Although both Θ_+ and J are unique functions of L and γ, the converse may not be true, for more than one set of L and γ may lead to the same set of observed $\Theta = |\Theta_+(L, \gamma)|$ and $J = |J(L, \gamma)|$. If many nearby trajectories lead to the same observed value of J, the orientation-averaged $J(L)$ will exhibit an extremum, and the scattered intensity will show a maximum at a given Θ as a function of J. This *rotational rainbow* can occur in inelastic scattering even for a purely repulsive potential, in contradistinction to the case of elastic scattering. Rotational rainbows also appear in the angular scattering distributions. Classical cal-

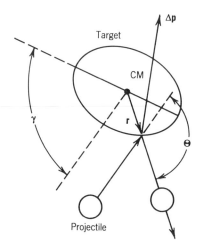

Figure 2-9-2. Rotational rainbow effect. An atom strikes a diatomic molecule (represented by a rigid ellipsoid) at an arbitrary orientation angle γ and is scattered at the angle Θ. The recoil momentum $\Delta \mathbf{p}$ imparts an angular momentum $\Delta \mathbf{J} = \mathbf{r} \times \Delta \mathbf{p}$ to the ellipsoid; $\Delta J(\gamma)$ must pass through a minimum as γ is varied, for a given constant Θ. Hence the classical, orientation-averaged cross section for rotational excitation shows a singularity at this value of $\Delta \mathbf{J}$. [Adapted from Faubel (1983, p. 365).]

culations predict a sharp edge, with a monotonic decline as Θ increases until the secondary rainbows appear, and the rainbow peak advancing toward larger Θ as ΔJ, the change of J in the collision, increases. Figure 2-9-3 illustrates these features. The rigid ellipsoid model discussed here is treated in greater depth in Note 2-4.

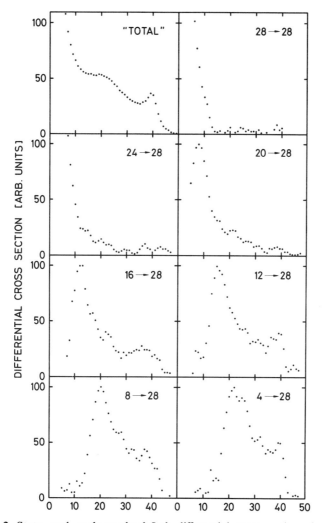

Figure 2-9-3. State- and angle-resolved Lab differential cross sections for rotational excitation of Na_2 by Ne at an impact energy of 175 meV. The cross sections are for transitions from $J = 4, 8, 12, 16, 20$, and 24 to $J' = 28$. The primary rotational rainbow appears at the left in each panel and unresolved secondary rainbows at the right. In experiments with low initial J values, Hefter et al. (1981) resolved the secondary rainbows. The elastic ($28 \rightarrow 28$) and total (all $J \rightarrow J' = 28$) differential cross sections are also shown. [Adapted from Bergmann et al. (1981, p. 63).]

E.   Simple Model Problems with Closed-Form Solutions

In his book, Child (1974) describes methods for solving the equations of motion that we have set up in Section 2-9-B. These equations can serve as the starting point for the solution of problems based on realistic interaction potentials and accurate wave functions. Such problems are necessarily complex, and they involve a great deal of computer computation. However, the equations of motion can also serve as the starting points for the solution of problems based on simple physical models that can be solved in closed analytic form. We now list the problems of this type that are solved by Child; the page numbers refer to his book.

1. One-dimensional potential coupling between two channels in exact resonance (pp. 91–96).
2. Inelastic scattering in a central field (pp. 96–100). Here Child points out that in a molecular scattering problem, two sources of angular momentum must be considered: the first is the rotational part associated with the internal motion (quantum number J); the second is the orbital part coming from the relative motion of the collision partners (quantum number L). In general, the two contributions are coupled, since only the magnitude of the total angular momentum and one of its components about a fixed axis are strictly conserved. However, if the interaction is spherically symmetric, so that $V = V(r, \rho)$, the rotational and orbital angular momenta are not coupled, and they are both constants of the motion.
3. The scattering of atoms by a rigid diatomic rotator (pp. 100–106).
4. The excitation of a rigid rotator (pp. 106–110).
5. The vibrational relaxation of a one-dimensional harmonic oscillator BC colliding with an atom A under an exponential repulsion (*Landau–Teller model*) (pp. 117–122).
6. Coupling induced by a constant interaction term between wave functions (pp. 122–123).

Serri et al. (1982) work out the equations for the rigid ellipsoid model of rotational excitation (see Fig. 2-9-2). They then extend the treatment to include simultaneous vibrational transitions, as outlined in Note 2-4.

The Fraunhofer diffraction model for elastic and inelastic collisions is discussed in Note 2-5. We have a great deal more to say about the forced oscillator model, which has proved very fruitful in discussions of vibrational excitation, in Note 2-6.

F.   Calculations on Realistic Systems: R, V, and E Excitation

A valuable source of information on the theory of various types of inelastic collisions is Bernstein (1979). In Chapter 6 of this edited volume, J. C. Light discusses the *general theory*, while in Chapter 7, D. J. Kouri treats *approximation methods*. There are three chapters devoted to *rotational excitation*: Chapter 8 by D. Secrest (the quantal treatment), Chapter 9 by D. J. Kouri (approximation methods), and Chapter 10 by M. D. Pattengill (classical trajectory methods). Two chapters on *vibrational excitation* follow: Chapter 11 by D. Secrest (the quantal treatment) and Chapter 12 by W. R. Gentry (classical and semiclassical methods). Finally, there is Chapter 13 by M. S. Child on *electronic excitation* (nonadiabatic transitions).

Bogdanov et al. (1989) have reviewed the theory of rotational and vibrational excitation of polyatomic molecules by heavy particle impact. Bobashev (1978) has discussed quasi-molecular interference effects in inelastic ion–atom collisions. Schinke (1983) has reviewed the theory of collision-induced rotational transitions, and Nakamura (1987) has treated the semiclassical theory of nonadiabatic transitions.

Kimura and Lane (1989) have discussed various kinds of low-energy, heavy particle, inelastic collision processes in a very general treatment of the theory. Recall that in Section 6-14 of the companion volume we discussed the excitation of the H atom by electron and proton impact in the Born and Bethe–Born approximations.

Finally, we call attention to the review by Brunner and Pritchard (1982) on *fitting laws*. These "laws" are mathematical relationships that represent the known rates for a particular atom–molecule system in terms of a few parameters.

PART $\mathscr{B}$.   COLLISIONAL AND FIELD DISSOCIATION

In Sections 5-5 and 7-3 of the companion volume, we discussed the dissociation of neutral molecules and positive molecular ions by electron impact. On pages 289 and 478, we listed the types of dissociative events that are possible with diatomic targets, and we saw how electron-impact dissociation can occur simultaneously with excitation, ionization, electron attachment, recombination, and ion-pair formation. This linkage complicates the interpretation of experiments, and the exposition of the subject as well!

When we turn to heavy particle impact, the situation is even more complicated. Although electron attachment is no longer a factor, charge transfer and chemical reactions enter the picture, while ion–ion recombination now replaces electron–ion recombination as a possibility (see Problem 2-17). Furthermore, *both* collision partners now have structure (except when H^+, D^+, or T^+ is involved). In this section we concentrate on simple dissociative processes, more complicated situations involving several kinds of simultaneous reactions being deferred until Chapters 4 and 5. We also discuss field dissociation here because of its connection with collisional dissociation in applications.

2-10. EXPERIMENT

The experimental studies of dissociation in heavy particle collisions may be divided into two types: measurements of cross sections and investigations of structure and dissociation mechanisms. We treat both types but restrict our attention to impact energies above about 1 keV, where most of the work has been done.

A. Cross-Section Measurements

One of the most intriguing approaches to controlled thermonuclear power production involves the establishment of a hot, dense plasma by transverse injection of high-energy molecular hydrogen ions into a dc magnetic mirror machine containing hydrogen gas at low pressure. If the fast injected ions are trapped in the mirror field in sufficient numbers and for sufficient time, they may be able to produce enough ionization to build up the required plasma temperature and density, but one of the major problems is that of efficiently trapping the incident ions. Particles entering the magnetic field from outside the machine have orbits that are topologically different from those of particles truly trapped within the machine, and injected ions will immediately be lost unless their charge-to-mass ratio is altered during their first circulation about the flux lines of the mirror field configuration. Changes in e/m may be effected by dissociating the ions in collisions with other particles inside the machine or by passing the incident ions through a strong electric or magnetic field. This fact has prompted a number of studies of collisional and field dissociation of molecular hydrogen ions of various isotopic composition.

Experiments on high-energy collisional dissociation have been conducted with both single beam/gas cell and crossed-beam instruments (McClure and Peek, 1972; Massey and Gilbody, 1974, pp. 2909–2936; Cooks, 1978; Montgomery and Jaecks, 1983). Some of these instruments are quite similar to those used in high-energy charge-transfer studies. Most cross-section measurements involve observations of fragments produced in a fast beam; it is difficult to quantify measurements on fragments formed in a static gas by projectiles passing through it. Data of relevance to controlled fusion research are presented in Figs. 2-10-1 through 2-10-3.

If a sufficiently strong external electric field $\mathbf{E}$ is applied to a molecular ion, the interatomic potential may be warped in such a way that the higher vibrational states become unstable and dissociation results even in the absence of a collision. The same effect may be produced by the equivalent field $\mathbf{E}' = \mathbf{v} \times \mathbf{B}$ caused by motion with velocity $\mathbf{v}$ through a magnetic field of flux density $\mathbf{B}$. The dissociation of H_2^+ and D_2^+ ions by static $\mathbf{E}$ and $\mathbf{B}$ fields has been measured by Riviere and Sweetman (1960, 1962) and by Kaplan et al. (1961, 1963). See Fig. 2-10-4 for data on H_2^+ dissociation.

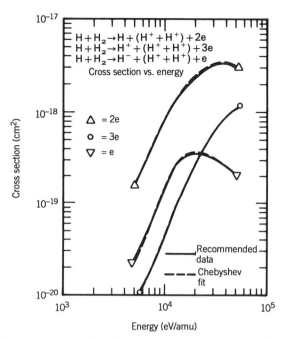

Figure 2-10-1. Integral cross sections for dissociative processes in collisions of H on H_2. [From Barnett (1990).]

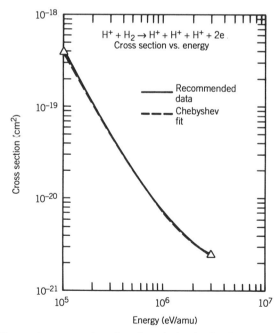

Figure 2-10-2. Integral cross sections for dissociative ionization in collisions of H^+ on H_2. [From Barnett (1990).]

149

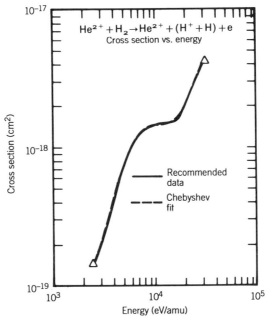

Figure 2-10-3. Integral cross sections for dissociative ionization in collisions of He^{2+} on H_2. [From Barnett (1990).]

B. Studies of Structure and Dissociation Mechanisms

Kanter and his colleagues at the Argonne National Laboratory and the Weizmann Institute of Science (Kanter, 1981) have used the *Coulomb explosion technique* for the determination of stereochemical structures of molecules and molecular ions. Two reviews (Kanter, 1983; Vager et al., 1989) describe the method and give references to applications to specific systems. In the usual case, a collimated beam of the structures to be studied is directed at MeV energy onto a thin foil (about 30 Å thick). Each dissociating projectile rapidly loses most or all of its electrons in sudden, violent collisions with the electrons in the target. The projectile then undergoes a "Coulomb explosion" as the highly charged constituent atoms abruptly repel one another. The dissociation fragments travel downstream in the Lab frame with their velocities shifted in both speed and direction from the incident beam velocity. The shifts typically amount to a few keV in energy and a few milliradians in angle. High-resolution measurements of the joint energy and angle distributions of the fragments provide the data from which structural information can be deduced. It is important to note that a projectile is usually stripped of electrons within about 10^{-16} s after it enters the foil. The characteristic times for the nuclear motions within the projectile are much longer (about 10^{-14} s for vibrations, and 10^{-12} s for rotations). If a tenuous gas target is used instead of a foil, the projectiles are dissociated in

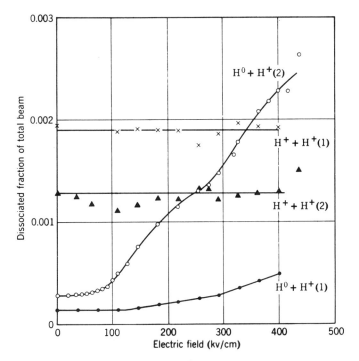

Figure 2-10-4. Fractional dissociation of H_2^+ ions as a function of the electric field intensity in the vacuum gap through which the ions are projected. Curves (1): H_2^+ ions direct from the ion source, with little vibrational energy. Curves (2): H_2^+ ions obtained from breakup of H_3^+ ions, with considerable vibrational energy. [From Riviere and Sweetman (1960).]

gentler collisions, and multielectron ionization is much less likely. Instead, many electrons are promoted to dissociative states that can decay into both neutral and charged fragments. Hence data obtained with gas targets give information about excited-state structures and are complementary to foil data.

Edwards and his colleagues at the University of Georgia have made a series of measurements in which they ionize molecules with a pulsed 1-MeV atomic ion beam and detect pairs of fragments of the dissociating molecular ions in coincidence. The kinetic energy spectra of the fragments are mapped using a "time-energy spectroscopic technique," wherein the ions released in the breakup are separated according to their mass-to-charge ratios and their kinetic energies recorded. In a study of He^+ impact on N_2 and the resulting dissociation of N_2^{2+} ions into pairs of N^+ fragments, Edwards and Wood (1982) reported three distinct dissociation channels, with total kinetic energy releases of 7.8, 10.2, and 14.8 eV. In later studies, Ezell et al. (1984) measured the angular correlation between the two N^+ ions ejected in the dissociation, and established that the recoil velocity of the parent molecular ions is considerably less than the mean thermal velocity of N_2 molecules at 300 K.

2-11. THEORY

The book edited by Bernstein (1979) contains two contributions on the theory of collisional dissociation: Chapter 20 by D. J. Diestler describes the quantal treatment, and Chapter 21 by P. J. Kuntz deals with trajectories and models. The theory of field dissociation is discussed by Hiskes (1961, 1962). Massey and Gilbody (1974) provide a useful summary of theoretical work on specific systems.

PART *𝒞*. ENERGY TRANSFER IN HEAVY PARTICLE COLLISIONS

2-12. INTRODUCTION

The subject under consideration here is the inelastic transfer of energy in heavy particle collisions at low energies (up to a few eV). This type of energy transfer provides the main mechanism for the passage of a molecular system from a nonequilibrium state to thermodynamic equilibrium, and experimental studies of energy transfer provide vital information about intermolecular potentials. Translational (T), rotational (R), vibrational (V), and electronic (E) energies are involved. Ideally, in measurements we would like perfect state resolution, with the internal states and momenta of the particles being known. Much of what we said about R, V, and E excitation and deexcitation processes in Part *𝒜* is relevant here, but the previous discussion dealt almost entirely with beam measurements, and they cannot be expected to furnish the enormous quantity of data required for the many applications of interest. Indeed, examination of Fig. 2-12-1, which relates only to V, R, and T energy, gives some indication of the immensity of the need and the difficulty of the problem. For example, state-to-state cross sections for an immense number of V–R transitions would be required, with all of the vibrationally excited reactants in selected states and their collision partners also in specified states. The principle of detailed balance (pp. 280 and 530–536 of the companion volume) is on hand to calculate inverse reactions, and we can allow some averaging over states, but evidently help is needed from other quarters. Here we concentrate on experiments of types different from those in Part *𝒜*, and look mainly at absorption cells and discharges probed with spectroscopic techniques. Excellent references of broad scope on the entire subject of molecular energy transfer are the books by Lambert (1977), Yardley (1980), Massey (1971), and Levine and Bernstein (1987, pp. 300–395), which should be consulted to supplement the present brief treatment based on their work.

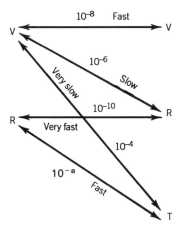

Figure 2-12-1. Comparison of typical rates of energy transfer processes in molecular collisions at thermal energies. The letters R, V, and T mean rotational, vibrational, and translational, respectively. The numbers alongside the double-ended arrow are values of $P\tau$(atm-s), the "bulk relaxation time". Transitions involving *electronic* (E) states are not shown here. Their rates vary widely and may be extremely fast for excited states (Gole, 1985). [Adapted from Flygare (1968).]

2-13. THERMODYNAMIC EQUILIBRIUM AND RELAXATION TOWARD IT

We begin by examining the Maxwell–Boltzmann distribution function that was discussed on pp. 50 and 74–75 of the companion volume, and that describes gas-phase equilibrium. We drop the dependence on kinetic energy to obtain the *Boltzmann distribution function*

$$p_i = \frac{n_i}{N} = \frac{g_i \exp(-E_i/kT)}{Q_I}, \tag{2-13-1}$$

where p_i is the relative population of molecules in the energy level E_i at temperature T, n_i the number of molecules in the level i, and N the total number of molecules. The degeneracy g_i gives the number of possible quantum states of the molecule corresponding to E_i, while Q_I is the internal partition function. Transient departures from the thermodynamic equilibrium state, which is discussed on pp. 43–44 of the companion volume, can be effected by a beam of particles or photons, by chemical reactions that are selective in their energy disposal, or by shock waves. By observing the subsequent changes in the population of the various energy levels, one can follow the "relaxation" to equilibrium and obtain information about the energy transfer processes.

Levine and Bernstein (1987, pp. 301–302) discuss the simple example of a two-state system that has been displaced from equilibrium and is undergoing relaxation back toward the original state. The equilibrium situation can be approximated by a mixture of HCl and a buffer rare gas at a temperature low enough that most of the HCl molecules are in the ground vibrational ($v = 0$) state. The only two processes assumed to be important are the V–T and T–V transfers indicated in the equation

$$HCl(v = 1) + M \leftrightharpoons HCl(v = 0) + M. \qquad (2\text{-}13\text{-}2)$$

Let n_i and n_0 be the number densities of the molecules in the $v = 1$ and $v = 0$ states, and n_B be the number density of the buffer gas, with which the HCl molecules mainly collide. Then, by definition of the rate coefficients k_{10} for deexcitation and k_{01} for excitation, we have

$$-\frac{dn_0}{dt} = \frac{dn_1}{dt} = -k_{10}n_B n_1 + k_{01}n_B n_0. \qquad (2\text{-}13\text{-}3)$$

At equilibrium this equation gives

$$\frac{k_{01}}{k_{10}} = \left[\frac{n_1}{n_0}\right]_{\text{equil}} = \exp\left(-\frac{E_1 - E_0}{kT}\right) = \exp\left(\frac{-\theta}{T}\right), \qquad (2\text{-}13\text{-}4)$$

where

$$\theta = \frac{E_1 - E_0}{k} = \frac{\hbar\omega}{k} \qquad (2\text{-}13\text{-}5)$$

is the *characteristic vibrational temperature*. Here $\omega = (E_1 - E_0)/\hbar$ is the angular frequency that a photon would have to have to bridge the gap between E_1 and E_0. If we let Δn_1 be the deviation of n_1 from its equilibrium value, we can show that the rate of relaxation from a nonequilibrium state is given by

$$\Delta n_1(t) = \Delta n_1(t = 0) \exp\left(\frac{-t}{\tau}\right), \qquad (2\text{-}13\text{-}6)$$

where

$$\tau = \frac{1}{n_B(k_{10} + k_{01})} = \frac{1}{n_B k_{10}[1 + \exp(-\theta/T]} \qquad (2\text{-}13\text{-}7)$$

is the *relaxation time* (Problem 2-21). The reader is invited to compare this discussion with that in Section 2-5 of the companion volume, which dealt with the relaxation of a gas by *elastic* collisions.

Note that in (2-13-7), $\tau \approx 1/n_B k_{10} \sim 1/P_B k_{10}$, where P_B is the pressure of the buffer gas. Hence $P_B \tau \sim 1/k_{10}$, and since $P_B \approx P$, the total gas pressure, we can write

$$P\tau \sim \frac{1}{k_{10}}. \tag{2-13-8}$$

We often express bulk relaxation data in terms of the quantity $P\tau$, where its reciprocal is a rate coefficient in units of 1/atm-s. Levine and Bernstein (1987, p. 303) show how to convert $(P\tau)^{-1}$ to more familiar units (L/mol-s).

We now reexamine Fig. 2-12-1, which gives approximate values of $P\tau$ for typical V–V′, V–R, R–R′, V–T, and R–T processes, along with their inverses at *thermal energies*. We can make a few useful generalizations. In general, a large translational energy defect is associated with an inefficient energy transfer process. R–R′ and R–T processes are very efficient because rotational levels are closely spaced. V–T energy transfers are much less efficient because of the large vibrational spacings, especially in diatomic molecules. (See Section 2-9-A on the adiabatic hypothesis.)

In Section 2-3 of the companion volume we worked out the collision frequency v for a molecule moving at random in a gas at thermal equilibrium. The calculation was based on the elastic sphere model and Maxwell's mean free path. The time interval between collisions, $T = 1/v$, may be used with the relaxation time τ to define the collision number Z_{A-B} for a particular relaxation process A–B:

$$Z_{A-B} \equiv \frac{\tau_{A-B}}{T_{A-B}}. \tag{2-13-9}$$

Roughly, the A–B process will occur once in every Z_{A-B} collisions. $Z_{R-R'}$ is typically between 1 and 10, whereas Z_{V-T} is on the order of 10^6. Hence R–R′ energy transfer occurs on nearly every collision, but the V–T process is about a factor of 10^6 less likely.

Levine and Bernstein (1987, p. 305) point out that although the relaxation of a gas involves many processes characterized by different rates, bulk relaxation studies may frequently be described accurately by a single relaxation time. The reason for this is that if the individual rates are widely different, the overall relaxation is determined by the slowest, or *rate-determining process*. Here there is a close analogy to sequential radioactive decays (Evans, 1955).

2-14. ADIABATICITY PARAMETER ξ AND RESONANCE FUNCTION $R(\xi)$

Here we extend the treatment of the adiabatic hypothesis presented in Section 2-9-A and discuss the rates of energy transfer processes involving rotational and vibrational states. Electronic states require special treatment and are considered at the end of this section. This discussion is based on Levine and Bernstein (1987, pp. 312–320, 368–383).

Let us first consider $T \leftrightarrows V$ *energy transfer* and work with the model that consists of an atom (A) colliding collinearly with a diatomic molecule (BC), B being located between A and C. The molecule is represented by a dumbbell harmonic oscillator, with natural frequency v. If A approaches BC with high velocity, or if the oscillator has a weak spring, the duration of the collision, t_c, is short compared with the oscillator period, t_v. When A strikes B, C barely moves and is said to act as a *spectator*. A change in the velocity of B is tantamount to a change in the relative velocity of B and C, and hence a change in the vibrational energy of BC. The inequality

$$t_c \ll t_v \tag{2-14-1}$$

means that the collision is in the *sudden limit*, where the energy transfer is efficient. On the other hand, if A approaches BC slowly, or if the oscillator spring is stiff, the separation distance B–C barely changes, and the collision is practically elastic. Hence T–V energy transfer is quite inefficient. The collision is now in the *adiabatic limit*, where the inequality above is reversed:

$$t_c \gg t_v. \tag{2-14-2}$$

We may say that the oscillator is able to adjust to the perturbation and is not strongly affected by it. It is useful to define the *adiabaticity parameter for* $T \leftrightarrows V$ *energy transfer as*

$$\xi_{T \leftrightarrows V} = \frac{t_c}{t_v} = \frac{a|\Delta E|}{hv}, \tag{2-14-3}$$

where a is the range of the interaction, $|\Delta E|$ the amount of energy transferred in the collision (here either $T \rightarrow V$ or $V \rightarrow T$), and v the initial velocity of A with respect to BC.

Similar considerations apply to $T \leftrightarrows R$ energy transfer. We introduce the *adiabaticity parameter for* $T \leftrightarrows R$ *energy transfer,*

$$\xi_{T \leftrightarrows R} = \frac{t_c}{t_r} = \frac{|\Delta E|}{h\omega}, \tag{2-14-4}$$

where t_r is the rotational period and $\omega = v/a$ is the orbital angular velocity of the relative projectile–target motion. Now $\Delta E \approx \hbar \omega_R$, where ω_R is the angular velocity of the molecule, so

$$\xi_{T \leftrightarrows R} = \frac{\omega_R}{\omega}. \tag{2-14-5}$$

In general, ξ is defined as

$$\xi = \frac{t_c}{t_{\text{char}}}, \tag{2-14-6}$$

the ratio of the collision duration to the appropriate characteristic period of the target motion. The adiabatic regime corresponds to $\xi \gg 1$.

The *resonance, or energy mismatch, function* $R(\xi)$ provides a more quantitative measure of the efficiency of energy transfer at a given value of ξ. It is defined by the equation

$$\frac{\langle\Delta E\rangle}{E_T} = \frac{\langle\Delta E\rangle_{\xi=0}}{E_T} R(\xi), \tag{2-14-7}$$

where E_T is the initial relative kinetic energy and $\langle\Delta E\rangle$ is the average energy transfer in the collision. The collision dynamics must be solved in order to determine $R(\xi)$, but for $\xi > 1$, the approximation

$$R(\xi) \approx \exp(-\xi) \tag{2-14-8}$$

is satisfactory.

In the harmonic oscillator model of T–V energy transfer discussed above, $R(\xi)$ is very small in the case of a stiff spring (inefficient energy transfer), and large if the spring is weak (efficient transfer). In general, at a given value of ΔE (the "energy gap"), $R(\xi)$ will be small at low impact velocities ($v \ll a \Delta E/h$) and will increase with increasing v. Levine and Bernstein (1987) make some useful general statements: Collisions with large impact parameters (and small scattering angles) tend to be more adiabatic (more elastic) than do closer collisions, because of the larger value of the range of interaction, a. The efficiency of inelastic energy transfer increases as the scattering angle increases, and backward scattering is the most inelastic. It follows that collisions leading to large changes in internal energy are mainly those with small impact parameters and hence rather small cross sections. Levine and Bernstein also point out that it is easier to transfer energy into or out of an oscillator in an excited state than one in the ground state.

As stated earlier, electronic energy transfer must be considered separately. Electronic excitation energies (typically, above 1 eV) are usually much larger than those for rotation and vibration, so on the basis of the adiabatic criterion, E–V and E–T energy transfer should be quite inefficient. However, this is not always the case, one reason being the occurrence of curve crossing in some instances. If the collision involves two potential curves, as in Fig. 1-13-1, the relevant energy gap ΔE is not the vertical distance ΔE_{asym} between the asymptotes in panel (*a*), but rather the much smaller vertical spacing ΔE_{cross} between the curves at the avoided crossing shown in panel (*b*) at the nuclear separation R_0. Levine and Bernstein (1987, pp. 365–378) discuss this effect in the context of the K–I system undergoing charge neutralization:

$$K^+ + I^- \rightarrow K + I. \tag{2-14-9}$$

In Fig. 1-13-1a, the asymptote at the right labeled $(A_2 + B_2)$ applies to the reactants $(K^+ + I^-)$, and the one designated $(A_1 + B_1)$ relates to the products $(K + I)$. Although $\Delta E_{asym} = $ I.E.(K) − E.A.(I) = 4.34 − 3.06 = 1.28\,eV, ΔE_{cross} is known to be only about 2.5 meV. Here I.E. represents the ionization energy, and E.A. the electron affinity. The discussion of curve crossing in Section 1-13-B-3 used the example of single charge transfer, with A being a projectile ion and B a neutral atom, but the result we obtained there for the probability Q of a net, overall transition from the $(A_1 + B_1)$ system to the $(A_2 + B_2)$ system is valid in the present example. Our result, the *Landau–Zener probability*, then is

$$Q = 2P(1 - P), \tag{2-14-10}$$

where P is the probability that no change occurs in the electronic states on a single pass of the system through the separation distance R_0. On the basis of the classical path method,

$$P \approx \exp(-\pi^2 \xi). \tag{2-14-11}$$

2-15. THEORY OF ENERGY TRANSFER PROCESSES

Of special interest here are the reviews by Bogdanov et al. (1989) and Uzer (1991). The references given in Section 2-9 are very useful, and, in fact, all of our previous discussion of the theory of inelastic collisions is relevant.

It is appropriate now to mention *propensity rules*, theoretical predictions that certain transitions, within a given class of collisions, are more strongly allowed than others. M. H. Alexander and his colleagues have worked out propensity rules for rotationally inelastic collisions involving various classes of molecules in various quantum states. References to their work are provided in papers by Dagdigian's group, which has undertaken extensive tests of the predictions (Jihua et al., 1986; Ali and Dagdigian, 1987; Dagdigian et al., 1988; Dagdigian, 1989). Propensity rules for vibrational transitions are discussed in the review by Krajnovich et al. (1987). In measurements of the first fully state-selected $(v_i, J_i, |M_i|) \rightarrow (v_f, J_f, |M_f|)$ differential cross sections for atom–molecule collisions, Mattheus et al. (1986) verified a strong $|\Delta M| \ll J$ propensity rule with respect to the direction of the linear momentum transfer. Here they studied Na_2–Ne $(0, 6, |M_i|) \rightarrow (0, 0, 0)$ rotationally inelastic scattering at a collision energy of 190 meV.

2-16.   EXPERIMENTAL STUDIES

Here we describe some examples of important modern work on various types of collisional energy transfer.

A.   Joint Institute for Laboratory Astrophysics

The HF molecule is important in chemical lasers, yet few measurements have been made on its relaxation from single high-J rotational states. Taatjes and Leone (1988) have used time-resolved, infrared, double-resonance spectroscopy to obtain rotational relaxation rates for $HF(v = 0, J = 13)$ colliding with the rare gases, H_2 and D_2. They used a two-laser pulse-and-probe scheme and took advantage of the fact that the $v = 1$ level of HF is nearly resonant with $(v = 0, J = 13)$ to form a significant population of molecules in the latter level. As shown in Fig. 2-16-1, the pump laser is used to excite the R(3) transition into $v = 1$, which then undergoes V–T, R relaxation to $(v = 0, J = 13)$ by collisions with unexcited HF. The probe laser is tuned to the R(13) transition, which

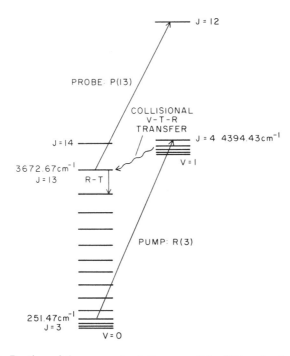

Figure 2-16-1. Portion of the energy-level diagram of the HF molecule, showing levels and transitions involved in the production of the $HF(v = 0, J = 13)$ rotationally excited state and the probing of this state to determine its population. [From Taatjes and Leone (1988).]

excites $(v = 0, J = 13)$ to $(v = 1, J = 12)$, and is used to monitor the $(v = 0, J = 13)$ population by transient absorption. The "pump laser" is actually composed of the pulsed Nd: YAG laser plus the LDS dye laser and optical components shown at the lower left of Fig. 2-16-2. The optoacoustic cell at the lower right is filled with HF and used to tune the pulsed laser to an HF absorption line. The probe is the CW F-center laser at the left center of the figure; it is pumped by the Kr ion laser at the right top and tuned to the desired wavelength by monitoring the absorption of the chopped beam through the heated absorption cell shown above the double resonance cell. The pump beam (5 mm diameter) and the probe beam (1.5 mm diameter) are counter propagated through the 65-cm stainless-steel "double resonance" cell shown in the center of the drawing. A transient-absorption signal is obtained from the time-resolved attenuation of the probe beam following each pump pulse, and these signals are stored in the internal memory of the signal averager near the right top of Fig. 2-16-2. The accumulated trace is then transferred to a microcomputer for storage and manipulation; it may also be displayed directly on a digital plotter. The R–R, T rate out of the $(v = 0, J = 13)$ probed level is much greater than the V–R, T rate that populates this level, so the rise of the probe signal is directly related to the relaxation rate out of the $(v = 0, J = 13)$ state. The relaxation rates are extracted from plots of reciprocal rise time, divided by the total pressure, versus the mole fraction of the collision partner.

The R–T relaxation rates measured for He, Ne, and Ar decrease dramatically in that order and may be described by a purely impulsive model. There is a substantial increase in progressing from Ar through Kr to Xe. This increase is ascribed to the increasing well depth of the attractive interaction, but cannot yet be explained quantitatively on the basis of any collision model. As indicated in Table 2-16-1, the rates for relaxation by H_2 and D_2 are nine times higher than that for He and dramatize the importance of the R–R energy transfer mechanism. In this table the accuracy of the rate constant determination ranges

Table 2-16-1   Rate Constants (k) and Cross Sections (σ) for the Rotational Relaxation of HF($v = 0, J = 13$) at 298 K

Collision Partner	$k (10^{-12} \, cm^3/molecule\text{-}s)$	$\sigma (10^{-16} \, cm^2)$
He	13	0.94
Ne	2.2	0.28
Ar	0.93	0.13
Kr	4.4	0.70
Xe	6.2	1.00
H_2	110	5.9
D_2	120	8.7

Source: (Taatjes and Leone (1988).

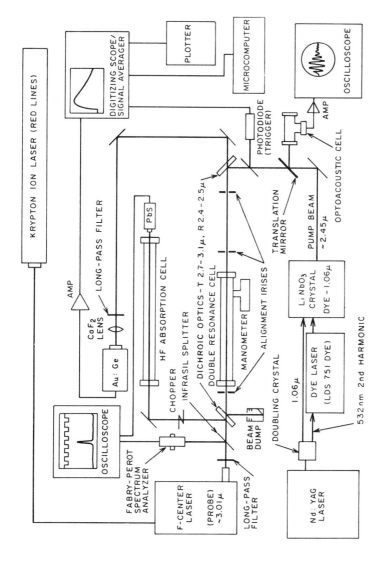

Figure 2-16-2. Apparatus used in Leone's laboratory for studies of the rotational relaxation of HF molecules in the $(v = 0, J = 13)$ state.

from about 15% for He, down to 10% for Ar, and then up to 17 to 18% for H_2 and D_2.

Another noteworthy experiment was performed by Yamasaki and Leone (1989), who used an absorption cell and laser spectroscopic techniques to measure state-specific total quenching rate constants for selected rotational levels of Br_2 under single-collision conditions with argon at 296 K. A 0.04-cm^{-1} bandwidth, etalon-narrowed pulsed dye laser excites ground-state $X^1\Sigma_g^+$ molecules in single ro-vibronic transitions to the $B^3\Pi(O_u^+)$ state, and fluorescence decays with and without the argon collision partner are analyzed at early times to obtain total quenching rate constants. Figure 2-16-3 illustrates the complexity of the problem by showing energy transfer and photochemical channels that Br_2 exhibits after being excited to a single ro-vibronic level in the B state. The rotational levels that are initially prepared are $J' = 26, 32, 37, 41, 46,$ and 58. For $J' \leqslant 41$, the total quenching rates are observed to decrease markedly with increasing J' (because of the decreasing probability of collision complex formation that results). For $J' > 41$, *negative* quenching rate constants

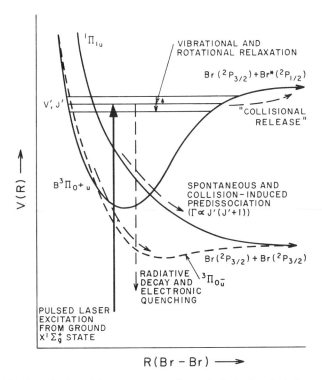

$R(Br - Br) \longrightarrow$

Figure 2-16-3. Complexity of the energy transfer and photochemical channels that Br_2 exhibits after being excited to a single rovibronic level in the B state. [From Yamasaki and Leone, (1989).]

are observed (i.e., the decay rates with Ar present are slower than those without Ar). This effect is believed to result from R–V energy transfer processes in which a high rotational level is transferred to a lower rotational level in the next higher vibration, the lower rotational state having a much slower spontaneous predissociation rate.

Wight et al. (1985) used an excimer laser to photodissociate H_2S in a photolysis cell and generate H atoms for studies of T–R, V energy transfer collisions with NO at 0.95 and 2.2 eV. A two-laser pulse-and-probe technique was employed.

Leone and his colleagues have also made a number of studies of the effect of orbital alignment in energy transfer events (Leone, 1988; Bussert and Leone, 1987; Schwenz and Leone, 1987; Bussert et al., 1987). Crossed-beam techniques were used in this work.

Wedding and Phelps (1988) have made an important measurement of the rate coefficients for the collisional quenching of $H_2(c^3\Pi_u^-)$ metastables by thermal energy H_2 molecules for various vibrational and rotational levels ($J = 1; v = 0$, 1, 2, 3 and $v = 1; J = 1, 2, 3$). The rate coefficients were observed to be independent of the vibrational and rotational quantum numbers and had a mean value of $(1.88 \pm 0.10) \times 10^{-9}$ cm^3/s at 300 K. Wedding and Phelps also determined destruction rate coefficients and radiative lifetimes of the $a^3\Sigma_g^+$ state, and studied collisional excitation transfer between the $J = 1$ levels of the $a^3\Sigma_g^+$ and $c^3\Pi_u^-$ states for $v = 0$ and 1. Wedding and Phelps used a pulsed discharge tube to produce the metastables and a laser absorption technique.

Next we introduce the term *energy pooling*, an excitation transfer process in which two excited atoms collide to produce one highly excited atom and one in the ground state. As in all energy transfer processes, spin change is an important consideration, and the transfer rate is normally enhanced when the electronic energy of the highly excited atom is close to the sum of the initial excitation energies. In A. Gallagher's laboratory, energy pooling has been studied in Sr vapor following pulsed optical excitation to the 5^3P_1 state (Kelly et al., 1988a). Energy pooling rate coefficients $k_{JJ'}$ were measured for the reactions

$$\text{Sr}(5^3P_J) + \text{Sr}(5^3P_{J'}) \rightarrow \text{Sr}(5^1S_0) + \text{Sr}(6^3S_1 \text{ or } 6^1S_0) \qquad (2\text{-}16\text{-}1)$$

by observing the radiative decay of the products via the cascade resonance line. The rate coefficients are large, almost gas kinetic. Kelly and his coworkers (1988a) also studied the dependence of the rates on spin, J, and the energy defect, ΔE. In this experiment the initial excitation is provided by a Nd:YAG pumped, pulsed dye laser directed through a gas cell. Fluorescence is detected at right angles to the laser beam and spectrally resolved by a monochromator. A Sr hollow cathode lamp sends focused light through the cell and into the monochromator so that individual lines can be isolated. This apparatus has also been used by Kelly et al. (1988b) to study collisional transfer with the $\text{Sr}(5^3P_J^0)$ multiplet that is produced by nearly adiabatic collisions with rare gases.

B. Massachusetts Institute of Technology

Pritchard and his colleagues at MIT have made a very extensive study of rotationally and vibrationally inelastic collisions of $Li_2^*(A^1\Sigma_u)$ with He, Ne, Ar, and Xe [see Magill et al. (1989) and the references therein]; of $Na_2^*(A^1\Sigma_u)$ with Xe (Smith et al., 1981); and of $I_2(B^3\Pi)$ with He and Xe (Dexheimer et al., 1983). Magill et al. studied the reactions

$$Li_2^*(v_i = 9, J_i) + X \rightarrow Li_2^*(v_f, J_f) + X, \qquad (2\text{-}16\text{-}2)$$

with $J_i = 8$, 22, and 42, and X = He, Ne, Ar, and Xe at $T \approx 600°C$. They reported *1088* level-resolved rate coefficients for vibrotationally inelastic collisions. The measurements are *rotationally resolved* in that the initial values of J and v are selected, and the final values are determined. A wide range of v_f and J_f are covered, and the error is typically only about 9%. The basic technique used here is laser-induced fluorescence (LIF). A ring dye laser is used to excite Li_2 in a heat pipe oven, which also contains the rare-gas target. Populations of the various (v_f, J_f) levels produced by collisional transfer out of the laser-populated (v_i, J_i) level of the $A^1\Sigma_u$ state are measured by their spectrally resolved fluorescence. The dependence of the population ratios $[f]/[i]$ on the pressure of the gas X is analyzed to determine the rate constants $k_{i \rightarrow f}$. The findings of Pritchard's group are too voluminous to be discussed in detail here; the reader should consult the references cited above and also Magill et al. (1988) for details.

C. University of Virginia

Gallagher's group at the University of Virginia has used laser spectroscopic techniques to study resonant collisional energy transfer between highly excited $K(ns)$ and $K([n-2]d)$ atoms which occurs when the levels are shifted into resonance with a weak electric field (Stoneman et al., 1987). At low relative velocities in an atomic beam, the collision times are so long ($\sim 0.17\,\mu s$) and the linewidths so narrow ($\sim 6\,MHz$) as to permit evaluation of the quantum defect of the K-atom p state with significantly improved accuracy. Thomson et al. (1989) have investigated resonant collisions of Na Rydberg atoms in an electric field and verified theoretical predictions concerning the differences in dipole–dipole energy transfer for the two cases $\mathbf{v} \parallel \mathbf{E}$ and $\mathbf{v} \perp \mathbf{E}$. Fu et al. (1989) have measured the n dependence of the microwave field intensity required to make the single-photon radiatively assisted collision

$$Na(ns) + Na(ns) + Nhv \rightarrow Na(np) + Na([n-1]p) + (N+m)hv, \qquad (2\text{-}16\text{-}3)$$

with $m = 1$, 10% as likely as the resonant collision process in which no photons are absorbed, $m = 0$. (A *radiatively assisted collision* is a resonant energy transfer collision between two structures in which the energy resonance requirement is met by the absorption or stimulated emission of photons.) Resonant collisional energy transfer between atoms has been reviewed by Gallagher (1992).

D. Bulk Measurements

By a bulk measurement we mean an observation of the behavior of a sample of gas (or a gas mixture) that is perturbed in such a way as to cause its behavior to depend on the rate of transitions into or out of excited states of the molecules. Some of the experimental methods that have been used successfully involve: (1) dispersion and absorption of high-frequency, (2) shock waves, (3) persistence of vibration in gas dynamics, (4) quenching of infrared fluorescence, (5) flash spectroscopy, (6) conversion of V to T energy by the optic-acoustic effect, and (7) thermal transpiration (analogous to thermal diffusion).

Bulk methods were employed long before beam methods, and they do not provide the detailed information obtainable by the latter. However, they have yielded an enormous amount of important data, and in many instances the bulk data are precisely what are wanted in applications! Massey (1971) has provided many useful tabulations of bulk data, from which the Tables 2-16-2 and 2-16-3 are abstracted.

2-17. NOTES

2-1. Normal Vibrations of Molecules. The material that we have presented on this subject is scattered throughout Chapters 5 through 8 of the companion volume and 1 and 2 of this volume, so it may be useful to take stock here. In Note 5-1 of the companion volume, we discuss briefly the technique of *normal-mode analysis*, definining *normal modes of vibration* and *normal coordinates*. On pp. 283–286 of the companion volume we describe experimental electron-impact studies of C_2H_2 and CO_2 and the excitation of fundamental normal modes of these molecules. The terms *infrared active* and *Raman active* are defined. Normal-mode excitation by heavy particle impact is discussed in Section 2-5, and experimental data are presented for

Table 2-16-2 Rotational Relaxation Times (τ_{relax}) and Probabilities of Deactivation per Collision (P_{deact}) for H_2 and D_2

Gas	Temp. (K)	τ_{relax} at 1 atm (10^{-8} s)	$P_{deact} \times 10^3$
H_2 (normal)	90		1.85
	207	1.2	4.65
	273	2.3	2.85
	298	1.9	3.9
H_2 (para)	90		2.8
D_2 (normal)	273	2.0	4.8

Source: (Experimental results abstracted from Massey (1971, p. 1567).

Table 2-16-3   Typical Values of the Probability of Vibrational Deactivation per Collision (P_{deact}) in Diatomic Gases

O_2	P_{deact}	N_2	P_{deact}	CO	P_{deact}	Cl_2	P_{deact}	I_2	P_{deact}
300	10^{-8}	550	1.6×10^{-8}	300	10^{-9}	300	2.6×10^{-5}	400	2.6×10^{-3}
500	10^{-7}	700	5×10^{-8}	1200	8×10^{-7}	440	8×10^{-5}	500	4.2×10^{-3}
1000	4×10^{-6}	1200	4.7×10^{-7}	2200	1.6×10^{-5}	900	3.2×10^{-3}		
1900	1.5×10^{-4}	2500	1.2×10^{-5}	4500	5×10^{-4}	1500	1.1×10^{-2}		
4000	5×10^{-3}	4200	1.5×10^{-4}						
8000	4×10^{-2}	5500	4.5×10^{-4}						

Source: Experimental values abstracted from Massey, (1971, p. 1521).

the excitation of CO_2, H_2O, and CF_4. The normal modes of vibration for tetrahedral molecules (such as CF_4) and octahedral structures (such as SF_6) are illustrated in Fig. 2-5-1, and the notation for the various modes is explained. Similar information for the CO_2 and N_2O molecules is presented in Fig. 2-5-7. Problems 2-4 through 2-6 involve the calculation of normal modes for various molecular geometries and for several macroscopic systems.

Finally, we should mention that Herzberg (1945), Landau and Lifshitz (1965), and Wilson et al. (1955) discuss the application of group theory to small oscillation theory. Also, Herzberg (1945) illustrates the normal-mode patterns for many types of molecules. Levine and Bernstein (1987, pp. 306–312) discuss $V–V'$ processes involving normal-mode vibrational states in polyatomic molecules and in the CO_2 laser.

2-2. *General Aspects of Energy-Loss Spectroscopy.* Energy-loss spectroscopy is the most general method for investigating inelastic collisions. We discussed its application to electron impact studies of rotational excitation on pp. 276–279 of the companion volume and vibrational excitation on pp. 282–286. The discussion was extended to heavy particle impact in Section 2-4 (rotational excitation) and Section 2-5 (vibrational excitation). Energy-loss spectra are displayed in Figs. 2-4-3 and 2-5-4.

We recall that the change in internal energy in an inelastic collision can be inferred from the observed decrease or increase in the translational energy, and that this change can be used to identify the transition. Buck (1988a, pp. 452–454) works out the kinematics for the inelastic collision of particles of "type 1" with particles of "type 2." The particles have initial Lab velocities v_1 and v_2, respectively, and the beams that they form intersect at a Lab angle α. Their initial CM velocities are u_1 and u_2, and the final CM velocities are u'_1 and u'_2. The asymptotic relative velocity is denoted by g before the collision, and by g' after the collision. Buck shows that

$$u'_1 = \frac{(m_2 g')}{m_1 + m_2} = u_1 \left(1 - \frac{\Delta E}{E_{tr}}\right)^{1/2}, \qquad (2\text{-}17\text{-}1)$$

where E_{tr} is the CM translational energy before the collision, and $\Delta E = E'_{int} - E_{int}$ is the gain in the internal energy of the collision system. When $\Delta E > 0$, excitation occurs, the collision is *inelastic*, and $u'_1 < u_1$. If $\Delta E < 0$, deexcitation has taken place and $u'_1 > u_1$; the collision is again *inelastic* but is frequently said to be *superelastic*.

If the collision is *elastic*, $\Delta E = 0$ and $u'_1 = u_1$. The tips of u_1 and u'_1 lie on the surface of a sphere whose radius u_1 is centered at the endpoint of the velocity of the center of mass, V_{CM}. We can easily obtain the magnitude of v'_1 from the equation

$$v'^2_1 - 2V_{CM} \cdot v'_1 - u^2_1 + c^2 = 0, \qquad (2\text{-}17\text{-}2)$$

and the CM scattering angle is given by

$$\cos \Theta = \frac{\mathbf{u}_1' \cdot \mathbf{u}_1}{u_1' u_1}. \qquad (2\text{-}17\text{-}3)$$

Depending on the relative sizes of u_1' and V_{CM}, there can be either one or two solutions to (2-17-2), the condition for two solutions being

$$(\lambda^2 - 1)\left(\frac{v_2}{v_1}\right)^2 + \lambda^2 - \left(\frac{m_1}{m_2}\right)^2 < 2\left(\frac{v_2}{v_1}\right)\left(\lambda^2 + \frac{m_1}{m_2}\right)\cos \alpha, \quad (2\text{-}17\text{-}4)$$

where $\lambda = u_1'/u_1$. This condition can be satisfied if the heavier mass is detected or if ΔE is large. If two solutions exist, the CM scattering angle is a double-valued function of the Lab angle, and the scattered particles are limited to a restricted range of the Lab angle (see p. 8 of the companion volume). The particles that are scattered throughout a range of 180° in the CM frame may be scattered throughout a much smaller Lab range, with the result that the Lab intensity is increased but the angular resolution is decreased for a given Lab acceptance angle of the detector.

A Newton diagram is shown in Fig. 2-17-1 for the case of a projectile

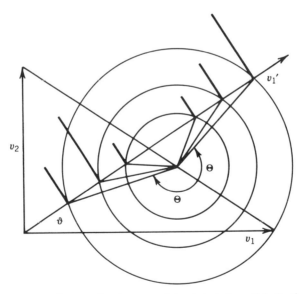

Figure 2-17-1. Newton diagram for the case of the detected particle (1, the "projectile") having a mass greater than that of its collision partner (2, the "target"). The circle of largest radius corresponds to elastic scattering, the smaller circles to inelastic transitions. The initial Lab velocities of the two particles are v_1 and v_2, and the final Lab velocities v_1' and v_2'. The Lab scattering angle is ϑ; the CM scattering angle is Θ. [Adapted from Buck (1988b, p. 526).]

that is heavier than the target. Note that the diagram is double valued and that in the case of elastic scattering, there are two CM contributions for one Lab angle. The discrete line spectra associated with the elastic and inelastic transitions are smeared out by the finite resolution of the apparatus.

Buck (1988a, b) also discusses two of the most commonly used beam arrangements: the *in-plane configuration*, in which the detector moves in the plane of v_1 and v_2, and the *out-of-plane configuration*, in which the detector moves in a plane that is perpendicular to the plane formed by v_1 and v_2. He points out that the behavior of the scattering angles is completely different in the two configurations and discusses the experimental consequences of this fact.

2-3. Molecular Orbital Theory. The motivation for the present discussion is the observation that during relatively slow collisions, the collision partners may be regarded as forming a molecule whose structure changes as the separation distance R changes, and that the outcome of the collision is determined by electronic transitions between molecular orbitals of this molecule. A *molecular orbital* (MO) is a wave function for a single-electron state of the molecule. Such a function can be constructed as a *linear combination of atomic orbitals* (LCAO), an *atomic orbital* being a single-electron wave function localized on one nucleus (Appendix I; Morrison et al., 1976; Seaton, 1962). The first matter to be settled is the choice of the most appropriate molecular wave function for a description of the collision under study. The second problem is to determine which mechanisms are responsible for the transitions that occur between energy levels of the quasi-molecule formed during the collision. Kessel et al. (1978) address both of these issues; we base our brief discussion on their treatment. First we look at the question of wave functions.

    *A. Wave Functions.** Let us assume initially that the Born–Oppenheimer approximation (pp. 290–293 of the companion volume) is valid, so that the nuclear and electronic motions can be separated. We are then dealing only with collisions for which the initial relative velocity is smaller than the orbital velocities under consideration. In an adiabatic atomic ion–atom or atom–atom collision (Sections 2-9-A and 2-9-B), for which the electronic motions are restricted to particular potential surfaces, the allowed electronic states at any fixed value of R are those of the quasi-molecule with that interatomic spacing. The molecular states change in an orderly fashion as R initially decreases from infinity to R_0 (the distance of closest approach) and finally back out to infinity. The

---

*Two of the most important methods of calculating electronic wave functions for molecular systems containing several electrons date back to 1927! The *Hund–Mulliken*, or *molecular orbital*, approach involves constructing the stationary-state wave functions as products of one-electron molecular orbitals. The alternative *Heitler–London*, or *valence bond*, approach is to approximate the wave function from orbitals based on the separated atom wave functions. See Herzberg (1950), Coulson and Lewis (1962), Morrison et al. (1976), and Bransden and Joachain (1983).

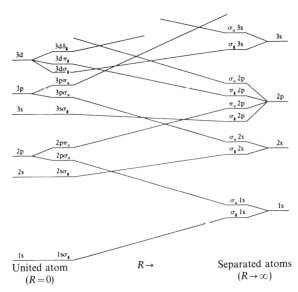

Figure 2-17-2. Correlation diagram for the one-electron H_2^+ molecular ion, showing how separated atom states connect to united atom states. [From Bransden and Joachain (1983, p. 413).]

allowed electronic energy at any R may be calculated, to a first approximation, by considering the nuclei as fixed, so that an *electronic energy curve* may be generated by making point-by-point calculations for many values of R. If the *Coulomb energy of repulsion between the nuclei* is added to this electronic energy, the result is the *potential energy function for the nuclear motion* in the Born–Oppenheimer approximation. In this fashion we can obtain a potential energy curve for each molecular electronic state and construct a set of *adiabatic molecular energy levels.* Herzberg (1950) and Bransden and Joachain (1983) provide rules for relating the atomic states of the isolated atoms $(R = \infty)$ to the molecular states of the quasi-molecule and to the atomic states of the united atom $(R = 0)$. A *correlation diagram* for the one-electron H_2^+ molecular ion is shown in Fig. 2-17-2; the molecular state notation is explained in Appendix II.

As Kessel et al. (1978) point out, adiabatic correlation diagrams may also be constructed for molecules having more than one electron. Here the 1929 von Neumann–Wigner noncrossing rule is operative; it forbids the crossing of adiabatic potential curves of the same electronic symmetry.* Multielectron adiabatic diagrams, however, generally do

---

*The crossing of potential curves is also discussed in Section 1-13-B-3 and Chapters 4 and 5. A central issue in all of these discussions is the symmetry character of the electronic wave function. In a diatomic molecule, a wave function that is symmetric with respect to interchange of the nuclei said to be *even*, or *gerade* (g). An antisymmetric wave function is *odd*, or *ungerade* (u). The von Neumann–Wigner noncrossing rule is discussed in Bransden (1983, pp. 374–375).

not explain inelastic collision phenomena satisfactorily. For example, Lichten (1963) observed that in (He$^+$, He) collisions at certain energies, the three electrons appear not to interact in pairs but to move separately in an average two-center potential produced by the two nuclei and the other electrons. Under completely adiabatic conditions, curves for states of the same parity cannot cross, but at finite velocities of approach there is a finite probability of crossing. Hence the intersecting curves are said to represent the *diabatic interactions* as opposed to the noncrossing adiabatic interactions [see Massey and Gilbody (1974, pp. 2546–2549)].

Lichten's approach is to ignore electron–electron interactions to first order, and to write the many-electron wave function as an antisymmetrized product of one-electron wave functions. It is assumed that the one-electron correlations between the united atom and separated atom limits are the same as in Fig. 2-17-2, which applies to the simplest system—the one-electron H$_2^+$ ion. Lichten's *diabatic correlations* and *diabatic energy-level diagrams* are useful in the analysis of both symmetric and antisymmetric systems. MO correlation diagrams for the latter type are shown in Kessel et al. (1978, pp. 152, 153). Many-electron molecular states are required for an accurate description of a collision, but discussions based on single-electron MOs can be useful.

An important difference between adiabatic and diabatic states is that the von Neumann–Wigner noncrossing rule does not apply to one-electron MOs or to antisymmetrized products of such states, the total energy of which is just the sum of the one-electron MOs. Now we briefly consider transitions.

B. *Transitions.* As two atoms begin to interact, they form some particular electronic state of the quasi-molecule, but a transition to some other molecular state may take place if the MOs for these states come sufficiently close together in energy for the states to be coupled. The first analysis of such behavior was performed by Morse and Stueckelberg (1929) on the H$_2^+$ system.

Actually, the use of diabatic orbitals, although in a less general fashion than Lichten's, dates back to Landau and Zener in 1932 in their discussions of nonadiabatic crossings of energy levels. Taking note of the limitations of static adiabatic states, they invoked the coupling between nuclear and electronic motions as a perturbation in the adiabatic approximation. The coupling was introduced through the radial component of **R**, and they found that if two adiabatic potential curves approach one another very closely at some value of R, this *radial coupling* can *ensure* that a transition will occur between states of the same symmetry, a sign of diabatic behavior. The main physical basis for radial coupling is the residual electrostatic interaction between pairs of electrons, associated with the breakdown of the independent electron model. Radial coupling is important when the two orbitals concerned have the same parity and the same component of electronic angular momentum along the internuclear axis (say, two Σ states). The radial coupling matrix element is displayed in (2-6-7), along with that for *rotational coupling*, which is associated with the rotation of the internuclear axis. Here the molecular axis of quantization for the electronic states changes with time, and states with values of Λ differing by $\hbar$ are coupled. Here Λ is the projection of the orbital angular momentum along the internuclear axis (see Appendix II). Thus rotational coupling can couple a Σ state with a Π state. An important paper by Russek (1971) on rotational coupling is discussed in Kessel et al. (1978, pp. 157–162) in their treatment of transitions between molecular orbitals.

When, in the coupling process, an electron is raised to a state of higher energy, *molecular orbital promotion* is said to have occurred. Lichten's original 1963 work has been extended by Fano and Lichten (1965), Lichten (1967), Barat and Lichten (1972), Eichler et al. (1976), Barat (1979), and others, and the model that has emerged is frequently known as the *Fano–Lichten molecular orbital promotion model*. Sidis and Dowek (1983) discuss the application of this method to atom–diatom collisions and display the procedure for constructing correlation diagrams based on MO surfaces for triatomic collision systems. Useful information on the matters discussed here is presented in the books by Bransden (1983) and by Bransden and Joachain (1983). Extensive research at various laboratories has demonstrated the utility of the MO promotion approach.

2-4. Rigid Ellipsoid Model of Collisional Excitation (see Section 2-9-D). Here we reproduce the development by Serri et al. (1982) of a classical hard ellipse, two-dimensional model that can reproduce the main features of rotationally and vibrationally inelastic collisions between an atom and a diatom. We begin by considering the case of rotational excitation only.

A. Vibrationally Elastic Collisions. The target diatomic molecule is represented by a hard ellipse that is initially stationary and nonrotating in the Lab frame. Its center is at the origin of the Lab coordinate system, as shown in Fig. 2-17-3. The major axis of the ellipse forms an angle γ with the x axis, and the ellipse is described by the equation

$$y^2(c^2\sin^2\gamma + a^2\cos^2\gamma) + 2xy\sin\gamma\cos\gamma(c^2 - a^2)$$
$$+ x^2(a^2\sin^2\gamma + c^2\cos^2\gamma) = a^2c^2, \tag{2-17-5}$$

where a and c are the semimajor and semiminor axis lengths, respectively. The projectile atom is represented by a point that

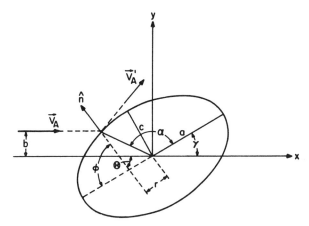

Figure 2-17-3. Excitation of a smooth, rigid ellipsoid by a point projectile proceeding from left to right with impact parameter b and velocity $\mathbf{V}_A$. [From Serri et al. (1982).]

approaches the target with velocity $\mathbf{V}_A$ parallel to the x axis; the impact parameter is b, and the x component of the point of impact is x_0. The scattering occurs in the plane of the ellipse. The impulse responsible for the rotational excitation is directed along the surface normal $\hat{n}$, and it makes an angle

$$\Theta = \tan^{-1} \frac{(a^2/c^2)[(b\cos\gamma - x_0\sin\gamma)/(x_0\cos\gamma + b\sin\gamma)]\cos\gamma + \sin\gamma}{\cos\gamma - (a^2/c^2)[(b\cos\gamma - x_0\sin\gamma)/(x_0\cos\chi + b\sin\gamma)]\sin\gamma}$$

(2-17-6)

with the x axis. Hence the momentum transfer is given by

$$p = m(V_{An} - V'_{An})\hat{n},$$
(2-17-7)

where m is the mass of the atom, and V_{An} and V'_{An} are the components of $\mathbf{V}_A$ along $\hat{n}$ before and after the collision, respectively. (The projectile velocity along the boundary is not altered in the collision.) The amount of angular momentum that is transferred is

$$L = pr,$$
(2-17-8)

where r is the moment arm. Conservation of energy requires that

$$\tfrac{1}{2}mV_{An}^2 = \tfrac{1}{2}mV'^2_{An} + \frac{\tfrac{1}{2}p^2}{M} + \frac{\tfrac{1}{2}L^2}{I},$$
(2-17-9)

where M is the mass and I is the moment of inertia of the molecule. The only nontrivial solution for p is

$$p = 2\left(\frac{1}{m} + \frac{1}{M} + \frac{r^2}{I}\right)^{-1} V_{An}. \qquad (2\text{-}17\text{-}10)$$

The Lab scattering angle is

$$\theta_{\text{Lab}} = \tan^{-1}\frac{V_{An}}{V_{As}} + \tan^{-1}\frac{V'_{An}}{V_{As}}, \qquad (2\text{-}17\text{-}11)$$

and the CM scattering angle is

$$\theta = \pi + \tan^{-1}\frac{p\sin\Theta/M|V_A|}{p\cos\Theta/M|V_A| - m/(m+M)}. \qquad (2\text{-}17\text{-}12)$$

The importance of the moment arm r as a parameter is evident: in all ellipse models, r determines the ratio L/p. We note that $r = 0$ if the point of impact is along the equator or the pole of the ellipse (either $\alpha = \pi/2$ or $\alpha = 0, \pi$). The maximum value of r is equal to

$$r_m = a - c. \qquad (2\text{-}17\text{-}13)$$

There is no first-order change in r with α near $r(\alpha) = r_m$, so there is a singular contribution to the differential scattering cross section, giving a rotational rainbow for which $L/p = r_m$. All of the remaining scattering occurs for smaller L/p, so the bright side of the rainbow extends to larger p (larger Θ) for measurements at fixed L, and to smaller L at fixed p (or Θ). For additional details of the rigid ellipsoid model as applied to rotational excitation, the reader is referred to Bosanac (1980).

B. *Vibrationally Inelastic Collisions.* Serri et al. (1982) account for the possibility of vibrational excitation by assuming that the target can absorb an amount of energy equal to the height of the first ($v = 1$) excited state above the ground ($v = 0$) vibrational state. They assume further that the probability of vibrational excitation is proportional to the square of the component of impulse ($p_\parallel = p\cos\phi$) along the major axis of the ellipse, which corresponds to the internuclear axis of the diatomic molecule. The angle ϕ has the value

$$\phi = \tan^{-1}\left[\frac{a^2}{c^2}\left(\frac{b\cos\gamma - \chi_o\sin\gamma}{b\sin\gamma + \chi_o\cos\gamma}\right)\right]. \qquad (2\text{-}17\text{-}14)$$

When an atom collides sharply with one atom of a homonuclear diatomic molecule, an amount of energy $\frac{1}{2}p_{\parallel}^2/\mu$ is transferred to vibration, where μ is the reduced mass of the diatomic. Now we take the probability of excitation $P_1(b, \gamma)$ to be proportional to this energy and write

$$P_1(b, \gamma) = Kp_{\parallel}^2, \tag{2-17-15}$$

where K is an adjustable parameter that determines the probability of a $\Delta v = 1$ transition. Hence the probability that the collision will be vibrationally elastic is $P_0(b, \gamma) = 1 - P_1(b, \gamma)$.

If vibrational excitation does occur, the equation of conservation of energy (2-17-9) is modified to

$$\tfrac{1}{2}m_A V_{An}^2 = \tfrac{1}{2}m_A V_{An}'^2 + \frac{\frac{1}{2}p^2}{M} + \frac{\frac{1}{2}L^2}{I} + \Delta E_{\text{vib}}, \tag{2-17-16}$$

where ΔE is the energy difference between the first excited and ground vibrational states. The decrease in postcollision energy that accompanies vibrational excitation thus affects the outcome of the collision by reducing the energy available for translation and rotation.

C. *Cross Sections.* To make this two-dimensional model useful for comparison with experimental data, it is extended to three dimensions by averaging over the azimuthal scattering angle. We require the major axis of the ellipse to rotate with the initial azimuthal angle in order that the scattering be confined to a plane. Then the classical differential cross section for scattering from an initial state (b, γ) into the state (Θ, J) is

$$\frac{d\sigma}{d\Omega}(o - j; \theta) \propto \frac{1}{\sin\theta} \sum b \sin\gamma \cdot \left|\frac{\partial(b, \gamma)}{\partial(\theta, j)}\right|, \tag{2-17-17}$$

where the sum is over all trajectories (b, γ) that yield a particular (Θ, J). Serri et al. (1982) describe a method for calculating cross sections. First, trajectories are computed for a large number of initial conditions (b, γ). Then the cross sections for vibrationally elastic ($v = 0$) and inelastic ($v = 1$) collisions are calculated by summing the relevant probabilities associated with trajectories with final $\Delta \Theta$ at Θ and ΔJ at J:

$$\frac{d\sigma}{d\Omega}(o - j, v; \theta) = \frac{1}{\sin\theta} \sum_i |b_i \sin\gamma_i| \cdot P_v(b_i, \gamma_i), \tag{2-17-18}$$

Of course, the final Θ and J depend on the process being studied. The calculation of P_0 requires the value of p to be calculated from (2-17-9), whereas (2-17-16) must be used in the determination of P_1.

Serri et al. (1982) state that their hard ellipse model should be useful for atom–diatom collisions characterized by (1) a potential that is mainly repulsive, (2) an impact that is rotationally sudden, and (3) a high impact energy relative to the inelasticity. They display predictions of the model and make comparisons with experimental data.

2-5. Fraunhofer Diffraction Model for Rotational Excitation. (Refer to Section 1-12-B, where this topic was introduced.) In the mid-1950s, nuclear physicists observed a $180°$ phase shift between the inelastic and elastic diffraction that occurs in the scattering of alpha particles from nuclei and that was explainable in terms of an optical Fraunhofer model. Fraunhofer diffraction theory leads to the following expression for the differential cross section for elastic scattering from a circular disk of radius R_0:

$$\frac{d\sigma}{d\omega} = (kR_0^2)^2 \frac{2F_1^2(\theta)}{\pi x^3} \cos^2(x + \tfrac{1}{4}\pi) \qquad (x \geqslant \pi/2), \qquad (2\text{-}17\text{-}19)$$

where $x = kR_0\theta$. and $F_1(\theta)$ is a form factor that accounts for the "fuzziness" of the potential (Toennies, 1982). For a target with no fuzziness, $F_1(\theta) = 1$. To describe inelastic scattering, we consider a deformed hard sphere, with surface $R = R_0[1 + \delta_2 P_2(\theta, \phi)]$, where δ_2 is a deformation parameter. For a transition $0 \to I$, we obtain

$$\frac{d\sigma}{d\omega}(0 \to I) = \begin{cases} \dfrac{\delta_I^2(kR_0)^2}{4\pi^2 x} 2F_1^2(\theta)\sin^2(x + \tfrac{1}{4}\pi), & I \text{ even} \\[2mm] \dfrac{\delta_I^2(kR_0)^2}{4\pi^2 x} 2F_1^2(\theta)\cos^2(x + \tfrac{1}{4}\pi), & I \text{ odd} \end{cases} \qquad (2\text{-}17\text{-}20)$$

for $x > I\pi/2$. Toennies (1982) points out that this approximation is similar to the DWBA (distorted wave Born approximation). We see that the Fraunhofer model predicts that for even transitions and deformations, the inelastic diffraction undulations will be shifted in phase from the elastic undulations. No phase shift results from odd transitions and deformations. The deformation parameter may be obtained from the relative amplitudes of the elastic and inelastic collisions.

In applications to rotational excitation of molecules, the full atom–molecule potential is approximated by an angle-dependent hard-wall repulsive potential with no attractive well. The sudden approximation is used for the inelastic collisions, and the full elastic scattering amplitude is replaced by the analytic Fraunhofer solution for the scattering of a wave by an obstacle. Approximate, analytic solutions for rotational cross sections can be obtained provided that the deformation of the target is not too great. Comparisons of the predictions of this model with experimental results and with other theoretical predictions appear in the papers by

Toennies (1985) and Buck et al. (1985). Faubel (1984) has discussed the theory of the Fraunhofer diffraction model in detail and applied it to the rotational excitation of small molecules by He atoms.*

2-6. *Forced Oscillator Model of Vibrational Excitation.* The forced harmonic oscillator has been discussed in detail in Rapp (1971, Chap. 24). As he points out, the goal here is to determine the final state of an oscillator that is initially in state i when it is subjected to a transient perturbation in a collision, but there are limitations on the information that we can expect to gain. In a single encounter, the oscillator may end in some final state f different from the initial state, but there is no procedure for predicting the outcome for a particular oscillator. Instead, we imagine a large ensemble of identical oscillators to be subjected to the same disturbance and calculate the *distribution of final states.* Alternatively, we phrase the result in terms of *transition probabilities* [i.e., the probabilities of transition $(P_{i \to f})$ from state i to each of the states f].

In a strictly classical treatment, only the total energy, ΔE, transferred to the oscillator is calculated, and transition probabilities do not enter the picture. Aside from variations produced by different phases of the oscillator, the projectile transfers to the target the calculated energy ΔE in every collision. If the target is oscillating initially, the amount of energy transferred depends on the phase, and the average amount $\langle \Delta E \rangle$ is the phase-averaged energy transfer per collision. If the target is not oscillating initially, ΔE is transferred to it in each collision.

By contrast, quantal and semiquantal analyses produce results phrased in terms of transition probabilities for some initial state i to final states f. One is limited to being able to say only that out of a large number N of collisions $N P_{i \to f}$ result in each of the states f. In a large number of collisions, the total amount of energy transferred to all N oscillators is

$$(\Delta E)_{\text{total}} = \sum N[(f - i)h\nu]P_{i \to f}, \qquad (2\text{-}17\text{-}21)$$

where the sum is over the final states f, starting with $f = 0$. Hence the average energy transferred to an oscillator per collision is

$$\Delta E = \sum (f - i)h\nu P_{i \to f}, \qquad (2\text{-}17\text{-}22)$$

*To show why the term *Fraunhofer diffraction* is applicable to the scattering of particles, let us consider the diffraction of light of wavelength λ from a circular aperture of radius R_0. If the distance of the source from the aperture (d_1) and the distance of the aperture from the detector (d_2) are large compared with R_0^2/λ, only a small fraction of a Fresnel zone appears through the aperture. This is the case of *Fraunhofer diffraction.* If, on the other hand, either d_1 or d_2 is $2R_0^2/\lambda$ or smaller, the aperture uncovers one or more Fresnel zones, and we have *Fresnel diffraction.* Similar considerations apply in the diffraction of light by a circular disk. [see Rossi, (1957, pp. 167, 177, 180)]. In an experiment on the scattering of *particles,* the relevant wavelength is $\lambda_{\text{deBroglie}}$ and R_0 refers to the effective radius of the target. Insertion of numerical values shows that the condition $d_1, d_2 \gg R_0^2/\lambda_{\text{deBroglie}}$ is satisfied in both nuclear and atomic collisions experiments, and the diffraction is of the Fraunhofer type.

which reduces to

$$\Delta E = \sum f h v P_{0 \to f} \tag{2-17-23}$$

if $i = 0$.

A. *Calculation of the Classical Energy Transfer.* Following Landau and Lifshitz (1960), we now derive a classical equation relating the energy transferred by an arbitrary external force $F(t)$ to a harmonic oscillator of mass m and natural frequency ω. We begin with the equation of motion

$$\ddot{x} + \omega^2 x = \frac{F(t)}{m}, \tag{2-17-24}$$

and rewrite it in the form

$$\frac{d\xi}{dt} - i\omega\xi = \frac{F(t)}{m}, \tag{2-17-25}$$

where

$$\xi = \frac{dx}{dt} + i\omega x \tag{2-17-26}$$

is a complex quantity. The solution is

$$\xi = \exp(i\omega t) \left[\int_0^t \frac{1}{m} F(t) \exp(-i\omega t) \, dt + \xi_0 \right], \tag{2-17-27}$$

where the constant of integration ξ_0 is the value of ξ at $t = 0$. The function $x(t)$ is given by the imaginary part of (2-17-26) divided by ω. Now we calculate the total energy transferred to the oscillator through all time, on the assumption that its initial energy is zero. Using the limits $-\infty$ and $+\infty$ in (2-17-27), with $\xi(-\infty) = 0$, we obtain

$$|\xi(\infty)|^2 = \frac{1}{m^2} \left| \int_{-\infty}^{\infty} F(t) \exp(-i\omega t) \, dt \right|^2. \tag{2-17-28}$$

The energy of the oscillator is

$$E = \tfrac{1}{2} m(\dot{x}^2 + \omega^2 x^2) = \tfrac{1}{2} m |\xi|^2, \tag{2-17-29}$$

so substituting $|\xi(\infty)|^2$ for $|\xi|^2$, we obtain

$$E = \frac{1}{2m} \left| \int_{-\infty}^{\infty} F(t) \exp(-i\omega t) \, dt \right|^2 \qquad (2\text{-}17\text{-}30)$$

for the energy transferred to the oscillator over all time. Note that E is the Fourier transform of $F(t)$, and that E is determined by the Fourier component of the force whose frequency is equal to the natural frequency of the oscillator.

If the duration of the force $F(t)$ is short compared with $1/\omega$, then

$$E = \frac{1}{2m} \left[\int_{-\infty}^{\infty} F(t) \, dt \right]^2. \qquad (2\text{-}17\text{-}31)$$

The force transfers a momentum $\int F \, dt$ without appreciably changing the displacement.

B. *Quantum Mechanical Treatment.* Here we merely quote two interesting and useful results of the quantal study of the forced oscillator. The first is that there is a correspondence identity between the classical and average quantum mechanical energy transfers:

$$\Delta E_{n,\,\text{class}} = \langle \Delta E_n \rangle_{\text{QM}}, \qquad (2\text{-}17\text{-}32)$$

where the subscript n denotes the vibrational mode being considered. This classical-quantal correspondence is exact only if the oscillator is forced by a time-dependent interaction potential that is linear in x. However, it holds approximately if the potential is approximately linear over the extent of the quantum oscillator wave packet during the collision. Then the wave packet will not spread, and it will remain centered on the classical value of $x(t)$, regardless of the time dependence of the driving force (Gentry, 1979).

The second result is that the transition probabilities obey a Poisson distribution (Treanor, 1965, 1966; Rapp, 1971, p. 453):

$$P_{0v} = \frac{\varepsilon^v e^{-\varepsilon}}{v!} \qquad (2\text{-}17\text{-}33)$$

Here v is the vibrational quantum number, and $\varepsilon = \Delta E / h\nu$ is the classical energy transfer in units of the vibrational quantum.

The use of the forced oscillator model in classical, semiclassical, and quantal studies of specific collision systems has been discussed by Gentry and Giese (1975), Toennies (1982), Ellenbroek and Toennies (1982), Skodje et al. (1977, 1983), Gentry (1984), Gierz et al. (1985a, b), and Niedner et al. (1987).

2-7. Decent and Indecent Methods. These methods,* developed by Gentry and Giese, involve semiclassical approaches based on classical–quantal correspondences (Gentry, 1979), and the present discussion is closely related to Note 2-6. DECENT starts with the calculation of a classical trajectory for each set of initial conditions (CM energy, impact parameter, relative orientation) on a nearly exact potential surface. One obtains a CM deflection angle for the projectile as well as an angular momentum and total energy transferred to the target. This total energy is then apportioned into rotation and vibrational energy for each trajectory. Next, Monte Carlo averaging over impact parameter and initial orientation, for fixed E_{CM}, yields the total differential cross section, the rotational energy transfer, and the vibrational energy transfer as a function of scattering angle. These results are all classical. Finally, the classical energy transfer is distributed among the quantum states of the target. In the case of vibrational excitation (the main application), the forced oscillator model prediction of a Poisson distribution is used. Hermann et al. (1978) discuss the distribution of both the vibrational and rotational energy transfer in their paper on (H^+, H_2) scattering.

The INDECENT method differs from DECENT in the choice of the initial translational boundary condition (dR/dT at $-\infty$). This difference and its consequences are discussed by Gentry (1979).

2-18. PROBLEMS

Solutions for many of the following problems appear in the references indicated.

2-1. Why do we not speak of rotational excitation of an atom?

2-2. Consider each of the following projectiles $(H^+, H, H_2, NaCl, CH_4)$ colliding with each of the following targets $(H_2, N_2, CO, CO_2, H_2O, HCl, CH_4, C_2H_2, SF_6)$. With which pairs of collision partners can

    (a) rotational excitation occur?

    (b) vibrational?

    (c) electronic? Give reasons for your answers.

2-3. **(a)** Use equations (2-2-1) and (2-2-2) to calculate E_v and E_J for the levels of H_2 and N_2 shown in Fig. 2-2-1, and compare your results with the values indicated in the figure.

*DECENT is the acronym for "Distribution of Exact Classical Energy Transfer," and INDECENT that for "Intermediate Speed Decent Approximation."

(b) Use the following equation from Problem 5-2 of the companion volume to calculate a few vibrational energies for these molecules:

$$E = -V_0 + \frac{L(L+1)\hbar^2}{2\mu d^2} + \hbar\omega_0\left(n+\frac{1}{2}\right) - \frac{3}{2}\frac{\hbar^3 L(L+1)(n+\frac{1}{2})}{\mu^2 d^4 \omega_0}.$$

(c) In naturally occurring H_2 and N_2, what are the abundance ratios of the ortho and para forms of the molecules; that is, what is $[o]/[p]$ for both H_2 and N_2? Can these ratios be altered in the laboratory, and if so, can experiments be carried out on each form separately? [See Rapp (1971, pp. 389–393).]

2-4. Solve each of the following problems by elementary means, not invoking formal small-oscillations theory (see Note 5-1 of the companion volume and Problem 2-6).

(a) A rigid, horizontal bar of length L and mass m is supported at each end by a vertical spring embedded at its lower end in a massive horizontal support. Each spring has a force constant k. Depress one end of the bar by a small distance d and release it from rest. Solve for the motion of the system, identifying the normal modes and frequencies. Sketch the normal modes. [See Chen (1974, p. 2).]

(b) Now the bar in part (a) is supported at its ends by two threads of length d fastened to a rigid ceiling. Calculate the frequencies and amplitudes of the normal modes if the bar receives a small impulse **p** at one end in a direction perpendicular to the rod and to the thread. [See Chen (1974, p. 3).]

(c) A block of mass M lies on a frictionless horizontal support and is connected by two springs (force constants k_1 and k_2) to rigid supports on each side. Displace the block slightly from its equilibrium position, and solve for the frequency of vibration. Now solve for the new frequency and amplitude of vibration if a mass m is dropped vertically on M as it passes through its equilibrium position. [See Chen (1974, p. 11).]

(d) Two particles of mass m are connected by a spring (k_1), and each particle is connected on its outboard side to a rigid wall by a spring (k_2). The two particles and the three springs are in a linear equilibrium configuration and are free to execute longitudinal one-dimensional oscillations. Find the normal modes of vibration, and illustrate the relative magnitudes and phases of the particle displacements with drawings. [See Chen (1974, p. 9).]

(e) Three rigid spheres, with masses $m, 2m, m$ from left to right, are equally spaced along a horizontal line. The center sphere is connected with a massless spring (k) to each of the outer spheres. Describe all of the normal modes of the system and say what you can about the relative frequencies. [See Cronin et al. (1979, p. 10).]

2-5. Solve the following problems by elementary means, as in Problem 2-4.

(a) Determine the normal-mode frequencies of vibration of a symmetric linear triatomic molecule ABA. Assume that the potential energy of the molecule depends only on the distances AB and BA and the angle ABA.

(b) Repeat part (a) for a nonlinear, symmetric, triangular molecule ABA. In the equilibrium position, B is located at the origin of coordinates, and AB and BA lie in the $X-Y$ plane, each line segment making an angle α with respect to the Y axis.

(c) Repeat part (a) for an unsymmetric linear molecule ABC. [See Landau and Lifshitz (1960, pp. 70–74).]

2-6. **(a)** Write out a summary of the classical small-oscillations theory as presented by Goldstein (1980, pp. 243–258). Then do the same for the quantal treatment as described by Wilson et al. (1955).

(b) Use the formal small-oscillations theory of classical mechanics to solve for the resonant frequencies and normal modes of the linear, symmetric, triatomic molecule ABA. [See Goldstein (1980, pp. 258–263).]

2-7. In Section 2-6-A-2 we distinguished between *violent collisions* ($\tau \equiv E\vartheta \geqslant 3\,\text{keV-deg}$) and *soft collisions* ($\tau < 3\,\text{keV-deg}$). Point out the physical differences between the two regimes of scattering, and show why the parameter τ is appropriate here. [See Andersen and Nielsen (1982).]

2-8. Refer to the paper by Kvale et al. (1985) discussed in Section 2-6-A-3. Write out an explanation of their data acquisition method and deconvolution techniques for obtaining excitation cross sections from the raw energy-loss data.

2-9. Derive equation (2-6-6) for the elastic energy loss.

2-10. Prepare a Grotrian diagram for the H atom, showing and labeling the Lyman, Balmer, Paschen, Brackett, and Pfund series of lines. Discuss the origin of the fine structure, the hyperfine structure, the Lamb shift, and the 21-cm line used in radio astronomy.

2.11. By taking cascading into account, convert the emission cross sections given in Van Zyl et al. (1985) to excitation cross sections.

2-12. Consider a beam of H atoms traveling with 50-keV energy in the laboratory and emitting Lyman-alpha radiation. Calculate the wavelength and frequency observed by a Lab observer at the following angles with respect to the beam: $\theta = 0°, 45°, 90°, 135°, 180°$. Solve the problem twice, using both the nonrelativistic and relativistic equations for the Doppler effect. (See pp. 612–613 of the companion volume.

2-13. Show that a 50-keV H atom moving across a 0.5-G magnetic field experiences a 1.5-V/cm electric field.

2-14. Use the adiabatic hypothesis (Section 2-9-A) to show that the probability of vibrational excitation of a molecule is small if

$$\frac{a}{d}\left(\frac{hvm}{\pi kTM}\right)^{1/2} \gg 1, \tag{2-18-1}$$

where the interaction range a may be taken to be on the order of gas-kinetic radii. Here v and d are the frequency and amplitude of the vibration that is excited, M the reduced mass associated with it, and m the mass of the lighter molecule involved in the collision. [See Massey (1971, p. 1465).]

2-15. Use the adiabatic hypothesis (Section 2-9-A) to show that rotational excitation of a molecule is unlikely if

$$\frac{ah}{8\pi^2 R^2 Mv} \gg 1, \tag{2-18-2}$$

where R is the molecular separation (same order of magnitude as a), M is the reduced mass of the internal molecular motion, and v is the velocity of approach. Note that the energy of the rotational quantum is $hv = I\omega^2/2$, while the moment of inertia of the molecule is $I = MR^2$. Except for the lightest molecules at very low temperatures, the wavelength h/Mv is very small, so the condition (2-18-2) is not satisfied. [See Massey (1971, p. 1466).]

2-16. Show that equation (2-9-27) for the rate coefficient is consistent with the following equation [(2-12-2) of the companion volume].

$$B_{ij}d^3v_i = \iint f_j(\mathbf{v}_j')f_i(\mathbf{v}_i')v_0'I_s(\Theta, v_0') \, d\Omega_{CM} \, d^3v_j' d^3v_i'$$

$$- \iint f_j(\mathbf{v}_j)f_i(\mathbf{v}_i)v_0I_s(\Theta, v_0) \, d\Omega_{CM} \, d^3v_j \, d^3v_i.$$

2-17. (a) List all of the reactions that can occur when an H atom collides with a DT molecule at impact energies below $100\,eV$.

(b) Repeat for H^+ projectiles.

(c) Outline methods for measuring the cross sections of all of the dissociative reactions in parts (a) and (b).

(d) Repeat (c) for a H_2 target.

2-18. A beam of H_2 molecules traveling through a vacuum at a Lab energy of

50 keV is undergoing dissociation that is isotropic in H_2 rest frame. What is the angular distribution of the dissociation products in the Lab frame? [See Sard (1970, pp. 168–181).] (*Note:* One of the difficulties in measurements of dissociation cross sections is the collection of "all" of the fragments from the dissociating structures.)

2-19. Show that Fig. 2-12-1 is consistent with the adiabatic hypothesis.

2-20. Work out the relations for $P\tau$ on p. 303 of Levine and Bernstein (1987).

2-21. Show that the relaxation of the two-state system discussed in the text is described by equations (2-13-6) and (2-13-7). [See Levine and Bernstein (1987, p. 302).]

2-22. **(a)** Justify (2-14-1) and (2-14-2) by discussing the physics of the harmonic oscillator model of vibrational energy transfer.

   (b) Consider the collinear collision of an atom A with a diatomic molecule BC represented by a harmonic oscillator. Show that in the spectator (or sudden) limit described in connection with (2-14-1), the vibrational energy acquired by BC is determined by a dimensionless mass ratio. [See Levine and Bernstein (1987, pp. 312–314).]

2-23. What is the probability that a molecule makes

   (a) exactly two collisions within a flight length x in a gas?

   (b) Exactly n collisions? [See Yamasaki and Leone (1989, pp. 967, 975–976).]

2-24. **(a)** Massey (1971, pp. 1540–1544) uses the distorted-wave method (see Section 3-23-F of the companion volume and an exponential interaction to calculate the deactivation probability per collision (P_{deact}) for head-on encounters of an atom A with a molecule BC. If BC is initially in its lowest excited state, an approximate result is easily obtained:

$$P_{deact}(1 \to 0) \sim \exp\left[\frac{-(54\pi^4 M v^2}{a^2 kT}\right)^{1/3}\right].$$

$$(2\text{-}18\text{-}3)$$

   Here M is the reduced mass for the relative motion, v the frequency associated with the transition, T the gas temperature, and a the range of the exponentially decreasing interaction. Show that this result provides a reasonable basis for plotting vibrational relaxation data in the form ($\log \tau_{relax}$) versus $T^{-1/3}$, and plot the data in Table 2-16-3 accordingly.

   (b) Use the adiabatic hypothesis to discuss the P_{deact} data. Explain why the values for I_2 are so much higher than those for the other gases and why P_{deact} rises so sharply with increasing T in each gas.

2-25. Derive the equations for the kinematics of inelastic collisions in Note 2-2.

2-26. Derive the equations for the rigid ellipsoid model in Note 2-4.

2-27. Fill in the intermediate steps in the derivation of the energy transfer in the forced oscillator model (Note 2-6).

2-28. A one-dimensional harmonic oscillator is at rest in its equilibrium position ($x = 0$) at $t = 0$, and it is being driven by a force $F(t)$. Calculate $x(t)$ for each of the following cases:

   (a) $F = F_0$ (a constant force);
   (b) $F = at$;
   (c) $F = F_0 e^{-\alpha t}$;
   (d) $F = F_0 e^{-\alpha t} \cos \beta t$. Here a, α, and β are constants. [See Landau and Lifshitz (1960, p. 64).]

2-29. Up to time $t = 0$, a one-dimensional oscillator is at rest at its equilibrium position $x = 0$. Calculate the final amplitude, x, under a force that is zero for $t < 0$, $F_0 t/T$ for $0 < t < T$, and F_0 for $t > T$. [See Landau and Lifshitz (1960, p. 64).]

2-30. Repeat Problem 2-29 for a constant force F_0 that is applied for a time T (i.e., for a rectangular pulse). [See Landau and Lifshitz (1960, p. 64).]

See also the following problems in the companion volume.

   Chapter 1: 5, 8, 11–13
   Chapter 2: 13–15, 18, 22
   Chapter 3: 23
   Chapter 4: 1–5, 21–25
   Chapter 5: 1–12, 15–17, 19–23
   Chapter 6: 2, 6–13, 21, 22
   Chapter 7: 1–6, 8–11
   Chapter 8: 1, 26–34, 36, 38, 39, 41–43

The following notes from the accompanying volume are also relevant to the present discussion.

   Chapter 2: 3, 6
   Chapter 4: 1
   Chapter 5: 1
   Chapter 6: 1, 2
   Chapter 7: 1, 2
   Chapter 8: 2, 4, 5

REFERENCES

Ali, A., and P. J. Dagdigian (1987), *J. Chem. Phys.* **87**, 6915.

Andersen, N., and S. E. Nielsen (1982), "Direct Excitation in Atomic Collisions: Studies of Quasi-One-Electron Systems," *AAMP* **18**, 265.

Andersen, N., M. Vedder, A. Russek, and E. Pollack (1980), *Phys. Rev. A* **21**, 782.

Andersen, N., J. W. Gallagher, and I. V. Hertel (1988), "Collisional Alignment and Orientation of Atomic Outer Shells. I. Direct Excitation by Electron and Atom Impact," *Phys. Rep.* **165**, 1.

Andersen, N., J. Broad, E. Campbell, J. W. Gallagher, and I. V. Hertel (1993a), "Collisional Alignment and Orientation of Atomic Outer Shells. II. Quasimolecular Excitation," *Phys. Rep.* to be published.

Andersen, N., K. Bartschat, J. Broad, J. W. Gallagher, and I. V. Hertel (1993b), "Collisional Alignment and Orientation of Atomic Outer Shells. III. Spin-Resolved Excitation,' *Phys. Rep.* to be published.

Andresen, P., D. Häusler, and H. W. Lülf (1984), *J. Chem. Phys.* **81**, 571.

Auerbach, D. J. (1988), "Velocity Measurements by Time-of-Flight Methods," Chap. 14 in G. Scoles, Ed., *Atomic and Molecular Beam Methods*, Vol. 1, Oxford University Press, New York.

Barat, M. (1979), *Commun. At. Phys.* **8**, 73.

Barat, M., and W. Lichten (1972), *Phys. Rev. A*, **6**, 211.

Barat, M., J. C. Brenot, D. Dhuicq, J. Pommier, V. Sidis, R. E. Olson, E. J. Shipsey, and J. C. Browne (1976), *J. Phys. B* **9**, 269.

Barnett, C. F. (1990), "Collisions of H, H_2, H_3, and Li Atoms and Ions with Atoms and Molecules," in *Atomic Data for Fusion.* Vol. 1, ORNL-6086, Oak Ridge National Laboratory, Oak Ridge, Tenn.

Beneventi, L., P. Casavecchia, F. Vecchiocattivi, G. G. Volpi, U. Buck, Ch. Lauenstein, and R. Schinke (1988), *J. Chem. Phys.* **89**, 4671.

Bergmann, K. (1988), "State Selection by Optical Methods," Chap. 12 in G. Scoles, Ed., *Atomic and Molecular Beam Methods*, Vol. 1, Oxford University Press, New York.

Bergmann, K., U. Hefter, and J. Witt (1980), *J. Chem. Phys.* **72**, 4777.

Bergmann, K., U. Hefter, A. Mattheus, and J. Witt (1981), *Chem. Phys. Lett.* **78**, 61.

Bernstein, R. B., Ed. (1979), *Atom–Molecule Collision Theory*, Plenum Press, New York.

Bernstein, R. B. (1982), *Chemical Dynamics via Molecular Beam and Laser Techniques*, Oxford University Press, New York.

Bischof, G., V. Hermann, J. Krutein, and F. Linder (1982), *J. Phys. B* **15**, 249.

Blum, K., and H. Kleinpoppen (1979), "Electron–Photon Angular Correlation in Atomic Physics," *Phys. Rep.* **52**, 203.

Blum, K., and H. Kleinpoppen (1983), "Angular Correlation Studies of Heavy-Particle Impact Excitation of Atoms," *Phys. Rep.* **96**, 251.

Bobashev, S. V. (1978), "Quasimolecular Interference Effects in Ion–Atom Collisions," *AAMP* **14**, 341.

Bogdanov, A. V., G. V. Dubrovskii, Yu. E. Gorbachev, and V. M. Strelchenya (1989), "Theory of Vibrational and Rotational Excitation of Polyatomic Molecules," *Phys. Rep.* **181**, 121.

Bosanac, S. (1980), *Phys. Rev. A* **22**, 2617.

Bransden, B. H. (1983), *Atomic Collision Theory*, 2nd ed., Benjamin/Cummings, Menlo Park, Calif.

Bransden, B. H., and C. J. Joachian (1983), *Physics of Atomic and Molecules*, Longman, Harlow, Essex, England.

Brenot, J. C., D. Dhuicq, J. P. Gauyacq, J. Pommier, V. Sidis, M. Barat, and E. Pollack (1975a), *Phys. Rev. A* **11**, 1245.

Brenot, J. C., J. Pommier, D. Dhuicq, and M. Barat (1975b), *J. Phys. B* **8**, 448.

Brunner, T. A., and D. E. Pritchard (1982), "Fitting Laws for Rotationally Inelastic Collisions," in K. P. Lawley, Ed., *Dynamics of the Excited State*, Wiley, New York, p. 589.

Buck, U. (1988a), "General Principles and Methods," Chap. 18 in G. Scoles, Ed., *Atomic and Molecular Beam Methods*, Vol. 1, Oxford University Press, New York.

Buck, U. (1988b), "Inelastic Scattering I: Energy Loss Methods," Chap. 21 in G. Scoles, Ed., *Atomic and Molecular Beam Methods*, Vol. 1, Oxford University Press, New York.

Buck, U., K. H. Kohl, A. Kohlhase, M. Faubel, and V. Staemmler (1985), *Mol. Phys.* **55**, 1255.

Bussert, W., and S. R. Leone (1987), *Chem. Phys. Lett.* **138**, 276.

Bussert, W., D. Neuschafer, and S. R. Leone (1987), *J. Chem. Phys.* **87**, 3833.

Carlson, T., M. O. Krause, and S. T. Manson (1991), *X-Ray and Inner-Shell Processes*, American Institute of Physics, Colchester, Vt.

Chen, M. (1974), *Berkeley Physics Problems with Solutions*, Prentice Hall, Englewood Cliffs, N.J.

Child, M. S. (1974), *Molecular Collision Theory*, Academic Press, New York.

Cooks, R. G., Ed., (1978), *Collision Spectroscopy*, Plenum Press, New York.

Coulson, C. A., and J. T. Lewis (1962), "Chemical Binding," in D. R. Bates, Ed., *Quantum Theory*, Vol. 2, Academic Press, New York.

Cronin, J. A., D. F. Greenberg, and V. L. Telegdi (1979), *University of Chicago Graduate Problems in Physics with Solutions*, Addison-Wesley, Reading, Mass.

Cruse, H. W., P. J. Dagdigian, and R. N. Zare (1973), *Faraday Discuss. Chem. Soc.* **55**, 277.

Dagdigian, P. J. (1988), "Inelastic Scattering II: Optical Methods," Chap. 23 in G. Scoles, Ed., *Atomic and Molecular Beam Methods*, Vol. 1, Oxford University Press, New York.

Dagdigian, P. J. (1989), *J. Chem. Phys.* **90**, 2617.

Dagdigian, P. J., and S. J. Bullman (1985), *J. Chem. Phys.* **82**, 1341.

Dagdigian, P. J., B. E. Forch, and A. W. Miziolek (1988), *Chem. Phys. Lett.* **148**, 299.

Demkov, Yu, N. (1964), *Sov. Phys. JETP* **18**, 138.

Dexheimer, S. L., T. A. Brunner, and D. E. Pritchard (1983), *J. Chem. Phys.* **79**, 5206.

Dickinson, A. S., and D. Richards (1982), "Classical and Semiclassical Methods in Inelastic Heavy Particle Collisions," *AAMP* **18**, 165.

Dowek, D., D. Dhuicq, J. Pommier, V. N. Tuan, V. Sidis, and M. Barat (1981), *Phys. Rev. A* **24**, 2445.

Dowek, D., D. Dhuicq, V. Sidis, and M. Barat (1982), *Phys. Rev. A* **26**, 746.

Dowek, D., D. Dhuicq, and M. Barat (1983), *Phys. Rev. A* **28**, 2838.

Düren, R. (1988), "Scattering Experiments with Laser-Excited Atomic Beams," Chap. 26 in G. Scoles, Ed., *Atomic and Molecular Beam Methods*, Vol. 1, Oxford University Press, New York.

Edwards, A. K., and R. M. Wood (1982), *J. Chem. Phys.* **76**, 2938.

Eichler, J., U. Wille, B. Fastrup, and K. Taulbjerg (1976), *Phys. Rev. A*, **14**, 707.

Ellenbroek, T., and J. P. Toennies (1982), *Chem. Phys.* **71**, 309.

Eriksen, F. J., D. H. Jaecks, W. de Rijk, and J. Macek (1976), *Phys. Rev. A* **14**, 119.

Eriksen, F. J., and D. H. Jaecks (1978), *Phys. Rev. A* **17**, 1296.

Evans, R. D. (1955), *The Atomic Nucleus*, McGraw-Hill, New York.

Ezell, R. L., A. K. Edwards, and R. M. Wood (1984), *J. Chem. Phys.* **81**, 1341.

Fano, U., and W. Lichten (1965), *Phys. Rev. Lett.* **14**, 627.

Fano, U., and J. H. Macek (1973), *Rev. Mod. Phys.* **45**, 553.

Faubel, M. (1983), "Vibrational and Rotational Excitation in Molecular Collisions," *AAMP* **19**, 345.

Faubel, M. (1984), *J. Chem. Phys.* **81**, 5559.

Faubel, M. (1985), "Low Energy Atom Collisions," in H. Kleinpoppen, J. S. Briggs, and H. O. Lutz, Eds., *Fundamental Processes in Atomic Collision Physics*, Plenum Press, New York, p. 503.

Faubel, M., and G. Kraft (1986), *J. Chem. Phys.* **85**, 2671.

Faubel, M., and J. P. Toennies (1978), "Scattering Studies of Rotational and Vibrational Excitation of Molecules," *AAMP* **13**, 229.

Faubel, M., K. H. Kohl, and J. P. Toennies (1980), *J. Chem. Phys.* **73**, 2506.

Faubel, M., K. H. Kohl, J. P. Toennies, K. T. Tang, and Y. Y. Yung (1982), *Faraday Discuss. Chem. Soc.* **73**, 205.

Faubel, M., K. H. Kohl, J. P. Toennies, and F. A. Gianturco (1983), *J. Chem. Phys.* **78**, 5629.

Flygare, W. H. (1968), *Acc. Chem. Res.* **1**, 121 (1968).

Friedrich, B., F. A. Gianturco, G. Niedner, M. Noll, and J. P. Toennies (1987a), *J. Phys. B* **20**, 3725.

Friedrich, B., G. Niedner, M. Noll, and J. P. Toennies (1987b), *J. Chem. Phys.* **87**, 5256.

Fu, P., J. D. Newman, D. S. Thomson, and T. F. Gallagher (1989), *Phys. Rev. A* **40**, 2745.

Fuchs, M., and J. P. Toennies (1986), *J. Chem. Phys.* **85**, 7062.

Gallagher, T. F. (1992), "Resonant Collisional Energy Transfer Between Rydberg Atoms," *Phys. Rep.* **210**, 319.

Gentry, W. R. (1979), "Vibrational Excitation II: Classical and Semiclassical Methods," in R. B. Bernstein, Ed., *Atom–Molecule Collision Theory*, Plenum Press, New York, p. 391.

Gentry, W. R. (1984), *J. Chem. Phys.* **81**, 5737.

Gentry, W. R. (1985), *ICPEAC XIV*, Stanford, Calif., p. 13.

Gentry, W. R. (1988), "Low-Energy Pulsed Beam Sources," Chap. 3 in G. Scoles, Ed., *Atomic and Molecular Beam Methods*, Vol. 1, Oxford University Press, New York.

Gentry, W. R., and C. F. Giese (1975), *J. Chem. Phys.* **63**, 3144.

Gentry, W. R., and C. F. Giese (1977), *J. Chem. Phys.* **67**, 5389.

Gierz, U., J. P. Toennies, and M. Wilde (1984), *Chem. Phys. Lett.* **110**, 115.

Gierz, U., M. Noll, and J. P. Toennies (1985a), *J. Chem. Phys.* **82**, 217.

Gierz, U., M. Noll, and J. P. Toennies (1985b), *J. Chem. Phys.* **83**, 2259.

Goldberger, A. L., D. H. Jaecks, M. Natarajan, and L. Fornari (1984), *Phys. Rev. A* **29**, 77.

Goldstein, H. (1980), *Classical Mechanics*, 2nd ed., Addison-Wesley, Reading, Mass.

Gole, J. L. (1985), "Probing Ultrafast Energy Transfer Among the Excited States of Small High Temperature Molecules," in A. Fontijn, Ed., *Gas Phase Chemiluminescence and Chemiionization*, Elsevier, New York, p. 253.

Greene, C. H., and R. N. Zare (1983), *J. Chem. Phys.* **78**, 6741.

Gundlach, G., E. L. Knuth, H. G. Rubahn, and J. P. Toennies (1988), *Chem. Phys.* **124**, 131.

Hall, G., K. Liu, M. J. McAuliffe, C. F., Giese, and W. R. Gentry (1984), *J. Chem. Phys.* **81**, 5577.

Hall, G., C. F. Giese, and W. R. Gentry (1985), *J. Chem. Phys.* **83**, 5343.

Hall, G., K. Liu, M. J. McAuliffe, C. F. Giese, and W. R. Gentry (1986), *J. Chem. Phys.* **84**, 1402.

Hasted, J. B. (1972), *Physics of Atomic Collisions*, 2nd ed., Elsevier, New York.

Hefter, U., and K. Bergmann (1988), "Spectroscopic Detection Methods," Chap. 9 in G. Scoles, Ed., *Atomic and Molecular Beam Methods*, Vol. 1, Oxford University Press, New York.

Hefter, U., P. Jones, J. Witt, K. Bergmann, and R. Schinke (1981), *Phys. Rev. Lett.* **46**, 915.

Hege, U., and F. Linder (1985), *Z. Phys. A* **320**, 95.

Hermann, V., H. Schmidt, and F. Linder (1978), *J. Phys. B* **11**, 493.

Herzberg, G. (1945), *Infrared and Raman Spectra of Polyatomic Molecules*, Van Nostrand Reinhold, New York.

Herzberg, G. (1950), *Spectra of Diatomic Molecules*, 2nd ed., D. Van Nostrand, Princeton, N.J.

Hiskes, J. R. (1961), *Phys. Rev.* **122**, 1207.

Hiskes, J. R. (1962), *Nucl. Fusion* **2**, 38.

Hoffbauer, M. A., S. Burdenski, C. F. Giese, and W. R. Gentry (1983), *J. Chem. Phys.* **78**, 3832.

Hollstein, M., J. R. Sheridan, J. R. Peterson, and D. C. Lorents (1969) *Phys. Rev.* **187**, 118.

Jaecks, D. H., C. Englehardt, and O. Yenen (1987), *ICPEAC XV*, Brighton, England, p. 403.

Jihua, G., A. Ali, and P. J. Dagdigian (1986), *J. Chem. Phys.* **85**, 7098.

Jones, P. L., U. Hefter, A. Matheus, J. Witt, K. Bergmann, W. Müller, W. Meyer, and R. Schinke (1982), *Phys. Rev. A* **26**, 1283.

Jones, P. L., E. Gottwald, and K. Bergmann, (1983), *J. Chem. Phys.* **78**, 3838.

Kanter, E. P. (1981), *Comments At. Mol. Phys.* **11**, 63.

Kanter, E. P. (1983), "The Role of Excited States of Molecular Ions in Structure Studies with High Energy Collisions," in J. Berkowitz and K.-O. Groeneveld, Eds., *Molecular Ions*, Plenum Press, New York.

Kaplan, S., G. A. Paulikas, and R. V. Pyle (1961), *Phys. Rev. Lett.* **7**, 96.

Kaplan, S., G. A. Paulikas, and R. V. Pyle, (1963), *Phys. Rev.* **131**, 2574.

Keller, H. M., M. Külz, R. Setzkorn, G. Z. He, K. Bergmann, and H. G. Rubahn (1992), *J. Chem. Phys.* **96**, 8819.

Kelly, J. F., M. Harris and A. Gallagher (1988a), *Phys. Rev. A* **38**, 1225.

Kelly, J. F., M. Harris, and A. Gallagher (1988b), *Phys. Rev. A* **37**, 2354.

Kessel, Q. C. (1969), "Coincidence Measurements," *Case Stud. At. Collision Phys.* **1**, 401.

Kessel, Q. C., and B. Fastrup (1974), "The Production of Inner-Shell Vacancies in Heavy Ion–Atom Collisions," *Case Stud. At. Collision Phys.* **3**, 137.

Kessel, Q. C., E. Pollack, and W. W. Smith (1978), "Inelastic Energy Loss: Newer Experimental Techniques and Molecular Orbital Theory," in R. G. Cooks, Ed., *Collision Spectroscopy*, Plenum Press, New York, p. 147.

Kimura, M., and N. F. Lane (1989), "Low Energy, Heavy Particle Collisions: A Close Coupling Treatment," *AAMOP* **26**, 79.

Knuth, E. L., H. G. Rubahn, J. P. Toennies, and J. Wanner (1986), *J. Chem. Phys.* **85**, 2653.

Krajnovich, D. J., C. S. Parmenter, and D. L. Catlett (1987), *Chem. Rev.* **87**, 237.

Krutein, J., and F. Linder (1977a), *J. Phys. B* **10**, 1363.

Krutein, J., and F. Linder (1977b), *Chem. Phys. Lett.* **51**, 597.

Krutein, J., and F. Linder (1979), *J. Chem. Phys.* **71**, 599.

Krutein, J., G. Bischof, F. Linder, and R. Schinke (1979), *J. Phys. B* **12**, L-57.

Kvale, T. J., D. G. Seely, D. M. Blankenship, E. Redd, T. J. Gay, M. Kimura, E. Rille, J. L. Peacher, and J. T. Park (1985), *Phys. Rev. A* **32**, 1369.

Lambert, J. D. (1977), *Vibrational and Rotational Relaxation in Gases*, Oxford University Press, Oxford.

Landau, L. D., and E. M. Lifshitz (1960), *Mechanics*, Addison-Wesley, Reading, Mass.

Landau, L. D., and E. M. Lifshitz (1965), *Quantum Mechanics: Nonrelativistic Theory*, 2nd ed, Pergamon Press, Oxford.

Leone, S. R. (1988), in J. C. Whitehead, Ed., *Selectivity in Chemical Reactions*, Kluwer Academic Publishers, Dordrecht, The Netherlands, p. 245.

Levine, R. D., and R. B. Bernstein (1987), *Molecular Reaction Dynamics and Chemical Reactivity*, Oxford University Press, New York.

Lichten, W. (1963), *Phys. Rev.* **131**, 229.

Lichten, W. (1967), *Phys. Rev.* **164**, 131.

Linder, F. (1979), *ICPEAC XI*, Kyoto, p. 535.

Liu, K., G. Hall, M. J. McAuliffe, C. F. Giese, and W. R. Gentry (1984), *J. Chem. Phys.* **80**, 3494.

Macek, J., and D. H. Jaecks (1971), *Phys. Rev. A* **4**, 2288.

Magill, P. D., B. Stewart, N. Smith, and D. E. Pritchard (1988), *Phys. Rev. Lett.* **60**, 1943.

Magill, P. D., T. P. Scott, N. Smith, and D. E. Pritchard (1989), *J. Chem. Phys.* **90**, 7195.

Martin, S. J., V. Heckman, E. Pollack, and R. Snyder (1987), *Phys. Rev. A* **36**, 3113.

Massey, H. S. W. (1971), *Electronic and Ionic Impact Phenomena*, 2nd ed., Vol. 3, Oxford University Press, Oxford.

Massey, H. S. W., and H. B. Gilbody (1974), *Electronic and Ionic Impact Phenomena*, 2nd ed., Vol. 4, Oxford University Press, Oxford.

Massey, H. S. W., E. W. McDaniel, and B. Bederson, Series Eds. (1982–1984), *Applied Atomic Collision Physics*, 5 vols., Academic Press, New York.

Mattheus, A., A. Fischer, G. Zeigler, E. Gottwald, and K. Bergmann (1986), *Phys. Rev. Lett.* **56**, 712.

McClure, G. W., and J. M. Peek (1972), *Dissociation in Heavy Particle Collisions*, Wiley, New York.

McDaniel, E. W. (1964), *Collision Phenomena in Ionized Gases*, Wiley, New York.

Meyer, G., and J. P. Toennies (1981), *J. Chem. Phys.* **75**, 2753.

Meyer, G., and J. P. Toennies (1982), *J. Chem. Phys.* **77**, 798.

Miller, D. R. (1988), "Free Jet Sources," Chap. 2 in G. Scoles, Ed., *Atomic and Molecular Beam Methods*, Vol. 1, Oxford University Press, New York.

Montgomery, D. L., and D. H. Jaecks (1983), *Phys. Rev. Lett.* **51**, 1862.

Morgenstern, R., M. Barat, and D. C. Lorents (1973), *J. Phys. B* **6**, L-330.

Morrison, M. A., T. L. Estle, and N. F. Lane (1976), *Quantum States of Atoms, Molecules, and Solids*, Prentice Hall, Englewood Cliffs, N.J.

Morse, P. M., and E. C. G. Stueckelberg (1929), *Phys. Rev.* **33**, 932.

Mueller, D. W., and D. H. Jaecks (1985), *Phys. Rev. A* **32**, 2650.

Nakamura, H. (1987), *ICPEAC XV*, Brighton, England, p. 413.

Niedner, G., M. Noll, and J. P. Toennies (1987), *J. Chem. Phys.* **87**, 2067.

Noll, M. (1988), *Phys. Scr.* **T-23**, 151.

Olsen, J. Ø., T. Andersen, M. Barat, Ch. Courbin-Gaussorgues, V. Sidis, J. Pommier, J. Agusti, N. Andersen, and A. Russek (1979), *Phys. Rev. A* **19**, 1457.

Pack, R. T. (1984), *J. Chem. Phys.* **81**, 1841.

Pack, R. T., E. Piper, G. A. Pfeffer, and J. P. Toennies (1984), *J. Chem. Phys.* **80**, 4940.

Park, J. T. (1983), "Interactions of Simple Ion–Atom Systems," *AAMP* **19**, 67.

Park, J. T., J. E. Aldag, J. L. Peacher, and J. M. George (1980), *Phys. Rev. A* **21**, 751.

Pauly, H. (1975), "Collision Processes: Theory of Elastic Scattering," in H. Eyring, D. Henderson, and W. Jost, Eds., *Physical Chemistry*, Vol. 6-B, Academic Press, New York, pp. 553–628.

Pauly, H., and J. P. Toennies (1965), "The Study of Intermolecular Potentials with Molecular Beams at Thermal Energies," *AAMP* **1**, 195.

Pauly, H., and J. P. Toennies (1968), "Neutral–Neutral Interactions: Beam Experiments at Thermal Energies," *MEP* **7-A**, 227.

Peacher, J. L., P. J. Martin, D. G. Seely, J. E. Aldag, T. J. Kvale, E. Redd, D. Blankenship, V. C. Sutcliffe, and J. T. Park (1984), *Phys. Rev. A* **30**, 729.

Phelps, A. V. (1990), "Cross Sections and Swarm Coefficients for H^+, H_2^+, H_3^+, H, H_2, and H^- in H_2 for Energies from 0.1 eV to 10 keV," *J. Phys. Chem. Ref. Data* **19**, 653.

Phelps, A. V. (1991), *J. Phys. Chem. Ref. Data* **20**, 557.

Phelps, A. V. (1992), *J. Phys. Chem. Ref. Data* **21**, 883.

Rapp, D. (1971), *Quantum Mechanics*, Holt, Rinehart and Winston, New York.

Ray, J. A., C. F. Barnett, and B. Van Zyl (1979), *J. Appl. Phys.* **50**, 6516.

Redd, E., T. J. Gay, D. M. Blankenship, J. T. Park, J. L. Peacher, and D. G. Seely (1987), *Phys. Rev. A* **36**, 3475.

Reuss, J. (1988), "State Selection by Nonoptical Methods," Chap. 11 in G. Scoles, Ed., *Atomic and Molecular Beam Methods*, Vol. 1, Oxford University Press, New York.

Riviere, A. C., and D. R. Sweetman (1960), *Phys. Rev. Lett.* **5**, 560.

Riviere, A. C., and D. R. Sweetman (1962), *Proc. 5th Int. Conf. Ionization Phenomena in Gases*, Munich, 1961, Vol. 2, p. 1236, North-Holland, Amsterdam.

Rosenthal, H. (1971), *Phys. Rev. A.* **4**, 1030.

Rosenthal, H., and H. M. Foley (1969), *Phys. Rev. Lett.* **23**, 1480.

Rossi, B. (1957), *Optics*, Addison-Wesley, Reading, Mass.

Rubahn, H. G., and K. Bergmann (1990), "Effect of Vibrational Bond Stretching in Molecular Collisions," *Annu. Rev. Phys. Chem.* **41**, 735.

Rubahn, H. G., and J. P. Toennies (1988), *Chem. Phys.* **126**, 7.

Russek, A. (1971), *Phys. Rev. A* **4**, 1918.

Russek, A. (1983), *ICPEAC XIII*, Berlin, p. 701.

Sard, R. D. (1970), *Relativistic Mechanics*, W. A. Benjamin, New York.

Schinke, R. (1983), *ICPEAC XIII*, Berlin, p. 429.

Schinke, R., H. Krüger, V. Hermann, H. Schmidt, and F. Linder (1977), *J. Chem. Phys.* **67**, 1187.

Schmidt, H., V. Hermann, and F. Linder (1978), *J. Chem. Phys.* **69**, 2734.

Schwenz, R. W., and S. R. Leone (1987), *Chem. Phys. Lett.* **133**, 433.

Seaton, M. J. (1962), "Complex Atoms," in D. R. Bates, Ed., *Quantum Theory*, Vol. 2, Academic Press, New York.

Serri, J. A., R. M. Bilotta, and D. E. Pritchard (1982), *J. Chem. Phys.* **77**, 2940.

Shamamian, V. A. (1989), Ph. D. thesis, University of Minnesota, Minneapolis, Minn.

Shimamura, I. (1984), in I. Shimamura and K. Takayanagi, Eds., *Electron–Molecule Collisions*, Plenum Press, New York.

Sidis, V. (1989), "Vibronic Phenomena in Collisions of Atomic and Molecular Species," *AAMOP* **26**, 161.

Sidis, V., and D. Dowek (1983), *ICPEAC XIII*, Berlin, p. 403.

Sidis, V., J. C. Brenot, J. Pommier, M. Barat, O. Bernardini, D. C. Lorents, and F. T. Smith (1977), *J. Phys. B* **10**, 2431.

Skodje, R. T., W. R. Gentry, and C. F. Giese (1977), *J. Chem. Phys.* **66**, 160.

Skodje, R. T., W. R. Gentry, and C. F. Giese (1983), *Chem. Phys.* **74**, 347.

Smith, N., T. A. Brunner, and D. E. Pritchard (1981), *J. Chem. Phys.* **74**, 467.

Stolte, S. (1988), "Scattering Experiments with State Selectors," Chap. 25 in G. Scoles, Ed., *Atomic and Molecular Beam Methods*, Vol. 1, Oxford University Press, New York.

Stoneman, R. C., M. D. Adams, and T. F. Gallagher (1987), *Phys. Rev. Lett.* **58**, 1324.

Taatjes, C. A., and S. R. Leone (1988), *J. Chem. Phys.* **89**, 302.

Tango, W. J., J. K. Link, and R. N. Zare (1968), *J. Chem. Phys.* **49**, 4264.

Thomas, E. W. (1972), *Excitation in Heavy Particle Collisions*, Wiley, New York.

Thomson, D. S., R. C. Stonemen, and T. F. Gallagher (1989), *Phys. Rev. A* **39**, 2914.

Toennies, J. P. (1976), "The Calculation and Measurement of Cross Sections for Rotational and Vibrational Excitation," *Annu. Rev. Phys. Chem.* **27**, 225.

Toennies, J. P. (1982), *Aust. J. Phys.* **35**, 593.

Toennies, J. P. (1985), "Low-Energy Atomic and Molecular Collisions," in J. Bang and J. de Boer, Eds., *Semiclassical Descriptions of Atomic and Nuclear Collisions*, North-Holland, Amsterdam, p. 29.

Treanor, C. E. (1965), *J. Chem. Phys.* **43**, 532.

Treanor, C. E. (1966), *J. Chem. Phys.* **44**, 2220.

Uzer, T. (1991), "Theories of Intramolecular Vibrational Energy Transfer," *Phys. Rep.* **199**, 73.

Vager, Z., R. Naaman, and E. P. Kanter (1989), *Science* **244**, 426.

van den Meijdenberg, C. J. N. (1988), "Velocity Selection by Mechanical Methods," Chap. 13 in G. Scoles, Ed., *Atomic and Molecular Beam Methods*, Vol. 1, Oxford University Press, New York.

Van Zyl, B. (1976), *Rev. Sci. Instrum.* **47**, 1214.

Van Zyl, B., and M. W. Gealy (1986), *Rev. Sci. Instrum.* **57**, 359.

Van Zyl, B., and M. W. Gealy (1987), *Phys. Rev. A.* **35**, 3741.

Van Zyl, B., and H. Neumann (1988), *J. Geophys. Res.* **93** (A-2), 1023.

Van Zyl, B., G. E. Chamberlain, G. H. Dunn, and S. Ruthberg (1976a), *J. Vac. Sci. Technol.*, **13**, 721.

Van Zyl, B., N. G. Utterback, and R. C. Amme (1976b), *Rev. Sci. Instrum.* **47**, 814.

Van Zyl, B., G. H. Dunn, G. Chamberlain, and D. W. O. Heddle (1980a), *Phys. Rev. A*, **22**, 1916.

Van Zyl, B., H. Neumann, H. L. Rothwell, and R. C. Amme (1980b), *Phys. Rev. A* **21**, 716.

Van Zyl, B., M. W. Gealy, and H. Neumann (1985), *Phys. Rev. A* **31**, 2922.

Van Zyl, B., M. W. Gealy, and H. Neumann (1986a), *Phys. Rev. A* **33**, 2333.

Van Zyl, B., H. Neumann, and M. W. Gealy (1986b), *Phys. Rev. A* **33**, 2093.

Van Zyl, B., M. W. Gealy, and H. Neumann (1987), *Phys. Rev. A* **35**, 4551.

Van Zyl, B., B. K. Van Zyl, and W. B. Westerveld (1988), *Phys. Rev. A* **37**, 4201.

Van Zyl, B., M. W. Gealy, and H. Neumann (1989), *Phys. Rev. A* **40**, 1664.

Wedding, A. B., and A. V. Phelps (1988), *J. Chem. Phys.* **89**, 2965.

Wight, C. A., D. J. Donaldson, and S. R. Leone (1985), *J. Chem. Phys.* **83**, 660.

Wilson, E. B., J. C. Decius, and P. C. Cross (1955), *Molecular Vibrations*, McGraw-Hill, New York.

Yamasaki, K., and S. R. Leone (1989), *J. Chem. Phys.* **90**, 964.

Yardley, J. T. (1980), *Introduction to Molecular Energy Transfer*, Academic Press, New York.

Zare, R. N. (1979), *Faraday Discuss. Chem. Soc.* **67**, 7.

Zen, M. (1988), "Accommodation, Accumulation, and Other Detection Methods," Chap. 10 in G. Scoles, Ed., *Atomic and Molecular Beam Methods*, Vol. 1, Oxford University Press, New York.

Ziegler, G., S. V. K. Kumar, H. G. Rubahn, A. Kuhn, B. Sun, and K. Bergmann (1991), *J. Chem. Phys.* **94**, 4252.

3

CHEMICAL DYNAMICS AND CHEMICAL REACTIONS

3-1. INTRODUCTION

We begin with some definitions. *Chemical kinetics* is the field of investigation of the rates of physical and chemical processes occurring at low energies (mainly below a few eV). Emphasis is placed on macroscopic bulk studies involving dense gases, liquids, and surfaces, and processes that are rapidly equilibrated, so that the rates of elementary reactions are not isolated in single collisions. *Molecular dynamics* is the subset of chemical kinetics that is concerned with the rates of microscopic molecular rearrangements, or "rate processes"; it deals with both intramolecular motions in isolated molecules and intermolecular collisions in the dilute gas phase. Further specialization leads to *chemical dynamics*, which centers on *chemical reactions*, processes that involve the rearrangement or association of colliding structures. Chemical dynamics and especially the microscopic study of individual chemical reactions are the subjects of this chapter, along with certain processes intimately involved with these reactions, as indicated in Section 3-2. We include no bulk studies, no investigations of clusters, and no surface impact phenomena. We also discuss no applications, but rather, refer the reader to Chapter 1 and Section 7-7 of Levine and Bernstein (1987) and to Massey et al. (1982–1984).

In one sense, chemical dynamics and chemical reactions constitute a branch of atomic collisions and hence must be included as one of many in this volume. In another sense, they constitute a large branch of chemistry that demands multivolume treatment for anything approaching adequate coverage. The best that we can do here is to try to define the scope of the subject, describe some of the most important experimental techniques, and discuss a few experiments and the conclusions drawn from them. The reader should supplement our sketchy treatment by consulting some of the numerous reviews and books on the subject, such as Herschbach (1987), Lee (1987), Polanyi (1987), Bernstein (1982), Levine and Bernstein (1987), and Scoles (1988).

In Part $\mathscr{A}$ of this chapter we deal with basic, general considerations, in Part

$\mathscr{B}$ we discuss reactions between neutral molecules, and in Part $\mathscr{C}$ we turn to ion–neutral reactions.

PART $\mathscr{A}$.   GENERAL CONSIDERATIONS

3-2.   DOMAIN OF CHEMICAL DYNAMICS

Following Bernstein (1982), we attempt to delineate the field of chemical dynamics by enumerating some topics that are germane to it. Let us consider the relatively simple case of an atom A colliding with a diatomic molecule BC and display some of the channels that are possible at low impact energies:

$$A + BC \rightarrow \begin{cases} A + BC & \text{elastic scattering} \\ A + BC(J) & \text{rotational excitation} \\ A + BC(J, v) & \text{vibrational and rotational excitation} \\ AB + C & \text{reaction} \\ A + B + C & \text{collision-induced dissociation} \\ A + BC(n) & \text{electronic excitation} \\ A + BC^+ & \text{ionization.} \end{cases} \qquad (3\text{-}2\text{-}1)$$

Similar results are obtained if the projectile atom A is replaced by the ion A^+, and if more complex collision partners are substituted for A and BC. Of course, several reactions in (3-2-1) can occur simultaneously. A matter of basic concern to us is what determines the probabilities of the various outcomes in (3-2-1) at a given CM impact energy.

We also need to know which factors determine whether the collision in the reaction channel is *direct*, with an interaction time of about one molecular vibration period ($\lesssim 10^{-13}$ s), or whether it proceeds through an *intermediate complex*, with a lifetime greater than a molecular rotation period ($\gtrsim 10^{-12}$ s).

We must ascertain the basis of the *energy partitioning* among the products of reactions and determine how the total available energy is distributed among the various internal degrees of freedom and the translational kinetic energy of the products. Of great interest is the physical basis for vibrational population inversions and for the nonequilibrium distributions of rotational states that frequently occur. We wish to understand how such distributions relax collisionally toward Boltzmann equilibrium distributions. It is important to know why so often highly excited states of nascent product molecules are formed in fast exoergic reactions, and why laser-excited reagents have enhanced reactivity in reactions with activation energy barriers.

We are also interested in the factors that operate in the one-body (unimolecular) decay of an intermediate product of a reaction. Is it possible to excite a particular bond selectively and thereby influence the branching ratio of the unimolecular decay (i.e., the way in which the intermediate dissociates)?

Another fundamental question relates to the molecular basis of three-body (termolecular) atomic or radical recombination and the inverse dissociation process.

In Section 8-17 of the companion volume we discussed infrared multiphoton dissociation, and visible and ultraviolet resonance-enhanced multiphoton ionization and dissociation. These subjects are of considerable importance in chemical dynamics, as are the internal energy states formed in single-photon ionization and dissociation (discussed in Sections 8-15 and 8-16 of the companion volume).

We must now define differential and integral cross sections and two-body (bimolecular) reaction rate coefficients for chemical reactions.

State-to-state reaction cross sections are of special interest here. How do such cross sections depend on the CM impact energy, the relative orientation of the reactants, and the rotational angular momentum of the reactants (and its polarization)?

The theoretical framework describing chemical reaction dynamics has steadily improved with the application of *ab initio* quantum mechanical calculations, computer simulations, classical and semiclassical computations, and information theory. A basic problem that is receiving a great deal of attention is that of obtaining information on potential surfaces from a combination of theory and experimental data. We discussed studies of this type that involve elastic and inelastic scattering in Chapters 1 and 2.

3-3. BASIC EXPERIMENTAL CONSIDERATIONS

Again following Bernstein (1982), we pose some questions related to the types of measurements required to investigate the topics discussed in Section 3-2:

For known and controlled reactant conditions, how can we measure the angular and recoil velocity distributions of the products of a chemical reaction?

How can we determine the relative populations of the internal states of the reaction products?

How can we measure state-to-state differential and integral cross sections for reactions and obtain the dependence on the impact energy and on the internal energies of the reactants?

How is it possible with molecular beam techniques to generate suitably intense beams of molecules with a given translational energy that may be varied from the meV region up to, say, 5 eV? How does one form beams of neutrals with energies ranging from about 1 eV up into the keV region using standard atomic collisions techniques? What about ions, radicals, and excited atoms and molecules?

By what techniques is it possible to form beams of state-selected molecules and beams that are polarized in angular momentum?

What kinds of molecules can be oriented with respect to their angular momentum, and how can molecules be oriented in a beam?

What methods are available for detecting beams of neutral molecules and for measuring their velocity distributions? How can the internal energy distributions of nascent products of reactions be determined?

All of these matters are discussed by Bernstein (1982) and by Levine and Bernstein (1987). Some of them, especially those related to beam formation and detection, have been discussed in Chapters 1 and 2. Others will be addressed in this and subsequent chapters.

PART ℬ.   NEUTRAL–NEUTRAL REACTIONS

3-4.   REACTION MECHANISMS

Before proceeding to laboratory methods and experiments, it is appropriate to expand our discussion of the mechanisms by which binary reactions can take place.

A.   Direct-Mode Reactions versus Collision Complex Formation

In Section 3-2 we made an important distinction between a reaction that occurs in a time less than about one vibrational period ($\lesssim 10^{-13}$ s) and one that takes place on a much longer time scale, in a time greater than a rotational period ($\gtrsim 10^{-12}$ s). In the former instance, the collision partners collide briskly and react impulsively, and the reaction is complete before the combined structure has time to rotate. In the latter case, a long-lived collision complex is produced, with the reactants forming a rotating structure that survives for one or more rotational periods. Measurements of the angular distributions of the reaction products can distinguish between these two extremes of behavior. The products of a direct-mode reaction have strongly anisotropic angular distributions in the CM frame. On the other hand, the products of a reaction involving complex formation are symmetrically distributed about the 90° CM scattering direction; they have forgotten the original direction of the relative velocity of the reactants. (A special case arises here if the reactants form a complex whose lifetime is longer than the transit time to the product detector. In that case, no dissociation of the complex is observed, and the angular distribution shows a sharp peak located at the centroid angle.)

A reaction taking place with the formation of a collision complex is alternatively referred to as a *compound reaction*. It is also said to involve a *sticky collision*.

B.   Spectator Stripping versus Rebound in Direct Reactions

Let us look within the class of direct reactions and take note of two limiting kinds of behavior, one involving *spectator stripping* and the other the *rebound mechanism* (Levine and Bernstein, 1987, pp. 10–13, 111–113). First consider the direct exchange reaction $Cl + HI \rightarrow I + HCl$. Here the very light H atom is transferred between two much heavier atoms, and the Cl and I nuclei are unable to move appreciably in the time required for the transfer. During the stripping of the H atom from the HI molecule, the I atom acts as a mere spectator, and its vector momentum is not significantly altered. The Cl atom also continues essentially in its original direction, with the H atom now attached to it. The angular distribution of the HCl product is highly anisotropic, indicating that no intermediate ClHI complex is formed.

The $K + CH_3I \rightarrow CH_3 + KI$ reaction provides an example of the rebound mechanism. Here the reaction occurs at small values of the impact parameter b, the reactants experience the short-range repulsive component of the interaction potential, and their deflections are similar to those for hard-sphere scattering. Accordingly, the reaction products are backscattered in the CM frame. Note that there will always be a rebound component in direct reactions, even in the spectator stripping limit, because of collisions at small impact parameters. Note also that the cross section for a rebound reaction is small because of the small target area associated with collisions having values of b restricted to small values.

C.   Harpoon Mechanism

Finally, we briefly describe the *harpoon mechanism*, which was proposed by M. Polanyi in the 1930s to account for the rapid rates of reactions between alkali atoms and halogen compounds. Let us illustrate it by the $Li + F_2 \rightarrow LiF + F$ reaction. The first step of the mechanism is the transfer of the valence electron of the Li atom to the F_2 molecule, this transfer being possible even when the reactants are separated by as much as about 7 Å. The resulting pair of ions are mutually attracted by the strong Coulomb force that is operative, and a stable LiF molecule and a free F atom are formed. The Li atom has used its valence electron as a harpoon to bring the reactants together; the interaction at large separation distances implies a large cross section. Examples of specific harpoon reactions and their interpretation are provided by Bernstein (1982) and Levine and Bernstein (1987).

3-5.   ADVANTAGES OF THE CROSSED-MOLECULAR-BEAM METHOD

We concentrate here on thermal and epithermal neutral–neutral chemical reactions, for the study of which the crossed-molecular-beam technique offers the important advantages listed below:

1. Each reactant molecule makes at most one collision, so the outcome of no collision is affected by a subsequent one.
2. The reactants are velocity selected, so the reaction cross section can be measured as a function of impact energy.
3. By supersonic nozzle expansion, one can produce beams of molecules that are predominantly in the ground rotational state and vibrationally cooled as well.
4. The angular distributions of the nascent products can be measured directly.
5. The recoil velocity distributions of the nascent products can also be measured to gain information on the energy partitioning of the reaction.
6. The use of inhomogeneous electric and magnetic fields permits the rotational and vibrational state selection of reactants and allows the study of the effect of reactant internal energy.
7. Molecular reactants may be oriented and polarized so that steric effects may be studied directly.
8. Molecular products may be analyzed with respect to rotational state and polarization, to elucidate energy and angular momentum disposal.
9. Lasers are easily coupled to crossed-beam experiments. Hence it is a straightforward matter both to excite the reactant molecules and to study the effect of this excitation by laser-induced-fluorescence (LIF) analysis on nascent product molecules (see Section 2-3). Furthermore, if a laser beam interacting with a reactant beam is polarized, one may measure the orientation dependence of the reaction cross section.

Two authoritative reviews that deal with experimental methods in crossed-molecular-beam research should be cited here. The first (Lee, 1988) is concerned with nonoptical methods, and the second (Dagdigian, 1988) with optical techniques.

3-6.   INSTRUMENTS AND EXPERIMENTS (NEUTRAL–NEUTRAL REACTIONS)

A.   Apparatus of Lee, McDonald, LeBreton, and Herschbach; Studies of Halogen Reactions

The Lee et al. instrument (Fig. 3-6-1), constructed in Herschbach's laboratory at Harvard (Lee et al., 1969), represented a major advance in experimental techniques. It contains a thermal dissociation atom beam source and a permanent gas oven source. The atom beam is chopped by a rotating paddle, and it intersects the permanent gas beam as shown at the lower right of Fig. 3-6-1. The beam sources are fixed in position to facilitate changing the source modules and to permit strong pumping of the source chambers. The detector

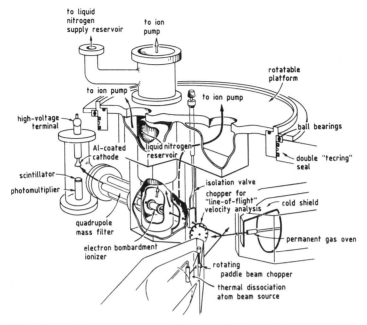

Figure 3-6-1. Crossed-beam apparatus of Lee, McDonald, LeBreton, and Herschbach. [From Herschbach (1987).]

unit (center and left in Fig. 3-6-1) is mounted on a rotatable lid. Product molecules headed in the direction at which the detector is set pass through a chopper for time-of-flight velocity analysis and into the detector assembly. A small fraction of these product molecules are ionized by electron bombardment, and the resulting ions that have the correct e/m ratio are transmitted by the quadrupole mass filter. The ion counting system is of the Daly type. Ions are accelerated onto an aluminum-coated stainless steel electrode maintained at $-30\,kV$. Secondary electrons (six to eight per ion) are accelerated back through the 30-kV potential difference and strike an aluminum-coated plastic scintillator, where they produce light pulses (three to six photons per electron) that are detected by a photomultiplier. The overall efficiency of the ion counter is 90 to 95%, with little dependence on the ion mass.

A vital feature of the design is the achievement of very low pressure in the electron bombardment ionizer [see Lee (1988)]. Here the ionizer is located within three nested chambers, each differentially pumped and provided with cryogenic traps. By this means, the background gas at the relevant mass that diffuses into the ionizer from the beam interaction chamber is reduced to a partial pressure that is no higher than that of the product molecules entering in free flight (typically about 10^{-14} torr). Also, the background arising from product molecules deposited on surfaces near the ionizer is minimized.

Herschbach (1987) points out that satisfactory reactive scattering data are often obtained with product molecule fluxes of only about 10^3 molecules/s.

The apparatus of Fig. 3-6-1 was first used to study the reaction

$$Cl + Br_2 \rightarrow BrCl + Br \qquad (3\text{-}6\text{-}1)$$

and other exchange reactions involving halogen atoms and molecules (Herschbach, 1987). The integrated reaction cross section is only about $10\,\text{Å}^2$ (less than the hard-sphere cross section), but the BrCl product angular distribution peaks strongly in the forward hemisphere, so one concludes that the main interaction is short range but attractive.

The next reaction studied (Herschbach, 1987) was

$$H + Cl_2 \rightarrow HCl + Cl, \qquad (3\text{-}6\text{-}2)$$

which has interesting similarities to the $K + CH_3I$ reaction and to the photodissociation of Cl_2. Angle-velocity flux contour maps for these three systems are shown in Fig. 3-6-2. For the (H, Cl_2) reaction, the angular distribution of the products is broad, but very anisotropic; the HCl recoils backward and the Cl forward with respect to the incident H atom. The product velocity is high, with about half of the available energy going into recoil of HCl and Cl. The remainder of the energy goes into vibrational and rotational energy of the HCl, as observed in infrared chemiluminescence. The shape of the angular distribution points to collinear H–Cl–Cl as being the preferred reaction geometry, and the large recoil energy indicates that strong repulsive forces are suddenly released. The contour map for the photodissociation process shown in Fig. 3-6-2c was generated from the continuous absorption spectrum of Cl_2. The angular distribution is determined by the electric dipole selection rule for absorption, so the transition probability varies as the square of the cosine of the angle between the Cl–Cl axis and the direction of the photon beam.

Interesting differences appear when Cl_2 in (3-6-2) is replaced by Br_2 and then by I_2 (see Fig. 3-6-3). The repulsive energy release becomes a smaller fraction of the release in photodissociation, and the molecular product angular distribution shifts from backward to the side with respect to the H-atom incident direction. Both of these effects are believed to be associated with the decrease in the halogen electronegativity, which favors a bent reaction geometry and reduces the repulsive energy release.

The reaction

$$O + Br_2 \rightarrow BrO + Br \qquad (3\text{-}6\text{-}3)$$

involves complex formation and no activation energy. The BrO product contour map displays prominent glory peaks. Arguments based on electronic structure suggest that the reaction proceeds mainly on the triplet O–Br–Br potential surface rather than on the more stable, singlet Br–O–Br surface.

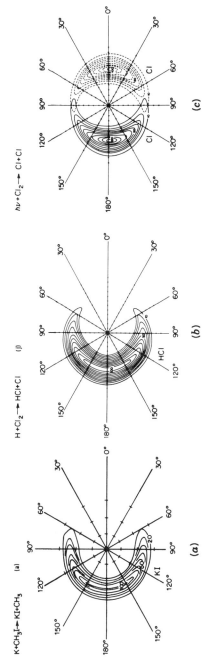

Figure 3-6-2. Angle-velocity flux contour maps for the three reactions indicated. The origin of each map is at the center of mass, and the horizontal axis is along the reactant relative velocity vector, with zero degrees corresponding to the direction of the incident atom or photon. Velocity is measured radially, with the markers along radial lines indicating velocity intervals of 200 m/s. The initial impact energy in (*b*) is about 0.44 eV. [From Herschbach (1987).]

K+CH₃I → KI+CH₃   (a)

H+Cl₂ → HCl+Cl   (J)

hν+Cl₂ → Cl+Cl

(*a*)

(*b*)

(*c*)

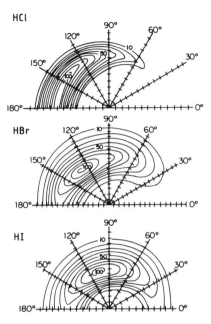

Figure 3-6-3. Angle-velocity flux contour maps for reactions of H atoms with Cl_2, Br_2, and I_2 molecules, the symmetrical portions below the horizontal axes not being shown. The markers on the radial lines correspond to velocity intervals of 200 m/s for Cl_2, and 100 m/s for Br_2 and I_2. [From Herschbach (1987).]

Similarly, the O + ICl reaction is expected to give mainly the products IO + Cl, rather than ClO + I, and this expectation is confirmed by experiment (Herschbach, 1987).

B. Lee's Apparatus Used in Studies of Reactive Scattering of F by H_2, D_2, and HD

The experimental arrangement shown in Fig. 3-6-4 has been used by Lee's group at Berkeley for the study of reactions between F atoms and H_2, D_2, and HD molecules (Neumark et al., 1985a, b).* The components are (1) an effusive F-atom beam source made of nickel and resistively heated; (2) a velocity selector; (3) a liquid nitrogen cold trap; (4) a molecular hydrogen nozzle expansion beam source; (5) a heater; (6) a liquid nitrogen feed line; (7) a skimmer; (8) a tuning fork chopper; (9) a synchronous motor; (10) a cross correlation chopper for time-of-flight velocity analysis; and (11) an ultrahigh vacuum, triply differentially pumped mass spectrometer detector chamber. The pressure (in torr) for each region is indicated.

Figure 3-6-5 shows a CM angle-velocity flux contour map for the

*Similar studies of the (D, H_2) reaction have been made by Continetti et al. (1990).

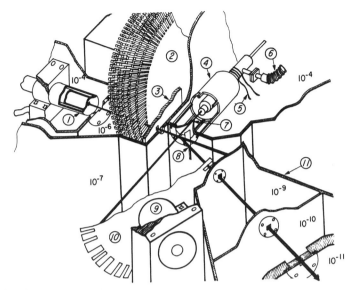

Figure 3-6-4. Lee's crossed-beam apparatus used in the study of reactions of F atoms with H_2, D_2, and HD. [From Lee (1987).]

$F + D_2 \rightarrow DF + D$ reaction. The F atoms move from right to left; the CM impact energy is 1.82 kcal/mol (0.079 eV). The contour map shows the probability of DF product molecules appearing at various angles and velocities. The angle $0°$ corresponds to the direction of the incident F beam, and the distance from the CM (at the center of the drawing) gives the CM velocity. The pronounced backward concentration of the DF product molecules indicates that not all of the collisions will lead to DF formation. Only those collisions in

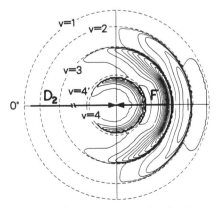

Figure 3-6-5. CM angle-velocity flux contour map for the $F + D_2 \rightarrow DF + D$ reaction at 1.82 kcal/mol. The F atoms are moving from right to left. [From Neumark et al. (1985b).]

which the F atom and the two D atoms are nearly collinear result in reactions and production of DF molecules. The appearance of DF in the indicated velocity bands can be explained by the fact that the DF molecules are produced in various vibrational states ($v = 1, 2, 3, 4, \ldots$). The total energy released in every reactive collision between F and D_2 is the same, so the energy available for translational motion will depend on the vibrational state of the DF product. Since the rotational energy spread of the products is smaller than the vibrational energy spacings, the recoil velocities of the separate vibrational states can easily be identified.

Time-of-flight velocity distributions for the reaction of F atoms with p-H_2 at an impact energy of 1.84 kcal/mol are shown in Fig. 3-6-6. The CM angle-

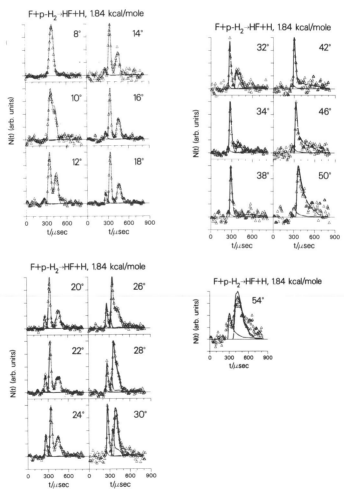

Figure 3-6-6. Time-of-flight spectra for the $F + p\text{-}H_2 \rightarrow HF + H$ reaction at 1.84 kcal/mol. [From Neumark et al. (1985a).]

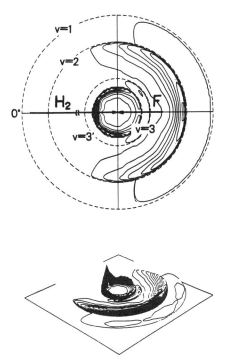

Figure 3-6-7. CM angle-velocity flux contour map for the $F + p\text{-}H_2$ reaction at 1.84 kcal/mol, with a three-dimensional perspective view shown below. [From Neumark et al. (1985a).]

velocity contour map derived from these distributions is displayed in Fig. 3-6-7. The forward peaking of the HF product in the $v = 3$ state shows that quasibound states are formed in the (F, H_2) reaction at 1.84 kcal/mol, and that these states decay into the $v = 3$ vibrational state of HF.

Some interesting results arise from quantal collinear calculations on the reaction $F + HD \rightarrow HF + D$. A sharp peak appears in the $HF(v = 2)$ reaction probability near threshold, and little else is produced at higher impact energies up to 0.2 eV. The conclusion is that HF formation in this reaction is dominated by resonance scattering, whereas the DF product is produced by direct scattering. These theoretical results are in agreement with experiment (Neumark, 1985b).

C.   Crossed-Beam Studies of the $D + H_2(v = 0, 1) \rightarrow HD + H$ Reactions in Toennies's Laboratory

The $H + H_2$ exchange reaction and its isotopic variants provide the simplest prototype for understanding bimolecular chemical reactions. The potential surfaces for these reactions are considered to be the most accurately known for any chemical reaction. The potential energy for the H_3 system is shown in Fig. 3-6-8 as a contour diagram and also plotted along the reaction coordinate for the collinear $D + H_2 \rightarrow HD + H$ reaction. In the lower drawing a comparison is made with the vibrational levels of both the reactants and the products. Experimental studies are difficult because of the small cross sections that result from the substantial energy barrier for the exchange process and because of the problems in generating and detecting fast hydrogen beams.

Götting et al. (1986) have studied the $D + H_2(v = 0)$ reaction in a crossed-

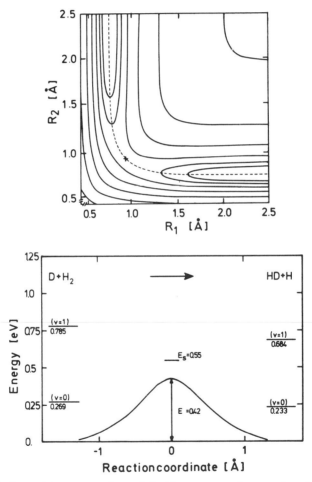

Figure 3-6-8. Potential energy surface of the H_3 system and energetics of the $D + H_2$ scattering experiment. A cut through the potential energy surface of the collinear configuration is shown in the lower drawing. The threshold of the reaction (E_s) and the vibrational states of the reactants and products are schematically shown. [From Götting et al. (1986).]

beam apparatus at a most probable CM impact energy of 1.5 eV. They employ an arc-heated primary beam source that provides a 3-eV beam of D atoms, which intersects a cold (45-K) collimated beam of H_2 molecules from an effusion source. The HD products are registered at a variable scattering angle in a plane perpendicular to the H_2 beam by a magnetic mass spectrometer arrangement and a particle counter. A pseudorandom chopper modulates the D-atom beam and makes it possible to obtain the time-of-flight distribution for the product HD molecules in the presence of the large background from the detector.

The experiment of Götting et al. (1986) was the first molecular beam-scattering experiment on the $D + H_2(v = 0) \rightarrow HD + H$ reaction with both angular and velocity resolution at a CM impact energy of 1.5 eV. The HD products were observed to be scattered into a broad angular region about the 90° direction; previous experiments at lower energies showed backward scattering in the CM frame. The average final translational energy amounted to about 52% of the total energy of 1.5 eV. Measurements of the differential cross sections at various scattering angles led to the assignment of the value $\langle \sigma \rangle = 1.7 \pm 0.8 \, \text{Å}^2$ for the integrated reaction cross section.

Additional studies of the (D, H_2) reaction, with the H_2 molecules in the vibrational state $v = 1$, have been made by Götting et al. (1987). Angular and time-of-flight product distributions were measured, and the absolute integral cross section for the reaction was determined to be $1.14 \pm 0.50 \, \text{Å}^2$ at the CM impact energy of 0.33 eV. The experimental results are in good agreement with the results of quasiclassical trajectory calculations (QCT) and the reactive infinite-order sudden approximation (RIOSA).

D.   Studies of the (H, D₂), (D, H₂), and (Ba, HF) Reactions in Zare's Laboratory

Rinnen et al. (1989) in Zare's laboratory at Stanford University have used thermal energy D_2 (about 298 K) and translationally hot H atoms to investigate the $H + D_2 \rightarrow HD + D$ reaction. Premixed D_2 and HI are introduced into a high-vacuum chamber through a capillary nozzle, and H atoms are produced 2 mm below the nozzle by pulsed laser photolysis of the HI at 266 nm. The (H, D_2) reaction is thereby initiated by the laser beam with H-atom CM collision energies of 1.3 and 0.55 eV, both of which are above the classical reaction barrier of 0.42 eV. After a delay of 55 ns, the HD product is state-selectively ionized by (2 + 1) resonance-enhanced multiphoton ionization (REMPI; see Section 8-17 of the companion volume). The resulting HD^+ ions are then selected and detected in a shuttered time-of-flight mass spectrometer. The populations of all the energetically accessible HD levels are measured, specifically ($v' = 0$, $J' = 0$ to 15), ($v' = 1$, $J' = 0$ to 12), and ($v' = 2$, $J' = 0$ to 8). The available energy is partitioned as follows: 73% goes into product translation, 18% into HD rotation, and 9% into HD vibration. The measured rotational and vibrational distributions are in good agreement with QCT calculations.

In Part $\mathcal{A}$ we stressed the importance of experimental studies of reactions in which the translational and internal energies of the reactants are specified, and the same quantities are measured for the products. The experiment by Rinnen et al. (1989), discussed above, employed quantum-state-specific detection of the products but used a distribution of reactant internal states. In a more recent experiment, Kliner and Zare (1990) [see also Kliner et al. (1991) and Adelman et al. (1992)] have succeeded in investigating the rovibronic-state-to-rovibronic-state dynamics of another isotopic variant of the hydrogen atom–hydrogen molecule exchange reaction. The reaction selected for this landmark study is $D + H_2(v, J) \rightarrow HD(v', J') + H$. The H_2 molecules are prepared in the $(v = 1, J = 1)$ state by stimulated Raman pumping (SRP), and the populations of the $HD(v', J')$ product states are measured by $(2 + 1)$ REMPI. The apparatus is similar to that used by Rinnen et al. (1989). Premixed DBr and H_2 flow effusively into a high-vacuum chamber, and this beam is crossed by the copropagating

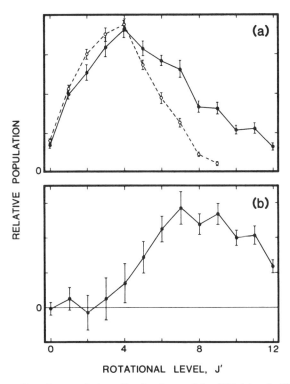

Figure 3-6-9. Ro-vibronic population distributions of the HD $(v' = 1, J')$ product from the $D + H_2$ reaction at about 1.0-eV CM impact energy. (a) The solid curve shows the pumped distribution following SRP of the H_2 reactant; the dashed curve shows the unpumped distribution from $D + H_2(v = 0$, thermal J). (b) The population distribution from the $D + H_2(v = 1, J = 1)$ reaction obtained by subtracting 87% of the unpumped distribution from the pumped distribution. [From Kliner and Zare (1990).]

SRP beams about 1 mm below the nozzle. The SRP beams pump about 19% of the $H_2(v = 0, J = 1)$ population (i.e., about 13% of the total population) into the $(v = 1, J = 1)$ state. A second laser (210 nm) is pulsed 20 ns later to photodissociate the DBr and produce fast D atoms. This laser also detects the HD product by REMPI. The impact energy (about 1.0 eV) is higher than the 0.42-eV classical reaction barrier, so the HD product is detected even in the absence of the SRP beams. Figure 3-6-9a displays the $HD(v' = 1)$ rotational distributions obtained with the SRP beams (solid curve) and without SRP (dashed curve). The contribution of the $D + H_2(v = 0, \text{thermal } J)$ reaction (i.e., the dashed curve) must be subtracted from the solid curve to obtain the HD distribution of the $D + H_2(v = 1, J = 1)$ reaction (Fig. 3-6-9b).

Gupta et al. (1980) have studied the $Ba + HF(v = 0) \rightarrow BaF + H$ reaction over a range of CM impact energy extending from 3 to 13 kcal/mol. In their apparatus, shown in Fig. 3-6-10, a seeded HF nozzle beam intersects a thermal Ba beam; the internal state distribution of the BaF product is monitored by laser-induced fluorescence as a function of impact energy. As shown in the

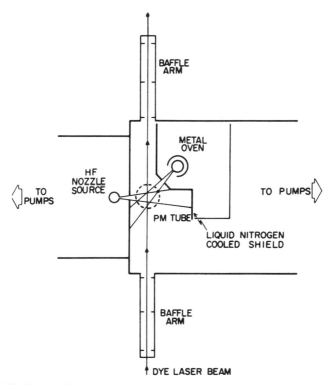

Figure 3-6-10. Crossed-beam apparatus used in Zare's laboratory for a study of the Ba + HF reaction. Laser-induced fluorescence is used to obtain the product BaF state distributions. [From Gupta et al. (1980).]

figure, the LIF dye laser beam enters and exits the apparatus through baffle arms to minimize scattered light. The fluorescence is observed by a photomultiplier tube, and LIF spectra for two different collision energies are shown in Fig. 3-6-11. The vibrational and rotational distributions of the product are determined from computer simulations of the excitation spectra. The reaction cross section peaks at CM impact energies close to 6-8 kcal/mol, the upper bound on the cross section being $15 \, \text{Å}^2$. The fraction of the available energy that goes into translation, vibration, and rotation is roughly constant over the range of impact energy covered; nearly half appears in translation while the remainder is divided about equally between vibration and rotation of the product BaF.

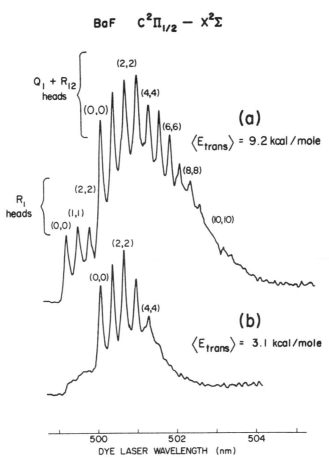

Figure 3-6-11. LIF spectra of the product BaF produced in the Ba + HF reaction at two difference average collision energies. [From Gupta et al. (1980).]

E. Lee's Apparatus for the Reactive Scattering of Electronically Excited Na Atoms by Molecules

Figure 3-6-12 shows the apparatus used in Lee's laboratory for the study of reactions of excited Na($3P$) atoms with HCl (Vernon et al., 1986). At impact energies slightly above the endoergicity of the reaction, the reactivity of the excited Na atoms is considerably greater than that of the Na($3S$) ground state. Measurements of the angular and velocity distributions of the NaCl product and comparison with theory indicate that the preferred transition state configuration, Na–Cl–H, is consistent with what would be predicted by a diffuse $3p$ orbital where the Na atom appears to be ionlike.

The apparatus of Fig. 3-6-12 employs two supersonic reactant beams that cross at 90° with an overlap volume of about 10^{-2} cm^3. The Na ground-state atoms are excited by a dye laser beam; because of the short lifetime of the Na($3P$) states, the excitation is made to take place in the beam interaction region. Product molecules are detected with a mass spectrometer that can rotate in the plane defined by the two reactant beams. The detector consists of an electron bombardment ionizer, a quadrupole mass filter, and a Daly ion counter that is enclosed in a triply differentially pumped vacuum chamber. The sodium beam is generated by a rare-gas seeded supersonic expansion from a two-chamber stainless steel oven, whereas the HCl beam is produced from a heated stainless steel tube with a nozzle on the end.

Lasers can also be used to control the alignment and orientation of electronically excited orbitals before they react. For instance, in the study of Na reacting with O_2, linearly polarized dye lasers can be used to excite Na atoms sequentially from the $3S$ to the $3P$ and finally, the $4D$ state, and the excited $4d$

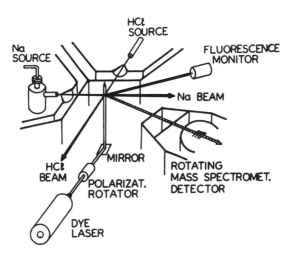

Figure 3-6-12. Apparatus for the study of reactive scattering of electronically excited Na atoms by HCl. [From Vernon et al. (1986).]

orbital can be aligned along the polarization direction of the **E** vector of the lasers. By this means, the effect of the alignment of the excited orbital on chemical reactivity can be investigated by rotating the polarization of the lasers with respect to the relative velocity vector. The dependence of chemical reactivity on molecular orientation can be determined from measurements of product angular distributions for many direct atom–molecule reactions that do not involve long-lived complexes. Control of molecular orientation is possible for symmetric top molecules (Note 3-4), and many careful studies of the effects of orientation on chemical reactivity have been made (Lee, 1987).

F.   Studies of Orientation of Reactant Molecules in Bernstein's Laboratory*

Because of their first-order Stark effect, polar symmetric-top molecules are good candidates for rotational state selection and orientation, and we restrict our attention here to that type of molecule. The method that we discuss for producing a beam of oriented reactant molecules utilizes an inhomogeneous electric field in a hexapole rod configuration to produce a $|JKM\rangle$* rotational state selection for a beam of polar symmetric-top molecules. During this process the molecules are oriented in the local inhomogeneous field. The beam then emerges from the hexapole and enters a tilted weak homogeneous electric field **E**, which orients the electric dipole moment **μ** (and the molecular frame) in the Lab frame (Note 3-4). The other beam of reactants collides with the oriented molecular beam at 90°; the orienting field **E** (and the average direction of **μ**) are arranged to be parallel to the relative velocity of the colliding beams. One may reverse the direction of **E**, and thereby the orientation of **μ**, and obtain two opposite orientations of the symmetric-top molecules in their interaction with the coreactants.

The rotational state of the symmetric-top molecule that is selected determines its average degree of orientation with respect to **E**. For molecules in a definite $|JKM\rangle$ state, the angle between **μ** and **E** is given by

$$\langle \cos \theta \rangle = \frac{KM}{J(J + 1)}. \qquad (3\text{-}6\text{-}4)$$

States with $K \approx M \approx J$ precess only slightly in the **E** field. A molecule in such a state collides with its dipole moment nearly parallel or antiparallel to the initial relative vector. By contrast, molecules with $K, M \ll J$ have $\langle \cos \theta \rangle \approx 0$, and they make collisions that are nearly broadside. In this case the effect of reversal of the **E** field on the reactivity is greatly diminished.

A revealing study of the $Rb + CH_3I \rightarrow RbI + CH_3$ reaction has been made in Bernstein's laboratory with the apparatus described above (Parker et al.,

---

*Many topics in this section are expanded in Notes 3-3 and 3-4.

*J, K, and M are, respectively, the quantum numbers for the total angular momentum, the component along the top axis, and the component along the **E** field (see Note 3-4).

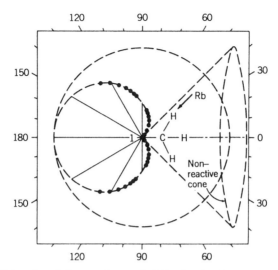

Figure 3-6-13. Polar plot of the reaction probability for CM backscattering as a function of the angle of attack for Rb on CH_3I. The points represent experimental data. [From Parker et al. (1981).]

1981). With oriented CH_3I reactants, they obtained the asymmetry in the reaction probability shown in Fig. 3-6-13 for a predominantly head-on ($b \approx 0$) collision. The reaction probability is highest for an approach from the I end of the molecular axis; it is practically zero for approach within the cone shown at the other end. Integrating over all orientations of the CH_3I molecules leads to a steric factor (at $b \approx 0$) of about 0.4. Other experiments on orientation (and alignment) are discussed in Levine and Bernstein (1987, pp. 478–496), Bernstein (1982, pp. 139–140), Gandhi and Bernstein (1990), and Friedrich et al. (1991).

G.   Chemiluminescence

This method of studying chemical reactions involves spectroscopic analysis of the radiation emitted from excited, nascent products (Fontijn, 1985) and the determination of the relative populations of the excited states. Much of the most detailed and useful results have come through the infrared chemiluminescence method, which was pioneered by J. C. Polanyi. Other useful data on vibrational state distributions have been obtained by observing visible and ultraviolet emission from electronically excited products, $XY^*(v', J')$.

Typically, an infrared chemiluminescence experiment is performed at very low pressure under steady-state conditions in a fast-flow system. In Fig. 3-6-14, jets of reactants merge, react, and are rapidly pumped away. The radiation from the interaction zone is collected and focused into an infrared spectrometer. Relative populations of the parent state can be derived from measurements of the relative intensities of the lines for the various (v, J) transitions and the

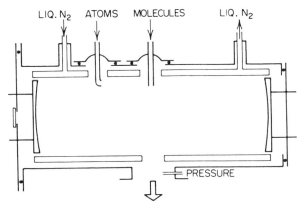

Figure 3-6-14. Apparatus for the study of chemiluminescence. Mirrors at either end of the reaction chamber collect the radiation and bring it to the sapphire window at the left. [From Polanyi (1987).]

relevant Franck–Condon factors. The results can be displayed in a "triangle plot," which shows how the energy release is distributed into vibration, rotation, and translation (Fig. 3-6-15).

A lengthy and detailed account of chemiluminescence (and laser-induced fluorescence as well) appears in the review by Dagdigian (1988). In this valuable study, attention is restricted to beam experiments and their analysis.

H. Studies of Ultrafast Reaction Dynamics in Zewail's Laboratory at Caltech

Thus far in this section we have examined *full collisions*, which are exemplified by the bimolecular reaction

$$A + BC \rightarrow [A \cdots B \cdots C] \rightarrow AB + C. \qquad (3\text{-}6\text{-}5)$$

The first "half" of the reaction consists of the encounter of the two heavy particles A and BC to form the transition state $[A \cdots B \cdots C]$; the second "half" involves the breakup of the transition state and the formation of the products $AB + C$. Here the reactants A and BC encounter each other on a time scale of nanoseconds $(1 \text{ ns} = 10^{-9} \text{ s})$ or microseconds $(1 \mu s = 10^{-6} \text{ s})$, and detection of the products AB and C requires a comparable amount of time. The transition state $[A \cdots B \cdots C]$ has a lifetime that is governed by the speed of the nuclear rearrangement and typically lies in the range 10 femtoseconds $(1 \text{ fs} = 10^{-15} \text{ s})$ to 10 picoseconds $(1 \text{ ps} = 10^{-12} \text{ s})$. The very-short-lived transition states have characteristics between those of the reactants and those of the products, and the continuum of the transition configurations along the reaction path determines the dynamics of the reaction.

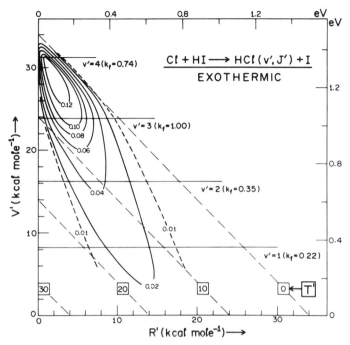

Figure 3-6-15. Triangle plot for the $Cl + HI \rightarrow HCl(v', J') + I$ reaction, showing the distribution of vibrational, rotational, and translational energy as contours of equal detailed rate constant $k(V', R', T')$. The ordinate gives the product vibrational energy, the abscissa the rotational energy. The total available energy is approximately constant (34 kcal/mol), and the translational energy T' is given by the broken diagonal lines. In the rate constant $k_f \equiv k(v')$, the subscript f denotes the forward, exothermic direction. [From Polanyi (1987).]

The transition state can also be produced in a *half-collision* (see Sections 8-14 and 8-15 of the companion volume), such as

$$hv + ABC \rightarrow [A \cdots B \cdots C] \rightarrow AB + C, \qquad (3\text{-}6\text{-}6)$$

wherein the transition state is produced not by a collision between two heavy particle reactants, but by a photon that deposits energy in a stable molecule ABC. Then a half-collision (a unimolecular dissociation reaction) takes place. In many cases it is advantageous to prepare the transition state by the half-collision method.

Because the nuclear and electronic motions in the transition regime determine the outcome of the reaction, there has been great interest in making direct studies of the transition state. However, until the mid-1980s the time resolution available to the experimenter was inadequate for this purpose, and direct studies were limited to the "before" and "after" regimes of reactions. Now, however, the latter investigations of the asymptotic regions of reactions are being complemented by direct observations of the transition state in real time (Gruebele and Zewail, 1990). The development of techniques to generate ultrafast laser pulses of suitable wavelength was the key factor in this advance.

The basic problem in chemical dynamics is to understand the process of nuclear rearrangement on reactive potential energy surfaces. The rearrangement occurs on the time scale of molecular vibrations (10 to 1000 fs), so the 10-fs laser pulses now available are adequate for direct real-time viewing of reactions as they proceed from reactants to transition state to products.

Figure 3-6-16 is a two-dimensional representation of a real-time femtosecond transition-state spectroscopy experiment. Here the half-collision approach is used. A femtosecond laser pulse (wavelength λ_1) pumps the molecule ABC to a transition state on the reactive surface V_1. (If the bimolecular reaction approach were used here, one of the two reactants, A or BC, would be activated.) The resulting molecular wave packet moves on V_1 according to quantum dynamics and eventually decays from the transition state to the products AB + C if undisturbed during the 10^{-11} s or less that is required. However, during the propagation of the molecular wave packet, it can be probed by a second

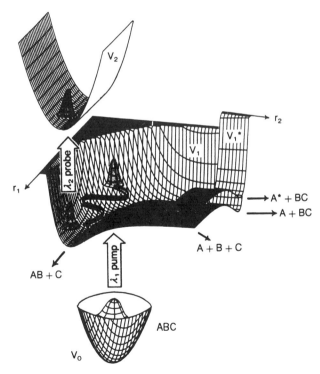

Figure 3-6-16. Two-dimensional portrayal of a real-time femtosecond transition-state spectroscopy experiment. [From Gruebele and Zewail (1990).]

femtosecond laser pulse (wavelength λ_2), after a variable delay Δt with respect to the first laser pulse. The second pulse elevates the wave packet to the surface V_2. By this means, a signal $I(\Delta t, \lambda)$ in the form of laser-induced fluorescence, ion counts, or some other observable is produced, and the motion of the dissociating wave packet can be imaged as a function of time. Depending on λ_2 and the detection scheme, it is possible to monitor the parent decay, the motion of the transition state, or the formation of the products. By polarizing the pump and probe pulses, one may also investigate the time dependence of the alignment and orientation of the molecules.

We now discuss a specific experiment in which these ideas are implemented, nanely the real-time clocking of the bimolecular reaction

$$H + CO_2 \rightarrow HOCO \rightarrow HO + CO$$

(Scherer et al., 1990). We start with the T-shaped van der Waals molecule $HI \cdot CO_2$, and photodissociate its HI component with a laser pulse of frequency v_1, as shown in panel (a) of Fig. 3-6-17. This pump laser pulse establishes the zero of time. The H atom is ejected with a velocity of about 20 km/s [panel (b)], and

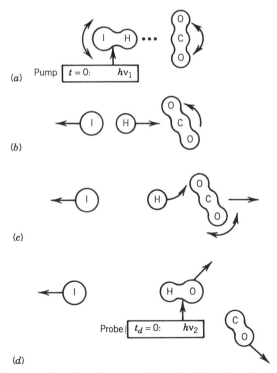

Figure 3-6-17. Pump-and-probe scheme used in the Caltech study of the bimolecular reaction $H + CO_2$. [From Scherer et al. (1990).]

quickly travels the 2-Å distance to the CO_2 component of the van der Waals molecule [panel (c)]. The H atom forms the HOCO complex, which undergoes vibrations and dissociates into the products OH and CO [panel (d)]. The OH (and CO) fragments are then detected at a variable time t_d by a probe laser of frequency v_2. The measurements by Scherer et al. (1990) show that the lifetime of the [HOCO] complex is about 5 ps. It should be noted that the method of establishing a zero of time here relies on the van der Waals forces holding the reactants in close proximity before the reaction is initiated. Establishing the zero time marker is not a problem in the half-collision approach, but it is usually difficult in the bimolecular reaction technique. Even if ultrashort laser pulses were used to produce species that could then react, the reactants would first have to find each other before they could react, and times in the nanosecond or microsecond range are required for this step.

Zewail and Bernstein (1988) point out that further progress in "femtochemistry" does not require further reduction in the width of laser pulses, as there is little to be gained by using pulses much shorter than those now obtainable. For example, with a wavelength $\lambda = 308$ nm and a pulse width $\Delta t = 7$ fs, the Heisenberg uncertainty principle requires an uncertainty of $\Delta E = \pm 0.13$ eV in the photon energy $E = 4.02$ eV. Hence when molecules are excited by such a short pulse, they are prepared with a significant spread of energies. We are at the point of having to choose between energy and time resolution.

Additional concepts in femtochemistry are discussed in several reviews (Zewail, 1988; Zewail and Bernstein, 1988; Sims et al. 1992), along with alternative experimental approaches. The theory of femtosecond transition state spectroscopy is the subject of Gruebele et al. (1990).

PART $\mathscr{C}$. ION–MOLECULE REACTIONS (BEAM STUDIES)

3-7. INTRODUCTION

We use the term *ion–molecule reaction* to refer to a heavy particle rearrangement or association reaction such as $A^+ + BC \rightarrow AB^+ + C$ or $A^- + B + C \rightarrow AB^- + C$. Other processes, such as charge transfer, ionization, and dissociation, may occur simultaneously or concurrently. By a *charge-transfer reaction* we mean a reaction in which an electron is transferred between the two colliding structures, as in $A^+ + B \rightarrow A + B^+$ or $A^- + BC \rightarrow A + BC^-$. These reactions are the subject of Chapter 5. Ionization by heavy particle impact is treated in Chapter 4, while dissociative processes are covered at various points through this volume. Here we concentrate on ion–molecule reactions occurring between simple structures and not involving other kinds of processes.

The fact that ion–molecule reactions can occur in gases has been known since early in this century, but until the 1950s such reactions excited comparatively little curiosity, and research proceeded at a slow pace. The first theoretical work of relevance here appeared in Langevin's famous mobility papers of 1903 and 1905, in which he considered the clustering of molecules about gaseous ions and orbiting collisions between ions and molecules (Section 1-5). More than 20 years elapsed before the next significant theoretical advance—the calculation of a rate constant for the reaction $H_2^{\ddagger} + H_2 \rightarrow H_3^{\ddagger} + H$ by H. Eyring, J. O. Hirschfelder, and H. S. Taylor in 1936. Experimental advances were characterized by the same slow time scale. J. J. Thomson observed ions of mass 3 in 1912 during experiments on hydrogen with his cathode-ray-tube apparatus, but the rate constant for their formation by the reaction displayed above was not measured until more than four decades later.

Starting in the late 1950s, however, it became apparent that a deep understanding of ion–molecule reactions is essential to our better comprehension of planetary atmospheres, interstellar molecules, radiation chemistry, combustion and flames, electrical discharges, and gas lasers. Consequently, there has been a great upsurge of interest in ion–molecule reactions, and today they are the subject of research in many laboratories.

There are strong parallels between ion–molecule reactions and the neutral–neutral reactions that we studied in Part $\mathscr{B}$. The goals in the study of the two kinds of processes are much the same, and many of the experimental techniques are common to both. However, most ion–molecule studies are performed at higher energies than the neutral–neutral studies, and a basic physical difference that has great consequences lies in the long-range potential that exists between an ion and a molecule.

Our study of ion–molecule reactions is divided into two parts. In this chapter we deal with beam studies, mostly performed between 1 and 500 eV impact energy. These investigations are closely related to the beam studies of neutral reactions in Part $\mathscr{B}$. In Chapter 7 we discuss experiments on thermal and epithermal ion–molecule reactions that are based on swarm techniques. These experiments are more closely related to electron–ion and ion–ion recombination (Chapter 9) and to ion transport through gases (Chapter 7). Our task is greatly facilitated by two books that have recently appeared. The first (Ng and Baer, 1992) deals with the experimental aspects of ion–molecule reactions, and the second (Baer and Ng, 1992) is concerned with theory.

3-8. INSTRUMENTS AND EXPERIMENTS (ION–MOLECULE REACTIONS)

We now examine several beam techniques used for the study of ion–molecule reactions and present a few of the data obtained by their use.

A. Armentrout's Guided Ion Beam Tandem Mass Spectrometer

Armentrout's instrument, shown in Fig. 3-8-1, has been used at the University of Utah for a large number of studies of various types. Ions are extracted from an ion source, accelerated to a known energy, and focused into a 60° magnetic sector momentum analyzer. The emerging mass-selected beam is then decelerated to the desired kinetic energy (which can range from about 0.05 to 500 eV Lab energy) and focused into an octopole ion beam guide that passes through a collision cell containing the neutral target. The pressure of the neutral reactant gas is sufficiently low that single-collision conditions apply. Both product and reactant ions drift from the gas cell to the end of the octopole, from which they are extracted and focused into a quadrupole mass filter. The emerging mass-selected ions are detected with a secondary electron scintillation ion detector and counted with digital electronics.

The octopole ion beam guide utilizes radio-frequency fields to form a potential well in the radial direction which guides ions along the axis. The octopole is 30 cm long and utilizes eight 0.8-mm-diameter stainless steel rods held on a 11.1-mm-diameter circumference. The ion energies along the octopole axis are perturbed to a negligible extent. The ion guide ensures efficient collection of all of the ionic products and also permits the determination of the absolute energy scale by retarding techniques.

Ion intensities are converted to absolute reaction cross sections in the manner described by Ervin and Armentrout (1985), who also describe the apparatus in detail. The uncertainties of the cross sections and the energy scale are believed to be generally less than 20% and 0.05 eV, respectively. In most cases laboratory energies are converted to CM energies by use of the stationary target

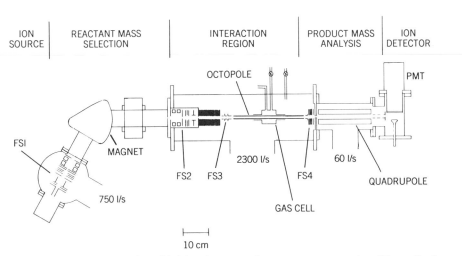

Figure 3-8-1. Armentrout's guided ion beam tandem mass spectrometer. [From Ervin and Armentrout (1985).]

approximation, but at very low beam energies, the thermal motion of the neutral reactant molecules is taken into account.

Primary ion production is carefully controlled to produce specific states or a known distribution of states of the desired ion. Three kinds of ion sources are most commonly used: surface ionization, electron impact, and high-pressure sources.

In one study made with this instrument, Ervin and Armentrout (1985) measured cross sections for the exothermic reactions of Ar^+ with H_2, D_2, and HD to form ArH^+ and ArD^+ over the CM energy range extending from thermal to 30 eV. Their data for D_2 and H_2 are shown in Fig. 3-8-2. The cross sections for H_2 and D_2 are nearly identical over this entire energy range when compared at the same CM energy. The total HD cross section is the same as for H_2 and D_2 at low energies but differs significantly above 4 eV CM, where product dissociation becomes important.

Sunderlin and Armentrout (1990) have studied the exothermic reactions of O^+ with HD as a function of translational energy and hydrogen temperature, with results shown in Figs. 3-8-3 and 3-8-4. While no change in the total cross section as a function of temperature is observed, the branching ratios for formation of OH^+ and OD^+ do depend on the HD rotational temperature.

Armentrout (1990) has accumulated cross-section data on the reactions of 44 atomic ions with molecular hydrogen. He finds pronounced diversity in behavior as he moves through the periodic table, revealing the fundamental changes that occur in the chemistry as the number of electrons is systematically

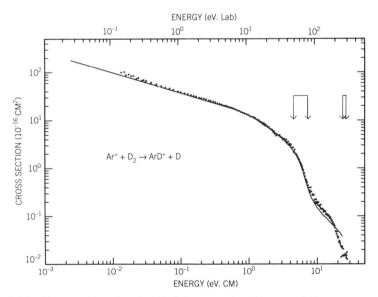

Figure 3-8-2. Cross section for the (Ar^+, D_2) reaction (dots) and the corresponding (Ar^+, H_2) reaction for comparison (solid curve). [From Ervin and Armentrout (1985).]

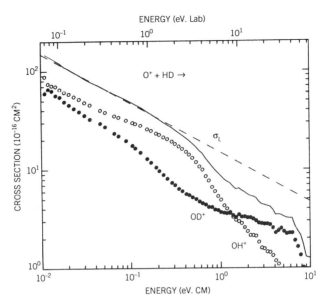

Figure 3-8-3. Cross sections for the reaction of O^+ with HD to form OH^+ and OD^+, and their sum (solid line) at 105 K. The dashed line is a plot of the Langevin cross section (Note 3-1). [From Sunderlin and Armentrout (1990).]

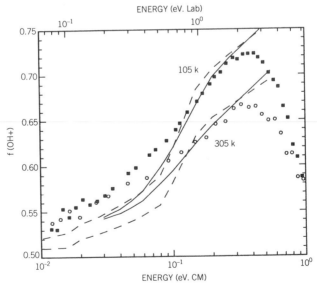

Figure 3-8-4. Fraction of OH^+ formed in the (O^+, HD) reaction at 305 K (open circles) and 105 K (closed squares). The dashed and solid lines are calculated values. In both cases the upper line corresponds to 100 K and the lower line to 300 K. [From Sunderlin and Armentrout (1990).]

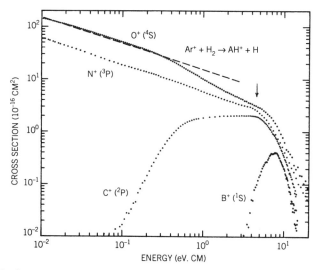

Figure 3-8-5. Cross sections for the reactions $A^+ + H_2 \rightarrow AH^+ + H$ with $A^+ = B^+(^1S)$, $C^+(^2P)$, $N^+(^3P)$, and $O^+(^4S)$. The dashed line gives the Langevin cross section (Note 3-1). [From Armentrout, (1990).]

varied. Armentrout's cross sections for several reactions, including the much studied $C^+ + H_2 \rightarrow CH^+ + H$ endothermic reaction, are shown in Fig. 3-8-5.

Fisher and Armentrout (1990) have measured the cross section for the reaction $O_2^+(^2\Pi_g, v = 0) + CH_4$ and used the results to obtain the heat of formation of HO_2 and the ionization energy of CH_4. In fact, over the years a large number of thermochemical data have been derived from the study of ion–molecule reactions.

Armentrout (1988) has published a brief review dealing with the dependence of ion–molecule reactions on kinetic, electronic, vibrational, and rotational energies. Information of this type is very useful in improving our understanding of reaction mechanisms.

It should be pointed out that the guided ion beam method was first developed by E. Teloy and D. Gerlich in 1974. The technique has been described in detail by Gerlich (1985).

B. Ng's Guided Ion Beam Instruments

Ng has constructed at the Iowa State University two instruments that have been used to study ion–molecule reactions, charge transfer, and dissociation. The first instrument is quite similar to Armentrout's; the second incorporates an additional octopole ion guide and quadrupole mass filter. In one study Flesch and Ng (1990) measured absolute integral cross sections for the reaction of $Ar^+(^2P_{3/2,1/2})$ with N_2 over the CM energy range 6.2 to 123.5 eV. Here dissociative charge transfer takes place as well as the ion–molecule exchange reaction. Flesch et al. (1990) have also measured the absolute state-selected and state-to-state integral cross sections for the $Ar^+(^2P_{3/2,1/2}) + O_2$ reactions over the CM energy range 0.044 to 133.3 eV. In this case charge transfer (dissociative and nondissociative) occurs in addition to the ion–molecule exchange reaction.

Liao et al. (1990) have made similar studies of the reactions of $H_2^+(X, v = 0$ to 4) + Ar, separating out the effects of charge transfer, dissociation, and the exchange reaction. The prominent features of the cross sections for the charge-transfer and ion–molecule reactions are controlled by the close resonance of the reactants $H_2^+(X, v = 2) + Ar$ and the charge-transfer products $H_2(X, v = 0) + Ar^+(^2P_{1/2})$. Below 3 eV CM, the charge-transfer cross section is peaked at $v = 2$, because of the preferential population of $Ar^+(^2P_{1/2})$. At $E > 5$ eV the intensity of the charge-transfer product $Ar^+(^2P_{3/2})$ exceeds that of $Ar^+(^2P_{1/2})$. On the other hand, the cross section for the ion–molecule reaction is only weakly dependent on the vibrational state of H_2^+. The ion–molecule reaction cross section decreases with increasing impact energy, and above 20 eV, it is negligible compared to the charge-transfer cross section.

Ng (1988) has published a useful review of ion–molecule reaction dynamics as studied by photoionization methods. It concentrates on measurements of state-selected and state-to-state ion–molecule reaction cross sections.

C. Lee's Studies of the Effects of Vibrational Excitation

Lee and his colleagues at Berkeley have used molecular beams and the guided ion beam technique to investigate the effects of translational and vibrational energy of reactant ions on the cross sections of ion–molecule reactions. In their first study, Anderson et al. (1981) examined several isotopic variants of the exoergic $H_2^+ + H_2$ reaction, which proceeds almost entirely by a direct mechanism without any observable barrier. They utilized a photoionization ion source to control the internal energy of the reagent ions. Some of their data on the proton transfer reaction

$$H_2^+ + D_2 \rightarrow D_2H^+ + H$$

and the atom transfer reaction

$$D_2^+ + H_2 \rightarrow D_2H^+ + H$$

are shown in Figs. 3-8-6 and 3-8-7. The CM impact energy was varied from 0 to 6 eV, and the vibrational state of the ion from $v = 0$ to $v = 4$. Figure 3-8-6 is a linear plot of the cross section versus energy for the proton transfer reaction. Results for the $v = 0$ and $v = 4$ states of the reactant H_2 ions are plotted separately. Since all of the cross-section data for the proton transfer and atom transfer reactions had approximately the same shape, the data could be plotted in Fig. 3-8-7 so as to emphasize the effects of reactant excitation. In the proton transfer reaction, vibrational energy decreases the cross section at low energies

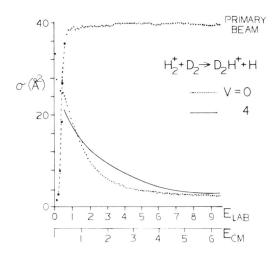

Figure 3-8-6. Cross sections for the $H_2^+ + D_2 \rightarrow D_2H^+ + H$ reaction, with the reactant ions in vibrational state $v = 0$ (lower dotted curve) and in $v = 4$ (solid curve.). The upper dotted curve shows the transmission function of the primary beam. [From Anderson et al. (1981).]

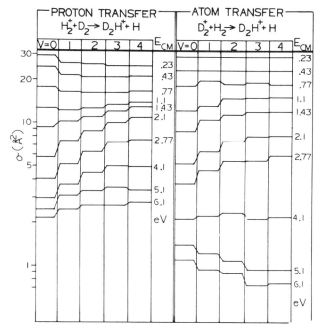

Figure 3-8-7. Vibrational and translational energy dependence of reactions forming D_2H^+ for reactant ions in the vibrational states $v = 0$ through $v = 4$. [From Anderson et al. (1981).]

227

but increases it at energies above about 1.4 eV. In the atom transfer reaction, vibrational excitation has essentially no effect at low energies, but there is considerable vibrational enhancement in the CM energy range 1 to 3 eV. At energies above 3 eV, the cross section is inhibited by vibrational excitation. It should be mentioned that Anderson et al. (1981) also studied collision-induced dissociation and charge transfer in the $H_2^+ + H_2$ system.

In Lee's laboratory, similar studies have also been made on the following systems:

$$H_2^+, D_2^+ + Ar \text{ (Houle et al., 1982)}$$

$$H_2^+ + N_2, CO, O_2 \text{ (Anderson et al., 1982)}$$

$$H_2^+, HD^+ + He \text{ (Turner et al., 1984)}$$

$$C_2H_2^+ + H_2 \text{ (Turner and Lee, 1984)}$$

D. Threshold Electron–Secondary Ion Coincidence Technique

The selection of specific internal states of reactant ions is an important feature of many modern measurements of ion–molecule reactions. This selection is often accomplished by the use of lasers, but the TESICO technique offers an interesting alternative. The basic physics of this technique may be outlined as follows. When ions are produced by photoionization, all energetically accessible internal states are generally populated, and each photoelectron that is emitted has an energy that indicates the internal energy of the corresponding photoion. Therefore, it is possible to study separately the reactions of ions in each of these populated states. The procedure is to measure reactant and product ions in coincidence with photoelectrons of a particular kinetic energy and to detect only the reactant ions in a single internal state and the product ions produced from reactant ions in that state.

The TESICO method is used in several laboratories and has produced a great deal of useful data. A review of the technique has been published by Koyano et al. (1985).

E. Multicoincidence Studies at Orsay

In the mid-1980s a group at Orsay began publishing the results of studies of negative ion–molecule reactions and other processes that they obtained with a powerful and sophisticated multicoincidence apparatus. Their instrument and techniques have been described in considerable detail by Brenot and Durup-Ferguson (1992). Here we give only a brief description, in the context of the Orsay studies of (S^-, H_2) collisions (Brenot et al., 1990). The following channels are possible:

Associative detachment (AD)	$SH_2 + e$
Associative detachment (AD)	$SH_2 + e$
H-transfer reaction (HT)	$SH^- + H$
Simple detachment (SD)	$S(^3p) + H_2 + e$
Reactive charge transfer (R)	$SH + H^-$
Reactive detachment (RD)	$SH + H + e$
Dissociative detachment (DD)	$S + H + H + e$
Dissociative charge transfer (DCT)	$S + H + H^-$

The apparatus is pictured in Fig. 3-8-8, in which a fast S^- beam intersects a cold H_2 beam that is produced by a supersonic expansion. The S^- ions are generated in an electrical discharge ion source containing a mixture of 0.1% of OCS in Ar buffer gas. After extraction, the negative ions are mass selected with a Wien filter to eliminate the SH^- component in the ion beam, and the S^- ions that remain in the beam are decelerated down to the desired energy (1.9 to 20 eV CM) immediately in front of the collision region. The fast neutral products (S, SH, and SH_2) and the slow charged products (e and H^-) are detected independently by two position-sensitive detectors containing microchannel plate assemblies (see Section 5-9 of the companion volume). The two correlated families of events are processed by a multicoincidence technique. The raw data consist of the number of events $N(yz \Delta t)$ associated with neutrals detected on an

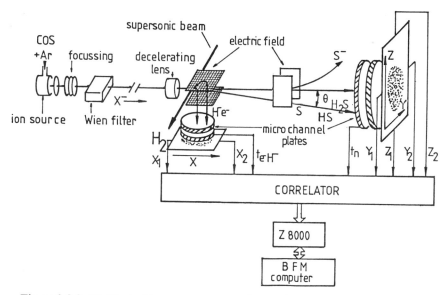

Figure 3-8-8. Multicoincidence apparatus at Orsay. [From Brenot et al. (1990).]

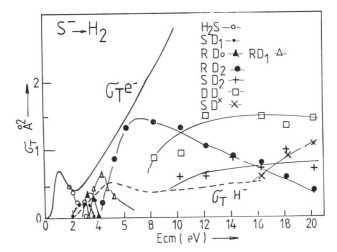

Figure 3-8-9. Integral cross sections as a function of CM collision energy for the various channels in (S^-, H_2) collisions. The data are normalized to the integral cross sections for e and H^- production. [From Brenot et al. (1990).]

elementary pixel (yz) of the microchannel plate and characterized by a difference of arrival time Δt between the neutral and charged products. A transformation from Lab to CM variables gives the CM differential cross sections, which are then integrated to obtain the integral CM cross sections for the various channels (Fig. 3-8-9). Brenot et al. (1990) display contour maps obtained at different energies that are very useful in the analysis of the dynamics of the collisions.

F.   Chemiluminescence Studies of Ion–Molecule Reactions

In Section 3-6-G we discussed chemiluminescence studies of neutral–neutral reactions; similar measurements are made on ion–molecule reactions by observing the radiation emitted from the products of the reactions. Emphasis is placed on the dynamic analysis of the reactions as revealed by their spectra. In an informative review, Ottinger (1984) compares ionic and neutral chemilumin-escence produced by ion–molecule reactions, discusses experimental techniques, and presents selected results. In a similar review, Leventhal (1984) discusses the emission of light from excited products of charge-transfer reactions.

3-9.   NOTES

3-1. Langevin Cross Section for Exothermic Ion–Molecule Reactions. The dashed lines in Figs. 3-8-3 and 3-8-5 are plots of the Langevin cross section σ_L for the indicated exothermic reactions. σ_L is the classical Langevin cross section for orbiting collisions between an ion and a molecule approaching one another under a pure $-1/r^4$ polarization potential with CM energy E.

It has the form (p. 91 of the companion volume; Su and Bowers, 1979, p. 87)

$$\sigma_L = \pi q \left(\frac{2\alpha}{E}\right)^{1/2}. \tag{3-9-1}$$

where α is the polarizability of the molecule and q is the ionic charge. The corresponding rate coefficient is

$$k_L = v_0\sigma_L = 2\pi q \left(\frac{\alpha}{M_r}\right)^{1/2}, \tag{3-9-2}$$

where v_0 is the relative velocity of approach and M_r is the reduced mass. We see that the pure polarization theory predicts that the cross section for capture of the ion is inversely proportional to v_0 (or $E^{1/2}$) and that the capture-rate coefficient is independent of v_0. The long-range attractive potential greatly increases the rate of collisions above what it would be if both structures were ground-state neutrals and is responsible for the very large cross sections (10^{-14} cm^2) of many ion–molecule reactions at thermal energies. The Langevin cross section closely approximates the measured cross section for many *exothermic* reactions but gives poor agreement in many others   the structure of the ion and molecule should be considered as well as additional terms in the interaction potential. Su and Bowers (1979) work out the theory of ion–dipole and ion–quadrupole collisions and predict the form of low-energy cross sections and rate coefficients for these interactions.

Armentrout (1990) discusses a number of methods of calculating cross sections for *endothermic* reactions for which the kinetic energy dependence of the cross sections is quite different. For general discussions of the theory of ion–molecule reactions, the reader is referred to the book by Baer and Ng (1992) and the review by Chesnavich and Bowers (1979).

3-2. *Paul Quadrupole Mass Filter; Paul and Penning Traps.* In the 1950s, W. Paul and colleagues at the University of Bonn were designing electric and magnetic multipoles to serve as lenses to focus atomic and molecular beams and to separate particles with different orientations of their dipole moments. Paul discovered that ions with different masses could be guided and separated by a radio-frequency (RF) electric quadrupole field upon which a dc electric field is superimposed (Fig. 3-9-1). Instruments of this type, known as *quadrupole mass filters*, are widely used in many kinds of research. A brief discussion is in order here.

In the mass filter the ions are subjected to an electric field **E** that is a linear function of the coordinates. This field is produced by a potential of the form

$$V(x, y, z, t) = V^*(t)(\alpha x^2 + \beta y^2 + \gamma z^2), \tag{3-9-3}$$

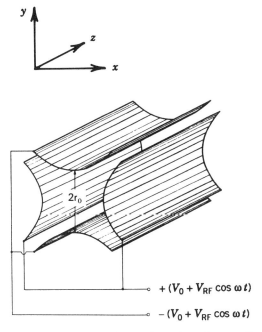

Figure 3-9-1. Hyperbolic electrodes for the Paul quadrupole mass filter. [Adapted from Paul et al. (1958).]

where $\alpha + \beta + \gamma = 0$ because $\nabla^2 V = 0$. In the present case we put $\alpha = -\beta = 1/r_0^2$ and $\gamma = 0$. The desired field is realized by using electrodes of hyperbolic cross section with spacing $2r_0$, and applying potentials $V^*(t) = \pm(V_0 + V_{RF}\cos\omega t)$, where V_0 and V_{RF} are constants (Fig. 3-9-1). Then

$$V(x, y, z, t) = (V_0 + V_{RF}\cos\omega t)\frac{x^2 - y^2}{r^2}\left(\frac{r}{r_0}\right)^2. \qquad (3\text{-}9\text{-}4)$$

Note that the equation $x^2 - y^2 = r^2$ describes a family of hyperbolas with vertices on the X axis. For a given value of $r \leqslant r_0$, the potential on the corresponding hyperbola is $V = (V_0 + V_{RF}\cos\omega t)(r/r_0)^2$. Now we rewrite (3-9-4) as

$$V(x, y, z, t) = -(V_0 + V_{RF}\cos\omega t)\frac{y^2 - x^2}{r^2}\left(\frac{r}{r_0}\right)^2, \qquad (3\text{-}9\text{-}5)$$

and note that the equation $y^2 - x^2 = r^2$ describes hyperbolae with vertices on the Y axis. Then for a given value of $r \leqslant r_0$, the potential of the associated hyperbola is $V = -(V_0 + V_{RF} \cos \omega t)(r/r_0)^2$. Equations (3-9-4) and (3-9-5) show that the potential can also be written in the form

$$V(r, \phi, t) = V^*(t) \left(\frac{r}{r_0}\right)^2 \cos 2\phi \qquad (3\text{-}9\text{-}6)$$

where ϕ is the azimuthal angle measured counterclockwise from the X axis. Then we see that the radial component of the electric field is

$$E_r = -\frac{2V^*(t)}{r_0^2} r \cos 2\phi, \qquad (3\text{-}9\text{-}7)$$

and that

$$\frac{\partial E_r}{\partial r} = -2V^*(t) \frac{\cos 2\phi}{r_0^2}. \qquad (3\text{-}9\text{-}8)$$

Singly charged ions that are injected along the Z axis oscillate under the influence of the alternating field according to the Mathieu equations (Mathews and Walker, 1965)

$$\ddot{x} + (a + 2q \cdot \cos 2\xi)x = 0$$
$$\ddot{y} - (a + 2q \cdot \cos 2\xi)y = 0, \qquad (3\text{-}9\text{-}9)$$

where

$$\omega t = 2\xi, \qquad (3\text{-}9\text{-}10)$$

$$a = \frac{8eV_0}{m\omega^2 r_0^2}, \qquad (3\text{-}9\text{-}11)$$

$$q = \frac{4eV_{RF}}{m\omega^2 r_0^2}. \qquad (3\text{-}9\text{-}12)$$

Here e is the elementary charge and m is the mass of the ion. There are two types of solutions to the equations of motion. In one case the orbits of the ions are stable. In the other case the amplitude of the oscillations grows exponentially with time, and the ions collide with the electrodes and are lost. The stability behavior of the ions is given by the two parameters a and q. An ion will be transmitted through the filter only if its (a, q) value lies within the bounded area shown in Fig. 3-9-2 [see Bassi (1988b)]. This value for (a, q) does not depend on the initial conditions of the ions. The ratio $a/q = 2V_0/V_{RF}$ does not depend on the mass, so ions of different masses all lie on a straight line passing through the origin of the stability diagram

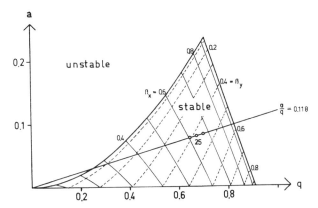

Figure 3-9-2. Stability diagram for the Paul quadrupole mass filter. [From Paul et al (1958).]

provided the field parameters are constant. If the ratio V_0/V_{RF} has a sufficiently high value, the stable mass interval may be narrowed to the point where only ions of one selected mass are transmitted. This situation makes it possible to obtain very high mass resolution ($> 10{,}000$), although only with extremely small currents. In atomic collisions research, a resolving power of 100 is usually sufficient, and a mass filter and its operating conditions are selected accordingly.

It should be noted that the hyperbolic electrodes shown in Fig. 3-9-1 are difficult to machine. Accordingly, circular rods are usually employed in their place; they provide an excellent approximation to the desired field (see Fig. 3-9-3).

The use of the quadrupole mass filter is illustrated many times in this volume (see, e.g., Figs. 3-6-1 and 3-8-1). It has largely supplanted the magnetic sector field spectrometer and other types of instruments for the identification of ions in atomic collisions experiments.

Paul has also developed a three-dimensional version of the mass filter to confine ions to a small region. An RF electric field is applied between the cap electrodes and the ring-shaped electrode at the center of the cylindrical trap shown in Fig. 3-9-4. Hyperbolic potentials are produced that make the ionic motion harmonic to first order. A Nobel Prize was awarded in 1989 for the development of the *Paul trap* and related work. H. Dehmelt also won a Nobel Prize in that year for his research with *Penning traps* (see Fig. 3-9-4). These devices have a static magnetic field applied parallel to the symmetry axis (perpendicular to the plane of the ring electrode), with a weak dc electric quadrupole field superimposed. The two kinds of traps described here have a wide range of intriguing and important applications in atomic collisions, spectroscopy, and other fields [see Paul (1990), Dehmelt (1990), and Ekstrom and Wineland (1980).]

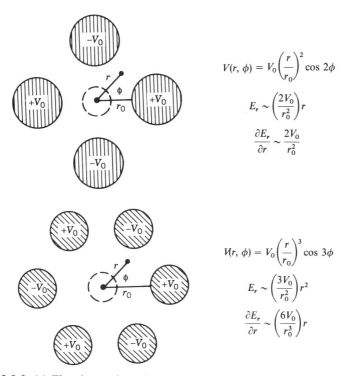

$$V(r, \phi) = V_0 \left(\frac{r}{r_0}\right)^2 \cos 2\phi$$

$$E_r \sim \left(\frac{2V_0}{r_0^2}\right) r$$

$$\frac{\partial E_r}{\partial r} \sim \frac{2V_0}{r_0^2}$$

$$V(r, \phi) = V_0 \left(\frac{r}{r_0}\right)^3 \cos 3\phi$$

$$E_r \sim \left(\frac{3V_0}{r_0^3}\right) r^2$$

$$\frac{\partial E_r}{\partial r} \sim \left(\frac{6V_0}{r_0^3}\right) r$$

Figure 3-9-3. (*a*) Electric quadrupole and (*b*) electric hexapole utilizing circular rods instead of hyperbolic electrodes. The potentials shown on the rods are for dc operation of the assemblies as state selectors. The shaded areas show the incoming beams, and the dashed profiles the maximum expected deflection of the beams. The displayed equations are for hyperbolic electrodes but are accurate to a good approximation for the circular rod systems near the central axis. [Adapted from Pauly and Toennies (1968).]

3-3. State Selection of Beam Molecules in Inhomogeneous Fields. As we shall show, it is possible to deflect certain classes of neutral molecules by passing them through static, inhomogeneous electric or magnetic fields. Furthermore, a suitable multipole device can select molecules in particular quantum states and also provide a focusing action. The selector acts like a lens, so that molecules in selected states coming from a small source are focused to a small image at the focal point of the lens. This focusing feature often plays a vital role in providing the beam intensity required in a given experiment. Here we restrict our attention to electrostatic quadrupole and hexapole fields, following the treatment by Bernstein (1982). Other types of multipole selectors are considered in the problems at the end of the chapter. Before proceeding, however, we must develop a general expression for the energy of an electric dipole in an electrostatic field and examine the force on the dipole in the field.

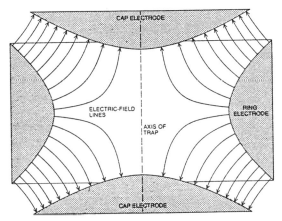

Figure 3-9-4. Electric quadrupole configuration in both the Paul and Penning traps. A radio-frequency field is employed in the Paul device, whereas the Penning trap utilizes both a dc quadrupole field and a constant, axial magnetic field. [From Ekstrom and Wineland (1980).]

A. *Force on a Polar Molecule in an Electrostatic Field.* Consider a molecule with a permanent electric dipole moment **μ** positioned within an electrostatic field **E** with its axis making an angle θ with respect to the local field lines. The energy of the molecule in a given rotational state as perturbed by the field is denoted by W_{JKM}. Here J is the total angular momentum quantum number, M is the orientation quantum number ($M = 0, \pm 1, \ldots, \pm J$) or the projection quantum number in the Lab frame, and K is the projection quantum number in the molecular frame ($K = 0, \pm 1, \ldots, \pm J$). If we put $P^2 = J(J+1)\hbar^2$, the component of the associated total angular momentum vector **P** along **E** is $M\hbar$, and the component of **P** along the figure axis of the molecule is $P_z = K\hbar$. [See the illustration on p. 53 of Bernstein (1982).]

The perturbation of the energy of the polar molecule can be expressed in a perturbation series as

$$W_{JKM} = W^{(0)} + W^{(1)} + W^{(2)} + \cdots = W_{JKM}^{(0)} + \lambda_1 E + \lambda_2 E^2 + \cdots. \quad (3\text{-}9\text{-}13)$$

Here $W_{JKM}^{(0)}$ is the energy of the rotor in the limit $E \to 0$, and λ_1 and λ_2 are the coefficients of the first- and second-order Stark terms, which split the degenerate rotational levels.* We then write

$$W = W^{(0)} - \boldsymbol{\mu} \cdot \mathbf{E} = W^{(0)} - \mu \cos \theta\, E \quad (3\text{-}9\text{-}14)$$

---

*The first-order Stark effect arises from the interaction of the permanent electric dipole moment of the molecule with the external electric field if the molecule has such a dipole moment. The second-order Stark effect is produced by the electric dipole moment that is induced by the field and interacts with it. The induced dipole moment has a strength directly proportional to E and is given by the equation $\mu_{\text{ind}} = \alpha \cdot \mathbf{E}$, where α is the tensor polarizability of the molecule (Townes and Schawlow, 1955, pp. 248–252).

and define the *effective dipole moment* of the molecule in the field to be

$$\mu_{\text{eff}} = \mu \langle \cos \theta \rangle = -\frac{\partial W}{\partial E}, \tag{3-9-15}$$

where $\langle \cos \theta \rangle$ is the quantum mechanical expectation value of the cosine of the orientation angle θ of the dipole moment. Using (3-9-13), we find that

$$\mu_{\text{eff}} = -\lambda_1 - 2\lambda_2 E + \cdots. \tag{3-9-16}$$

Then the radial force on the molecule located at some point **r** in the field is

$$F_r = -\frac{\partial W}{\partial r} = -\frac{\partial W}{\partial E} \frac{\partial E}{\partial r} = \mu_{\text{eff}} \frac{\partial E}{\partial r}, \tag{3-9-17}$$

and by (3-9-16) we obtain

$$F_r = -\lambda_1 \frac{\partial E}{\partial r} - 2\lambda_2 E \frac{\partial E}{\partial r} + \cdots \tag{3-9-18}$$

as the radial force on the polar molecule.

We now consider as a special case a polar symmetric-top molecule, which exhibits a first-order Stark effect. Here

$$W_{JKM} \approx W_{JKM}^{(0)} + W^{(1)}, \tag{3-9-19}$$

where

$$W^{(1)} = -\mu_{\text{eff}} E = -\mu \langle \cos \theta \rangle E = \lambda_1 E. \tag{3-9-20}$$

Quantum mechanics gives the first-order Stark energy to be

$$W^{(1)} = -\mu \frac{KM}{J(J+1)} E \tag{3-9-21}$$

(Problem 3-17), and comparison with (3-9-20) shows that

$$\langle \cos \theta \rangle = \frac{KM}{J(J+1)}. \tag{3-9-22}$$

Note that in the present first-order limit, $\langle \cos \theta \rangle$, which is a measure of the alignment of the dipole, is independent of the strength of the field, E.

We now have the radial force on a polar symmetric-top molecule:

$$F_r \approx -\lambda_1 \frac{\partial E}{\partial r} = \frac{KM}{J(J+1)} \mu \frac{\partial E}{\partial r}. \tag{3-9-23}$$

If we assume that the molecule is in an electrostatic hexapole field, whose properties are given in Fig. 3-9-3, we obtain

$$F_r = \frac{KM}{J(J+1)} \mu \frac{6V_0}{r_0^3} r = -k_{S-H} r, \tag{3-9-24}$$

where

$$k_{S-H} = -6 \frac{V_0}{r_0^3} \mu \frac{KM}{J(J+1)} \tag{3-9-25}$$

is the radial force constant.

Next we consider a polar diatomic (or linear polyatomic) molecule, which exhibits no first-order Stark effect, so that $\lambda_1 = 0$ in (3-9-13) (see Problem 3-15). Here

$$W_{JM} = W_{JM}^{(0)} + \lambda_2 E^2 + \cdots, \tag{3-9-26}$$

where

$$W_{JM}^{(0)} = BJ(J+1) \tag{3-9-27}$$

for a rigid rotator, with $B = \hbar^2/2I$, where I is the moment of inertia. For this class of molecules,

$$\lambda_2 = f(J, M) \frac{\mu^2}{B}, \tag{3-9-28}$$

where

$$f(J, M) = \frac{J^2 + J - 3M^2}{2(J^2 + J)(2J - 1)(2J + 3)} \tag{3-9-29}$$

except for $J = M = 0$, for which $f(0,0) = -\frac{1}{6}$ (Problem 3-16). Then, by (3-9-16) and (3-9-28),

$$\mu_{\text{eff}} \equiv \mu \langle \cos \theta \rangle \approx -2\lambda_2 E = -2 \frac{\mu^2}{B} f(J, M)E, \tag{3-9-30}$$

so that

$$\langle \cos \theta \rangle = -2f(J, M) \frac{\mu E}{B} + \cdots . \tag{3-9-31}$$

(Here the alignment for a given state does depend on the field strength.) Thus we obtain

$$F_r \approx -2\lambda_2 E \frac{\partial E}{\partial r} = -2f(J, M) \frac{\mu^2}{B} E \frac{\partial E}{\partial r} \tag{3-9-32}$$

for the radial force on the molecule [cf. eq. (10-8) on p. 251 of Townes and Schawlow (1955)]. If we now assume the field to be of the electric quadrupole type (see Note 3-2 and Fig. 3-9-3), the radial force on a polar diatomic molecule becomes

$$F_r = -2f(J, M) \frac{\mu^2}{B} \frac{4V_0^2}{r_0^4} r = -k_{D-Q} r, \tag{3-9-33}$$

where

$$k_{D-Q} = 8 \frac{V_0^2}{r_0^4} \frac{\mu^2}{B} f(J, M). \tag{3-9-34}$$

B. *State Selection and Focusing of Molecules in Electrostatic Multipoles.* Let the Z axis coincide with the axis of the electrostatic multipole under discussion, and let the length of the device be l. A beam of neutral molecules is injected axially with speed v, the molecules making small angles with respect to the Z axis. The multipole field acts on the small radial component of the velocity, but not on the large longitudinal component. The radial equation of motion is

$$F_r = -kr = m\ddot{r}, \tag{3-9-35}$$

with a sinusoidal solution if $k > 0$, or an exponential solution if $k < 0$. If we take $k > 0$, along with the initial condition $r = 0$ at $z = 0$, the solution for the radial motion with $z = l$ is

$$r = \sin\left[\left(\frac{k}{m}\right)^{1/2} \frac{l}{v}\right]. \tag{3-9-36}$$

In this case the molecule will execute a sinusoidal orbit and return to the axis when

$$\left(\frac{k}{m}\right)^{1/2}\frac{l}{v} = n\pi \qquad (n = 1, 2, 3, \ldots). \tag{3-9-37}$$

For a polar symmetric top in an electrostatic 6-pole, the condition for single-loop ($n = 1$) focusing is, by (3-9-25),

$$-6\frac{V_0}{r_0^3}\mu\frac{KM}{J(J+1)} = \frac{\pi^2 mv^2}{l^2}, \tag{3-9-38}$$

or

$$V_0 = \frac{\pi^2}{6}\frac{r_0^3}{l^2}\frac{mv^2}{\mu}\left[\frac{-KM}{J(J+1)}\right]^{-1}, \tag{3-9-39}$$

where V_0 is the magnitude of the voltage applied to the hexapole electrodes. Only states with $\langle\cos\theta\rangle < 0$ can be focused.

On the other hand, for a polar diatomic in an electrostatic 4-pole, (3-9-34) gives as the condition for single-loop ($n = 1$) focusing

$$\frac{8V_0^2\mu^2 f(J, M)}{r_0^4 B} = \frac{\pi^2 mv^2}{l^2}, \tag{3-9-40}$$

$$V_0^2 = \frac{\pi^2}{8}\frac{r_0^4}{l^2}\frac{mv^2}{\mu^2/B}\frac{1}{f(JM)}. \tag{3-9-41}$$

Only states with $f(J, M) > 0$ (i.e., $\mu_{\text{eff}} < 0$) will be focused.

We see that if we are given the molecular properties m, B, μ, and v, as well as the multipole dimensions l and r_0, we can calculate by (3-9-39) and (3-9-41) the rod voltage V_0 at which focusing occurs (see Problem 3-25).

In practice, the direct unfocused beam ($n = 0$) and the even-loop trajectories ($n = 2, 4, \ldots$) are blocked by a beam stop placed at the midpoint of the multipole axis. Molecules following the odd trajectories with $n = 3, 5, \ldots$ diverge rapidly and do not interfere appreciably with the image of the desired ($n = 1$) molecules.

Bernstein (1982, pp. 50–54) discussed experiments that have been performed on polar diatomics with electrostatic 4-poles and on symmetric-top molecules with electrostatic 6-poles. As mentioned in Note 1-6, Toennies (1965) has utilized two electric 4-poles in tandem to measure cross sections for rotational transitions in polar T*l*F molecules colliding with a variety of targets. As indicated in Fig. 3-9-5, the first

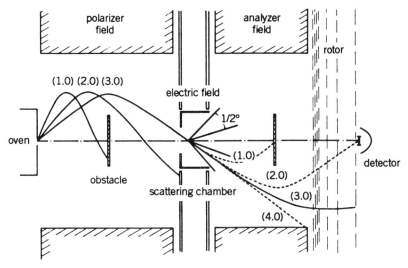

Figure 3-9-5. Side view of Toennies's apparatus for measuring inelastic cross sections. Trajectories for various (J, M) molecular states are shown. The polarizer (selector) field produced by the first quadrupole allows only molecules in the $(3, 0)$ state to enter the scattering chamber. The analyzer field produced by the second quadrupole permits only $(2, 0)$ molecules to arrive at the detector. The uniform electric field perpendicular to the velocity prevents randomization of the M states and provides polarization of the state-selected molecules [see Stolte (1988).] [From Pauly and Toennies (1968).]

quadrupole serves as a selector for (J, M) states in the incident beam, the second as an analyzer of the final (J, M) states. Toennies was the first to perform experiments of this type.

3-4. Orientation of Molecular Beams by Inhomogeneous Electric Fields. Before proceeding it is appropriate to review the concepts of orientation and alignment, taking note of the difference between them (Levine and Bernstein, 1987). Let us consider an ensemble of diatomic molecules (rotational state J) in a weak electric field. If the ensemble is *randomly oriented* (i.e., *unpolarized*), all of the orientation quantum numbers $M_J = 0, \pm 1, \ldots, \pm J$ are equally probable. If, however, the population distribution varies linearly with M_J^2, we have the case of *pure orientation*. Finally, if the distribution depends only on M_{J^2}, the ensemble is said to be *purely aligned*. These three cases are illustrated in Fig. 5-10-2 of the companion volume); there are many intermediate situations. The *polarization* of the ensemble is the difference between the actual multipole mean value and that for the randomly oriented ensemble.

Leone (1984) expresses the distinction between alignment and orientation in graphic terms. Consider first a linear molecule whose ends are unlike each other. Alignment permits the experimenter to position the linear axis either parallel or perpendicular to a collision partner. Orientation permits the system to be situated so that either of the two different ends of the linear molecule can point toward the collision partner.

The following discussion is based on the paper by Harland et al. (1990), which describes studies of reactions between K atoms and CF_3I and CH_3I molecules. Harland et al. point out that a polar symmetric-top molecule such as CH_3I rotates in an electric field in the same manner as a toy top in a gravitational field. The symmetry axis precesses about an applied **E** field, and the electric dipole moment does not average to zero.* In a collision-free molecular beam, each symmetric-top molecule is oriented with respect to a weak applied **E** field. However, all orientations are represented, so an ensemble of the molecules has no net orientation. An oriented ensemble may be produced by removing those molecules with unwanted orientations, and this can be done by passing the beam through an inhomogeneous **E** field that deflects each molecule according to its orientation with respect to the local field direction.

As we saw in Note 3-3, the energy of interaction of a symmetric-top molecule with the **E** field is given by the first-order Stark effect:

$$W^{(1)} = -\mu E \frac{KM}{J(J+1)}, \qquad (3\text{-}9\text{-}21)$$

or, by (3-9-22),

$$W^{(1)} = -\mu \langle \cos\theta \rangle E. \qquad (3\text{-}9\text{-}42)$$

Here μ is the electric dipole moment of the molecule; $J, K,$ and M are quantum numbers defined in Note 3-3; and $\langle \cos\theta \rangle$ is interpreted classically as the angle between the symmetry axis of the molecule and the local field lines. In an inhomogeneous **E** field, each molecule experiences a force $\mathbf{F} = -\nabla W$ that tends to minimize its energy, so molecules with negative $\langle \cos\theta \rangle$ can be separated from those with positive $\langle \cos\theta \rangle$. Any inhomogeneous **E** field that is sufficiently strong will accomplish this objective, but as we showed in Note 3-3, an electric hexapole field will also *focus* symmetric-top molecules with negative $\langle \cos\theta \rangle$.

Harland et al. (1990) use an electric 6-pole that is 137 cm long, with 6.4-mm-diameter stainless steel rods mounted on a 15.9-mm-diameter circle. The rod voltage, V_0, is typically about 5 kV. Inside the 6-pole, molecules with negative values of $\langle \cos\theta \rangle$ are oriented with respect to the local

*On the other hand, a polar diatomic molecule rotates like a two-blade propellor in a plane normal to the angular momentum vector, so its electric dipole moment averages to zero in the absence of an electric field. In the presence of a field, the molecule has an electric dipole moment whose value depends on the direction in which its angular momentum vector is pointing, as well as on the quantity $(\mu E)^2$ (see Problem 3-15). Diatomic molecules exhibit second-order Stark effects.

nonuniform field. These molecules pass out of the hexapole into a region of uniform **E** field, which provides a uniform axis of orientation for studies of collisions.

As we have stressed, the state selection is actually according to the algebraic sign of $\langle \cos \theta \rangle$, but if only a few (J, K, M) states are populated, it is possible to select individual (J, K, M) states.

3-10. PROBLEMS

Solutions for many of these problems appear in the references indicated.

3-1. The chemical identities, masses, and internal energies of particles may change in a reaction of the type $A + B \rightarrow C + D$. Suppose that B is at rest in the Lab frame. Let $T_{CM,i}$ be the initial kinetic energy in the CM frame, and Q be the decrease in the total internal energy of the system occurring in the collision. Show that the CM and Lab differential cross sections for observation of the product C in a given direction are related by

$$\frac{d\sigma_{\text{Lab}}}{d\Omega_{\text{Lab}}} (\theta_{\text{Lab}}, \varphi_{\text{Lab}}) = \frac{(1 + x^2 + 2x \cos \theta_{\text{CM}})^{3/2}}{|1 + x \cos \theta_{\text{CM}}|} \frac{d\sigma_{\text{CM}}}{d\Omega_{\text{CM}}} (\theta_{\text{CM}}, \varphi_{\text{CM}}),$$

where

$$x = \left(\frac{m_A m_C}{m_B m_D} \frac{T_{CM,i}}{T_{CM,i} + Q} \right)^{1/2}.$$

[See Joachain (1975, pp. 20–21).]

3-2. Consider an unstable particle with internal energy E_0 moving with velocity **V** in the Lab frame. It undergoes *spontaneous dissociation* into two fragments with masses m_1 and m_2 and internal energies E_1 and E_2. The velocities of the fragments make angles θ_1 and θ_2 with respect to **V** in the Lab frame. Show that the relation between θ_1 and θ_2 is

$$\frac{m_2}{m_1} \sin^2\theta_2 + \frac{m_1}{m_2} \sin^2\theta_1 - 2 \sin \theta_1 \sin \theta_2 \cos(\theta_1 + \theta_2)$$

$$= \frac{2\Delta E}{(m_1 + m_2)V^2} \sin^2(\theta_1 + \theta_2),$$

where $\Delta E = E_0 - E_1 - E_2$. [See Landau and Lifshitz (1960, p. 43).]

3-3. In Problem 3-2, consider one of the fragments, and let v_L and v_C be its velocities in the Lab and CM frames. The angle between v_L and **V** is labeled θ here. Show that the angular distribution of the fragments in the Lab frame is

$$\frac{1}{2} \sin \theta \, d\theta \left[2 \frac{V}{v_C} \cos \theta + \frac{1 + (V^2/v_C^2) \cos 2\theta}{[1 - (V^2/v_C^2) \sin^2\theta]^{1/2}} \right] \quad (0 \leqslant \theta \leqslant \pi) \quad \text{if } v_C > V$$

and

$$\sin \theta \, d\theta \, \frac{1 + (V^2/v_C^2) \cos 2\theta}{[1 - (V^2/v_C^2) \sin^2\theta]^{1/2}} \quad (0 \leqslant \theta \leqslant \theta_{max}) \quad \text{if } v_C < V.$$

[See Landau and Lifshitz (1960, p. 44).]

3-4. In Problem 3-2 assume that $v_{2C} > v_{1C}$ applies for the speeds of the fragments in the CM frame, and show that the range of possible values of the angle between the velocities of the fragments $(\theta_t = \theta_1 + \theta_2)$ is

$$0 < \theta_t < \pi \qquad \text{if } v_{1C} < V < v_{2C};$$
$$\pi - \theta_0 < \theta_t < \pi \qquad \text{if } V < v_{1C};$$
$$0 < \theta_t < \theta_0 \qquad \text{if } V > v_{2C},$$

where

$$\sin \theta_0 = \frac{V(v_{1C} + v_{2C})}{V^2 + v_{1C}v_{2C}}.$$

[See Landau and Lifshitz (1960, p. 44).]

3-5. Study the discussion in Scoles (1988) of the seeded beam technique for generating neutral beams with energies in the eV region.

3-6. Study the discussion of surface ionization detectors in Bassi (1988a).

3-7. Thermal detectors can operate either at room temperature (thermistors) or at low temperatures (cryogenic bolometers). Discuss the principles of operation of these devices. [See Zen (1988).]

3-8. In 1955, Taylor and Datz published an account of the first successful crossed-beam measurement on a gas-phase chemical reaction (Taylor and Datz, 1955). Taking into account the advances in experimental techniques that have occurred since that time, how would you modify their landmark experiment if you were to do it now?

3-9. Plot the pure polarization potential for the (H^+, CH_4) system. Then, on the same graph, roughly indicate the effects of other contributions to the true potential that are neglected if only the $-1/r^4$ polarization potential is used.

3-10. Apply classical elastic scattering theory to ion–molecule interactions, assume a pure polarization potential, and obtain (3-9-1) for the Langevin cross section and (3-9-2) for the reaction rate. [See Su and Bowers (1979).]

3-11. Study the treatment of ion (permanent)–dipole and ion (permanent)–quadrupole scattering theory as presented by Su and Bowers (1979).

3-12. Consider a beam of H^+ ions (0.1 eV Lab energy) intersecting a beam of N_2 molecules (0.1 eV Lab energy) at an angle of 45°. Calculate the Langevin cross section and the corresponding reaction rate (Note 3-1). What is the cross section for H^+ ions of 1 eV energy?; of 10 eV?; of 100 eV?

3-13. A quadrupole mass filter with length of 1 m and a radius (r_0) of 2 cm is operated with $V_0 = 166.8$ V, $V_{RF} = 1000$ V, and an RF frequency of 1.0 MHz (see Note 3-2). Ions of mass 32 amu are injected with an energy of 8 eV. Will they be transmitted? (Incidentally, the resolving power of this mass filter under the conditions stated is about 250.)

3-14. What are the advantages and disadvantages of the Stern–Gerlach arrangement in state selection and analysis? [See Reuss (1988).]

3-15. Consider a rotating polar diatomic molecule whose field-free values of the electric dipole moment, the angular momentum, the angular velocity, and the moment of inertia are μ, $\mathbf{P}$, ω_0, and I, respectively, and represent this molecule by a rotating dipole.

(a) Fix the dipole in a uniform electric field $\mathbf{E}$, with $\mathbf{P}$ perpendicular to $\mathbf{E}$. During half of each rotation of the dipole, the field produces an angular acceleration such that the angular velocity ω has its maximum value when the dipole points in the direction of $\mathbf{E}$. During the other half of the revolution, the field produces an angular deceleration, and ω has its minimum value when the dipole is antiparallel to $\mathbf{E}$. The dipole has a component antiparallel to $\mathbf{E}$ for a longer time (t_A) than it has a component parallel to $\mathbf{E}$ (t_P). Show that the fractional difference between t_A and t_P is

$$f \propto \frac{\mu E}{\frac{1}{2}I\omega_0^2}, \qquad (3\text{-}10\text{-}1)$$

the ratio of the energy of the dipole in the field to the rotational energy when $\mathbf{E} = 0$. Then show that the change in energy produced by the field is the positive quantity

$$\Delta W = f\mu E \propto \frac{(\mu E)^2}{hBJ(J+1)}, \qquad (3\text{-}10\text{-}2)$$

where B is the rotational constant of the molecule.

(b) Now consider the dipole to be rotating with $\mathbf{P}$ parallel or antiparallel to **E**. Show that its energy is decreased by an amount proportional to the same factor that appears on the right side of (3-10-2). The energy change averaged over random orientations of the dipole equals zero. (The direction of the dipole is taken to be from the negative end to the positive end.)

3-16. Work out the Stark shift for a polar diatomic molecule. [See Townes and Schawlow (1955, pp. 248–252) and Landau and Lifshitz (1965, p. 316).]

3-17. Work out the Stark shift for a symmetric-top molecule. [See Townes and Schawlow (1955, pp. 248–252).]

3-18. Explain how the requirement in Note 3-2 that the electric field be a linear function of the coordinates led to the use of electrodes with hyperbolic cross sections to produce the desired electric quadrupole field (see Fig. 3-9-3a). Use similar arguments to derive the equation for $V(r, \phi)$ displayed in Fig. 3-9-3b for the electric hexapole.

3-19. Study the discussion of the electric octupole beam guide in Ervin and Armentrout (1985, main text and appendix).

3-20. Read Gordon (1955), Gordon et al. (1955), and Feynman et al. (1965, Vol. III, Chap. 9) for discussions of the use of an electric quadrupole to separate the upper and lower inversion states in the Townes ammonia maser.

3-21. **(a)** Show that the magnetic quadrupole lens can focus a beam of *charged* particles passing through it nearly parallel to its longitudinal (Z) axis, the plane in which the focusing occurs depending on the sign of the charge of the particles.

(b) Show that the magnetic hexapole lens (Fig. 4-6-5 of the companion volume) is the simplest magnetic lens that can perform a similar function with *uncharged* particles having magnetic dipole moments polarized parallel or antiparallel to the X axis.

[See Cronin et al. (1979, p. 20).]

3-22. The magnetic field pattern of two parallel wires carrying current in opposite directions has been used as a guide in the design of "Columbia magnets" for state selection. Calculate the magnetic equipotentials of the two-wire system, showing them to be cylinders passing through the wires. Demonstrate how the two wires may be replaced by a magnet whose pole pieces are shaped so as to produce exactly the same magnetic field as the wires. One of the pole pieces is convex, the other concave. [See Reuss (1988).]

3-23. Show that in a magnetic quadrupole, the magnetic scalar potential Φ depends on r and ϕ in the same manner that the potential V in an electric quadrupole depends on these variables (see Fig. 3-9-3a). Calculate the magnetic induction B and its gradient. [See Reuss (1988), Stolte (1988), and Iannotta (1988).]

3-24. Repeat Problem 3-22 for the magnetic hexapole (refer to Fig. 3-9-3*b*).

3-25. In Note 3-3 we derived the focusing condition for a polar diatomic molecule in an electric quadrupole [equation (3-41)] and for a symmetric top in an electric hexapole field [equation (3-39)]. In the case of the diatomic molecule, it is possible to isolate successively several of the lower rotational states. Note that in the case of the symmetric top, the state selection is according to the combination of quantum numbers $KM/J(J + 1)$, which is equal to $\langle \cos \theta \rangle$. Discuss what is observed as the rod voltage V_0 is varied, and interpret the results. [See Bernstein (1982, pp. 50–54).]

3-26. Discuss the trapping mechanisms in the Paul and Penning traps. [See Paul (1990, Dehmelt (1990), and Ekstrom and Wineland (1980).]

REFERENCES

Adelman, D. E., N. E. Shafer, D. A. V. Kliner, and R. N. Zare (1992), *J. Chem. Phys.* **97**, 7323.

Anderson, S. L., F. A. Houle, D. Gerlich, and Y. T. Lee (1981), *J. Chem. Phys.* **75**, 2153.

Anderson, S. L., T. Turner, B. H. Mahan, and Y. T. Lee (1982), *J. Chem. Phys.* **77**, 1842.

Armentrout, P. B. (1988), *Comments At. Mol. Phys.* **22**, 133.

Armentrout, P. B. (1990), *Int. Rev. Phys. Chem.* **9**, 115.

Baer, M., and C. Y. Ng, Eds. (1992), "*State-Selected and State-to-State Ion–Molecule Reaction Dynamics: Theory*", *Adv. Chem. Phys.* **82**, Part II.

Bassi, D. (1988a), "Ionization Dectors I: Ion Production," Chap. 7 in G. Scoles, Ed., *Atomic and Molecular Beam Methods*, Vol. 1, Oxford University Press, New York., (1988).

Bassi, D. (1988b), "Ionization Detectors II: Mass Selection and Ion Detection," Chap. 8 in G. Scoles, Ed., *Atomic and Molecular Beam Methods*, Vol. 1, Oxford University Press, New York.

Bernstein, R. B. (1982), *Chemical Dynamics via Molecular Beam and Laser Techniques*, Oxford University Press, New York.

Brenot, J. C., and M. Durup-Ferguson (1992), "Multicoincidence Detection in Beam Studies of Ion–Molecule Reactions: Technique and Applications to $X^- + H_2$ Reactions," in C. Y. Ng and M. Baer, Eds., *Adv. Chem. Phys.* **82**, Part I.

Brenot, J. C., M. Durup-Ferguson, J. A. Fayeton, K. Goudjil, Z. Herman, and M. Barat (1990), *Chem. Phys.* **146**, 263.

Chesnavich, W. J., and M. T. Bowers (1979), "Statistical Methods in Reaction Dynamics," in M. T. Bowers, *Gas Phase Ion Chemistry*, Vol. 1, Academic Press, New York, p. 119.

Continetti, R. E., B. A. Balko, and Y. T. Lee (1990), *J. Chem. Phys.* **93**, 5719.

Cronin, J. A., D. F. Greenberg, and V. L. Telegdi (1979), *University of Chicago Graduate Problems in Physics with Solutions*, Addison-Wesley, Reading, Mass.

Dagdigian, P. J. (1988), "Reactive Scattering II: Optical Methods," in G. Scoles, Ed., *Atomic and Molecular Beam Methods*, Vol. 1, Oxford University Press, New York, pp. 596–630.

Dehmelt, H. (1990), *Rev. Mod. Phys.* **62**, 525.

Ekstrom, P., and D. Wineland (Aug. 1980), *Sci. Am.* **243**, 105.

Ervin, K. M., and P. B. Armentrout (1985), *J. Chem. Phys.* **83**, 166.

Feynman, R. P., R. B. Leighton, and M. Sands (1965), *The Feynman Lectures on Physics*, Addison-Wesley, Reading, Mass.

Fisher, E. R., and P. B. Armentrout (1990), *J. Phys. Chem.* **94**, 4396.

Flesch, G. D., and C. Y. Ng (1990), *J. Chem. Phys.* **92**, 2876.

Flesch, G. D., S. Nourbakhsh, and C. Y. Ng (1990), *J. Chem. Phys.* **92**, 3590.

Fontijn, A., Ed. (1985), *Gas-Phase Chemiluminescence and Chemi-Ionization*, North-Holland, Amsterdam.

Friedrich, B., D. P. Pulman, and D. R. Herschbach (1991), *J. Phys. Chem.* **95**, 8118.

Gandhi, S. R., and R. B. Bernstein (1990), *J. Chem. Phys.* **93**, 4024.

Gerlich, D. (1985), *ICPEAC XIV*, Stanford, Calif., p. 541.

Gordon, J. P. (1955), *Phys. Rev.* **99**, 1253.

Gordon, J. P., H. J. Zeiger, and C. H. Townes (1955), *Phys. Rev.* **99**, 1264.

Götting, R., H. R. Mayne, and J. P. Toennies (1986), *J. Chem. Phys.* **85**, 6396.

Götting, R., V. Herrero, J. P. Toennies, and M. Vodegel (1987), *Chem. Phys. Lett.* **137**, 524.

Gruebele, M., and A. H. Zewail (May 1990), *Phys. Today* **43**, 24. See also *J. Phys. Chem.* **95**, 7973 (1991); *J. Chem. Phys.* **95**, 7763 (1991); *J. Chem. Phys.* **96**, 198 (1992).

Gruebele, M., G. Roberts, and A. H. Zewail (1990), *Philos. Trans. R. Soc. London* **A-332**, 223.

Gupta, A., D. S. Perry, and R. N. Zare (1980), *J. Chem. Phys.* **72**, 6237.

Harland, P. W., H. S. Carman, L. F. Phillips, and P. R. Brooks (1990), *J. Chem. Phys.* **93**, 1089.

Herschbach, D. R. (1987), "Molecular Dynamics of Elementary Chemical Reactions" (Nobel lecture), *Angew. Chem.* **26**, 1221.

Houle, F. A., S. L. Anderson, D. Gerlich, T. Turner, and Y. T. Lee (1982), *J. Chem. Phys.* **77**, 748.

Iannotta, S. (1988), "Experiments with Spin-Polarized Beams," Chap. 27 in G. Scoles, Ed., *Atomic and Molecular Beam Methods*, Vol. 1, Oxford University Press, New York.

Joachain, C. J. (1975), *Quantum Collision Theory*, North-Holland, Amsterdam, pp. 20–21.

Kliner, D. A. V., and R. N. Zare (1990), *J. Chem. Phys.* **92**, 2107.

Kliner, D. A. V., D. E. Adelman, and R. N. Zare (1991), *J. Chem. Phys.* **95**, 1648.

Koyano, I., K. Tanaka, and T. Kato (1985), *ICPEAC XIV*, Stanford, Calif., p. 529.

Landau, L. D., and E. M. Lifshitz (1960), *Mechanics*, Addison-Wesley, Reading, Mass.

Landau, L. D., and E. M. Lifshitz (1965), *Quantum Mechanics: Nonrelativistic Theory*, Pergamon Press, Oxford.

Lee, Y. T. (1987), "Molecular Beam Studies of Elementary Chemical Processes" (Nobel lecture), in *Les Prix Nobel en 1986*, Norstedts Tryckeri, Stockholm, pp. 168–206. Also published in *Chem. Scr.* **27**, 215.

Lee, Y. T. (1988), "Reactive Scattering I: Nonoptical Methods," in G. Scoles, Ed., *Atomic and Molecular Beam Methods*, Vol. 1, Oxford University Press, New York.

Lee, Y. T., J. D. McDonald, P. R. LeBreton, and D. R. Herschbach (1969), *Rev. Sci. Instrum.* **40**, 1402.

Leone, S. R. (1984), *Ann. Rev. Phys. Chem.* **35**, 109.

Leventhal, J. J. (1984), "The Emission of Light from Excited Products of Charge Exchange Reactions," in M. T. Bowers, Ed., *Gas Phase Ion Chemistry*, Vol. 3, Academic Press, New York, p. 309.

Levine, R. D., and R. B. Bernstein (1987), *Molecular Reaction Dynamics and Chemical Reactivity*, Oxford University Press, New York.

Liao, C. L., R. Xu, G. D. Flesch, M. Baer, and C. Y. Ng (1990) *J. Chem. Phys.* **93**, 4818.

Massey, H. S. W., E. W. McDaniel, and B. Bederson, Series Eds. (1982–1984), *Applied Atomic Collision Physics*, 5 vols., Academic Press, New York.

Mathews, J., and R. L. Walker (1965), *Mathematical Methods of Physics*, W. A. Benjamin, New York.

Neumark, D. M., A. M. Wodtke, G. N. Robinson, C. C. Hayden, and Y. T. Lee (1985a), *J. Chem. Phys.* **82**, 3045.

Neumark, D. M., A. M. Wodtke, G. N. Robinson, C. C. Hayden, K. Shobatake, R. K. Sparks, T. P. Schafer, and Y. T. Lee (1985b), *J. Chem. Phys.* **82**, 3067.

Ng, C. Y. (1988), "State-Selected and State-to-State Ion–Molecule Reaction Dynamics by Photoionization Methods," in J. M. Farrar and W. Saunders, Eds., *Techniques for the Study of Ion–Molecule Reactions*, Wiley, New York.

Ng, C. Y., and M. Baer, Eds. (1992), "State-Selected and State-to-State Ion-Molecule Reaction Dynamics: Experiment," *Adv. Chem. Phys.* **82**, Part I.

Ottinger, Ch. (1984), "Electronically Chemiluminescent Ion–Molecule Exchange Reactions," in M. T. Bowers, Ed., *Gas Phase Ion Chemistry*, Vol. 1, Academic Press, New York, p. 250.

Parker, D. H., K. K. Chakravorty, and R. N. Bernstein (1981), *J. Phys. Chem.* **85**, 466.

Paul, W. (1990), *Rev. Mod. Phys.* **62**, 531.

Paul, W., H. P. Reinhard, and U. von Zahn (1958), *Z. Phys.* **152**, 143.

Pauly, H., and J. P. Toennies (1968), "Neutral–Neutral Interactions: Beam Experiments at Thermal Energies," *MEP*, **7-A**, 227.

Polanyi, J. C. (1987), "Some Concepts in Reaction Dynamics" (Nobel lecture), in *Les Prix Nobel en 1986*, Norstedts Tryckeri, Stockholm, pp. 208–258.

Reuss, J. (1988), "State Selection by Nonoptical Methods," Chap. 11 in G. Scoles, Ed., *Atomic and Molecular Beam Methods*, Vol. 1, Oxford University Press, New York.

Rinnen, K. D., D. A. V. Kliner, and R. N. Zare (1989), *J. Chem. Phys.* **91**, 7514.

Scherer, N. F., C. Sipes, R. B. Bernstein, and A. H. Zewail (1990), *J. Chem. Phys.* **92**, 5239.

Scoles, G., Ed. (1988), *Atomic and Molecular Beam Methods*, Vol. 1, Oxford University Press, New York.

Sims, I. R., M. Gruebele, E. D. Potter, and A. H. Zewail (1992), *J. Chem. Phys.* **97**, 4127.

Stolte, S. (1988), "Scattering Experiments with State Selectors," Chap. 25 in G. Scoles, Ed., *Atomic and Molecular Beam Methods*, Vol. 1, Oxford University Press, New York.

Su, T., and M. T. Bowers (1979), "Classical Ion–Molecule Collision Theory," in M. T. Bowers, Ed., *Gas Phase Ion Chemistry*, Vol. 1, Academic Press, New York, p. 84.

Sunderlin, L. S., and P. B. Armentrout (1990), *Chem. Phys. Lett.* **167**, 188.

Taylor, E. H., and S. Datz (1955), *J. Chem. Phys.* **23**, 1711.

Toennies, J. P. (1965), *Z. Phys.* **182**, 257.

Townes, C. H., and A. L. Schawlow (1955), *Microwave Spectroscopy*, McGraw-Hill, New York, Chap. 10.

Turner, T., and Y. T. Lee (1984), *J. Chem. Phys.* **81**, 5638.

Turner, T., O. Dutuit, and Y. T. Lee (1984), *J. Chem. Phys.* **81**, 3475.

Vernon, M. F., H. Schmidt, P. S. Weiss, M. H. Covinsky, and Y. T. Lee (1986), *J. Chem. Phys.* **84**, 5580.

Zen, M. (1988), "Accomodation, Accumulation, and Other Detection Methods," Chap. 10 in G. Scoles, Ed., *Atomic and Molecular Beam Methods*, Vol. 1, Oxford University Press, New York.

Zewail, A. H. (1988), *Science* **242**, 1645.

Zewail, A. H., and R. B. Bernstein (1988), *Chem. Eng. News* **66**, 24.

4

IONIZATION BY HEAVY PARTICLE IMPACT

4-1. INTRODUCTION

Ions—that is, atoms or molecules bearing an electric charge—are formed in collisions by either of two processes. In the first, one or more electrons may be transferred between the target and the incident particle. Transfer of an electron to the projectile will leave an initially neutral target positively charged while a transfer from the projectile results in a negatively charged target. These are both examples of charge transfer, a process treated in the following chapter. The second process, the direct ejection of one or more electrons into unbound or continuum states, is the subject of this chapter. The term *ionization* has been variously used in the literature to mean the production of a positive ion, a free electron, or an ion–electron pair. At a sufficiently high energy the cross section for charge transfer is negligibly small and then the cross sections for the three are essentially equal. However, at lower energies, they may be quite different. We use the term in this chapter to mean the ejection of a free electron unless otherwise specified.

The ejection of secondary electrons during fast collisions of ions or atoms with other heavy particles generally has a higher probability than any of the other elementary atomic collision processes. Furthermore, it is the atomic process involving the greatest transfer of energy which, in this case, is the sum of the binding energy and the kinetic energy given to the electron or electrons ejected. In the case of outer-shell electrons, the binding energies are usually in the range 5 to 25 eV and the average kinetic energy given to an electron is about 30 eV. For inner shells these quantities may be much larger. As a result, 80 to 90% of the energy transferred in ion–atom collisions at impact energies per unit mass above about 50 keV/u is due to ionization.

Ionization has important practical applications. Most types of particle detectors utilize the ionization produced by the passage of charged particles through matter. Radiation damage of biological tissues and property modification of solids by ion bombardment result largely from ionization. Thermonuclear and plasma research, auroral and other upper atmospheric studies, and modeling of solar and stellar processes require knowledge of the systematics of ionization. Unlike collisional excitation, which is a two-body problem, even the simplest ionization event is a three-body problem and therefore its description poses a challenge to theorists who attempt to understand and describe it.

General references that have dealt with this subject are the books by McDaniel (1964), Hasted (1972), and Massey et al. (1974) and the review papers of Ogurtsov (1972), Rudd and Macek (1972), Stolterfoht (1978b), Toburen and Wilson (1979), Berényi (1981a), and Toburen (1982, 1990). Rudd et al. (1976, 1979), Barnett et al. (1977), and McDaniel et al. (1977–1982) have published data compilations that include ionization cross sections. The books by Mott and Massey (1965) and by McDowell and Coleman (1970) contain descriptions of theoretical methods used to describe ionization by heavy particle impact.

In Part $\mathscr{A}$ we describe the methods that have been used to measure ionization cross sections, both integral and differential, and give examples of the experimental data to illustrate the systematics of the process. In Part $\mathscr{B}$ we look at some of the theoretical and semiempirical methods that have been used to describe the ejection of electrons. Finally, in Part $\mathscr{C}$ some of the mechanisms by which ionization takes place are described and some specialized topics related to ionization are discussed.

PART $\mathscr{A}$.   MEASUREMENT OF IONIZATION CROSS SECTIONS

The probabilities that certain processes take place during the collisions of two particles is described by the cross section, a concept introduced and defined in Section 1-4 of the companion volume. Usually, one particle, referred to as the target particle, is at rest or nearly at rest in the laboratory frame of reference and the other, referred to as the incident particle or projectile, carries the energy. Because many more measurements have been made with incident ions than with neutral projectiles, we will often refer to the incident particle as an ion.

4-2.   INTEGRAL IONIZATION CROSS SECTIONS

As in the case of electron impact, the designation *integral cross section* refers to the total ionization resulting from the ejection of any number of electrons at any energy and in any direction. Three basic methods of measuring integral ionization cross sections have been used, each of which involves the passage of a beam of ions through the target gas at low pressure. The earliest and most direct is the condenser method, also called the transverse-field method. The other two, the energy-loss method and the method of integration of differential cross sections, are each able to yield cross sections of this type as by-products of more elaborate measurements designed primarily for other purposes. Most of the available data were produced by the condenser method.

A. Condenser Method

This method was used in the 1930s by Smith (1930) and by Tate and Smith (1932) to measure ionization from electron impact and was first applied to ion impact by Goldman (1932), who made an unsuccessful attempt to measure an ionization cross section while measuring electron transfer. The first successful proton ionization measurement was made by Keene (1949).

The general arrangement of the apparatus in this method is shown schematically in Fig. 4-2-1. The ion beam enters the target gas cell GC from the left, travels between the ion and electron collection electrodes IP and EP, leaves the gas cell, and is caught in a Faraday cup FC. Ions and electrons produced along the beam path between the electrodes are forced by a transverse electric field to the corresponding electrode, and the currents are recorded at the same time as the primary beam current is measured. A suitably biased grid G is placed in front of the ion collection electrode to prevent loss of the secondary electrons formed when ions strike the plate.

If the target gas density n is low enough to ensure single-collision conditions, the cross sections for production of positive or negative charge are given by

$$\sigma_{\pm} = \frac{I_{\pm}}{nlI_0},$$

(4-2-1)

where I_+ and I_- are the positive (ion) and negative (electron) currents, respectively, l the length of the collection electrodes parallel to the beam

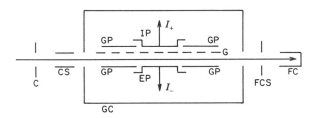

Figure 4-2-1. Apparatus for measuring integral ionization cross sections by the condenser method. C, beam collimator; CS, suppressor for secondary electrons from collimator; GC, target gas cell; EP, electron collection plate; IP, ion collection plate; GP, guard plates; G, secondary electron suppression grid; FC, Faraday cup; FCS, suppressor for electrons from Faraday cup. [From Rudd et al. (1985a).]

direction, and I_0 the primary beam current. Alternatively, and more accurately, the currents are integrated over time so that the total positive and negative charges produced corresponding to a given collected beam charge are used to calculate the cross sections. The number density n is usually determined by measuring the pressure and temperature of the target gas and applying the ideal gas law.

In practice, corrections are necessary for such effects as the neutralization of the beam as it traverses the target gas (see Section 5-2), failure to collect all of the electrons or slow ions, and the presence of various kinds of spurious currents. These are discussed in detail by Rudd et al. (1985a).

The cross section σ_- is usually considered to be the ionization cross section, although it also includes contributions from slow negative ions which may be formed in some collisions. This contribution, however, is generally very small. The cross section σ_+ combines the contributions from electron ejection and electron capture, the latter generally being dominant at low energies.

Some of the groups who have published comprehensive sets of data using this method were McDaniel's group [see, e.g., McDaniel et al. (1961) and Hooper et al. (1961)], Gilbody's group [see, e.g., Gilbody and Lee (1963)], De Heer et al. (1966), Pivovar and Levchenko (1967), McNeal (1970), and Rudd et al. (1983)].

B.   Energy-Loss Method

If a beam of charged particles traverses a target gas, the energy loss of the particles making collisions is a signature of certain inelastic collision processes; for example, beam particles that have lost an energy greater than the ionization potential may be assumed to have caused ionization provided that single-collision conditions obtain. By knowing the density and path length of the beam in the target and the fraction of particles suffering an energy-loss characteristic of a given process, the cross section for that process can be determined. The advantage of this method is that since it involves measurements made only on the primary beam, it avoids the problems inherent in the detection of secondary products from the collision. Park (1978) has reviewed this method.

This method has been used extensively with electrons as incident particles, but it is much more difficult to apply in the case of ions. Because of their larger masses, ions must have correspondingly higher energies to attain the velocities needed for most inelastic reactions to take place. For example, the maximum in the ionization cross section of hydrogen with electron impact is at about 70 eV, but for proton impact it comes at 130,000 eV. So, for example, to distinguish among inelastic processes a few eV apart requires a resolution of only about 1 to 2% for electrons but about 1 part in 10^5 for ions. The difficulties involved in attaining a resolution of that magnitude were solved by Park and Schowengerdt (1969), who accelerated beams of ions to energies up to 200 keV, passed the beams through the target gas, and then decelerated them again to a relatively low energy before energy analyzing them. Deconvolution techniques were also used to improve the resolution further.

The apparatus used by Park and his collaborators at the University of Missouri at Rolla for excitation measurements has been discussed in Chapter 2. An earlier version used in their ionization measurements is shown schematically in Fig. 4-2-2. Ions from the source IS are accelerated at AC, pass through the target gas in the collision chamber CC at ground potential, and are magnetically analyzed at M. They are then decelerated at DC to an energy (typically, 2 keV) selected by the offset voltage ΔV before being energy analyzed by an electrostatic analyzer EA. By using the same voltage supply for both the acceleration and deceleration, they were able to obtain resolutions of 2 eV or better with ion beam energies up to 200 keV. Sweeps of the energy-loss spectrum with and without target gas yield cross sections as a function of energy loss. Integration of these spectra over all energy losses greater than the ionization potential gives the integral ionization cross section.

A source of error in this method results from the fact that in the collision, some beam particles are scattered away from the acceptance aperture of the detection system and are therefore not detected. This is a particular problem when investigating cross sections for processes resulting from small-impact parameters. To correct for this error and also to enable measurements to be made of the angular distributions of scattered ions from a number of other processes, the Rolla group later modified their apparatus to allow rotation of the entire accelerator and associated apparatus about the collision center [see Park et al. (1978)].

Another source of error arises from the fact that ionizing collisions which result in a change in the charge state of the projectile are not counted. This is an especially serious cause of discrepancies at low energies where the probability of

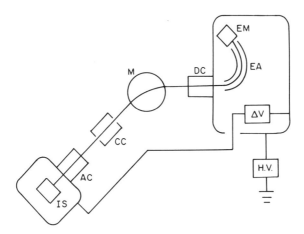

Figure 4-2-2. Apparatus for making energy-loss measurements. IS, ion source; AC, acceleration column; CC, collision chamber; M, deflection magnet; DC, deceleration column; EA, energy analyzer; EM, electron multiplier detector; HV, high-voltage source; ΔV, offset voltage source. [From Rudd et al. (1985a).]

a charge transfer associated with ionization is large. An additional discrepancy may result from the fact that the energy loss does not distinguish between collisions in which one, two, three, or more electrons are ejected.

C.   Integration of Differential Cross Sections

By energy analyzing and detecting electrons ejected in various directions, cross sections differential in energy and angle may be obtained. The apparatus and method for doing this are discussed in Section 4-3. By integrating these doubly differential cross sections (DDCSs) over angle and energy, the integral cross sections may be obtained. Since the cross sections fall off rapidly with increasing secondary electron energy, the low-energy part of the spectrum dominates the integral. Unfortunately, measurements of low-energy electrons are usually subject to large uncertainties with this type of apparatus. The measured cross sections may be too large because of spurious electrons reaching the detector, or they may be too small if stray electric and magnetic fields deflect the low-energy electrons from their expected trajectories from the collision center to the detector.

To obtain accurate integrals, not only must the low-energy part of the spectra be accurate, but also the angular and energy steps must be small enough to reproduce any structure in the distributions. Such structure results, for example, from autoionization or Auger peaks (see Section 4-11) in the energy spectrum. The area under these peaks, however, is generally a small fraction of the total area. Other structures are the binary encounter peak in the angular and energy spectrum and the forward peak due to electron capture to the continuum (see Section 4-8).

D.   Mass/Charge Analysis

To obtain more detailed information about the ionization process, the slow-recoiling target ions have, in some cases, been analyzed magnetically, yielding the distribution of m/q values. This technique was used to distinguish between dissociative and nondissociative ionization by, among others, Keene (1949), who was the first to measure ion impact ionization cross sections. He analyzed the products from H^+, H_2^+, and He^+ collisions with hydrogen and helium gases. Using a 20-V transverse extraction potential which directed the recoil ions into a 180° magnetic analyzer, he found a few percent of H^+ ions along with a large fraction of H_2^+ recoil ions from 15-keV collisions of H_2^+ with hydrogen. Because of the large energy given to the H^+ fragment ion during dissociation, this early measurement may have missed most of the H^+ ions (see Section 4-13). Afrosimov et al. (1958) performed a similar experiment and determined the partial cross sections for production of slow H^+ and H_2^+ ions from hydrogen bombarded by H^+, H_2^+, and H_3^+ at energies of 5 to 180 keV.

Mass/charge analysis has also been used to determine the charge-state distribution of recoil ions. Wexler (1964), for example, studied proton ionization of the rare gases up through krypton with a mass spectrometer capable of resolving the naturally occurring isotopes as well as the charge states. Relative cross sections for production of charge states up to $5+$ were measured for argon and krypton, $3+$ for neon, and $2+$ in helium at proton energies from 0.8 to 3.75 MeV. Hvelplund et al. (1980) and DuBois et al. (1984a) used time-of-flight spectrometers to measure cross sections for production of individual charge states. These experiments are discussed further in Section 4-12.

E. Coincidence Measurements

Analyzing the final charge state of only one of the colliding particles gives less than complete information since only an average over the possible final states of the unobserved particle results. Afrosimov and coworkers (1963, 1964, 1968) and Everhart and Kessel (1965) were the first to utilize coincidence methods to detect the final charges of both scattered and recoil particles, thereby giving more complete information on the exact processes. These experiments are discussed further in Section 4-10.

Ionization and charge-transfer processes are conveniently described by a notation suggested by Hasted (1960). An atomic reaction such as

$$A^{i+} + B^{j+} \rightarrow A^{m+} + B^{n+} + (m + n - i - j)e^- \tag{4-2-2}$$

is specified by the notation $(ijmn)$. The cross section for the process is $_{ij}\sigma_{mn}$. Thus, in a collision of a singly charged projectile with a neutral target atom, a simple ionization would be written (1011) and a simple electron transfer would be (1001), while a combination of electron transfer and ionization would be (1002). In the common situation of a singly charged ion incident on a neutral target, the i and j subscripts are often omitted. However, in studies of charge transfer where the initial and final charges are determined only for the projectile, the j and n subscripts are omitted and the cross section is written σ_{im}.

The apparatus used by Afrosimov et al. (1963, 1964) is shown in Fig. 4-2-3. Ions accelerated in the accelerator tube T are mass selected by the magnet M and make collisions at the scattering center C_3, the scattered primary particles being charge analyzed by a magnet at A_3 and detected at D_3 if charged and at D_4 if neutral. The recoil particles are analyzed by a magnet at A_2 and detected similarly at D_1 or D_2. The angles were variable over a wide range by the use of flexible bellows. Seven high-vacuum pumps placed at strategic locations kept the densities of the gas in the paths of the particle trajectories low to minimize charge-changing collisions between the collision center and the detectors. Collimators, which defined the incident beam as well as the detected product particles, were adjustable from outside the vacuum system. This system provided rather complete flexibility to detect recoil particles of any charge in coincidence with scattered beam particles of any charge, thus yielding cross sections for each combination [i.e., for each elementary process $(ijmn)$]. The

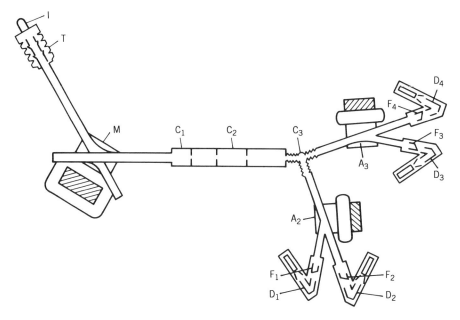

Figure 4-2-3. Apparatus used to detect recoil ions and scattered projectile ions in coincidence. I, ion source; T, acceleration column; M, magnetic mass spectrometer; C_1 and C_2, chambers for charge exchange of beam and measurements of integral cross sections; C_3, scattering center; A_2 and A_3, analyzers for recoil and scattered beam particles; $D_1 - D_4$, particle detectors; $F_1 - F_4$, detectors for neutral and charged particles. [Adapted from Afrosimov et al. (1963, 1964). For a more complete description of this apparatus, consult those references.]

apparatus of Everhart and Kessel (1965), which provided a similar capability, is discussed in Section 4-10.

F. Data: Incident Protons or Other Bare Nuclei

The low-, intermediate-, and high-energy regions are defined as those for which the projectile velocity is, respectively, less than, approximately equal to, or higher than the orbital velocity of the least tightly bound electron in the target atom or molecule. As with electron impact, the general dependence of the integral cross sections with impact energy is rather simple. In the low-energy region, the cross section increases with the energy of the incident ion and is approximately proportional to a low power (typically, 0.5 to 1.5) of the excess of the energy over the threshold energy. At an intermediate energy the cross section reaches its maximum and at high energies it decreases again, approaching a $(1/E_1) \log E_1$ dependence, where E_1 is the projectile energy. Little, if any, structure is seen in the integral ionization cross-section curves for ion impact.

Rudd et al. (1985a) have critically reviewed all measured proton ionization cross-section data available in 1985 and have given recommended values of the integral cross sections for 13 targets over a wide energy range. An additional target, water vapor, was subsequently measured (Rudd et al., 1985b). The recommended data were given in the form of parameters in the semiempirical equation

$$\sigma_- = (\sigma_l^{-1} + \sigma_h^{-1})^{-1}, \tag{4-2-3}$$

where

$$\sigma_l = 4\pi a_0^2 C x^D \tag{4-2-4}$$

$$\sigma_h = \frac{4\pi a_0^2 [A \ln(1 + x) + B]}{x}. \tag{4-2-5}$$

Here $x = \lambda E_1/R$, where λ is the ratio of the electron to the proton mass and R is the Rydberg of energy, 13.6 eV. A, B, C, and D are dimensionless adjustable fitting parameters. Values of these parameters for the 14 targets are given in Table 4-2-1.

Cross sections described by this equation are typically expected to be reliable to within 10% in the high-energy region, 10 to 25% at intermediate energies, but to have somewhat greater uncertainties in the low-energy range because of larger variations among the reported values. Examples of the integral cross-section dependence on energy are shown for proton impact on several targets in Fig. 4-2-4.

Table 4-2-1.   Values of Parameters for (4-2-3)

Target	A	B	C	D
H	0.28	1.15	0.44	0.907
He	0.49	0.62	0.13	1.52
Ne	1.63	0.73	0.31	1.14
Ar	3.85	1.98	1.89	0.89
Kr	5.67	5.50	2.42	0.65
Xe	7.33	11.1	4.12	0.41
H_2	0.71	1.63	0.51	1.24
N_2	3.82	2.78	1.80	0.70
O_2	4.77	0.00	1.76	0.93
CO	3.67	2.79	2.08	1.05
CO_2	6.55	0.00	3.74	1.16
NH_3	4.01	0.00	1.73	1.02
CH_4	4.55	2.07	2.54	1.08
H_2O	2.98	4.42	1.48	0.75

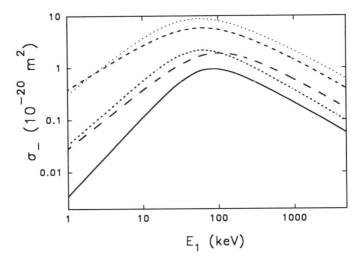

Figure 4-2-4. Integral ionization cross sections for proton impact on various target gases. Solid line, helium; long dashed line, neon; medium dashed line, argon; short dashed line, molecular hydrogen; dotted line, carbon dioxide. Calculated from (4-2-3) and the parameters in Table 4-2-1.

At high energies, the cross sections for ionization by bare ion projectiles of charge Z_1 are approximately Z_1^2 times the cross sections for proton impact at the same velocity. This Z_1^2 scaling is predicted by several theoretical treatments, but it is not a universal rule. Figure 4-2-5 shows cross sections for ionization of H_2 by H^+, He^{2+}, and Li^{3+} measured by Shah and Gilbody (1982). Since they measured electrons in coincidence with recoil ions, their cross sections are for production of ion–electron pairs. When the cross sections are divided by Z_1^2 and plotted against the energy per unit projectile mass, the cross sections approach a common value at high energies but diverge at lower energies. This simple Z_1^2 scaling holds only for the integral cross sections, however; the differential cross sections show deviations from that rule, even at high energies.

Coincidence data taken by Afrosimov et al. (1967) for protons incident on argon yielded cross sections for the elementary processes $(10mn)$, where m ranged from -1 to 1 and n from 1 to 3. These results are shown in Fig. 4-2-6 as functions of incident energy from 5 to 50 keV. It is possible to obtain the cross sections for electron transfer and ionization by summing appropriate elementary cross sections. But these data also give us further insight into the particular elementary processes that contribute to those general processes. Notice, for example, that the $(100n)$ cross sections, which involve a combination of electron transfer and electron ejection, dominate over the corresponding $(101n)$ processes, which involve only electron ejection, thus violating the rough rule of thumb that the most likely processes are those involving the fewest electrons.

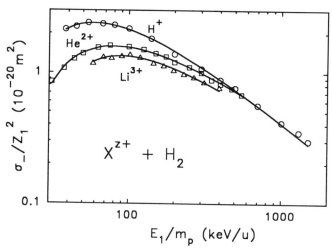

Figure 4-2-5. Scaled cross sections for ionization of hydrogen by H^+, He^{2+}, and Li^{3+} projectiles. Note the convergence at high energies. Data from Shah and Gilbody (1982) are the sum of dissociative and nondissociative ionization.

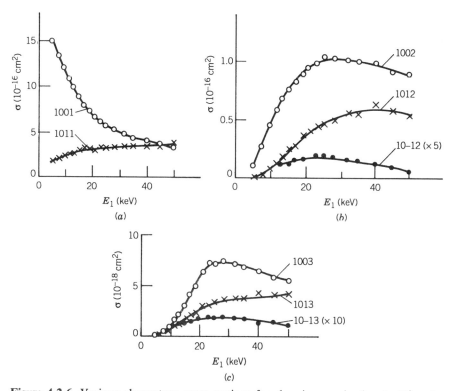

Figure 4-2-6. Various elementary cross sections for slow ion production in $H^+ + Ar$ collisions. (a) Cross sections of the type (10m1) in which Ar^+ is formed. (b) cross sections of the type (10m2) in which Ar^{2+} is formed. The (10–12) cross sections have been multiplied by 5. (c) Cross sections of the type (10m3) in which Ar^{3+} is formed. The (10–13) cross sections have been multiplied by 10. [Adapted from Afrosimov et al. (1967).]

G.   Data: Incident Dressed Ions or Neutral Atoms

The term *dressed ions* refers to those atoms that are not completely ionized and therefore carry one or more electrons. When such ions or neutral atoms are used as incident particles, the emitted electrons may come from the projectile as well as from the target. The fewer and the more tightly bound the electrons carried by the projectile, the more the collision resembles one by a bare projectile of charge $Z_1 - N$, where Z_1 is its nuclear charge and N the number of electrons on it.

Figure 4-2-7 shows the variation with energy per unit mass of the total electron production cross section for H^+, He^0, He^+, and He^{2+} incident on helium. There is Z_1^2 scaling between the H^+ and He^{2+} projectiles at high energies, as discussed earlier, but there is no simple scaling law relating the cross sections for the other two incident particles.

The amount of screening provided by the electrons carried with the incident particle depends on the distance of closest approach during the collision. For close collisions the effective charge of the projectile approaches that of its nuclear charge Z_1, while for distant collisions the effective charge approaches $Z_1 - N$. The differences in the systematics between ionization by bare and dressed ions are most noticeable in the differential cross sections discussed in Section 4-3.

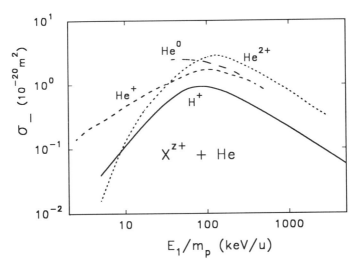

Figure 4-2-7. Comparison of cross sections for production of electrons in collisions of various projectiles incident on helium. H^+, data of Rudd et al. (1985a); He^{2+}, data of Rudd et al. (1985c); He^0 and He^+, data from Barnett (1990).

4-3.   DIFFERENTIAL IONIZATION CROSS SECTIONS

After an ionizing collision there are at least three particles leaving the collision center: the scattered projectile, the recoiling target, and one or more secondary electrons. While a complete characterization of such a collision would involve the direction, velocity, and charge of each particle (and for some purposes also the spin), experimental work to date has measured only one particle at a time or in a few cases two or three particles in coincidence. Most of the measurements on electrons have not involved coincidence, but a great deal of information about the ionization process has been gained from a study of the energy spectra and the angular distributions of electrons in ionizing collisions.

A.   Energy and Angular Distribution of Electrons

Blauth's (1957) measurements of secondary electrons from proton collisions with He, Ne, Ar, Kr, H_2, and N_2 at 8.8 to 49 keV were the first studies of the energy spectra in such collisions. His apparatus, however, was limited to a single angle of emission, namely 54.5°, and only relative cross sections were measured. Kuyatt and Jorgensen (1963) and Rudd and Jorgensen (1963) reported on measurements of both the energy and angular dependence of cross sections for ejection of secondary electrons in $H^+ + H_2$ and $H^+ + He$ collisions. Measurements of this type have been continued up to the present time. These doubly differential cross sections (DDCSs) have not only been valuable in understanding the mechanisms by which electron ejection takes place, but such detailed cross sections also provide a far more stringent test of proposed theoretical descriptions than do the integral cross sections. Thus their availability has stimulated a great deal of new theoretical work. The DDCSs may be numerically integrated over all directions of ejection to obtain the singly differential cross section $\sigma(\varepsilon)$ which gives the overall energy distribution of electrons ejected in collisions. These SDCSs are of basic interest in studies of radiation damage.

1.   *Apparatus*

In principle, the measurement of DDCSs is straightforward and a schematic diagram of a generic apparatus for it is illustrated in Fig. 4-3-1. A collimated ion beam traverses a target gas, the latter being kept at a low enough density (typically, at a pressure of a few times 10^{-4} torr) that single-collision conditions obtain. The beam is caught in a Faraday cup and integrated. If a neutral beam is used, a thermal or a secondary emission detector may be employed in place of the Faraday cup. Electrons originating from collisions within a small length of beam path l and emitted within a small range of solid angle $\Delta\Omega$ enter an electrostatic analyzer. Those within a narrow range of energies $\Delta\varepsilon$ around a selected energy ε exit the analyzer and go to a particle detector, where they are

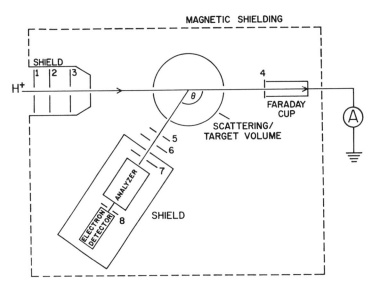

Figure 4-3-1. Schematic diagram of apparatus for measuring doubly differential cross sections for ejection of electrons. Beam enters from left and is collimated by apertures 1 and 2. Aperture 3 is a secondary electron suppressor. Beam traverses collision region and enters Faraday cup with suppressor aperture 4. Electrons ejected at angle θ are collimated by apertures 5 and 6, are preaccelerated by 7, and enter an energy analyzer. Electrons passed by the analyzer are detected. [From Rudd et al. (1992).]

counted for the same period of time that the beam integrator collects charge from the Faraday cup. The analyzer voltage is then set to pass a different energy and the process is repeated until the entire energy region of interest is surveyed. The angle is changed and the procedure repeated enough times to obtain the angular distribution. In modern versions of the experiment the energy distribution measurement is controlled by a computer, but the angle is still changed manually. Exceptions are the spectrometers described by Gibson and Reid (1984) and by Berényi (1981b, 1982) by means of which data may be taken simultaneously at several different angles.

It is generally necessary to take a background count, N_B, which is subtracted from the count N with target gas present. This procedure corrects not only for electrons from the residual gas in the target region but also for any other spurious counts, such as electronic noise, provided that they are independent of the target gas. It is also necessary to determine the efficiency of the electron detector as a function of energy. This is usually done by arranging to read absolutely the current to the detector, used as a Faraday cup, and comparing this current to the count rate from the same detector used as a particle detector. It is generally necessary to reduce the current for the latter measurement by a known factor to avoid overloading the counter. This can be done, for example, by using two apertures of a known area ratio provided that the current density is the same over both. A quite different and more convenient method of determining the detector efficiency was given by Cheng et al. (1989b). This method assumes a Poisson distribution of electron multiplication at each dynode and requires only an integral pulse height measurement. However, its use is limited to discrete dynode electron multipliers.

If N_0 is the number of primary beam particles associated with the electron count N, then the DDCS, $\sigma(\varepsilon, \theta)$, is obtained from the equation

$$N - N_B = \sigma(\varepsilon, \theta)N_0(nl\,\Delta\Omega)_{\text{eff}}\,\Delta\varepsilon. \tag{4.3.1}$$

The quantity $(nl\,\Delta\Omega)_{\text{eff}} = \int n(x)\,\Delta\Omega(x)\,dx$, where $n(x)$ is the target gas density and $\Delta\Omega(x)$ is the solid angle subtended by the analyzer–detector system from points on the beam near the collision center. In general, both $n(x)$ and $\Delta\Omega(x)$ are functions of position x along the beam path, but if a static gas target is used, $n(x)$ is independent of x and can be taken outside the integral. The remaining quantity, $(l\,\Delta\Omega)_{\text{eff}}$, is determined only by the slit geometry and is given in a relatively simple equation (Kuyatt, 1968). The energy range passed by the analyzer is the product of a geometrically determined constant and the pass energy for which the analyzer is set. The constant depends on the type of analyzer used, and expressions for various types are given, e.g., by Sevier (1972).

The majority of the DDCS data available have been provided by three experimental groups: at the University of Nebraska at Lincoln (UNL), Pacific Northwest Laboratories (PNL) in Richland, Washington, and the Hahn–Meitner Institute (HMI) in Berlin, Germany. In the measurements at UNL a static gas target was used. This has the advantage that the density of the target is constant over the length of path viewed, and therefore the determination of $n(l\,\Delta\Omega)_{\text{eff}}$ is a simple matter. A disadvantage is that electrons are produced over the entire length of the beam in the gas, posing problems in discriminating against spurious electrons that may reach the detector. In addition, a correction must be made for the absorption (or scattering) of electrons by the target gas before reaching the detector and for the neutralization of the beam in traversing the gas cell. The latter effect and the equation needed to correct for it are discussed in Section 5-2.

In the PNL apparatus, shown in Fig. 4-3-2, a small target cell was used which has a slot to facilitate measurements at various angles. This feature results in a much shorter path length in the gas for the beam and the ejected electrons, but because of the nonuniform target gas density distribution in the cell, corrections have to be made in the measurements of the angular distributions. Later measurements [e.g., Toburen et al. (1990)] at PNL used a target formed by diffusion of the target gas through a multichannel plate. In the HMI apparatus, gas is admitted through a capillary tube directly above the collision center. This also improves the ratio of the target gas pressure at the collision center to that elsewhere in the chamber but poses similar problems in measuring the angular distributions. Usually, these problems are solved by normalizing at each angle to measurements made with a static gas.

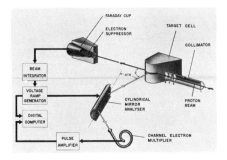

Figure 4-3-2. Apparatus for measuring doubly differential cross sections for electron ejection in ion–atom collisions. [From Toburen, (1971).]

An important improvement made on most PNL measurements was the addition of a time-of-flight (TOF) system for energy analysis of electrons (Toburen and Wilson, 1975). The incident beam was chopped by a 3.33-MHz 40-kV peak-to-peak square-wave transverse electric field, which provided beam pulses shorter than 0.7 ns. With a delay cable, the signal from the chopper provided the stop pulse to a time-to-pulse-height converter which was started by the detected electron. In this way a time spectrum of electrons was obtained that could be converted into an energy spectrum. The TOF system has the advantage of being a more open system with fewer disturbing influences on the electron trajectories. Therefore, it is able to operate well to as low as 1 eV and lower. However, data from the TOF system have to be normalized to data taken at higher electron energies by the electrostatic analyzer.

2. Data: Incident Protons or Other Bare Nuclei

Generally, the DDCS falls off with increasing electron energy and increasing angle, as shown in Fig. 4-3-3, but for angles of less than 90° and for sufficiently high primary energies a binary peak appears which becomes more prominent as the incident particle energy increases. This is shown in Fig. 4-3-4. The binary peak comes at the combination of energy and angle at which conservation of momentum and energy would predict initially stationary electrons to be ejected in a purely billiard ball–type binary collision. The energy ε_B of the binary peak is given by

$$\varepsilon_B = 4T\cos^2\theta - \chi, \tag{4-3-2}$$

where θ is the angle of ejection relative to the direction of the incident beam, $T = m_e v_1^2/2$, v_1 the projectile velocity, m_e the mass of the electron, and χ the ionization potential or binding energy of the electron in the target. The width of the binary encounter peak is due to the distribution of momenta of the electron before the collision, the *Compton profile*. Measurements of the shape of the binary encounter peak have, in fact, been used by Böckl and Bell (1983) to determine the Compton profile (see Section 4-14). When the DDCSs are

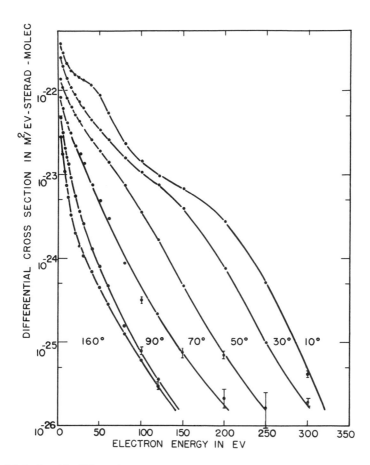

Figure 4-3-3. Doubly differential cross sections for ejection of electrons at various angles in 100-keV $H^+ + H_2$ collisions. The broad hump in the 10° curve at about 200 eV is the binary encounter peak and the one at 50 eV is due to electron capture to the continuum. [From Rudd and Jorgensen (1963).]

integrated over angle, the binary peak disappears since the electron energy at which it appears is a continuously varying function of the angle.

In a purely binary collision of fast ions and stationary electrons, the more probable distant collisions result in the ejection of slow electrons at 90°. However, when the target electron is in an atom, the field of the nucleus affects its trajectory. Very slow electrons are therefore scattered essentially in random directions and form a nearly isotropic distribution, while faster electrons retain more of their initial binary collision direction and are deflected less by the field of the nucleus. Figure 4-3-5 illustrates this pattern for 2-MeV $H^+ + Xe$ collisions. Notice, however, that even at 1500-eV some electrons are found at angles far greater than the one given by (4-3-2) for a binary collision.

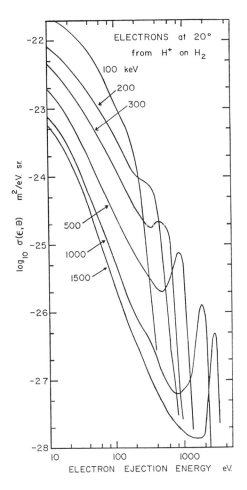

Figure 4-3-4. Energy distribution of electrons ejected at 20° from $H^+ + H_2$ collisions at various incident energies. Note increasing prominence of the binary peak at higher energies. Data at 100 to 300 keV are from Rudd et al. (1966a) and at 500 to 1500 keV are from Toburen and Wilson (1972). [From Rudd and Macek (1972).]

In the apparatus described above, electrons from all shells of the target are detected. To obtain DDCS data for electrons ejected from a specific shell, Sarkadi et al. (1983) measured continuum electrons in coincidence with Auger electrons characteristic of the L shell in 350-keV $H^+ + Ar$ collisions. More measurements of this type are needed.

The Z_1^2 dependence of the total cross section for high-energy fully stripped projectiles described in Section 4-2 does not always hold for the differential cross sections. Figure 4-3-6 concerns doubly differential cross sections for ejection of electrons from helium by 25-MeV/u Mo^{40+} and H^+ ions at two angles, 20° and 150°. The ratio of the cross sections due to the heavy, highly charged projectiles to the suitably scaled proton cross sections is plotted versus ejected electron energy. If Z_1^2 scaling were operative, all data points would have the value unity, as shown by the dashed line. In fact, the ratio for 20° is larger while that for 150° is smaller. This is attributed to the two-center effects on the

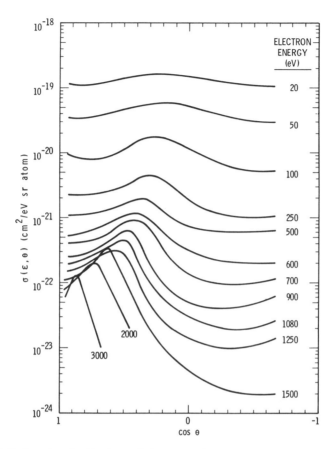

Figure 4-3-5. Angular distributions of electrons of various energies ejected from xenon by 2-MeV protons. [From Toburen (1974).]

electron trajectories. The fast, highly charged Mo^{40+} ion exerts a forward attractive force on the ejected electron as it leaves the collision site. This enhances the forward cross sections, especially for faster-moving electrons, but reduces the cross section in the backward directions. The authors find that the continuum distorted wave-eikonal initial state (CDW-EIS) approximation of Fainstein and Rivarola (1987) yields ratios consistent with the experimental results.

There are also scaling laws relating cross sections for different targets. Wilson and Toburen (1975) showed from their data for 0.3-, 1.0-, and 2.0-MeV proton impact on methane, ethane, ethylene, acetylene, and benzene and later (Lynch et al., 1976) for ammonia, monomethylamine, and dimethylamine that the DDCSs scale according to the numbers of weakly bound electrons in the molecule. That is, the cross section $\sigma(\varepsilon, \theta)$ per weakly bound electron is approximately independent of target. Cheng et al. (1989a, b) found a way to scale DDCS data

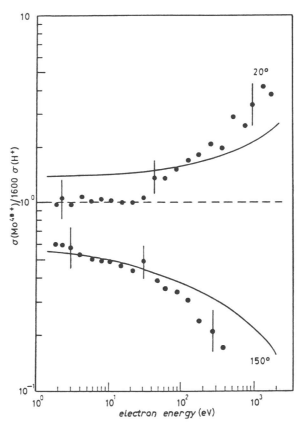

Figure 4-3-6. Ratio of doubly differential cross sections for electron emission from 25 MeV/u Mo^{40+} and H^+ impact on He. Solid lines are the calculated values using the CDW-EIS theory. [From Stolterfoht et al. (1987).]

for low-energy protons ($T < \chi$) on N_2, O_2, CO_2, Ne, Ar, and Kr reasonably well by plotting the ratio $f(\theta) = \sigma(\varepsilon, \theta)/\sigma(\varepsilon)$ versus θ. The angular dependence of $f(\theta)$ was found to be largely independent of target, proton energy, and ejected electron energy. This scaling was fairly good in the forward hemisphere but less accurate in the backward directions. The scaling for helium was also poor.

Fine structure in the form of peaks in the energy distribution, especially noticeable at large angles where the continuum background is small, is due to autoionization and the emission of Auger electrons. These effects are considered in Section 4-11. Another kind of structure in the ejected electron spectrum is a peak seen at the point where the velocity of the secondary electron is equal to the projectile velocity. This peak, which appears only at very small angles, is the result of electron ejection by the mechanism known as electron capture to the continuum, a subject discussed in Section 4-8.

3. Data: Incident Dressed Ions or Neutral Atoms

In the case of ionization produced by dressed ions or by neutral projectiles, an even more prominent peak not confined to small angles also appears in the energy spectrum at the equal velocity point. This effect, first seen by Wilson and Toburen (1973) for molecular hydrogen ions and by Burch et al. (1973) for atomic oxygen ions, is due to electrons that are detached from the projectile during the collision and is known as electron loss to the continuum (ELC). The ELC peak is apparent in data taken by the PNL group using a beam of H_2^+ incident on H_2 and is shown in Fig. 4-3-7. The electron-loss process is actually the same as ionization, when viewed from the projectile frame of reference, and

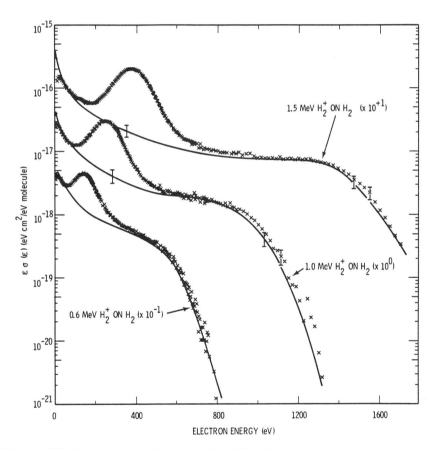

Figure 4-3-7. Energy spectra of electrons. Line, $H^+ + H_2$ multiplied by 2; $\times$, $H_2^+ + H_2$. All cross sections have been multiplied by the electron energy. Prominent peaks are due to electron loss from the projectile. [From Wilson and Toburen (1973).]

can be treated as such theoretically. The results are then transformed to the laboratory frame using the relation

$$\frac{d^2\sigma}{d\varepsilon\, d\Omega} = \frac{\varepsilon}{\varepsilon'} \frac{d^2\sigma'}{d\varepsilon'\, d\Omega'}, \tag{4-3-3}$$

where the primes indicate quantities measured in the projectile frame and where the energy in the laboratory frame is given by

$$\varepsilon = \varepsilon' + T + 2(\varepsilon' T)^{1/2} \cos\theta', \tag{4-3-4}$$

where T is the electron-equivalent energy of the projectile and θ' is the ejection angle of electrons in the projectile frame.

This process has been studied by several investigators, for example, Duncan and Menendez (1976) and Kövér et al. (1983), who measured the angular dependence of the position, width, and intensity of the equal velocity peak for He$^+$ and H$_2^+$ incident on argon. Schneider et al. (1983) studied projectile ionization in fast heavy-ion collisions and compared experimental loss peaks to ones calculated using the binary encounter approximation integrated over the initial velocity distribution and transformed to the laboratory frame. One of their findings was that the shape of the ELC peak was not very sensitive to the wave function chosen to represent the initial velocity distribution of the electrons in the projectile. Wang et al. (1991) studied electrons detached from the projectile which appeared in the backward direction in collisions of H^0 and He$^+$ with Ar and showed that those electrons were due primarily to elastic scattering. A review of projectile ionization has been given by Stolterfoht (1978b).

A dressed-ion or neutral atomic projectile, especially one with many electrons, can be thought of as a cloud of more-or-less loosely bound electrons being carried along by the nuclear core. One or more of these electrons, especially the outermost ones, may make collisions with the target and contribute independently to the ionization cross section. Because of this, some features of the collision are similar to those that would be caused by electron impact at the same velocity.

Consider, for example, Fig. 4-3-8, which shows the secondary electron spectra at 10° from helium bombarded by three different primary particles of nearly the same velocity: 150-keV protons, 150-keV neutral hydrogen, and 100-eV electrons. Since these are not of exactly the same velocity, their features can be better compared by plotting versus ε/T, where T is the equivalent electron energy of the projectile. As described in the companion volume, the lower-energy part of the electron-impact spectrum results from the ejection of true secondary electrons from the target, while the rise at higher energies is from the corresponding scattered primary electrons, which have lost a small amount of energy in the same collision. The spectrum with the H^0 beam has a very similar

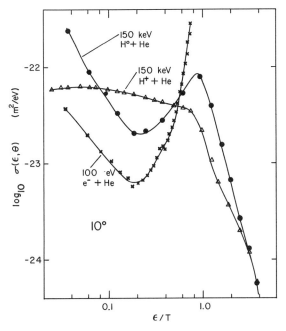

Figure 4-3-8. Cross sections for electron ejection at $10°$ from helium by 150-keV H^+ and H^0, and by 100-eV electrons. [From Rudd et al. (1980).]

shape to the electron-impact spectrum up to $\varepsilon/T = 1$, while the H^+ spectrum is quite different.

The shape of the H^0 curve can be understood qualitatively in some detail. The highest secondary energies result from close collisions for which the nucleus of the incident particle is not screened by the orbital electron. Consequently, the H^0 curve approaches the H^+ curve for large ε. At $\varepsilon/T = 1$, there is a maximum in the H^0 curve where the secondary electron velocity is equal to the projectile velocity. This is the ELC peak; that is, the electron brought along with the neutral projectile is detached during the collision and scatters elastically from the target. At lower energies, where the impact parameter is larger, the neutral curve drops below that of the proton because of screening of the nucleus by the electron. Then, at the very lowest energies where the impact parameter is largest, the neutral curve again rises, largely because of the ionization caused by the electron carried by the projectile. Wilson and Toburen (1973) also demonstrated the similarity in the angular distribution of electrons between dressed ion and electron impact.

The variation of the screening with electron energy, also noted by Toburen et al. (1980) in comparing He^+ and He^{2+} collisions with water vapor, is described in the calculations of Manson and Toburen (1981). Using the Born approximation, they show that the screening function for an incident He^+ projectile is given by

$$F_s = 2 - \frac{1}{[1 + (Ka_0/4)^2]^2}, \qquad (4\text{-}3\text{-}5)$$

where K is the momentum transfer and a_0 is the Bohr radius. The screening function varies from 1 for small values of K to 2 for large K. For example, applying (4-3-5) to fast collisions of $He^+ + He$, they obtain the results shown in Fig. 4-3-9 for electrons ejected at $60°$. At small electron energies, K is small, the screening function approaches 1, and the cross section approaches that of proton impact. At high energies, where K is large, F_s approaches 2, and the cross section is the same for He^+ as for He^{2+}. In Fig. 4-3-10 the angular distribution of 218-eV electrons is shown for 2-MeV $He^+ + He$ collisions. The experimental cross section is reasonably well represented by the sum of the cross sections for electron ejection from the target $(C + D)$ and from the projectile $(A + B)$.

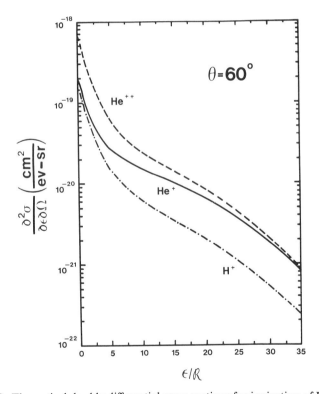

Figure 4-3-9. Theoretical doubly differential cross sections for ionization of He by equal velocity H^+, He^{2+}, and He^+ at $60°$. The incident velocity corresponds to 0.5-MeV H^+. [From Manson and Toburen (1981).]

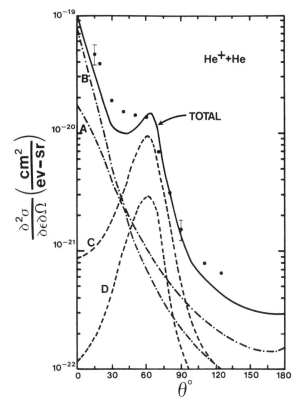

Figure 4-3-10. Doubly differential cross section for ejecting electrons of 218-eV energy in 2-MeV He$^+$ + He collisions. A, projectile ionization with the target remaining in the ground state; B, projectile ionization with simultaneous target excitation; C, target ionization with the projectile remaining in the ground state; D, target ionization with simultaneous projectile excitation. [From Manson and Toburen (1981).]

Simultaneous excitation of the projectile (D) and of the target (B) are important contributions.

The dependence of various processes in heavy ion–atom collisions on angle and on initial projectile charge are illustrated by data of Stolterfoht et al. (1974) shown in Figs. 4-3-11 and 4-3-12. The large number of slow electrons from soft collisions appears near zero energy for emission from the target (T) and in the electron-loss peak for emission from the projectile (P). The electron-loss peak decreases in size as the initial charge on the projectile increases because there are fewer electrons to be detached. The binary collision peak from the target shifts with angle, as expected, and is largely insensitive to primary beam charge. The oxygen K-Auger peak comes at about 500 eV when emitted from the stationary target, but when coming from the projectile the energy at which it appears depends on angle because of the Doppler shift (see Section 4-11). As expected,

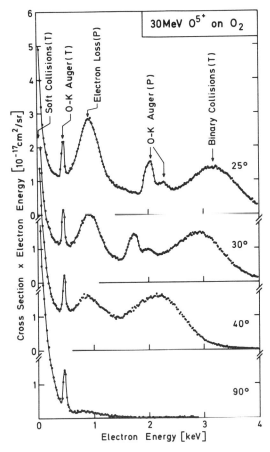

Figure 4-3-11. Cross section multiplied by electron energy for electrons ejected at various angles in 30-MeV $O^{5+} + O_2$ collisions. [From Stolterfoht et al. (1974).]

the projectile Auger peak decreases in size as the number of outer-shell electrons in the projectile decreases.

DuBois and Manson (1990) made a combined experimental and theoretical study of 400- to 750-keV/u He^+ collisions with helium and argon targets. The experimental method used coincidence between ejected electrons and projectiles that were charge analyzed following the collision, which produced energy spectra of electrons at various angles associated with processes in which the projectile did or did not lose an electron. The theoretical calculations using the plane-wave Born approximation (PWBA) yielded reasonably good results for the total cross sections for ionization of the target and of the projectile; the agreement was much worse for the differential cross sections. In particular, the simultaneous ionization of target and projectile was seriously underestimated. Similar comparisons for He^0, He^+, and H^+ on helium by Heil et al. (1991, 1992)

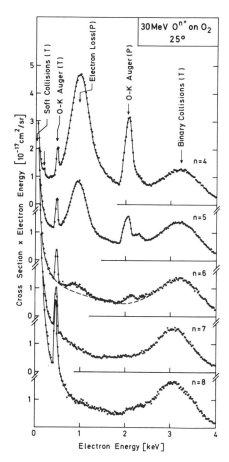

Figure 4-3-12. Cross section multiplied by electron energy for electrons ejected at 25° in 30-MeV $O^{n+} + O_2$ collisions. [From Stolterfoht et al. (1974).]

indicated that the PWBA was adequate at 0.5 and 1.0 MeV/u to describe individual ionization of target or projectile and that the previous poor agreement for $He^+ + Ar$ was probably due to inadequate wave functions describing the argon target.

B. Energy and Angular Distribution of Recoil Ions

The ratio of the geometric cross section of an atom to that of the nucleus is on the order of 10^8. Therefore, the vast majority of colliding particles interact only with the outer parts of the atom (i.e., the electrons) rather than with the nucleus, which acts like a spectator in the collision. It is not surprising, then, that very little energy is transferred to the atom as a whole during a typical collision, and therefore the residual ions in ionizing collisions receive little recoil energy.

One way to obtain the distribution of energy given to the recoiling ions is to calculate the elastic scattering cross section using a suitable scattering potential.

This must then be multiplied by the ionization efficiency, which is the fraction of the recoil atoms that are ionized. The use of the Coulomb potential yields a reasonable differential cross section for elastic scattering except near zero recoil energy, where it approaches infinity. Therefore, when integrated over all possible energies, the resulting cross section is infinite. A more realistic potential is the Bohr or screened Coulomb potential mentioned briefly in Section 3-17 of the companion volume. This representation has the advantage over more complex potential models of being expressible in an analytic equation. For two atomic particles (charged or uncharged) with nuclear charges Z_1 and Z_2 a distance r apart, this potential may be written $V(r) = (Z_1 Z_2 e^2/r) \exp(-r/a)$, where a is the screening radius given by

$$a = \frac{a_0}{\zeta}, \tag{4-3-6}$$

where $\zeta = (z_1^{2/3} + z_2^{2/3})^{1/2}$, where z_1 and z_2 are the numbers of electrons screening the atomic particles 1 and 2, respectively. Although Bohr (1941) did not make a distinction between the nuclear charge Z and the number of screening electrons z, the equation in this form allows its use for neutral–neutral, neutral–ion, or ion–ion interactions. Using the Born approximation (Born, 1926) with this potential, Schiff (1949) has calculated an expression for the electron scattering cross section per unit solid angle as a function of the scattering angle of the incident particle which can be generalized and modified to apply to atomic scattering (see Problem 4-18). By applying conservation of momentum and energy, the resulting modified equation can be transformed to the cross section per unit recoil energy ε_r as a function of that energy. The result may be written

$$\frac{d\sigma}{d\varepsilon_r} = \frac{\pi r_0^2 \varepsilon_m/4}{(\varepsilon_r + \varepsilon_0)^2}, \tag{4-3-7}$$

where

$$\varepsilon_0 = \frac{m_e}{m_t} R\zeta^2 \tag{4-3-8}$$

and where the quantity

$$r_0 = \frac{2a_0 R Z_1 Z_2 (1 + \gamma)}{\gamma E_1} \tag{4-3-9}$$

is the distance of closest approach for the case of zero screening. The maximum recoil energy is the quantity

$$\varepsilon_m = \frac{4\gamma E_1}{(1 + \gamma)^2}, \tag{4-3-10}$$

$R = 13.6 \, \text{eV}$, $\gamma = m_t/m_p$, and m_e, m_t, and m_p are the masses of the electron, the target atom, and the projectile, respectively. The corresponding expression for a Coulomb potential may be obtained by setting $\varepsilon_0 = 0$ in (4-3-7).

If the ionization efficiency can be calculated or estimated, the cross section for production of ions can be determined. Figure 4-3-13 shows a comparison of cross sections calculated with the Bohr and Coulomb potentials and the experimental results of Crooks (1974) for $H^+ + Ar$ collisions at three primary energies. Since the cross section is expected to be proportional to $1/E_1$, the quantity σE_1 is plotted to yield a universal curve for all impact energies. The falloff of the experimental values below about $5 \, \text{eV}$ may be partly instrumental, resulting from a reduction in the transmission of the analyzer at low energies, and partly due to a decrease in the ionization efficiency. A comparison of the measured and calculated curves in the straight-line region yields an ionization efficiency of roughly 50%.

From (4-3-7) it is easy to show that the fraction f of recoils with energies greater than a given value ε_r' is equal to

$$f = \frac{\varepsilon_0}{\varepsilon_r' + \varepsilon_0}. \tag{4-3-11}$$

The quantity ε_0 is quite small; for example, for $H^+ + Ar$, $\varepsilon_0 = 5.2 \times 10^{-4} \, \text{eV}$, and therefore only about 0.5% of the recoil ions have energies greater than $0.1 \, \text{eV}$.

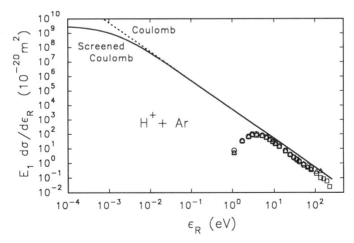

Figure 4-3-13. Elastic cross sections for production of recoils of target atoms in $H^+ + Ar$ collisions calculated using screened Coulomb and Coulomb potentials. These are compared with experimental cross sections of Crooks (1974) for recoil ion energy distributions. O, 50 keV; $\triangle$, 100 keV; $\square$, 200 keV.

The angular distribution of recoil ions can also be calculated for the Bohr potential. The result is

$$\frac{d\sigma}{d\Omega} = \frac{(r_0^2/4)\cos\phi}{(\cos^2\phi + \varepsilon_0/\varepsilon_m)^2}.$$ (4-3-12)

Since $\varepsilon_0/\varepsilon_m$ is very small, the angular distribution of the recoiling target atoms reaches a maximum very near $\phi = 90°$, where ϕ is measured from the beam direction.

Equation (3-8-8) of the companion volume relates the scattering angle θ to the impact parameter b, at least for Coulomb collisions. A measurement, then, of the projectile scattered at an angle θ in coincidence with the resulting detected signal from any process allows one to know the value of b for that event. However, as noted earlier, typical values of θ are very small and may be difficult to measure. Ullrich et al. (1988) have devised a method using conservation of energy and momentum in which a measurement of the transverse component of the recoil ion momentum provides the desired projectile scattering angle. In collisions of 5.9-Mev/u U^{65+} + Ne they were able by this method to determine scattering angles in the range of 2×10^{-6} to 4×10^{-5} rad. Because thermal motion is a limiting factor, cooling the gas target is expected to allow measurements of scattering angles of $\theta \approx 10^{-9}$ rad to be made for 1-GeV uranium ions scattering from neon atoms.

If the target is a molecule, dissociation can produce angular and energy distributions of the recoil ions quite different from those for atomic targets. The collision can put the molecule into a high-energy repulsive state, resulting in two (or more) fragments that fly apart with considerable energy. Many aspects of this so-called Coulomb explosion have been investigated by the group at Argonne National Laboratory [see, e.g., Kanter et al. (1979); see also Sections 5-5 and 6-10 of the companion volume]. Figure 4-3-14 shows a spectrum of positive ions detected at 90° from 50-keV H^+ + H_2 collisions measured by G. W. Kerby and Y.-Y. Hsu at the University of Nebraska (private communication, 1991). The peaks at 6 to 9 eV are due to H^+ ions from the Coulomb explosion of the H_2 molecule. Dissociative ionization is discussed further in Section 4-13. A review of the experimental and theoretical aspects of the production of recoil ions with emphasis on highly charged recoils was recently published by Cocke and Olson (1991).

PART ℬ.   THEORY OF ION IMPACT IONIZATION

Ionization is a difficult process to treat theoretically because even in the simplest case of a proton ionizing a hydrogen atom, one must consider the mutual interactions of three particles in the final state: the incident projectile, the electron, and the residual ion. Since it is impossible in principle to solve a three-

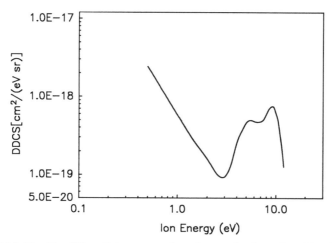

Figure 4-3-14. Doubly differential cross sections for production of positive ions at 90° from 50-keV $H^+ + H_2$ collisions. [Data from G. W. Kerby and Y.-Y. Hsu (private communication, 1991).]

body problem exactly, either classically or quantum mechanically, assumptions and/or approximations must be made.

In some ways, a theoretical description of ion-induced ionization is simpler than that for ionization by electrons. One does not need to consider electron exchange since the incident and ejected particles are dissimilar. Furthermore, in the vast majority of ion collisions there is little deflection of the projectile because it has a much greater mass than the electron with which it interacts. A simplifying argument for both electron and ion impact is that since the target atom itself is given such a minute amount of recoil energy in most collisions, it can be considered to remain at rest.

Early treatments of ionization assumed that the interaction between the projectile and the target atom took place rapidly, so that the projectile gave an impulse to a single electron and then left the scene without further influence on the electron trajectory. This is an appropriate approximation if the projectile is fast and the ejected electron is not too fast. A further simplification also neglects the influence of the residual ion on the electron trajectory. In this case, many of the electrons are ejected into the binary encounter peak, which is a sharp maximum expected when the DDCS is plotted as a function of angle or energy. In its most elementary form this approximation gives rise to the Rutherford or Thomson equation in which the electron is assumed to be initially at rest. In the somewhat more realistic binary encounter approximation (BEA) treatment, the distribution of initial orbital electron velocities is taken into account, thus giving a breadth to the binary encounter peak. (See Note 4-4 for a brief discussion of the determination of the Compton profile from the shape of the binary encounter peak.) But otherwise, the fields of the projectile and the residual ion are assumed to have no effect on the outgoing electron. This may be called the *zero-center* case, since no centers of forces are assumed to influence the trajectory of the electron after the initial collision. The Born approximation, in which the final state of the ejected electron is represented by a plane wave, is a quantum mechanical zero-center treatment.

The next stage of approximation would be to consider the effect of the residual ion on the trajectory but still to neglect the influence of the projectile. This *one-center* process is embodied in the first Born approximation, with the final state centered on the target nucleus (Bates and Griffing, 1953). The low-energy, soft-collision part of the spectrum results from dipole transitions in this treatment, but the binary-encounter, hard-collision part is also included.

If the projectile has an appreciable effect on the outgoing electrons, it is necessary to treat the problem as *two-center electron emission* (TCEE). Stolterfoht et al. (1987) first suggested this "*n*-center" nomenclature and classification and stressed the importance of two-center effects. Fainstein et al. (1991) have reviewed the theoretical treatments of TCEE. The 1970 discovery of a forward peak in electron spectra at the velocity of the projectile (see Section 4-8) and the more recent finding of significant number of electrons ejected in the forward direction with a velocity approximately half that of the projectile (see Section 4-9) serve as indications that under some conditions electrons are emitted in the combined fields of the ion and the projectile.

When the target contains more than one electron, an additional complication known as electron correlation arises due to the interactions between the ejected electron and the remaining orbital electrons. However, a precise definition of the concept of electron correlation is difficult to formulate. In the usual independent particle model the ejected electron is considered to move in a potential that is the sum of the potential due to the nucleus and the average potential of the other electrons. It is not a simple matter to separate the specific correlation with a given electron from that electron's contribution to the average potential. In fact, this is a question that has not yet been completely settled.

4-4.   CLASSICAL BINARY ENCOUNTER APPROXIMATION

For many collisions the interaction of the charged projectile with a single electron in an atom dominates the collision process, in which case the effect of the target nucleus and the other electrons may be neglected except for the binding energy which they provide. Treatments of the collision problem based on this assumption, known as binary encounter approximations (BEA), were discussed for electron impact in Chapters 3 and 6 of the companion volume.

Rutherford (1911) and Thomson (1912) studied the collision of a charged projectile with a charged target particle initially at rest. Rutherford was interested in the scattered projectile and using classical physics obtained an expression for the probability of scattering as a function of scattering angle. However, he did not relate this to Q, the energy transfer to the target particle.

This was done by Thomson, who derived an equation for electron–electron collisions from which the Thomson ionization cross section of (6-13-4) in the companion volume can easily be derived. For ion impact the projectile energy E_1 is generally much greater than the ionization energy χ, allowing us to rewrite that equation neglecting the last term. Writing the equation for a single-target electron (i.e., for $N = 1$) and using a slightly different notation, we have what has come to be called the *Rutherford equation*:

$$\sigma_R = \frac{4\pi a_0^2 R^2}{T\chi}. \qquad (4\text{-}4\text{-}1)$$

The differential form of the equation is

$$\sigma_R(Q) = \frac{d\sigma}{dQ} = \frac{4\pi a_0^2 R^2}{TQ^2}, \qquad (4\text{-}4\text{-}2)$$

where $Q = \chi + \varepsilon$ and ε is the kinetic energy of the emitted electron.

Both the differential and integral forms of the Rutherford equation are often used as references, and cross sections are sometimes stated in terms of a quantity Y, defined as the ratio of a cross section to the Rutherford cross section.

However, the Rutherford (or Thomson) ionization cross-section equation does not reproduce the experimental data very accurately. The inaccuracy inherent in the assumption that the target electron is initially at rest was corrected by Williams (1927) and by Thomas (1927), both of whom took account of its initial orbital velocity. Letting U_2 stand for the initial orbital energy in the target, Williams's equation is

$$\sigma(Q) = \sigma_R(Q)\left(1 + \frac{4U_2}{3Q}\right), \qquad Q \geqslant \chi. \qquad (4\text{-}4\text{-}3)$$

However, this is correct only for the part of the energy range for which momentum and energy conservation allow electrons to be ejected at all angles. Thomas's treatment was more general and can be written

$$\sigma(Q) = \begin{cases} \sigma_R(Q)\left(1 + \dfrac{4U_2}{3Q}\right), & \chi \leqslant Q \leqslant Q_- \quad (4\text{-}4\text{-}4) \\[3mm] \sigma_R(Q)\dfrac{U_2}{6Q}\left[\left(\dfrac{4T}{U_2}\right)^{3/2} + \left(1 - \sqrt{1 + \dfrac{Q}{U_2}}\right)^3\right], & Q_- \leqslant Q \leqslant Q_+, \quad (4\text{-}4\text{-}5) \end{cases}$$

where $Q_\pm = 4T \pm 4(TU_2)^{1/2}$. These are generally taken to be the correct BEA equations for proton ionization. Unfortunately, Thomas's work was forgotten for many years while attention was focused on the development of quantum mechanical methods. Gryzinski (1957, 1959) revived interest in the classical theory of collisions by a series of two papers in which he derived a classical equation based on the work of Chandrasekhar (1941) on colliding stars. Gryzinski, however, made additional assumptions in these and in his later series of three papers (Gryzinski, 1965). Gerjuoy (1966) and Vriens (1967), independently, arrived at BEA equations that were equivalent to Thomas's equation for the collision of a heavy ion with an electron.

Atomic or molecular targets with more than one subshell can be handled by the BEA if the binding energies, orbital energies, and numbers of electrons in each subshell are known. Calculations are made separately for each subshell and then added to obtain the cross section.

Three flaws are inherent in the Thomas BEA equation: (1) There is a discontinuity in the slope of the cross section as a function of secondary electron energy at the point where both parts of the equation are valid (i.e., at $Q = Q_-$), (2) the cross section drops to zero at $Q = Q_+$ (the measured cross section, while small and decreasing rapidly above that energy, is still nonzero) and (3) the integral cross section has a $1/T$ dependence at high energies instead of the well-verified $(1/T)\ln T$ behavior.

The first two of these problems can be overcome by integrating the BEA equation over a distribution of initial velocities of the target electron. Using the equation as it stands is equivalent to the assumption of a delta-function velocity distribution. Fock (1935) derived quantum mechanically the exact velocity distribution for electrons of any principal quantum number in the hydrogen atom. His normalized distribution of orbital velocities v_2 may be written

$$f(v_2) = \frac{32 v_{av}^5 v_2^2}{\pi (v_{av}^2 + v_2^2)^4}, \tag{4-4-6}$$

where v_{av} is the velocity associated with the average orbital kinetic energy U_2. Gregoire [see Rudd et al. (1971)] performed the integration using this velocity distribution and obtained an analytical expression for the resulting cross section in terms of the quantities $\alpha = \varepsilon/U_{av}$, $\phi = v_1/v_{av}$, and $\beta = (\alpha/4\phi - \phi)^2$, where v_1 is the incident ion velocity and Q is the energy transfer in the collision. The cross section is

$$\sigma(\varepsilon) = \begin{cases} S_A + S_B, & \chi \leqslant Q \leqslant 4T \\ S_B & Q \geqslant 4T \end{cases} \tag{4-4-7}$$

where

$$S_A = \sigma_R(Q) \frac{U_2}{\pi Q} \left[\frac{32\beta^{3/2}\alpha}{3(1+\beta)^3} + \left(\frac{4}{3} + \alpha\right)(\pi - 2R_1) \right], \tag{4-4-8}$$

$$S_B = \sigma_R(Q) \frac{U_2}{\pi Q} \left[\frac{16}{3(1+\beta)^3} \left(\frac{4\phi^3}{3} - \beta^{3/2}\alpha - \frac{\alpha(\alpha+\beta)^{3/2}}{1-\alpha} \right) \right.$$
$$\left. + \left(\frac{4}{3} + \alpha\right) R_1 - \left(\frac{4}{3} - \frac{\alpha}{1-\alpha}\right) R_2 \right], \tag{4-4-9}$$

and

$$R_1 = \tan^{-1} \beta^{-1/2} + \frac{\beta^{1/2}}{(1+\beta)^3}\left(1 + \frac{8}{3}\beta - \beta^2\right),$$

$$R_2 = R_3 + (1-\alpha)^{-3/2} \tan^{-1}\left(\frac{1-\alpha}{\alpha+\beta}\right)^{1/2}, \qquad \alpha < 1$$

$$= R_3 + (\alpha-1)^{-3/2} \ln\frac{(\alpha+\beta)^{1/2} - (\alpha-1)^{1/2}}{(1+\beta)^{1/2}}, \qquad \alpha > 1$$

$$R_3 = \left(2 + \frac{14}{3}\beta + \frac{8}{3}\alpha\right)\frac{(\alpha+\beta)^{1/2}}{(1+\beta)^3} - \frac{(\alpha+\beta)^{1/2}}{(1+\beta)(1-\alpha)}. \qquad (4\text{-}4\text{-}10)$$

At $\alpha = 1$, the equations above converge to the expression

$$S_B = \sigma_R(Q)\frac{U_2}{\pi Q}$$

$$\times \left\{\frac{16}{3(1+\beta)^3}\left[\frac{4}{3}\phi^3 - \beta^{3/2} - \frac{4}{3}(1+\beta)^{3/2} + \frac{3}{5}(1+\beta)^{1/2}\right] + \frac{7}{3}R_1\right\}, \qquad \alpha = 1.$$

$$(4\text{-}4\text{-}11)$$

A different initial velocity distribution was suggested by Gryzinski (1965), but the results differ little from those using the Fock distribution except at the highest ejected electron energies. Figure 4-4-1 shows a comparison of the SDCS calculated with the BEA with experimental data for 0.3 to 2-MeV protons incident on xenon. The BEA cross sections were calculated separately for each subshell and added.

As an approximation, the average kinetic energy U_2 needed in the BEA may be taken to be equal to the ionization energy χ of the subshell. More accurate values can be calculated quantum mechanically or, as suggested by Robinson (1965), by using Slater's rules. In Fig. 4-4-1 the BEA results are shown both with the kinetic energy taken to be equal to χ and calculated by Slater's rules. The quantity ρ is the ratio of the average initial orbital energy to the ionization potential for a given shell. At low secondary energies the data favor the results using Slater's rules, while at high energies $\rho = 1$ seems to give better agreement. Typical calculations of proton ionization with the BEA yield results that agree with experiment to within about 40% for primary particle velocities greater than the orbital velocity. The accuracy of the approximation is not as good for lower incident energies, however.

Bonsen and Vriens (1970) have extended the BEA method to calculate cross sections differential in angle as well as electron energy. Instead of the Fock velocity distribution, which is correct only for hydrogen atoms, they used more accurate wave functions and calculated the velocity distributions for helium

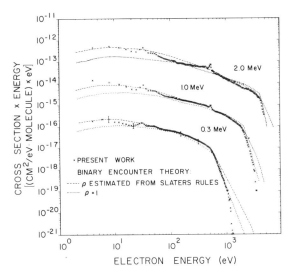

Figure 4-4-1. Energy distributions of electrons from 0.3- to 2.0-MeV H^+ + Xe collisions. $\times$, Experimental data; lines, binary encounter approximation with ρ, the ratio of the average orbital kinetic energy to the ionization potential, approximated by two methods. [From Toburen (1974).]

atoms and for hydrogen molecules. The results for 300-keV protons incident on helium are shown in Fig. 4-4-2 compared to Born approximation calculations (using hydrogen wave functions) and to experiment. The comparison indicates that the BEA yields results very similar to those of the Born approximation except at angles greater than 90°, where the BEA results are much too small.

4-5. QUANTUM MECHANICAL METHODS

The quantum mechanical approximation proposed by Born (1926) has been used widely in calculations of ionization cross sections. The Born approximation has been discussed in the companion volume as applied to scattering in Sections 3-16 and 3-17, and to excitation and ionization by electron impact in Section 3-20. Some of the variations of the method were treated in Sections 3-23 and 6-14. A review of the theory of fast ion–atom collisions with emphasis on Born theory has recently been given by Briggs and Macek (1991).

The three particles (the incident ion, the interacting electron, and the residual ion) are described in different ways depending on the problem being considered. In some cases a classical description of the ion trajectory (the impact parameter approximation) suffices, while in other situations a complete quantum description of the projectile motion as well as the electron motion is necessary. As mentioned in the introduction to Part $\mathscr{B}$, the correlation or interaction between electrons also may be important when more than one active electron is involved.

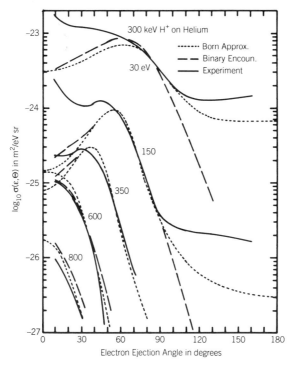

Figure 4-4-2. Angular distributions of electrons of various energies from 300-keV H^+ + He collisions. Solid line, experiment (Rudd et al., 1966a); dashed line, binary encounter approximation calculations of Bonsen and Vriens (1970); dotted line, Born approximation calculations using (4-5-4). [From Rudd and Macek (1972).]

A. Plane-Wave Born Approximation

The basic assumption in the plane-wave Born approximation (PWBA) is that the deflection of the projectile results only from interaction with the electron. Consequently, the outgoing wave describing the projectile motion is distorted only slightly during the collision, and the deflection of the projectile is very small. The interaction, then, can be treated as a perturbation. This assumption is valid when $Z_1/Z_2 \ll v_1/v_0$, where Z_1, Z_2, v_1, and v_0 are the projectile and target nuclear charges and the projectile and Bohr velocities, respectively. For protons interacting with outer-shell electrons, this assumption generally restricts the validity of the PWBA to energies greater than about 500 keV/u.

An additional approximation is that the electron–nucleus interaction is considered only in the final state. However, ionization is a three-body process and since, as pointed out by Briggs and Macek (1991), the Coulomb force is long range, the three particles can never be considered to be free, and there is always a final-state interaction between the electron and the projectile. The reason that the PWBA gives as good results as it does is that the large majority of collisions are distant ones for which the assumptions are good approximations. The total cross sections are dominated by these collisions, and only when one examines the details of the momentum distributions of the electrons does one see the results of the relatively rare close collisions.

Under the assumptions of the plane-wave Born approximation, the cross section for excitation or ionization depends only on the charge and velocity of the projectile. The only effect the mass of the projectile has is through the reduced mass of the target atom and the projectile and the motion of the center of mass. Otherwise, the cross sections for proton impact and electron impact are identical, and therefore the derivations and most of the results given in the sections in the companion volume listed above may also be used for ion impact.

In the Born approximation, it is necessary to determine the form factor

$$F(q) = \left| \int \psi_\kappa \exp(i\mathbf{q} \cdot \mathbf{r}) \psi_0 \, d^3r \right|^2 \tag{4-5-1}$$

where q is the momentum transfer, κ the momentum of the ionized electron, and ψ_0 and ψ_κ are the initial ground-state and final-state continuum wave functions. The integral in this equation can be solved exactly only for hydrogen wave functions; additional approximations must be made for all other atomic targets. A convenient approach was suggested by Bates and Griffing (1953) in which the wave function of the atom in question is approximated by a hydrogenic wave function scaled according to the ionization potential χ of the atom. The hydrogen 1s wave function is used with an effective charge given by

$$\mu = \sqrt{\frac{\chi}{R}}, \tag{4-5-2}$$

where R is the Rydberg of energy. This type of scaling, of course, does not reproduce features that result from peculiarities of more complex orbitals (e.g., those having nodes in their wave functions).

Mott and Massey (1965) give an explicit expression for the ionization cross section in the Born approximation using hydrogenic wave functions. In atomic units the cross section differential in the directions of the scattered ion and in the direction and energy ε of the ejected electron is given by

$$d\sigma = n \, \frac{2^8 \mu^6 m_p^2}{\pi q^2} \, \frac{\exp\{-(2\mu/\kappa)\arctan[2\mu\kappa/(\mu^2 + q^2 - \kappa^2)]\}}{[1 - \exp(-2\pi\mu/\kappa)](\mu^2 + q^2 + \kappa^2 - 2q\kappa\cos\delta)^4}$$

$$\times \frac{(q - \kappa\cos\delta)^2 + \mu^2\cos^2\delta}{(\mu^2 + q^2 - \kappa^2)^2 + 4\mu^2\kappa^2} \, d\Omega_p \, d\Omega_e \, d\varepsilon, \tag{4-5-3}$$

where κ is the momentum of the ejected electron, k the initial momentum of the projectile, q the momentum transfer, m_p the mass of the projectile, n the number of electrons, and δ is defined by the equation $\cos\delta = q \cdot \kappa / q\kappa$.

To obtain the doubly differential cross section (DDCS) for electron ejection as a function of the angle and energy of electron ejection, it is necessary to integrate this equation over Ω_p, the projectile solid angle. Kuyatt and Jorgensen (1963) were able to perform the integral over azimuthal angle analytically. The remaining integral over the polar angle was converted into an integral over the momentum transfer q. Their result is

$$\frac{d\sigma}{d\Omega_e\, d\varepsilon} = n \frac{2^8 m_p^2}{k^2} \int_{q_m}^{q_{max}} \frac{1}{q} \frac{\mu^6 \exp\{-(2\mu/\kappa)\tan^{-1}[2\kappa\mu/(q^2 - \kappa^2 + \mu^2)]\}}{[(q+\kappa)^2 + \mu^2][(q-\kappa)^2 + \mu^2][1 - \exp(-2\pi\mu/\kappa)]}$$
$$\times \frac{CD^3 + 4CDE^2 - 4BD^2E - BE^3 + 2AD^3 + 3ADE^2}{(D^2 - E^2)^{7/2}}\, dq, \qquad (4\text{-}5\text{-}4)$$

where

$$A = q^2 - 2q_m\kappa\cos\theta + (\kappa^2 + \mu^2)\left(\frac{q_m}{q}\right)^2 \cos^2\theta,$$

$$B = 2(q^2 - q_m^2)^{1/2}\kappa\sin\theta - (\kappa^2 + \mu^2)\left(\frac{2q_m}{q^2}\right)(q^2 - q_m^2)^{1/2}\sin\theta\cos\theta,$$

$$C = (\kappa^2 + \mu^2)\frac{q^2 - q_m^2}{q^2}\sin^2\theta,$$

$$D = q^2 - 2q_m\kappa\cos\theta + \kappa^2 + \mu^2,$$

$$E = 2\kappa(q^2 - q_m^2)^{1/2}\sin\theta,$$

θ = angle of ejection of the electron,

q_m = minimum value of $q \approx (m/2)(\kappa^2 + \mu^2)/k$,

k = wave vector for the incident proton in the Lab system,

$$q_{max} \approx 2k. \qquad (4\text{-}5\text{-}5)$$

Cross sections are obtained from this equation by numerical integration over the one remaining variable, q.

To find the DDCS from this equation for a given value of the effective charge μ, given by (4-5-2), we need to use a slightly more complicated scaling rule. It can be shown (see Problem 4-6) that

$$\sigma(\varepsilon, E_1, \mu) = \left(\frac{1}{\mu^6}\right)\sigma\left(\frac{\varepsilon}{\mu^2}, \frac{E_1}{\mu^2}, 1\right). \qquad (4\text{-}5\text{-}6)$$

Therefore, calculations made for atomic hydrogen ($\mu = 1$) for secondary electrons of energy ε/μ^2 and protons of energy E_1/μ^2 may be divided by μ^6 to obtain the cross section for a target with an effective charge μ for energies ε and E_1 [see Rudd et al. (1966a)].

There are two main features in the DDCS obtained from the PWBA calculations. There is a binary peak, which occurs at a combination of angles and electron energies given by (4-3-2). In a three-dimensional plot this forms the so-called *Bethe ridge*. In addition, there is a peak at zero energy due to recoil of electrons emitted in soft collisions.

As experimental DDCS data for proton impact became available during the 1960s and 1970s, comparisons were made with the results of the Born hydrogenic calculation described above. It was found that although the Born calculations gave a good account of integral cross sections at high energies, the much more detailed DDCS revealed large discrepancies, especially in the angular distributions of electrons. The example given in Fig. 4-5-1 for 300-keV protons on helium shows that while there is agreement over an intermediate range of angles, the calculations seriously underestimate the cross sections in the forward and backward directions. Madison (1973) showed that the discrepancy in the backward direction could be largely eliminated by the use of the more accurate Hartree–Fock wave functions in place of the hydrogenic ones. Figure 4-5-1 illustrates this.

The discrepancy in the forward direction, however, resulted from the neglect

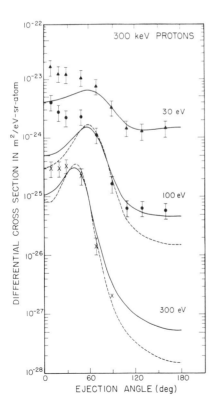

Figure 4-5-1. Comparison of calculated differential ionization cross sections for 300-keV $H^+ + He$ with experimental values. Solid lines, plane-wave Born approximation with wave functions obtained from a Hartree–Fock potential; dashed lines, scaled Born approximation using hydrogenic wave functions; data points, experimental measurements by Rudd et al. (1966a). [From Madison (1973).]

of electron–projectile interaction in the final state and led to the discovery of the mechanism of electron capture to the continuum, discussed in Section 4-8.

B. Distorted-Wave Born Approximation

Most calculations have used the PWBA, which ignores the interaction between the projectile and the nucleus and thus is strictly applicable only for collisions in which the projectile scattering angle is very small. As noted above, the results of this approach are useful only when describing collisions with small momentum transfer. The distorted-wave Born approximation (DWBA) described in Section 3-23 of the companion volume, takes account of the change in the projectile wave function caused by the interaction with the nucleus. DWBA calculations have been made, for example, by Madison (1973), Manson et al. (1975), Rudd and Madison (1976), and Madison and Manson (1979).

C. Bethe Theory

As described in Sections 3-23 and 6-14 of the companion volume, a modification of the Born approximation was proposed by Bethe (1930) in which the exponential term is expanded and integrated, yielding terms corresponding to the electric dipole, quadrupole, and so on, atomic transition moments. For the singly differential cross sections the result may be written

$$\sigma(Q) = \frac{4\pi a_0^2 R}{T} \left[a(Q) \ln \frac{T}{R} + b(Q) + c(Q) \frac{R}{T} + \cdots \right], \qquad (4\text{-}5\text{-}7)$$

where $a(Q)$, $b(Q)$, and $c(Q)$ depend on the energy transfer $Q = \chi + \varepsilon$ but not on the primary energy. The first of these constants is related to the differential oscillator strength df/dQ by the equation

$$a(Q) = \frac{R}{Q} \frac{df}{dQ}. \qquad (4\text{-}5\text{-}8)$$

The oscillator strength is a quantity that can be obtained from measurements of photoionization cross sections and thus provides an important link between photon and ion (or electron) impact ionization. Equation (4-5-7) may be integrated over Q to obtain the integral cross section. The resulting expression is identical to the right-hand side of (4-5-7) except that constants a, b, and c replace the functions $a(Q)$, $b(Q)$, and $c(Q)$.

Bethe's approximation is valid only under the conditions specified for the PWBA, namely for large projectile velocities and small momentum transfers. Inokuti (1971) has discussed the Bethe theory in detail.

A very important result deduced from the Bethe theory is the prediction that at high energies, cross sections for all allowed transitions should asymptotically approach a $(1/T)\ln T$ dependence on the projectile energy. For ionization of neutral targets by ion impact, this prediction has been so well verified that it is often used as a test of the accuracy of experimental cross-section measurements. However, one must remember that whether an incident energy is high or low depends on the binding energy of the electron shell in question. A projectile energy may be "high" for interaction with an electron in an outer shell, but at the same time "low" for an interaction with a tightly bound inner-shell electron. At some projectile energy the asymptotic dependence just described may be reached for ionization of the outer shell of an atom, but not yet attained for other shells. Usually, the outer shell supplies most of the cross section for ionization, but for the differential cross sections the inner shells may become important (in some cases even dominant) at high enough incident energies and for certain ranges of parameters, such as large angles or large values of electron ejection energy.

Dropping the higher-order terms in (4-5-7) and multiplying by T, we get

$$T\sigma(Q) = 4\pi a_0^2 R \left[a(Q) \ln \frac{T}{R} + b(Q) \right]. \tag{4-5-9}$$

A plot of $T\sigma(Q)$ versus $\ln(T/R)$ will be a straight line for any given value of Q. The slope of that line is easily calculated from $a(Q)$ in the case of differential cross sections or from the quantity a for integral cross sections. The parameter a is related to the integral of the optical oscillator strength in a fashion similar to that indicated in (4-5-8) for differential oscillator strengths. Such a plot, first suggested by Fano (1954), is known as a Fano plot. In Fig. 4-5-2 experimental data for the integral ionization cross section for $H^+ + H_2$ collisions are compared on a Fano plot. Relatively small discrepancies at high energies are magnified in such a plot. Because this behavior of the Born approximation is so well established for high projectile and low secondary energies, any divergence of experimental data from the expected straight line constitutes strong evidence that the data are unreliable.

D.   Electron Correlation

Most theoretical treatments of ionization of atoms more complex than hydrogen assume that the electron being ejected departs in a field that is the sum of the field of the residual nucleus and the average (usually spherically symmetric) fields due to the remaining electrons. Interactions between the electrons that are not attributable to this average field are known as correlations. When doubly excited states of helium were identified (Madden and Codling, 1963) it was soon realized that such states represented examples of correlation in stationary states and that the independent particle model was inadequate to describe them. Under certain conditions double (or multiple) ionization provides an example of

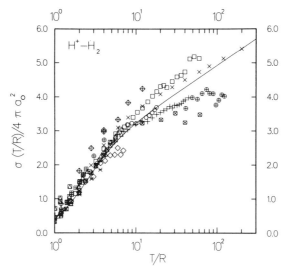

Figure 4-5-2. Fano plot of cross sections for ionization of H_2 by proton impact. The line gives the values recommended by Rudd et al. (1985a). Experimental data: $\triangle$, Afrosimov et al. (1958, 1969); $\boxplus$, Schwirzke (1960); $+(T/R > 4)$, Hooper et al. (1961); $\diamondsuit$, Solov'ev et al. (1962); $\oplus$ $(T/R < 4)$, Kuyatt and Jorgensen (1963); $\oplus$, Rudd and Jorgensen (1963), Rudd et al. (1966a), and Rudd (1979b); $\bigcirc$, Gilbody and Lee (1963); $\boxtimes$, Gordeev and Panov (1964); $+(T/R < 4)$, Hollricher (1965); $\boxtimes$ De Heer et al. (1966); $*$, Desequelles et al. (1966); $\oplus(T/R > 40)$, Piovovar and Levchenko (1967); $\otimes$, Toburen and Wilson (1972); $\square$, Shah and Gilbody (1982); $\times$, Rudd et al. (1983). [From Rudd et al. (1985a).]

dynamic correlation. An initial ionizing interaction occurs between the incoming projectile and one electron, which may be followed by an interaction of the outgoing electron with a second electron, which is ionized as well. However, multiple ionization is also exhibited in independent, noncorrelated events such as the interaction of the projectile with two different electrons in the same atom, a process that can be described by the independent electron model. Other two-electron processes, such as the Auger effect, transfer ionization, and transfer excitation, also involve electron correlation.

When correlation is treated theoretically, the already difficult three-body problem becomes an even more complex four-body problem. As with the three-body problem, assumptions have to be made in order to reduce the problem to a solvable one. Treatments of correlation have been given by McGuire (1987), Stolterfoht et al. (1989), and Stolterfoht (1990, 1991). The papers from the Symposium on Correlation Effects in Ion Induced Electron Emission at the 15th International Conference on the Physics of Electron and Atomic Collisions, held in Brighton, England in 1987, are also useful [e.g., Stolterfoht (1988)].

4-6.   CLASSICAL-TRAJECTORY MONTE CARLO METHOD

The three-body character of the collision is explicitly taken into account in the classical-trajectory Monte Carlo (CTMC) method. This is accomplished by following all three bodies through the collision, taking account of all inter-actions, and integrating the equations of motion step by step through a large number of increments of the motion until the three particles are well separated. Abrines and Percival (1966) developed the method, and Bonsen and Banks (1971) used it to calculate proton DDCS cross sections. Olson and Salop (1977) and Olson (1979, 1983) have applied the method to calculate a wide variety of cross sections.

In this treatment the initial state of the electron in the target atom is represented by spatial and momentum distributions obtained from quantum mechanical wave functions. The residual target ion is represented by a model

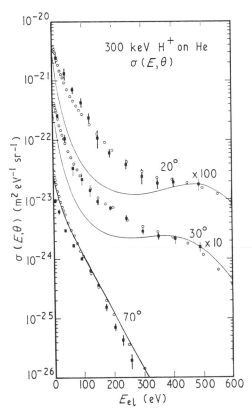

Figure 4-6-1. Comparison of experiment to two calculations of doubly differential cross sections for electron ejection by 300-keV H^+ + He collisions. $\bigcirc$, Experimental values of Rudd et al. (1966a); $\bullet$, calculations by the CTMC method; lines, calculations with the Born approximation. [From Bonsen and Banks (1971).]

potential such as the screened Coulomb potential. The impact parameter is chosen randomly and the initial electron position and momentum are selected randomly from the distributions. After following the collision until the projectile is far from the collision center, the final state is determined from the relative positions and velocities of the three particles. The procedure is repeated for a large number of projectile trajectories (typically, 10^4 to 10^5), and the cross sections determined. In the case of very small cross sections, computations on a very large number of trajectories are required to obtain statistical accuracy. Recently, a calculation to determine DDCSs for ejection of high-energy electrons from $H^+ + H$ collisions (Schultz et al., 1991) required approximately 10^7 trajectories to obtain reasonably good statistics at some angles. Sometimes the range of allowed initial conditions can be judiciously limited in order to focus on a specific process. In the CTMC method it is also possible to calculate a cross section as a function of the impact parameter.

The CTMC model suffers from one of the defects of the BEA, namely that the high-energy cross section varies as $1/T$ rather than as $(1/T) \ln T$. Therefore, this method yields best results at intermediate energies, with the accuracy falling off both at high and low energies. However, in other respects the CTMC method is superior to the BEA in that it takes explicit account of the three-body nature of the final state and therefore simulates the actual experimental conditions about as well as classical mechanics allows. Figure 4-6-1, showing the DDCS calculated by the CTMC method and by the plane-wave Born approximation, indicates the better agreement of the former with experimental results.

The CTMC method has also been used to study multiple ionization of He, Ne, and Ar by highly stripped ions (Olson, 1979), the production of saddle-point electrons (see Section 4-9) in ion–atom collisions (Olson, 1987; Gay et al., 1988) and in antimatter–matter collisions (Olson and Gay, 1988).

4-7. SEMIEMPIRICAL MODELS

Since none of the *ab initio* theoretical models developed to date yields accurate ionization cross sections for all atomic and molecular targets over a wide energy range, there is a need for algorithms that yield the cross sections required in various applications. This need is sometimes met by the use of semiempirical models. The better models are based on firm theoretical ground, but as the name implies, they also depend on experiment to determine certain parameters.

Green and McNeal (1971) presented an equation for fitting integral ionization cross sections which utilizes four adjustable fitting parameters:

$$\sigma_- = \frac{(Z_2 a)^\Omega E_1^\gamma}{J^{(\Omega + \nu)} + E_1^{(\Omega + \nu)}}, \tag{4-7-1}$$

where a, J, Ω, and ν are the fitting parameters, Z_2 is the total number of electrons in the target atom or molecule, and E_1 is the incident particle energy. The authors have taken Ω to be 0.75.

This equation does not have the desired $(1/E_1)\ln E_1$ dependence at high energies, however. A fitting equation that has the correct asymptotic form and yields a somewhat better fit to experimental data was given by Rudd et al. (1983). Another version of this model (Rudd et al., 1985a) was presented as (4-2-3).

A scaling relation for ionization cross sections based on the Bethe equation has been developed by Gillespie (1982). For incident particles of charge q the cross section is given by

$$\sigma_- = q^2 f \frac{q^{1/2}\alpha}{\beta} \sigma_{\text{Bethe}}(\beta), \qquad (4\text{-}7\text{-}2)$$

where β is the relativistic velocity ratio v/c, α is the fine-structure constant, and where

$$f\left(\frac{q^{1/2}\alpha}{\beta}\right) = \exp\left[-\lambda\left(\frac{q^{1/2}\alpha}{\beta}\right)^2\right] \qquad (4\text{-}7\text{-}3)$$

with $\lambda = 0.76$. The relativistic form of the Bethe equation may be written

$$\sigma_{\text{Bethe}}(\beta) = \frac{4\pi a_0^2 \alpha^2}{\beta} \left\{ A\left[\ln\left(\frac{\beta^2}{1-\beta^2}\right) - \beta^2\right] + C\right\}. \qquad (4\text{-}7\text{-}4)$$

A and C are constants, and as before, A can be determined from the oscillator strength.

Semiempirical models have also been proposed for the differential cross sections describing the energy distributions of the electrons ejected. Four of these models are described next.

A. Model of Khare and Kumar

Khare and Kumar (1980) modified an electron impact model by Jhanwar et al. (1975) to apply to proton impact. For the case where the proton energy E_1 is much greater than the ionization energy χ, their equation may be written

$$\sigma(\varepsilon) = \frac{4\pi a_0^2 R^2}{T} \left[\frac{(df/dQ)\ln(1+CT)}{(\varepsilon+\chi)(1+\chi/T)^{1/2}} + \frac{N\varepsilon}{\varepsilon^3 + \varepsilon_0^3} \right], \qquad (4\text{-}7\text{-}5)$$

where ε is the energy of the ejected electron, df/dQ is the differential optical oscillator strength for absorption of a photon of energy Q, and N is the number of electrons in the atom, not counting the K electrons. C and ε_0 are adjustable parameters which have values 0.07 and 60 eV, respectively, for nitrogen as a target. This model has been shown to fit 50- to 250-keV $H^+ + N_2$ data reasonably well up to the kinematic cutoff.

B. Miller Model

A model by Miller et al. (1983) is based on the Bethe equation (4-5-7). Dropping the third- and higher-order terms, that equation may be rewritten

$$\sigma(\varepsilon) = \frac{4\pi a_0^2 R}{T}\left[\frac{R}{Q}\frac{df}{dQ}\ln\frac{4T}{R} + b(\varepsilon)\right] \qquad (4\text{-}7\text{-}6)$$

where $Q = \chi + \varepsilon$ is the energy loss in the collision. The first term in the equation is the "soft-collision" term. In the second, or "hard-collision" term, $b(\varepsilon)$ is independent of projectile properties and may, in principle, be determined from experimental data at one incident energy by subtraction of the first term from the measured cross section at that energy. The value of the function $b(\varepsilon)$, once determined for one incident energy, may be used for all other energies. In practice, the binary encounter approximation is also used to help determine $b(\varepsilon)$. Based as it is on the Bethe approximation, Miller's model holds only for high-energy collisions. An example of the fit to experimental data is shown in Fig. 4-7-1.

C. Rudd Model

A model given by Rudd (1988) for the singly differential ionization cross sections is useful for proton impact at all energies. It is based on the molecular promotion model of Fano and Lichten (1965) at low primary and high secondary energies and on the Williams equation (4-4-3), modified to agree with Bethe–Born theory at high primary and low secondary energies. Three parameters suffice to fit the energy distributions at a single primary energy, but a total of 10 parameters are needed to fit cross sections at all combinations of primary and secondary energies. The cross section is the sum of the partial cross sections for the various subshells. The model is stated in terms of the dimensionless variables $w = \varepsilon/\chi$ and $v = (T/\chi)^{1/2}$. The cross section for each subshell is

$$\sigma(w) = \frac{S}{I}\frac{F_1 + F_2 w}{(1 + w)^3\{1 + \exp[\alpha(w - w_c)/v]\}}, \qquad (4\text{-}7\text{-}7)$$

where $w_c = 4v^2 - 2v - R/4\chi$, $S = 4\pi a_0^2 N(R/\chi)^2$, χ is the ionization potential, and N is the number of electrons in the subshell. F_1 is given by

$$F_1 = L_1 + H_1, \qquad (4\text{-}7\text{-}8)$$

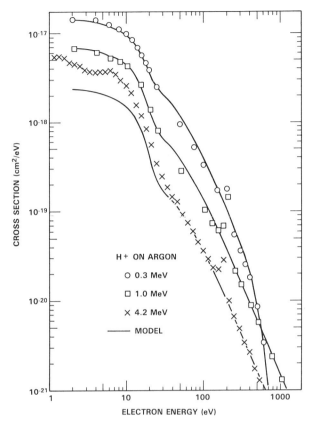

Figure 4-7-1. Singly differential cross sections for electron ejection in H^+ + Ar collisions. Lines, Miller model; experimental data points at 0.3 and 1.0 MeV, Toburen et al. (1978); data points at 4.2 MeV, Gabler (see Rudd et al., 1979). [From Miller et al. (1983).]

where $L_1 = C_1 v^{D_1}/(1 + E_1 v^{D_1 + 4})$ and $H_1 = A_1 \ln(1 + v^2)/(v^2 + B_1/v^2)$, and F_2 by

$$F_2 = \frac{L_2 H_2}{L_2 + H_2}, \qquad (4\text{-}7\text{-}9)$$

where $L_2 = C_2 v^{D_2}$ and $H_2 = A_2/v^2 + B_2/v^4$. The quantities $A_1 \cdots E_1$, $A_2 \cdots D_2$, and α are the 10 fitting parameters which, when determined, are sufficient to give the singly differential cross sections at all primary and secondary energies. Rudd et al. (1992) have fitted this equation to experimental data for protons on 10 target gases to determine the fitting parameters. Figure 4-7-2 shows calculations

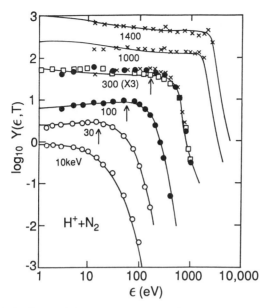

Figure 4-7-2. Singly differential cross sections for $H^+ + N_2$ plotted as $Y(\varepsilon, T)$, the ratio of the cross section to the Rutherford cross section, versus electron energy. Lines, Rudd model; +, Toburen (1971); ○, Rudd (1979); □, Stolterfoht (1971); ■, Crooks and Rudd (1971). Arrows indicate the positions of the peaks due to electron capture to the continuum. [From Rudd et al. (1992).]

of SDCSs from the model for protons on nitrogen compared with experimental values by various investigators.

D. Kim Model

Kim has devised a model [see Rudd et al. (1992)] that is also based on the Bethe equation. As with the Miller model, it requires as input differential optical oscillator strength spectra, but instead of using those data directly, the spectra are fitted either by a series of Gaussian functions or by a series of inverse powers of the energy transfer ε. The fitting equations are

$$\frac{df}{d(Q/R)} = \frac{R}{Q} \sum_i a_i \exp\left[-\left(\frac{R/Q - b_i}{c_i}\right)^2\right]\left(\frac{R}{Q}\right)^{d_i} \qquad (4\text{-}7\text{-}10)$$

and

$$\frac{df}{d(Q/R)} = \frac{R}{Q} \sum_i e_i \left(\frac{R}{Q}\right)^i, \qquad (4\text{-}7\text{-}11)$$

where a_i, b_i, c_i, d_i, and e_i are fitting parameters. The model gives Y, the ratio of the cross section to the Rutherford cross section, (4-4-2):

$$Y = \frac{\sigma(Q)}{\sigma_R(Q)} = Q\frac{df}{dQ}\frac{\ln(4fT/R)[1 + gR/T + h(R/T)^2] + Y_{BE} + N}{1 + \exp[k(Q - 4T)/R]},$$

$$(4\text{-}7\text{-}12)$$

where

$$Y_{BE} = \frac{NU_2}{Q}\left[1 - \left(\frac{U_2}{\varepsilon + U_2}\right)^p\right] \qquad (4\text{-}7\text{-}13)$$

and where f, g, h, k, and p are additional parameters, determined by fitting to experimental data. N is the occupation number, and U_2 is the average kinetic energy of the electrons in a given shell. Values of U_2, which may be calculated from nonrelativistic Hartree–Fock wave functions, are available for the various subshells of some targets (Kim, 1975; Rudd et al., 1992). The additional terms in the Kim model allow it to be used over a somewhat wider primary energy range than the Miller model, but so far the parameters have been evaluated only for two targets, argon and nitrogen [see Rudd et al. (1992)]. Figure 4-7-3 shows the model results for nitrogen.

Although these SDCS models have not yet been tested for applicability to

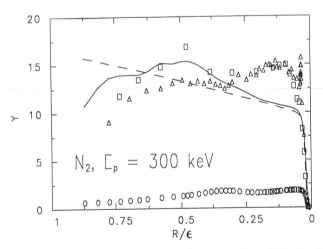

Figure 4-7-3. Singly differential cross sections for ionization of N_2 by 300-keV H^+ plotted as Y, the ratio of the cross section to the Rutherford cross section, versus reciprocal of the electron energy ε. Solid line, Kim model; dashed line, Rudd model; □ and △, experimental data; ○, calculated contribution from the $2\sigma_g$ state. [From Rudd et al. (1992).]

bare ion projectiles heavier than protons, Z_1^2 scaling could probably be used at high energies. No similar models are known for heavy dressed ion or neutral impact, nor have models yet been published for the angular distributions.

PART $\mathscr{C}$. MECHANISMS AND SPECIAL TOPICS IN IONIZATION

4-8. ELECTRON CAPTURE TO THE CONTINUUM

Ionization and electron capture were long considered to be separate processes, but in the 1960s and 1970s a new mechanism of electron ejection known as electron capture to the continuum* was identified through a combination of experimental and theoretical investigations. This effect is both true ionization and true electron capture and therefore is important as a link between these two processes, in addition to its own intrinsic interest.

Reviews of this mechanism have been given by Rudd and Macek (1972), Sellin (1982), Groeneveld et al. (1984), Jakubassa-Amundsen (1983), Lucas et al. (1984), and Briggs (1989).

The first indication of such a mechanism appeared in 1963 in the measurements by Rudd and Jorgensen (1963) of angular and energy distributions of electrons ejected by protons, where an unexplained peak appeared (see Fig. 4-3-3) in the energy distribution of electrons ejected at 10°. When the DDCSs were plotted against angle, the effect was also visible as a forward peak (Rudd and Macek, 1972), as shown in Fig. 4-4-2. Oldham (1967) suggested that the peak, which was found to occur when the velocity of the outgoing electron matched the velocity of the proton, resulted from a strong interaction between the ejected electron and the scattered proton.

Salin (1969a, b) approached the problem by considering the process to be an ionization in which the effective charge of the target is modified by the influence of the projectile. This gave a velocity-dependent effective charge and led to the multiplication of the Born approximation expression for ionization by the factor

$$N(\gamma)^2 = \frac{2\pi\gamma}{1 - e^{-2\pi\gamma}}, \tag{4-8-1}$$

*The mechanism has been variously labeled "charge exchange to a continuum state (CEC)" (Macek, 1970b), "charge exchange into the continuum" (Harrison and Lucas, 1970), "charge transfer into continuum states (CTC)" (Crooks and Rudd, 1970), "production of convoy electrons" (Brandt and Ritchie, 1977), "continuum electron capture (CEC)" (Vane et al., 1978), "electron capture to the continuum (ECC)" (Shakeshaft and Spruch, 1978), "electron transfer to the continuum (ETC)" (Yu and Lapicki, 1987), and "capture ionization" (Briggs and Macek, 1991). Since there is no exchange involved, and the expression "transfer to the continuum" could be taken to mean direct transfer not involving capture, one of the labels using the word "capture" would seem to be more descriptive. We use the term *electron capture to the continuum* (ECC). The term *convoy electrons* has come to be associated with this and related processes in solids.

where

$$\gamma = \frac{1}{|\mathbf{v} - \mathbf{v}_1|} - \frac{1}{v},$$

(4-8-2)

and where $\mathbf{v}_1$ and $\mathbf{v}$ are the projectile and ejected electron velocities, respectively.

Macek (1970a, b), on the other hand, interpreted the forward peak as an electron capture by the proton. However, instead of capture into the ground state or into one of the bound excited states, he suggested that this was the previously overlooked case of capture into a continuum state of the projectile. Macek treated the problem mathematically by using the first term in the Neumann expansion of Faddeev's equations (Fadeev, 1960), an approach that treats the three outgoing particles on an equal basis. His calculated angular distributions agree well in shape with the experimental results but overestimate the absolute values by a factor of 2. Figure 4-8-1 shows a comparison with experimental angular distributions (Toburen and Wilson, 1972) at various incident energies from 0.3 to 1.5 MeV. In each case the electron energy was chosen at the equal-velocity point. Godunov et al. (1983) have obtained better agreement using a modification of the Fadeev equations.

Since this treatment involves the addition of direct and exchange amplitudes, it follows that there should be interference terms. However, no structure attributable to interference has been observed (Dettmann et al., 1974; Duncan et al., 1977). Furthermore, Macek's results do not approach the Born approximation expression in the limits of very small and very large electron energies.

A prediction of both Salin's and Macek's treatments was that the cross section should go to infinity for electrons ejected at zero angle with a velocity $v = v_1$. Of course, the necessarily finite angular acceptance angle of any experimental apparatus would limit the measured peak to a finite height. Energy distributions at zero angle were measured for protons impinging on a foil target by Harrison and Lucas (1970) and on a gaseous target by Crooks and Rudd (1970). Both showed a large cusp-shaped peak in the energy distribution, a shape that is unusual for the description of physical processes. Figure 4-8-2 shows energy spectra for electrons ejected at $0°$ for H^+ and H_2^+ incident on helium at 100 keV. The cusp-shaped peak appears in each case at the energy for which the electron velocity equals that of the incident particle. It became evident later that most of the peak in the H_2^+ case is due to electron loss from the projectile, which also comes at the equal-velocity point. In Fig. 4-8-3 data from the gas target are compared with Macek's calculations: one calculated for zero angle of emission and one for an angle equal to the largest angle accepted by the experimental apparatus. As expected, the measured peak falls between the two.

That an infinite peak is to be expected was shown by Macek (Rudd and Macek, 1972) using a very general argument based on a transformation of the cross section from the projectile (primed) frame to the laboratory (unprimed) frame and on the continuity of the cross section per unit energy interval across

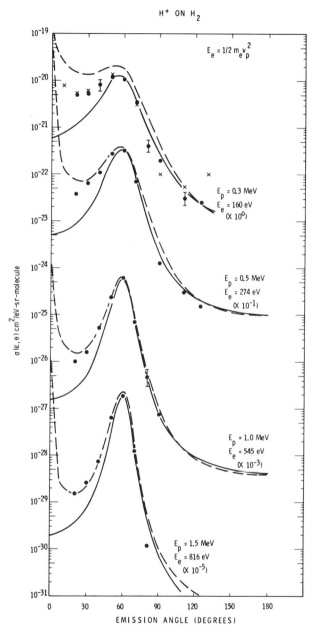

Figure 4-8-1. Angular distributions of electrons from $H^+ + H_2$ collisions at various energies. The electron energies were chosen such that the electron velocity is equal to that of the projectile in each case. ×, Rudd et al. (1966a); ●, Toburen and Wilson (1972); dashed lines, theory of Macek (1970a, b); solid line, hydrogenic Born approximation calculations. [From Toburen and Wilson (1972).]

303

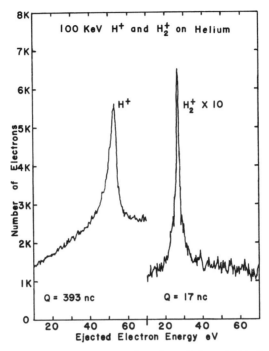

Figure 4-8-2. Energy spectra of electrons ejected at $0°$ from 100-keV H^+ and H_2^+ on helium, showing the cusp-shaped peaks where the electron velocity is equal to the incident particle velocity. [From Crooks and Rudd (1970).]

the ionization threshold. For a hydrogen atom the energies of the stationary states are given by $E_n = -R/n^2$, where R is the Rydberg of energy. Then the density of states near the continuum is

$$\frac{\Delta n}{\Delta E_n} = \frac{n^3}{2R}. \tag{4-8-3}$$

The cross section for excitation of the states in an interval Δn is proportional to $\Delta n/n^3$, so

$$\frac{\Delta \sigma}{\Delta n} \propto n^{-3}. \tag{4-8-4}$$

Then $\Delta \sigma/\Delta E_n$ is a constant independent of n, say $\bar{\sigma}$. But the cross section per unit energy interval is continuous across the ionization threshold, so just above the threshold,

$$\frac{d\sigma}{dE} = \bar{\sigma}(E), \tag{4-8-5}$$

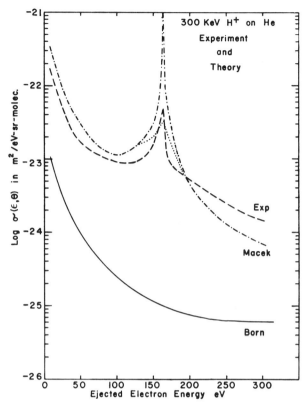

Figure 4-8-3. Cross sections for ejection of electrons at $0°$ for 300-keV H^+ incident on helium. Dashed curve, experimental data (at $0° \pm 1.4°$), Crooks and Rudd (1970); dash-dot curve, theory of Macek (1970a, b) for $0°$; dotted curve, the same for $1.4°$; solid curve, Born approximation. [From Rudd and Macek (1972).]

where $\bar{\sigma}(E)$ is a slowly varying function of E. In the projectile reference frame the cross section for electron capture to continuum states of the projectile obeys the same equation:

$$\frac{d\sigma'}{dE'} = \bar{\sigma}'(E'), \qquad (4\text{-}8\text{-}6)$$

where $\bar{\sigma}'(E')$ is a slowly varying function of E', the energy in the projectile frame. This can be extended to the doubly differential cross section for ejection into a solid angle $d\Omega'$:

$$\sigma'(E', \theta') = \frac{d^2\sigma'}{dE' \, d\Omega'} = \bar{\sigma}'(E', \theta'). \qquad (4\text{-}8\text{-}7)$$

Since the signal detected is the same regardless of the frame of reference in which it is described,

$$\sigma(E, \theta)\, dE\, d\Omega = \sigma'(E', \theta')\, dE'\, d\Omega'. \tag{4-8-8}$$

The element of volume in velocity space is the same in both frames (i.e., $d\mathbf{v} = d\mathbf{v}'$), so

$$v\, dE\, d\Omega = v'\, dE'\, d\Omega'. \tag{4-8-9}$$

Then

$$\sigma(E, \theta)\, dE\, d\Omega = \sigma'(E', \theta')\, \frac{v}{v'}\, dE\, d\Omega, \tag{4-8-10}$$

so finally,

$$\sigma(E, \theta) = \bar{\sigma}'(E', \theta')\, \frac{v}{v'} = \bar{\sigma}'(E', \theta')\, \frac{v}{|\mathbf{v} - \mathbf{v}_1|}, \tag{4-8-11}$$

where $\mathbf{v}_1$ is the velocity of the projectile. Since $\bar{\sigma}'(E', \theta')$ is finite at $E' = 0$, $\sigma(E, \theta)$ diverges for $\mathbf{v} = \mathbf{v}_1$.

In physical terms, one may think of the process as a capture of the electron by the projectile immediately followed by a reemission at a very small velocity in the projectile frame. Since the energy distribution of ejected electrons is largest at zero energy in the emitter's frame, the cusp may thus be thought of as a Doppler-shifted version of the zero-energy peak. Drepper and Briggs (1976) have shown this in a way that separates the effects due to a frame transformation from those attributable to dynamical processes.

Another way of picturing the process is as a Thomas double scattering. Thomas (1927) suggested this mechanism to explain electron capture (see Section 5-6), but it can apply equally well to capture to the continuum. The incoming ion first makes an elastic collision with an electron that recoils at 60°. At this angle the electron recoils at the same speed as the projectile ion. Subsequently, the electron goes into a hyperbolic orbit around the nuclear core and emerges from the atom in approximately the same direction as the nearly undeflected projectile. The field of the projectile further tends to focus the electron trajectory in the forward direction.

Other theoretical treatments that have been applied to ECC are the classical-trajectory Monte Carlo (CTMC) method (Bonsen and Banks, 1971), the second Born approximation (Dettman et al., 1974), a linear combination of atomic orbitals (Band, 1974), the impulse approximation (Jakubassa-Amundsen, 1983, 1989; Miraglia and Macek, 1991), a density matrix description (Burgdörfer, 1984), a multiple scattering theory (Garibotti and Miraglia, 1983), the continuum distorted wave (CDW) approximation (Belkic, 1978, 1980), and the distorted-wave strong-potential Born approximation (Brauner and Macek, 1992). A completely satisfactory theoretical treatment of electron capture to the continuum is not yet available.

Neelavathi et al. (1974) showed that a fast charged particle would cause wavelike variations in electron density and therefore potential when traveling in a solid. The moving potential minima could trap electrons and sweep them along in the forward direction. They proposed that this "wake-riding" mechanism could account for the forward cusp in the electron distribution from solids. However, measurements by Duncan and Mendenez (1976) of energy profiles, angular distributions, and variations of the widths of the cusps with angle for 0.175- to 1-MeV H^+ and H_2^+ on carbon foils tended to agree with expectations of the ECC mechanism rather than with those of the wake-riding mechanism. Nevertheless, the origin of the forward peak from solids is still a matter of controversy.

The same investigators found that the yields of velocity matching electrons were independent of foil thickness, which seemed to indicate that these electrons are produced at or very near the foil exit surface. However, Sellin et al. (1985), using 15.2-MeV/u Ni^{24+} and Ni^{26+} ions in aluminum and carbon foils, did find an increase in the yield with thickness up to about 50 $\mu g/cm^2$. However, when the thickness exceeds the mean free path, saturation occurs. Yamazaki and Oda (1984) concluded from experiments involving 0.9 to 2-MeV/u hydrogen beams on carbon foils that the cusp results primarily from a process in which secondary electrons traveling with the projectile ion are detached as they pass the potential step at the surface of the foil. They were able to distinguish that mechanism from the contribution to the cusp due to the ECC mechanism.

Koschar et al. (1987), who measured absolute numbers of convoy electrons when 1.2 to 3.0-MeV H^+ and H^0 passed through carbon foils with thicknesses of 2 to 20 $\mu g/cm^2$, found that the yield was the same for the two projectiles for thicknesses greater than 15 $\mu g/cm^2$, but for very thin foils it was as much as 1000 times greater for the neutral hydrogen projectiles than for protons. They account for the dependence on thickness by calculating the charge distribution as a function of depth and taking the number of convoy electrons to be proportional to the sum of the electrons lost from H^0 and captured by H^+. The electron distribution includes not only the convoy electron peak but also an associated broad peak that shifts to a lower energy for thicker targets. The shape of the distribution has recently been described successfully by a Coulomb transport simulation method developed by Reinhold et al. (1992).

Because of the negative charge of antiprotons, the spectrum of electrons ejected by them would be expected to show an inverted cusp or dip in the spectrum. Such a dip was predicted by Garibotti and Miraglia (1980). It was also suggested (Crooks and Rudd, 1970) that this effect could explain the fact that positive pions experience greater energy loss in emulsions than do negative pions, a fact observed by Heckman and Lindstrom (1969).

Because the cross section for ionization is continuous with that of excitation across the ionization threshold, it is possible to obtain cross sections for excitation from measurements of ECC. This method, which was exploited by Rødbro and Anderson (1979), avoids the problems associated with analyzing, detecting, and normalizing optical spectra.

An interesting experiment related to ECC but involving hydrogen atoms instead of electrons was done by Breinig et al. (1983). A beam of Ar^+ (or Kr^+) at 100 to 300 eV/u was incident on a CH_4 target. By the Thomas double-scattering mechanism a hydrogen atom was captured into a state in the vibrational continuum of the electronic ground state of the temporary ArH^+ molecular ion resulting in protons emerging into a narrow cone around the forward direction. Such a process had been analyzed by Bates et al. (1964) and by Shakeshaft and Spruch (1980).

4-9. SADDLE-POINT ELECTRONS

It had long been assumed that the electrons ejected from a target atom in an ion–atom collision had a distribution centered on the residual target ion. With the discovery of the mechanism of electron capture to the continuum, it came to be recognized that electrons ejected in a collision could, instead, be associated with the receding projectile.

During the calculation of differential cross sections for proton–hydrogen atom collisions at intermediate energies using the CTMC method (see Section 4-6), Olson (1983) discovered that in addition to ion-centered and projectile-centered electrons, there was also a group of electrons ejected with velocities close to $v_1/2$. He interpreted this enhancement to the "stranding" of electrons on the transitory saddle-shaped region of the electric potential between the moving proton and the residual hydrogen ion. Since the electron is influenced simultaneously by both centers of attraction, this mechanism is a good example of "two-center" ionization. For a review of two-center effects in ionization, see Fainstein et al. (1991).

Meckbach et al. (1986) measured electron spectra at various angles near 0° for H^+ and H^0 collisions with helium, which showed a ridge sharply focused in angle but broadly distributed in energy, which they interpreted as being due to electrons propagating along the saddle point in the potential. Later, the same group (Bernardi et al., 1988) showed that the sharpness of the ridge had been enhanced artificially by an extended gas target effect which they corrected in subsequent measurements. Even after correction, the ridge (Fig. 4-9-1) indicates that a considerable fraction of the electron emission is influenced by both attracting centers. Similar data with H^0 impact do not have the ridge.

In a paper on $H^+ + He$ collisions, Olson et al. (1987) made CTMC calculations indicating that up to about 200 keV the saddle-point electron flux was actually dominant over that associated with the target or the projectile alone. In the same paper they presented new experimental data for 60- to 200-keV collisions, showing that electrons ejected at 17° (the smallest angle available

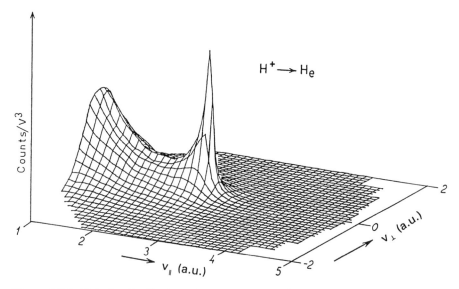

Figure 4-9-1. Velocity distribution of electrons from H^+ collisions with He. In addition to the peak at the equal-velocity point, there is a ridge concentrated at zero transverse electron velocity attributed to the saddle-point mechanism. [From Bernardi et al. (1988).]

in their apparatus) exhibited a definite peak at an intermediate velocity when plotted versus velocity (see Fig. 4-9-2).

Meckbach et al. (1986) made quantum mechanical calculations of a Gaussian-shaped charge cloud, initially concentrated on the saddle, which showed that as time went on, the velocity distribution of the cloud spreads out in the direction of the relative motion of the two positive charge centers but narrows in the perpendicular direction, in agreement with intuitive expectations and with measurements.

Gay et al. (1990) investigated the saddle-point mechanism using projectiles with a charge greater than unity. In their measurements of electrons at 10° and 20° with H^+ and He^{2+} bombarding He, Ne, and Ar targets, they found saddle-point maxima which depended on the ratio of the projectile to the target atom charge roughly as predicted by the equation

$$\frac{v_{sp}}{v_1} = \frac{1}{1 + (q_1/q_2)^{1/2}} \tag{4-9-1}$$

which gives the velocity at the expected position of the saddle-point potential energy maximum in terms of the velocity v_1 of the projectile and the charges q_1 and q_2 of the projectile and target. Similar measurements by Bernardi et al. (1990), however, failed to find a dependence of the position of the maximum on the projectile charge.

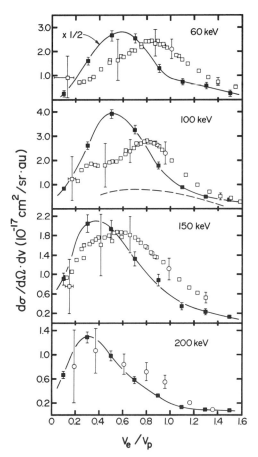

Figure 4-9-2. Velocity spectra for electrons ejected at $17°$ in $H^+ + He$ collisions. ◯ (100 and 150 keV), Rudd and Jorgensen (1963); ◯ (200 keV), Rudd et al. (1966a); ☐, Olson et al. (1987); ■, CTMC calculations; dashed line, plane-wave Born approximation. [From Olson et al. (1987).]

4-10. ELECTRON PROMOTION

When the velocity of an incident ion in a collision is comparable to or less than the orbital velocity of an electron in the target, the situation no longer involves a sudden transition from an initial to a final state as described in the Born approximation. If the period of the orbiting electrons is short compared to the time the two colliding nuclei are within a few atomic radii of each other, a temporary molecule is formed as an intermediate state. A different mechanism of excitation and ionization comes into play, which usually becomes dominant at low energies. In fact, the cross section for this mechanism for inner-shell vacancy production is so large as to be essentially equal to the geometric cross section.

This mechanism, known as electron promotion, was suggested by Fano and Lichten (1965). The changeover takes place approximately at a projectile velocity equal to the orbital velocity of the electron involved. A 25-keV proton, for example, has the same velocity as the orbital electron in a ground-state hydrogen atom. However, for tightly bound inner shells of heavier atoms, the critical energy may be many MeV. Thus a projectile at a given velocity may be "fast" for an interaction with the outer shell of an atom but be "slow" when considering processes involving inner shells of the same atom. While in high-energy collisions, as described by the Born approximation, the velocity is the relevant parameter, the distance of closest approach is a more important quantity in the promotion mechanism.

Reviews of work involving promotion, molecular orbital theory, and inelastic energy loss have been written by Kessel (1969), Kessel and Fastrup (1973), Garcia et al. (1973), Fastrup (1975, 1977), Datz (1977), Stolterfoht (1978a, b), and Kessel et al. (1978). See also Chapter 2 of this volume.

A. Energy-Level Diagrams

A diagram showing energy levels as a function of the internuclear separation of the two colliding atoms is useful in discussing the electron promotion model. Figure 4-10-1 shows such a diagram, which was drawn by Fano and Lichten (1965) for an argon–argon pair. Levels at the right for large internuclear distances are the atomic levels of the separate atoms, while those on the left for small separations represent the atomic levels of the krypton atom, which is what results when two argon atoms are forced together. In between are the levels appropriate to the temporary Ar_2 molecule at various separations. The levels at the opposite extremes are connected according to the rules of molecular structure involving angular momentum and parity [see, e.g., Herzberg (1950), Lichten (1967), Barat and Lichten (1972), or Kessel and Fastrup (1973)].

In a collision, one starts at the right in this diagram with the atoms well separated. As the collision proceeds, the internuclear separation r decreases to a minimum value and then increases again as the atoms separate. Since levels cross, it is possible that the atoms which are in one state coming into the collision branch off and are "promoted" into a different, sometimes higher, final state. For slow collisions the electron promotion mechanism often has a much higher probability than other mechanisms of excitation or ionization. The $4f\sigma$ orbital is especially noteworthy in this regard, since it rises steeply at the critical internuclear distance $r_c \simeq 0.25$ Å to put a $2p$ electron of argon into either a higher unfilled orbital or into a continuum state, thus resulting in an L vacancy followed by the emission of an Auger electron or an X-ray. Cacak et al. (1970) showed that this resulted in a nearly constant cross section for Auger emission as a function of the incident energy provided that the energy was great enough to ensure that the nuclei penetrated to within the critical distance. Figure 4-10-2 shows cross-section data for Auger emission combined with relative cross

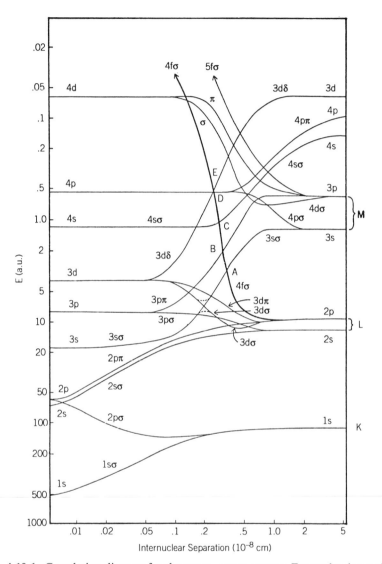

Figure 4-10-1. Correlation diagram for the argon–argon system. Energy levels are those of krypton at the united atom limit at the left. The argon energy levels are at the right. [Adapted from Lichten (1967).]

sections for X-ray emission. The failure of the X-ray cross section to remain constant above the knee is probably due to changing fluorescence yields.

The experimental data are compared to calculations on a simple but effective model suggested by Kessel [see Cacak et al. (1970)] that assumes a step-function increase in the probability P of the process from zero to unity (actually, the probability has an effective value of 2, since two electrons are ejected in the same

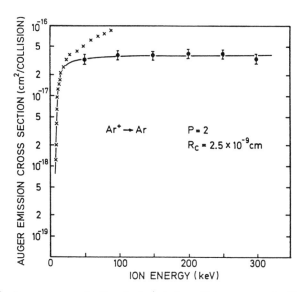

Figure 4-10-2. *L*-vacancy production in Ar$^+$ + Ar collisions versus incident ion energy. ●, Measured *L*-Auger electron emission cross sections of Cacak et al. (1970); ×, measured *L* X-ray data of Saris and Onderdelinden (1970) normalized to the Auger data at 20 keV; solid line, calculations from (4-10-2). [From Kessel and Fastrup (1973).]

collision) as the closest internuclear separation r_0 reaches the critical value r_c. Using a screened Coulomb potential with an atomic screening radius a given by (4-3-6), the relation between the impact parameter p and the distance of closest approach r_0 is

$$p = r_0 \left[1 - \frac{b}{r_0} \exp\left(-\frac{r_0}{a} \right) \right]^{1/2}, \tag{4-10-1}$$

where $b = 2Z_1 Z_2 a_0 R/E_1$. The cross section follows as

$$\sigma = 2\pi r_c^2 \left[1 - \frac{b}{r_c} \exp\left(-\frac{r_c}{a} \right) \right]. \tag{4-10-2}$$

Calculations using this equation are also shown in Fig. 4-10-2.

B.   Coincidence Measurements

While the determination of total cross sections for inner-shell vacancy production from Auger or X-ray measurements yields some information about the promotion mechanism, it involves an integration over impact parameters and final charge states and thus conceals important information. A more detailed type of measurement was made in the 1960s at the Ioffe Institute in Leningrad and at the University of Connecticut. Both groups used coincidence methods to obtain differential cross sections for inner-shell vacancy production as a function of the scattering angle and final charge states of the colliding atomic particles.

The apparatus used by Afrosimov et al. (1963, 1964) was shown in Fig. 4-2-3. The system studied with this apparatus was $Ar^{0,+1} + Ar$ at 12.5 and 50 keV. Another study of the same system from 3 to 400 keV was made at about the same time by Everhart and Kessel (1965) and by Kessel et al. (1965). Their apparatus is shown in Fig. 4-10-3. Target gas fills the main chamber while three "boxes," or wedge-shaped chambers, are evacuated. These chambers are the scattered box, the recoil box, and an inlet port for the incident ion beam. Both the scattered and recoil boxes are rotatable around the collision center and both contain electrostatic analyzers and detectors. Both neutral and ionized particles can be detected. A similar but larger chamber was later used by Kessel (1969) for ion–atom collision research at energies up to 10 MeV.

Both the Leningrad and Connecticut groups measured relative values of the probabilities for collisions involving various values of the inelastic energy loss

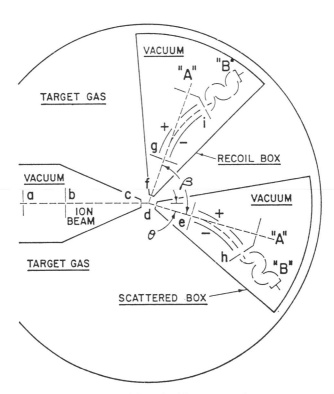

Figure 4-10-3. Schematic diagram of the coincidence scattering apparatus used by Kessel and Everhart (1966).

during (10mn) and (00mn) collisions. One result to come out of the measurements was that there were characteristic energy losses Q which depended on m and n. However, after subtracting the ionization energies associated with the ejected electrons, the energy losses Q^* were independent of m and n, as well as being approximately independent of the distance of closest approach of the colliding particles and of the primary beam energy. The values obtained for the three energy loss peaks in argon were $Q_I^* = 53 \pm 14\,\text{eV}$, $Q_{II}^* = 263 \pm 16\,\text{eV}$, and $Q_{III}^* = 475 \pm 22\,\text{eV}$. The first of these peaks is due to the production of M-shell vacancies, the second results from a single L-shell vacancy accompanied by M-shell vacancies, while the third is from two L-shell vacancies in addition to M-shell vacancies. From the measured deflection of the scattered beam particle the distance of closest approach r_0 can be determined. The relative probabilities of these characteristic energy losses varied with the distance of closest approach as shown in Fig. 4-10-4.

In Fig. 4-10-5 the average inelastic energy loss $\bar{Q}$ is plotted versus r_0. By comparing this figure with the molecular correlation diagram the specific excitation mechanisms involved with each increase in $\bar{Q}$ could be deduced. The triple peaks described above come at values of r_0 in the vicinity of 0.24 Å. The promotion of electrons on the diabatic $4f\sigma$ molecular orbital (MO) accounts for the single or double L-vacancy production. Fano and Lichten (1965) in proposing the promotion model showed that the promotion probabilities for this case would be close to unity because of the many excited orbitals crossed by the $4f\sigma$ orbital. They also predicted that the probability would be velocity independent.

At $r_0 \simeq 0.10$ Å there is an additional rise in $\bar{Q}$, due to additional L vacancies.

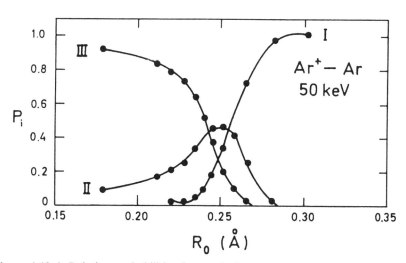

Figure 4-10-4. Relative probabilities for producing zero, one, or two (I, II, III) $L_{2,3}$ vacancies in 50-keV Ar$^+$ + Ar collisions versus distance of closest approach. [From Fastrup et al. (1971).]

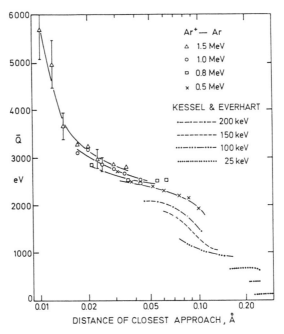

Figure 4-10-5. Average inelastic energy loss versus distance of closest approach for $Ar^+ + Ar$ collisions at various energies. [From Kessel and Fastrup (1973).]

Since the K shells of argon begin to interpenetrate between 0.01 and 0.02 Å, there is an additional rise in the average energy loss in that region resulting from the production of K vacancies.

C. Energy Spectra of Electrons from Slow Collisions

When integrated over all angles, the cross section for emission of electrons tends to approach an exponential decrease with electron energy for ion velocities smaller than the orbital velocities. Macek developed a model to explain this behavior [see Rudd (1979)] for low-energy proton impact, which is illustrated in the energy-level diagram in Fig. 4-10-6. As the proton approaches the target, a charge-transfer transition with an energy difference ΔE_1 takes place from the initial $H^+ + X$ state to the $H(1s) + X^+$ state. At the distance of closest approach a rotational coupling promotes the system to one of several possible excited states of either the projectile or target. Upon separation, this brings the system to a point where a transition energy ΔE_2 of only a few electron volts is sufficient to raise it to a continuum state. Applying an expression derived by Meyerhof (1973), an exponential expression of the form

$$\sigma(\varepsilon) \propto \left\{ 1 + \exp\left[\frac{\alpha(\varepsilon + \Delta E_2)}{(\chi T)^{1/2}} \right] \right\}^{-1} \tag{4-10-3}$$

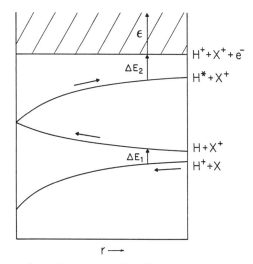

Figure 4-10-6. Energy-level diagram used to illustrate promotion of target to excited state and transition into the ionization continuum. [From Rudd (1979).]

results. Here α is a dimensionless constant near unity which is related to the size of the target. This expression is also incorporated into the semiempirical model by Rudd (see Section 4-7) where it also describes the quasi-exponential dependence on electron energy above the kinematic cutoff for high-energy collisions. In that case, ΔE_2 is replaced by $-\varepsilon_c$, where ε_c is the electron energy at the kinematic cutoff.

Woerlee et al. (1981) and Gordeev et al. (1981) have applied a direct-coupling-to-the-continuum mechanism to collisions between heavy ions and atoms (e.g., $Kr^{n+} + Kr$) and obtained a similar exponential expression.

4-11.   AUTOIONIZATION AND THE AUGER EFFECT

Certain excited states of atoms, molecules, or ions can deexcite by the emission of electrons rather than photons. This general process, known as *autoionization* (AI), was first observed by Auger (1925) for excited states involving inner-shell vacancies. For this reason, such a process is known as the Auger effect, but it is actually only a special case of autoionization. States that autoionize may be excited by photons, electrons, or ions. Since autoionization produced by electron impact was taken up in Sections 6-6, 7-1, and elsewhere in the companion volume, in this section we deal primarily with studies of AI and the Auger effect involving excitation by ion impact. This subject has a large literature and various aspects of it have been reviewed by Mehlhorn (1969, 1978, 1985), Sevier (1972), Ogurtsov (1972), Rudd and Macek (1972), Wille and Hippler (1986), Lin (1986), Stolterfoht (1987a, b), and Niehaus (1990).

A. Outer-Shell Autoionization

For autoionization to be energetically possible, there must be a continuum state of the same energy as the excited state (i.e., the excited state must have a higher energy than the ground state of the ion of the next-higher charge state). If the energy requirement is met and certain selection rules are satisfied, auto-ionization proceeds very rapidly. Typical lifetimes are about 10^{-13} s, although some states have lifetimes against autoionization as long as 10^{-5} s.

Autoionization can take place from a multiply excited state or from a state with a single excitation plus a rearrangement of the core electrons. An example of a doubly excited state is the $2s2p$ state of helium, which can decay to the $1s$ ground state of the He$^+$ ion. An example of the second case occurs in the oxygen atom, the ground state of which may be written $2p^3(^4S)2p(^3P)$. This atom can be excited to $2p^3(^2D)4s(^3D)$, which is at an energy higher than the first ionization potential of oxygen. In the case of molecules, a single electronic excitation coupled with a vibrationally excited state may put the molecule into a high-enough energy state to autoionize.

Peaks in the energy spectrum of electrons due to autoionization were first identified by Moe and Petsch (1958), who used low-energy K$^+$ collisions with rare gases, and by Berry (1961), who studied collisions of Ar$^+$ + Ar and He$^+$ + He. In the latter case, a single peak of about 3.5 eV FWHM was seen centered at about 31 eV, an energy which indicated that a possible autoionizing transition was taking place. Rudd (1964), with 75-keV H$_2^+$ incident on helium, produced an electron spectrum with a high-enough resolution (0.36 to 0.66 eV) that several individual AI transitions could be identified by comparison to spectra obtained by photoabsorption (Madden and Codling, 1963) and by electron energy-loss measurements (Simpson et al., 1964). Using this method, Rudd (1965), with a resolution improved to 0.1 eV, also identified two previously unobserved series in helium, the $2sns(^1S)$ and $2snp(^3P)$.

Many other target and projectile combinations have been used. For example, Edwards and Rudd (1968) studied the neon spectrum with proton impact, Ogurtsov and Bydin (1969) studied the AI of alkali atoms induced by collisions with rare gases, Bruch et al. (1975) produced doubly and triply excited states of LiI and LiII using beam-foil techniques, and Yagishita et al. (1978) used lithium ions on helium.

Mention should be made of an experiment performed at Freiburg by Kessel et al. (1979), who measured AI electron spectra from 2-keV He$^+$ + He collisions in coincidence with ions scattered at an angle of 6°. By scanning the azimuthal scattering angle the population amplitudes for various magnetic sublevels of the $(2p^2)^1D$ and $(2s2p)^1P$ states could be determined. It was found that the ^1D state was excited with a 97% probability and that the $m = 0$ sublevel was the main one excited. By this means the shape of the electron cloud during excitation could be mapped out. The mechanism for the excitation was deduced to be the "blow up" of the electron cloud caused by promotion of two electrons via the

$2p\sigma_u$ molecular orbital. At an internuclear separation of about $0.5a_0$, the electron cloud is no longer able to follow the rotation of the internuclear axis.

B.   Selection Rules

For atoms, the selection rules governing autoionization transitions are the conservation of parity and $\Delta J = 0$. For LS coupling the additional rules $\Delta L = 0$ and $\Delta S = 0$ must also be satisfied. The excitation process is also subject to selection rules, but for electron or ion collisions these are generally much less restrictive than for photon excitation. For atoms subject to LS coupling, the multiplicity of the target can change at most by twice the number of electrons in the incident particle. This is because a spin change can be caused only by an electron exchange. If LS coupling is not appropriate, the only requirement is that the total angular momentum be conserved.

Since the angular momentum and spin of the emitted AI electron is generally not measured, it is sometimes useful to recast the selection rules in a form that relates the initial excited state of an atom to the final state of the ion of the next-higher charge state rather than to the final continuum state of the original atom (see Problem 4-15).

C.   Doppler Effect in Electron Spectra

In a collision, either the incident ion or the target atom or both may be excited to an autoionizing state. When the projectile emits an electron with a characteristic energy in its moving frame of reference, its energy measured in the laboratory frame is shifted due to the motion of the source. This kinematic shifting, which is analogous to the Doppler shift in acoustical and optical spectra, was first noted by Rudd et al. (1966b) in the spectra from $Ar^+ + Ar$ collisions. The lines in the AI spectrum that shifted with incident energy were identified as coming from the moving projectile, while others, which did not change with energy, came from the target. A simple velocity vector triangle relates the shifted energy ε' to the unshifted energy ε and the angle θ at which the ejected electrons are detected. The relation is

$$\varepsilon = \varepsilon' - 2(\varepsilon'T)^{1/2}\cos\theta + T \qquad (4\text{-}11\text{-}1)$$

where $T = \frac{1}{2}m_e v_1^2$. Figure 4-11-1 shows the shifted and unshifted peaks observed at $160°$. Note that if $\varepsilon \geqslant T\sin^2\theta$, (4-11-1) yields two possible values of the shifted energy for a given value of the angle θ, depending on the direction of emission in the projectile frame.

Elaborations on the influence of the Doppler effect on the shift and on the broadening of discrete and continuous electron spectra have been made by Gordeev and Ogurtsov (1971), Rudd and Macek (1972), Ogurtsov (1972), and Dahl et al. (1976).

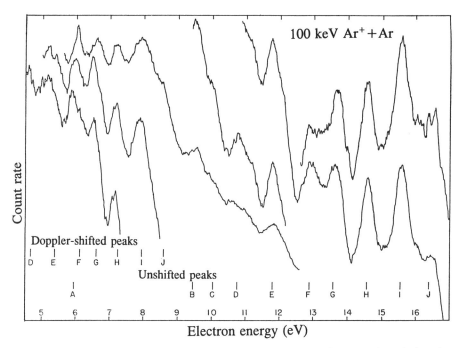

Figure 4-11-1. Energy spectrum of electrons observed at 160° from 100-keV Ar⁺ + Ar collisions. Unshifted autoionizing peaks are due to electrons from the stationary target atoms; Doppler-shifted peaks, from the moving projectiles. [From Rudd et al. (1966c).]

D.   Line Shapes and Widths

The Fano theory of autoionization (Fano, 1961) accounts for the asymmetric shapes of the lines seen in the electron spectra under some conditions. This theory is presented in Chapter 8 of the companion volume. In the reaction $A^+ + B \rightarrow A^+ + B^+ + e^-$, the electron may be ejected from B by either of two mechanisms: direct ionization or the formation of an unstable doubly excited state B** followed by an autoionizing transition to the final continuum state. Since the final state may be the same by both routes, the amplitudes for the two mechanisms interfere, giving rise to the asymmetric line shape. The shape of the line as a function of the electron energy ε is described by the Fano parameters q, ε_0, and Γ in the equation

$$\sigma = \sigma_0 + \sigma_a \frac{(q + \delta)^2}{1 + \delta^2}, \qquad (4\text{-}11\text{-}2)$$

where

$$\delta = \frac{\varepsilon - \varepsilon_0}{\Gamma/2}. \qquad (4\text{-}11\text{-}3)$$

The quantity ε_0 is the energy of the resonance, Γ is its width, q is a parameter that describes the asymmetry, σ_0 is the continuum cross section, and σ_a is the autoionization cross section.

Shore (1967) has also described resonant line shapes, using a different parametrization. His equation is

$$\sigma = \sigma_0 + \frac{a\delta + b}{1 + \delta^2},\qquad(4\text{-}11\text{-}4)$$

where a and b are the Shore parameters.

Schowengerdt and Rudd (1972), Stolterfoht et al. (1972a, b), and Bordenave-Montesquieu et al. (1973, 1975, 1976) have investigated the rapid variation of the shape parameters of the helium autoionization lines with emission angle and projectile energy for proton impact. Figure 4-11-2 shows the change in the lineshape of the autoionizing transition from the $2s^2(^1S)$ state in helium as the incident energy is changed. In Fig. 4-11-3 the variation in the shape of the line profiles with angle is shown for the same transition. Note that in the 20° direction the transition appears as a dip, changes at 30° to an asymmetric shape with a dip followed by a peak, and becomes a more symmetric peak at larger angles. Figure 4-11-4 shows the corresponding variation of the asymmetry parameter q with proton energy for the same transition observed at 10°.

Just as lines in the optical spectra are Doppler broadened because of the thermal motion of the emitting atoms, AI peaks from atomic collisions are Doppler broadened due to any motion of the source particles, whether they be target or projectile. For example, a large fraction of the target particles recoil at angles close to 90° from the beam direction, but all azimuthal angles are possible. Thus AI electrons from the recoiling targets will have their energies shifted by different amounts depending on the particular azimuthal angle. Rudd and Macek (1972) analyzed the problem and found that under some conditions a double peak may be obtained from a single AI transition. They also derived an approximate equation for the distribution and width. Dahl et al. (1976) gave a more accurate expression for the Doppler width. Assuming that the source particles (whether recoiling target particles or incident beam particles) have a fixed velocity u and that the cone into which they are projected has a fixed angle β, the Doppler width from their analysis is

$$\Delta\varepsilon_D = 4\left(\frac{E}{\varepsilon}\right)^{1/2} \varepsilon' \sin\beta \sin\theta,\qquad(4\text{-}11\text{-}5)$$

where $E = m_e u^2/2$, ε and ε' are the energies of the electron measured in the source frame and the laboratory frame, respectively, and θ is the angle of observation relative to the axis of the cone. In the case of the AI spectrum from recoil particles, the cone angle β is nearly 90°, while for scattered incident beam particles, β is generally very small.

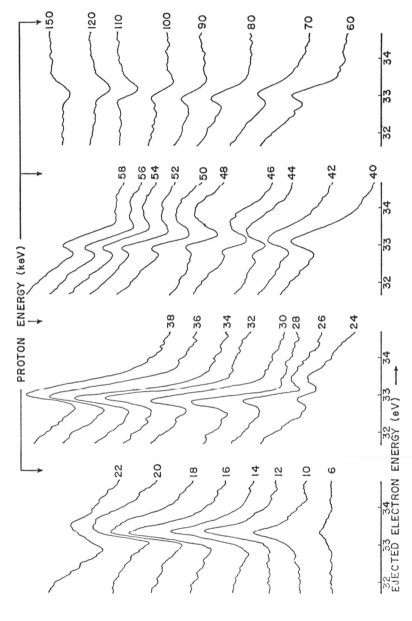

Figure 4-11-2. Electron spectra at $10°$ from $H^+ + He$ collisions at various energies in the region of the $2s^2(^1S)$ autoionizing state. [From Schowengerdt and Rudd (1972).]

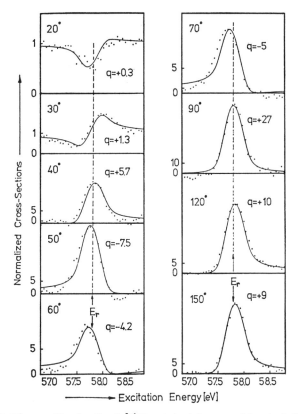

Figure 4-11-3. Line profiles for the $2s^2(^1S)$ autoionizing transition in helium excited by 100-keV proton impact. Data points fitted by (4-11-2), where q is the asymmetry parameter. [From Stolterfoht et al. (1972b).]

The fact that the Doppler width is proportional to $\sin \theta$ has led investigators to study AI electron spectra at zero angle to avoid Doppler broadening and thereby to obtain better resolution. Elston (1981) first used this technique along with coincident detection of selected final charge states of projectiles to study multiple inner- and outer-shell ionization and excitation. Zero-degree Auger spectroscopy was further exploited by Itoh et al. (1983, 1985) and Itoh and Stolterfoht (1985) for studying AI spectra from fast, highly charged projectiles.

Autoionization peaks are also broadened by a post-collision interaction first described by Barker and Berry (1966). Electrons emitted before the incident ion has left the scene of the collision find themselves in the electric field of the passing ion and are therefore detected with a smaller energy. Berry showed that for slow incident ions the distribution of energies due to this effect took the form $f(\Delta\varepsilon) = (b/\Delta\varepsilon^2)\exp(-b/\Delta\varepsilon)$, where $\Delta\varepsilon$ is the change of the electron energy due to

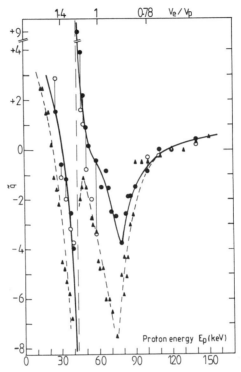

Figure 4-11-4. Variation of the Fano q parameter with proton energy for the ejection of electrons at $10°$ from the $2s^2(^1S)$ autoionizing state of helium excited by protons. △, Data of Schowengerdt and Rudd (1972); ○ and ●, data of Bordenave-Montesquieu et al. (1975) obtained by two different methods. [From Bordenave-Montesquieu et al. (1975).]

the field of the ion. This is given by $\Delta\varepsilon = q/(v_1 t)$, where q is the charge, v_1 the velocity of the incident particle, and t the time between excitation and electron emission. The quantity b is the value of $\Delta\varepsilon$ for $t = \tau$, the mean lifetime of the AI state. Arcuni (1986) has studied, both experimentally and theoretically, the opposite situation, in which the incident particle is faster than the emitted electron and finds in this case that the change of energy of an emitted electron is strongly angle dependent and is given by

$$\Delta\varepsilon = \frac{q}{vt}\frac{B\cos\theta}{1 - B\cos\theta},\qquad(4\text{-}11\text{-}6)$$

where B is the ratio of the electron's initial velocity to the projectile ion velocity, q the charge of the projectile, and θ the emission angle relative to the beam direction.

E. Energy Levels

Calculations of the energies of doubly excited states of helium were made as early as 1934 (Fender and Vinti, 1934) by variational methods. Since helium is the prototype for double excitation, most of the theoretical work has been directed toward that atom. More recent theoretical calculations include those of Burke et al. (1963), Altick and Moore (1965), O'Malley and Geltman (1965), Burke and McVicar (1965), and Lipsky and Russek (1966). Lipsky et al. (1977) have given a catalog of energy levels and classifications for doubly excited states in two-electron systems.

Measurements of X-ray and Auger spectra combined with outer-shell energy levels known from optical spectroscopy yield the energy levels of inner shells. A cross-check between atoms may be made with Moseley diagrams. Since most X-ray measurements are made with solids, the energy levels are often given relative to the Fermi level and a conversion must be made to obtain free-atom energy levels referred to the vacuum level. Lotz (1979) has tabulated electron binding energies in free atoms for all subshells of atoms up to $Z = 108$.

Conversely, if the energy levels are known, a matrix with initial inner-shell vacancy-state energies along the top of the table and the possible final-state configuration energies in the left-hand column enables one to calculate expected Auger electron energies by subtraction. Table 4-11-1 shows such a matrix for a single K-shell vacancy in neon, and Table 4-11-2 is a similar matrix for initial vacancies in both the K and L shells. Spectral lines from single inner-shell vacancies are called diagram lines and those from combined inner and outer shell vacancies are nondiagram or satellite lines. Figure 4-11-5 shows the energy spectrum of electrons from neon in which the Auger spectra induced by protons, electrons, and photons are seen to be virtually identical. However, satellite lines are enhanced in Auger spectra induced by heavy ions, especially those carrying electrons.

Table 4-11-1. Matrix for Computation of K-Auger
Transition Energies (eV) for Diagram Lines in Neon[a]

		$K(^2S)$ 870.0
$L_{2,3}^2(^3P)$	62.7	PL
$L_{2,3}^2(^1D)$	65.8	804.2
$L_{2,3}^2(^1S)$	69.6	800.4
$L_1L_{2,3}(^3P)$	88.0	782.0
$L_1L_{2,3}(^1P)$	98.5	771.5
$L_1^2(^1S)$	121.9	748.1

[a]PL, transition that would violate the parity-L value selection rule.

Table 4-11-2. **Matrix for Computation of K-Auger Transition Energies (eV) for Satellite Lines in Neon**[a]

		$KL_{2,3}(^3P_0)$ 917.3	$KL_{2,3}(^1P_0)$ 921.9	$KL_1(^3S_e)$ 950.3	$KL_1(^1S_e)$ 955.0
$L_{2,3}^3(^4S_0)$	126.4	PL	S, PL	PL	S, PL
$L_{2,3}^3(^2D_0)$	131.5	785.8	790.4	PL	PL
$L_{2,3}^3(^2P_0)$	134.1	783.2	787.8	816.2	820.9
$L_1L_{2,3}^2(^4P_e)$	149.3	768.0	S	PL	S, PL
$L_1L_{2,3}^2(^2D_e)$	157.9	759.4	764.0	792.4	797.1
$L_1L_{2,3}^2(^2S_e)$	163.5	753.8	758.4	786.8	791.5
$L_1L_{2,3}^2(^2P_e)$	166.1	751.2	755.8	PL	PL
$L_1^2L_{2,3}(^2P_0)$	186.5	730.8	735.4	763.8	768.5

[a]PL, transition that would violate the parity-L value selection rule; S, transition that would violate the spin rule.

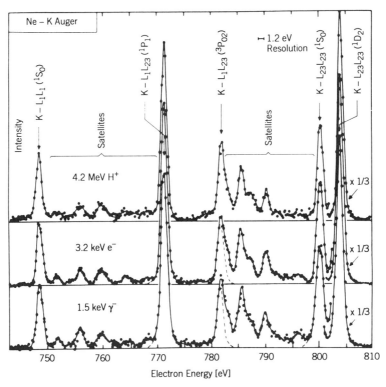

Figure 4-11-5. Neon K-Auger spectrum produced by 4.2-MeV protons, 3.2-keV electrons, and 1.5-keV photons. [From Stolterfoht et al. (1973).]

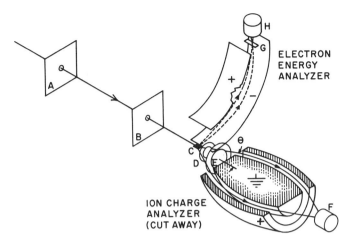

Figure 4-11-6. Schematic diagram of apparatus to measure Auger electrons in coincidence with charge-analyzed incident particles that have been scattered through an angle θ. [From Thomson et al. (1970).]

An apparatus used by Thomson et al. (1970) is shown in Fig. 4-11-6. Incident ions scattered through a narrow range of polar angles around $21°$, but over the entire 2π azimuthal angular range, are charge analyzed and detected in coincidence with L-Auger electrons emitted at $90°$. $Ar^+ + Ar$ collisions at 10 to 30 keV were studied in this way. The low-resolution Auger spectra in Fig. 4-11-7 show that higher final projectile charge states are associated with Auger spectra that have been shifted to lower energies. Since Auger satellite lines come at lower energies than diagram lines, this observation is consistent with the expectation that the additional electrons removed are from the outer shells.

F.   Inner-Shell Vacancy Production

For light atoms the fluorescence yield (i.e., the fraction of excited atoms that deexcite by emission of photons) from inner-shell vacancies is very small, and therefore the measurement of Auger yields is an effective way of measuring inner-shell vacancy production cross sections, thus supplementing X-ray measurements, which are more useful for heavier targets. Garcia et al. (1973) have reviewed both the theory and measurement of inner-shell vacancy production in ion–atom collisions. Data on cross-section measurements of K-shell ionization by light ion impact were compiled by Rutledge and Watson (1973) and by Gardner and Gray (1978). Paul and Muhr (1986) and Paul and Sacher (1989) have updated those reviews with more recent experimental values.

The Born approximation was applied to inner-shell vacancy production by Merzbacher and Lewis (1958), while the impact parameter method has been used by Bang and Hansteen (1959). Garcia (1970) successfully applied the BEA

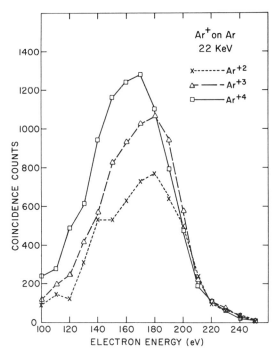

Figure 4-11-7. Electron spectra in the region of the *L*-Auger lines in coincidence with scattered incident ions of various charge states for 22-keV Ar⁺ + Ar collisions. [From Thomson et al. (1970).]

to inner-shell cross-section calculations. In a series of papers, the New York University group [see, e.g., Basbas et al. (1971, 1973) and Brandt and Lapicki (1974, 1981)] made successive corrections to the Born approximation which accounted for (1) the Coulomb deflection and change of velocity of the projectile in the field of the target nucleus, (2) the increase in the binding energy due to the presence of the incident ion during the collision, (3) the influence of the projectile on the electron orbit, (4) relativistic effects, and (5) the energy loss of the projectile during inner-shell excitation. The cross sections calculated with these corrections, labeled σ_{ECPSSR}, are in good agreement with experiment over a wide range of conditions. This is shown in Fig. 4-11-8, where the ratio $\bar{s} = \sigma_1/\sigma_{\text{ECPSSR}}$ is plotted versus the scaled velocity $\xi = (2/\theta)(v_1/v_K)$. Here v_1 is the incident ion velocity, v_K is the hydrogenic velocity of the K electron in the target, and $\theta = \chi_K/(Z_2 - 0.3)^2$, where χ_K is the ionization energy of the K shell and Z_2 is the nuclear charge of the target. The quantity θ is the K-shell binding energy modified by the shielding by the outer electrons. Because of the large amount of experimental data in the literature, the range of scaled velocities has been divided into intervals, and average cross sections in each interval given for three ranges of Z_2. Except for the lowest energies, most of the data are within 15% of

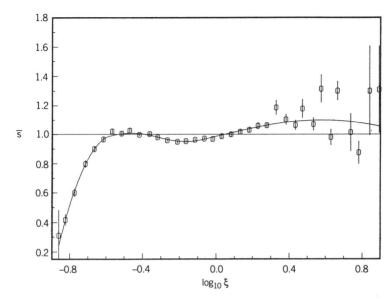

Figure 4-11-8. Measured K-shell ionization cross sections for proton impact divided by the ECPSSR-corrected Born approximation calculations plotted versus scaled proton impact energy. Measured data from many targets have been averaged. [From Paul and Muhr (1986).]

the values predicted by the ECPSSR theory. This is quite remarkable considering the fact that the numerical values of the cross sections vary 1 million-fold in the region described.

DuBois et al. (1984b) measured L-Auger electrons from magnesium bombarded by 30- to 80-keV protons in coincidence with the charge state of the incident particles to show that the contribution of electron capture to the production of $2p$ vacancies was substantial, being 35 to 50% in the region measured.

4-12.   MULTIPLE IONIZATION

Experimental measurements of charge-state distributions of slow ions produced in ion–atom collisions have accumulated for many years. Examples of such investigations are those of Fedorenko and Afrosimov (1956), who studied various rare gas ions incident on rare gas atoms; Solov'ev et al. (1962), who used protons and hydrogen atoms on H_2, N_2, He, Ne, Ar, and Kr; Wexler (1964), who bombarded He, Ne, Ar, and Kr with 0.8 to 3.75-MeV protons; Haugen et al. (1982), who measured multiple ionization of noble gases by fully stripped ions of H, He, Li, B, C, and O at energies of 1.44 and 2.3 MeV/u; and DuBois et al. (1984a), who used time-of-flight measurements to obtain cross sections for multiple ionization of the rare gases by H^+ and He^+.

With a time-of-flight spectrometer Hvelplund et al. (1980) measured the production of He^+ and He^{2+} from helium targets bombarded by various multicharged ions and correlated them with the final charge state of the incident particles. They found that the single ionization cross sections follow a universal curve when the cross section divided by the charge state of the incident ion is plotted versus the energy per unit mass per unit charge of the ion. This is in agreement with results of the CTMC model. They also found that for multiply charged incident ions the cross sections depend only on the charge state of the ion and not on its atomic number.

Little effort was made to understand or separate out the contributions to multiple ionization from various channels until Manson et al. (1983) used their own and earlier data for protons on neon to show at least for that target that direct multiple outer-shell ionization was far more probable in this energy range than initial K-shell ionization followed by the emission of one or more Auger electrons.

Dubois (1984b) and DuBois and Manson (1987) have made analyses of the channels resulting in the production of multiply ionized slow ions when protons are incident on rare gases. These channels are: (1) direct multiple outer-shell ionization, (2) multiple electron capture, (3) single or multiple electron capture with simultaneous ionization, and (4) inner-shell ionization followed by Auger electron emission. An additional mechanism, known as *shakeoff*, may result when a single electron is ejected from an atom by the projectile. In this process the ensuing readjustment to the new potential may cause one or more of the remaining electrons to be emitted. For some channels, specific cross sections were obtained by subtraction of known cross sections, while for others only sums were found.

Measurements were also made (DuBois and Toburen, 1988) with various projectiles from protons to oxygen atoms incident on helium over a wide energy range to find the effect of charge state and impact velocity on the production of double ionization. They found that the independent ionization of both target electrons was the dominant mechanism for charged-particle impact above 100 keV/u. On the other hand, neutral projectiles interact with only one target electron, even in double ionization (DuBois and Kövèr, 1989).

DuBois (1984a) also utilized coincidence measurements to elucidate the mechanisms of multiple ionization. H^+ and He^+ ions with energies from 15 to 100 keV were charge analyzed after making collisions with Ne, Ar, or Kr targets. The scattered beam particles gave the start pulses for a time-of-flight system while the slow recoil ions of various charge states extracted from the collision center provided the stop pulses. In this way the cross sections could be determined for both the capture and the ionization channels (i.e., for processes in which the incident particles did or did not capture one of the electrons from the target, respectively). He concluded that higher-order capture-plus-ionization processes can be considerably more important in free-electron production than can lower-order direct ionization processes. In Hasted's notation, he found that for He^+ collisions, for example, $_{10}\sigma_{02} > {}_{10}\sigma_{01}$.

A comparison of the double ionization produced by positive and negative incident particles was made when Haugen et al. (1982) measured R, the ratio of the double to the single ionization cross sections for helium using protons and equal-velocity electrons. They found that R is approximately twice as large for electrons as for protons of the same velocity. Anderson et al. (1986) repeated the experiment using antiprotons in place of electrons and found an even greater difference. Their data are shown in Fig. 4-12-1, where it is seen that the value of R for antiprotons approaches the value for electrons of equal velocity at high energies, and both approach a constant value, equal to the value for protons, at even higher energies. At lower energies the value of R for electrons decreases as it approaches the threshold for double ionization. Reading and Ford (1987) were able to calculate R for protons and antiprotons using the forced impulse approximation. Their results, also shown in Fig. 4-12-1, both come to the same constant value at high energies but are 35% lower than experiment. Attributing

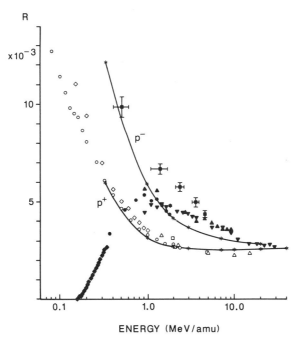

Figure 4-12-1. Plots of the ratio R of the double- to single-ionization cross sections for p^-, p^+, and e^- colliding with He. ■, Antiproton data of Anderson et al. (1986); ●, electron data of Adamczyk et al. (1966); ▲, electron data of Nagy et al. (1980); ▽, electron data of Schram et al. (1966); ■, electron data of Stephen et al. (1980); □, proton data of Anderson et al. (1986); △, proton data of Knudsen et al. (1984); ○, proton data of Shah and Gilbody (1985); ◇, proton data of Puckett and Martin (1970). Lines, which show the calculations using the forced impulse method, have been scaled by the factor 1.35. [From Reading and Ford (1987).]

this to the absence of *d* states in their basis set, they arbitrarily multiplied all of their values of *R* by 1.35. So normalized, their calculations are in good agreement with measured values for protons and in fair agreement for antiprotons.

Using the CTMC method, Olson (1987) has also calculated values of the ratio *R* which are higher for antiprotons than for protons. While the agreement with experiment was not as good as in Reading's work, the CTMC calculations were able to elucidate two effects which cause the antiproton cross sections to be higher than those for protons. The first is that the antiproton can force one electron toward the other at a larger impact parameter. The proton must, in fact, pass between two electrons to force them closer together. The more important effect, however, was that at small impact parameters the antiproton screens the helium nucleus, causing a Coulomb explosion, while the proton increases the binding of the electrons.

4-13.   DISSOCIATIVE IONIZATION

Dissociative ionization induced by electron collisions was discussed briefly in Sections 5-5 and 7-3 of the companion volume. The same process results from ion impact.

As noted in Section 4-3, the distribution of energies of recoils from nondissociative collisions, as described by the screened-Coulomb model, yields a $1/\varepsilon_R^2$ dependence down to very low energies. Thus a large majority of the recoil ions from ionization of atoms have energies much less than 1 eV. This is not the case with ionization of molecules, where the target may be excited into a highly repulsive molecular ion state that flies apart with considerable energy. This so-called Coulomb explosion produces protons with energies up to 9 or 10 eV in the case of dissociation of H_2 [see Section 6-10 of the companion volume and Kanter et al. (1979)]. In the condenser method of measuring total ionization cross sections, the collecting field must be quite large to avoid losses of such high-energy ions from the interaction region. Measurements using mass/charge analysis are especially prone to this difficulty. That some early workers failed to collect all of the fragmentation ions was shown by Browning and Gilbody (1968) who did such measurements on H_2, N_2, O_2, CO and CH_4 bombarded by 5 to 45-keV protons and, with careful collection of all energetic fragments, measured cross sections as much as five times larger than previous values.

The University of Georgia group (Wood et al., 1976) devised a technique for simultaneously measuring the time of flight (TOF) and the energy of positive ions from the decay of excited molecules. This technique enabled them to measure the energy spectra of selected molecular fragments. A schematic diagram of their system is shown in Fig. 4-13-1. It consists of an electrostatic analyzer, the back plate of which is driven by a triangular wave voltage to scan the energies of emitted ions. By keeping the potential difference between the front and back plates fixed during the sweep, all ions were analyzed at the same

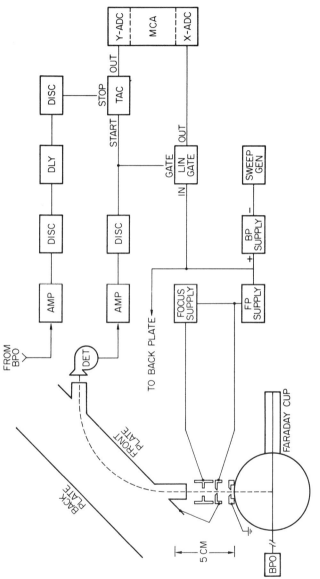

Figure 4-13-1. Schematic diagram of the time–energy spectrometer and electronic instrumentation used to study dissociative ionization. [From Wood et al. (1976).]

energy and therefore with the same energy resolution. The beam from the accelerator was pulsed every $16\,\mu s$ with a 250-ns pulse. A capacitive beam pickoff provided a signal that served as the time reference for the TOF measurement. The output signals were recorded by a multichannel analyzer operating in the two-parameter mode with a 64×64 array displayed on an oscilloscope screen as in Fig. 4-13-2. In the case of $O^+ + O_2$, they find that the recoil energy spectrum of the O^{4+} extends as high as 90 eV (Steuer et al., 1977).

Knowing the energy distributions of the fragment ions has enabled the Georgia group to determine the relative contributions from various molecular states in a number of cases. For example, Wood et al. (1977) and Edwards et al. (1977) have identified six different states in H_2 the contributions from which combine to yield a good fit to their experimental data taken with various incident particles at energies of 0.5 to 4 MeV. The states are $1s\sigma_g$, $2p\sigma_u$, $2p\pi_u$, $2s\sigma_g$, H^+H^+, and an assumed autoionizing state. The shapes of the kinetic energy distributions for each state are calculated by the reflection approximation (Coolidge et al., 1936). In this approximation the relative intensity as a function of dissociation energy is taken to be proportional to the overlap integral between the H_2 ground state and the dissociative state. The data are shown in Fig. 4-13-3 along with the fit by this procedure.

Lindsay et al. (1987) also studied H_2 but at lower primary energies and were able to fit their data using only the first five of the states listed above. They also measured the angular distribution of protons from the various states and found that they could be fitted by the expression

$$I(\theta) = \frac{\sigma}{4\pi} \left[1 + \beta P_2(\cos\theta)\right], \tag{4-13-1}$$

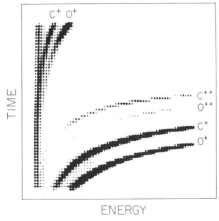

Figure 4-13-2. Oscilloscope screen view of data from the dissociation of CO measured by the time–energy spectrometer. [From Wood et al. (1976).]

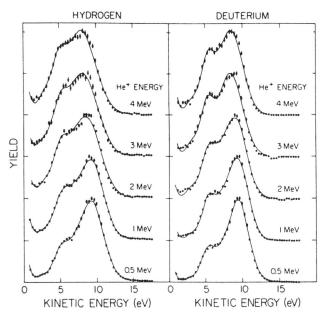

Figure 4-13-3. Energy spectra of H^+ and D^+ ions from dissociation of H_2 and D_2 by 0.5 to 4.0-MeV He^+. Points, experiment; lines, calculations using the reflection approximation. [From Wood et al. (1977).]

where

$$P_2(\cos\theta) = \frac{3\cos^2\theta - 1}{2} \qquad (4\text{-}13\text{-}2)$$

and where β is a parameter that can range from -1 to $+2$.

4-14. NOTES

4-1. Ionization of Atoms in Rydberg States. Atoms in which one electron has a large principal quantum number n are known as Rydberg atoms. Such atoms have some interesting properties, as described in Section 6-17 of the companion volume. While much of the interest in Rydberg atoms has to do with changes in excitation states rather than with ionization, and while most of the studies of ionization of Rydberg atoms have been done with photons or electrons, collisions with heavy particles may have applications in studies of nebulae and interstellar gas clouds, where Rydberg atoms and charged particles coexist. Gilbody (1986) has reviewed electron removal from highly excited H atoms.

Because the excited electron in a Rydberg atom is so far from the nucleus and the core electrons, a particle in collision with it cannot interact simultaneously with both the electron and the core. Thus the excited electron can be considered to be quasi-free. Collisions with free electrons can therefore be studied by observing collisions with Rydberg atoms. The "free" electron model, originally proposed by Fermi (1934), has been developed by Flannery (1973) and by Matsuzawa (1980, 1983).

Ionizing interactions of Rydberg atoms with molecules can result in either simple ionization:

$$A^* + BC \rightarrow A^+ + e^- + BC$$

or associative ionization:

$$A^* + BC \rightarrow AB^+ + e^- + C.$$

Cross sections for ionization involving Rydberg atoms approach the geometrical size of the atom. Since the radius scales as n^2, the cross section goes as n^4 and can therefore be very large for large principal quantum numbers. Hotop and Niehaus (1967) have measured cross sections between 10^{-13} and 10^{-12} cm^2, for example.

4-2. Penning Ionization. In addition to processes in which the kinetic energy of the colliding particles is used to eject electrons into the continuum, there is also the process of Penning ionization, in which electron ejection occurs by a transfer of internal excitation energy from one structure to the other, as in the process

$$A + B^* \rightarrow A^+ + B + e^-.$$

This process can occur at thermal energies when an excited atom collides with an atom or molecule whose ionization energy is less than the excitation energy of the other atom. The cross sections at thermal energies are usually somewhat larger than gas kinetic cross sections and the process is generally quite efficient. See also Fontijn (1985) and Massey (1971).

The term *Penning ionization* has also been applied to ionizing collisions in which the two colliding atoms are of the same type. Ionization can take place in that case if both atoms are excited, as in the reaction

$$A^* + A^* \rightarrow A^+ + A + e^-.$$

4-3. Ionization of Ions. Since ion–ion collisions require the use of intersecting beam techniques, they are much more difficult and time consuming than ion–neutral collisions, and very little work has been reported. The Belfast group (Mitchell et al., 1977; Angel et al., 1978a, b; Dunn et al., 1979) detected only the doubly charged products and thus measured the sum of the cross sections for the charge-transfer process

$$A^+ + A^+ \rightarrow A^{2+} + A$$

and for ionization

$$A^+ + A^+ \rightarrow A^{2+} + A^+ + e^-.$$

Their measurements were in the center-of-mass energy range 40 to 400 keV, and in the case where the interacting atoms were cesium the cross sections were in the range 1 to 3×10^{-16} cm^2. Such cross sections are of special interest in connection with certain schemes for heavy ion fusion (Kim and Magelssen, 1979). In addition, ion–ion cross sections are useful in modeling processes that occurred in the early universe when it consisted largely of charged particles. In this connection, Poulaert et al. (1978) have measured the cross section for the reaction

$$H^+ + H^- \rightarrow H_2^+ + e^-.$$

Reviews of ion–ion collision measurements have been given by Dolder and Peart (1985, 1986).

4-4. *Compton Profiles.* At the time of the discovery of the Compton effect (Compton, 1923), it was found that the inelastically scattered photon line had a significant width. This broadening, which was due to the distribution of velocities of the electrons in the target, was later exploited to determine the Compton profile, the name given to that distribution. Such profiles have also been determined from inelastic scattering of electrons [see, e.g., Barlas et al. (1978)]. Bell et al. (1983) and Böckl and Bell (1983) have shown that since the width and shape of the binary peak in the secondary electron spectra from fast ion–atom collisions (see, e.g., Fig. 4-3-4) are determined by the distribution of initial electron velocities, the Compton profile could also be extracted from measurements of such spectra. This method has important advantages over electron scattering. The use of light, fast ions yields high counting rates, there is no problem of distinguishing primary from secondary particles, and the method is less sensitive to multiple scattering.

Bell et al. showed that the cross section for the binary peak can be written

$$\frac{d^2\sigma}{d\varepsilon\,d\Omega} = \frac{1}{2v_1} \left(\frac{d\sigma}{d\Omega}\right)_R J_T(p_z) \qquad (4\text{-}14\text{-}1)$$

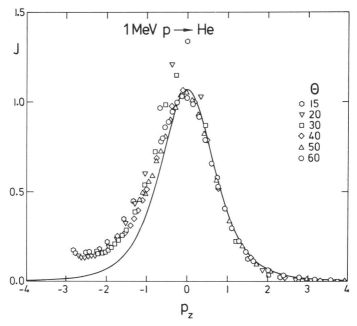

Figure 4-14-1. Compton profile of helium from 1-MeV protons at different electron emission angles. Points, calculations from experimental data of Manson et al. (1975); solid line, theoretical Compton profile based on Hartree–Fock calculations of Biggs et al. (1975). [From Böckl and Bell, 1983.]

where

$$J_T(p_z) = \int d^2 p_\perp \, |\psi_i(p_\perp, p_z)|^2 \qquad (4\text{-}14\text{-}2)$$

is the Compton profile and where $(d\sigma/d\Omega)_R$ is the Rutherford cross section for scattering a free electron into a solid angle $d\Omega$. Figure 4-14-1 shows an example of Compton profiles extracted from 1-MeV H^+ + He secondary electron spectra at various angles compared with the profile based on Hartree–Fock calculations.

4-5. *Symbols and Notation.* Because we use some symbols in this chapter that are not used elsewhere in this book, we give here a list of the more common ones.

a_0	Bohr radius (5.29×10^{-9} cm)
v_0	Bohr velocity (2.19×10^8 cm/s)
R	Rydberg of energy (13.6 eV)
m_p, m_t, m_e	mass of projectile, target atom or molecule, and electron, respectively

v_1, E_1	velocity and energy of incident particle, respectively
v_2, U_2	velocity and orbital kinetic energy of target electron, respectively
Z_1, Z_2	nuclear charge of incident and target particle, respectively
χ	binding energy of electron
$\varepsilon, \varepsilon_R$	energy of ejected electron and recoil atom or ion, respectively
T	energy of an electron with same velocity as the incident particle ($T = \frac{1}{2}m_e v_1^2$)
Q	energy transfer (e.g., $Q = \chi + \varepsilon$)
θ	angle of ejection of electron relative to the incident beam direction
ϕ	angle of recoil of target particle relative to the incident beam direction

4-6. References and Data on Ionization by Heavy Particle Impact. McDaniel et al. (1985) and McDaniel and Mansky (1993) have published two extensive bibliographies on atomic collisions that cover the period 1965 through mid-1992. These documents, which contain a total of 3200 entries, deal mainly with conventional bibliographies, data compilations, and reviews. Valuable sources of information on both experimental and theoretical aspects of ionization and the other areas of atomic collisions arc cited.

4-7. Relationship to Electron and Photon Projectiles. Many of the concepts, experimental techniques, and theoretical methods described in this chapter correspond closely to the discussions of ionization by electron and photon impact in Chapters 6, 7, and 8 of the companion volume. Cross-reference these topics and note the similarities.

4-15. PROBLEMS

4-1. In designing a transverse-field system (see Figure 4-2-1) to measure ionization cross sections, a compromise must be made in setting the potential between the parallel plates used to measure ion and electron currents. The potential must be high enough to produce saturation in the current measurements but not so high that the primary beam is deflected away from the Faraday cup monitoring it. Suppose that the cup has a diameter of 10 mm and is 30 cm from the center of the parallel plates. Each pair of plates is 10 cm long and has a 2-cm spacing. A potential difference of 30 V is needed for saturation. Assuming the beam of protons to be of negligible diameter and centered on the Faraday cup in the absence of a transverse field, how low a beam energy can be used before the beam misses the cup?

4-2. Consider the apparatus of Fig. 4-2-1.

(a) The electrodes labeled CS and FCS are often provided with positive biases. If slow positive ions are formed in the vicinity of those electrodes, what effect could they have on the resulting cross-section measurements?

(b) What are some possible sources of spurious currents at the electrodes IP and EP? [See Rudd et al. (1983).]

4-3. It is desired to measure the $n = 2$ and $n = 3$ excitation cross sections of the hydrogen atom using the energy-loss apparatus of Fig. 4-2-2. The two states are at energies of 10.2 and 12.7 eV, respectively, and the analyzer has a resolution of 0.1%. To what energy must the beam be decelerated in order to separate these states?

4-4. Using the apparatus of Fig. 4-3-1, how small a doubly differential cross section can be measured if it is necessary to obtain at least 100 counts while collecting 1 μC of beam charge? Assume a negligible background count and the following parameters: detector efficiency is 0.9, $(l\Delta\Omega)_{eff} = 10^{-6}$ m-sr, target gas pressure 10^{-4} torr, and $\Delta\varepsilon = 20$ eV.

4-5. In most experimental arrangements to measure cross sections, the beam must travel through the target gas and be caught in a Faraday cup. Some of the beam, however, is scattered elastically by the gas and misses the cup. Make a calculation to see if this is an important effect. Assume cylindrical symmetry, a beam of negligible diameter, and consider the cup to be a circular aperture of radius r. Use the small-angle approximation and the Rutherford scattering cross section, $\sigma(\theta) = K/\sin^4(\theta/2) \simeq 16K/\theta^4$, where $K = (Ra_0 Z_1 Z_2 / E_1)^2$. Assume the target gas density to be n between two points on the beam path that are at the distances x_1 and x_2 from the cup and that the density is zero elsewhere. Show that the fraction of the beam that is scattered outside the cup is given by

$$f = \frac{16\pi Kn}{3r^2}(x_2^3 - x_1^3). \tag{4-15-1}$$

Calculate the value of f for 5000-eV protons going through krypton gas at a density $n = 3 \times 10^{19}$ m^{-3} when $r = 0.01$ m, $x_1 = 0.1$ m, and $x_2 = 0.2$ m.

4-6. From the Born approximation expression of (4-5-4), show that if k is replaced by μk and κ by $\mu\kappa$, one obtains a cross section $\sigma(\mu)$ that is μ^{-6} times the cross section $\sigma(\mu = 1)$. Since $\mu = (\chi/R)^{1/2}$, $\varepsilon \propto \kappa^{1/2}$, and $E_1 \propto k^{1/2}$, show that the scaling equation (4-5-6) follows.

4-7. Show that for electrons detected in the forward direction, the cusp-shaped peak described by (4-8-1) is approximately symmetric in v.

4-8. Prove (4-9-1).

4-9. **(a)** Show that at $Q = Q_- = 4T - 4(TU_2)^{1/2}$ the expressions of (4-4-4) and (4-4-5) have the same value but different slopes.

(b) Show that $Q = Q_+ = 4T + 4(TU_2)^{1/2}$, the cross section given by (4-4-5) goes to zero.

4-10. Write a computer program to calculate $N(\gamma)^2$ from (4-8-1) as a function of electron energy for an angle of 5° and a proton impact energy of 100 keV. Plot the results.

4-11. Using (4-10-2), find the threshold energy for producing L-vacancies in $Ar^+ + Ar$ collisions. Take r_c to be 0.25 Å. Compare with Fig. 4-10-2.

4-12. Using a velocity vector diagram, verify (4-11-1). Use this equation to show that if $\varepsilon \geqslant T\sin^2\theta$, there are two Doppler-shifted peaks at different energies. Calculate those energies.

4-13. Suppose that a beam of 100-keV helium ions is deflected into a cone angle of 1° in making collisions with heavier target atoms, resulting in Al electrons being ejected from the projectiles with an energy of 35 eV in the projectile frame of reference. The Doppler-shifted electrons are detected at an angle of 150° from the beam direction. Find their energy in the laboratory frame of reference and the width of the peak due to Doppler broadening.

4-14. Show that the Fano line shape of (4-11-2) is the same as that given by the Shore parameterization of (4-11-4) for small values of δ. What are the relationships between the parameters?

4-15. The selection rules for autoionization are given in Section 4-11. They may be restated in a form that relates the excited state to the possible final states of the residual ion rather than the ion plus the continuum electron [see Rudd and Smith (1968)]. Show that for LS coupling, the rules for parity and L-values may be given as follows:

1. Transitions between states other than S states are allowed.

2. S states combine only with $S, D, G, \ldots$ states if the parities are the same.

3. S states combine only with $P, F, H, \ldots$ states if the parities are opposite.

Also, show that the spin rule may be restated as:

4. The multiplicities $2S + 1$ of the two states must differ by unity.

4-16. Verify that the transitions in Tables 4-11-1 and 4-11-2 either satisfy these selection rules or violate them as indicated in the tables.

4.17. Describe the apparatus used by the University of Georgia group (Wood et al., 1976). In particular, explain the oscilloscope display shown in Fig. 4-13-2.

4-18. The differential cross section for the angular distribution of particles of initial momentum k scattered at an angle θ_1 from an atom represented by a screened Coulomb potential, generalized from the result given by Schiff (1949), is

$$\frac{d\sigma}{d\Omega} = f(\theta_1)^2 \qquad (4\text{-}15\text{-}2)$$

where

$$f(\theta_1) = \frac{2\mu Z_1 Z_2 e^2}{h^2(K^2 + a^{-2})}, \qquad K = 2k \sin\frac{\theta_1}{2} \qquad (4\text{-}15\text{-}3)$$

and where μ is the reduced mass. Use this to derive (4-3-7) for the energy distribution of recoils.

REFERENCES

Abrines, R., and I. C. Percival (1966), *Proc. Phys. Soc. London* **88**, 861.

Adamczyk, B., A. J. H. Boerboom, B. L. Schram, and J. Kistemaker (1966), *J. Chem. Phys.* **44**, 4640.

Afrosimov, V. V., R. N. Il'in, and N. V. Fedorenko (1958), *Zh. Eksp. Teor. Fiz.* **34**, 1398 [transl. in *Sov. Phys. JETP* **34**, 968].

Afrosimov, V. V., Yu. S. Gordeev, M. N. Panov, and N. V. Fedorenko (1963), *Compt. Rend. VI Conf. Int. Phenomena d'Ionisation dans les Gaz SERMA*, Paris, **I**, 111.

Afrosimov, V. V., Yu, S. Gordeev, M. N. Panov, and N. V. Fedorenko (1964), *Zh. Tekhn. Fiz.* **34**, 1613 [transl. in *Sov. Phys. Techn. Phys.* **9**, 1248 (1965)].

Afrosimov, V. V., Yu. A. Mamaev, M. N. Panov, and V. Uroshevich (1967), *Zh. Tekhn. Fiz.* **37**, 717 [transl. in *Sov. Phys. Tech. Phys.* **12**, 512].

Afrosimov, V. V., G. A. Leiko, Yu. A. Mamaev, and M. N. Panov (1968), *Zh. Eksp. Teor. Fiz.* **56**, 1204 [transl. in *Sov. Phys. JETP* **29**, 648 (1969)].

Afrosimov, V. V., G. A. Leiko, Yu. A. Mamaev, and M. N. Panov (1969), *Zh. Eksp. Teor. Fiz.* **56**, 1204 [transl. in *Sov. Phys. JETP* **29**, 648].

Altick, P. L., and E. N. Moore (1965), *Phys. Rev. Lett.* **15**, 100.

Anderson, L. H., P. Hvelplund, H. Knudsen, S. P. Møller, K. Elsener, K.-G. Rensfelt, and E. Uggerhøj (1986), *Phys. Rev. Lett.* **57**, 2147.

Angel, G. C., K. F. Dunn, E. C. Sewell, and H. B. Gilbody (1978a), *J. Phys. B: At. Mol. Phys.* **11**, L49.

Angel, G. C., E. C. Sewell, K. F. Dunn, and H. B. Gilbody (1978b), *J. Phys. B: At. Mol. Phys.* **11**, L297.

Arcuni, P. W. (1986), *Phys. Rev. A* **33**, 105.

Auger, P. (1925), *J. Phys. Radium* **6**, 205.

Band, Y. B. (1974), *J. Phys. B* **7**, 2557.

Bang, J., and J. M. Hansteen (1959), *K. Dan. Vidensk. Selsk. Mat. Fys. Medd.* **31**(13).

Barat, M., and W. Lichten (1972), *Phys. Rev. A* **6**, 211.

Barker, R. B., and H. W. Berry (1966), *Phys. Rev.* **151**, 14.

Barlas, A. D., W. H. E. Rueckner, and H. F. Wellenstein (1978), *J. Phys. B: At. Mol. Phys.* **11**, 3381.

Barnett, C. F., Ed. (1990), *Atomic Data for Fusion*, Vol. 1, *Collisions of H, H_2, He and Li Atoms and Ions with Atoms and Molecules*, ORNL-6086/V1, Oak Ridge National Laboratory, Oak Ridge, Tenn.

Barnett, C. F., J. A. Ray, E. Ricci, M. I. Wilker, E. W. McDaniel, E. W. Thomas, and H. B. Gilbody (1977), *Atomic Data for Controlled Fusion Research*, ORNL-5206, Oak Ridge National Laboratory, Oak Ridge, Tenn.

Basbas, G., W. Brandt, and R. Laubert (1971), *Phys. Lett.* **34A**, 277.

Basbas, G., W. Brandt, and R. Laubert (1973), *Phys. Rev. A* **7**, 983.

Bates, D. R., and G. W. Griffing (1953), *Proc. R. Soc. London* **66A**, 961.

Bates, D. R., C. J. Cook, and F. J. Smith (1964), *Proc. Phys. Soc. London* **83**, 49.

Belkic, D. R. (1978), *J. Phys. B: At. Mol. Phys.* **11**, 3529.

Belkic, D. R. (1980), *J. Phys. B: At. Mol. Phys.* **13**, L589.

Bell, F., H. Böckl, M. Z. Wu, and H.-D. Betz (1983), *J. Phys. B: At. Mol. Phys.* **16**, 187.

Berényi, D. (1981a), *Adv. Electron. Electron Phys.* **56**, 411.

Berényi, D. (1981b), *Acta Phys. Hung.* **51**, 157.

Berényi, D. (1982), *Proc. Int. Seminar on High Energy Ion–Atom Collisions*, Debrecen, Hungary, D. Berényi and G. Hock, Eds., Akadémiai Kiadó, Budapest, p. 131.

Bernardi, G., S. Suárez, P. Focke, and W. Meckbach (1988), *Nucl. Instrum. Methods Phys. Res.* **B33**, 321.

Bernardi, G., P. Fainstein, C. R. Garibotti, and S. Suárez (1990), *J. Phys. B: At. Mol. Opt. Phys.* **23**, L139.

Berry, H. W. (1961) *Phys. Rev.* **121**, 1714.

Bethe, H. (1930) *Ann. Phys.* **5**, 325.

Biggs, F., L. B. Mendelsohn, and J. B. Mann (1975), *At. Data Nucl. Data Tables* **16**, 201.

Blauth, E. (1957), *Z. Phys.* **147**, 228.

Böckl, H., and F. Bell (1983), *Phys. Rev. A* **28**, 3207.

Bohr, N. (1941), *Phys. Rev.* **59**, 270.

Bonsen, T. F. M., and D. Banks (1971), *J. Phys. B* **4**, 706.

Bonsen, T. F. M., and L. Vriens (1970), *Physica* **47**, 307.

Bordenave-Montesquieu, A., P. Benoit-Cattin, M. Rodiere, and A. Gleizes (1973), *8th Int. Conf. Physics of Electronic and Atomic Collisions*, Abstracts, B. C. Čobić and M. V. Kurepa, Eds., p. 523.

Bordenave-Montesquieu, A., P. Benoit-Cattin, M. Rodiere, A. Gleizes, and H. Merchez (1975), *J. Phys. B: At. Mol. Phys.* **8**, 874.

Bordenave-Montesquieu, A., P. Benoit-Cattin, A. Gleizes, and H. Merchez (1976), *At. Data Nucl. Data Tables* **17**, 157.

Born, M. (1926), *Z. Phys.* **37**; 863, **38**, 803.

Brandt, W. and G. Lapicki (1974), *Phys. Rev. A* **10**, 474.

Brandt, W., and G. Lapicki (1981), *Phys. Rev. A* **23**, 1717.

Brandt, W., and R. H. Ritchie (1977), *Phys. Lett.* **62A**, 374.

Brauner, M. and J. H. Macek (1992), *Phys. Rev.* **46**, 2519.

Breinig, M., G. J. Dixon, P. Engar, S. B. Elston, and I. A. Sellin (1983), *Phys. Rev. Lett.* **51**, 1251.

Briggs, J. S. (1989), *Comments At. Mol. Phys.* **23**, 155.

Briggs, J. S., and J. H. Macek (1991), *Adv. At. Mol. Opt. Phys.* **28**, 1.

Browning, R., and H. B. Gilbody (1968), *J. Phys. B (Proc. Phys. Soc.)* **1**, 1149.

Bruch, R., G. Paul, and J. Andrä (1975), *Phys. Rev. A* **12**, 1808.

Burch, D., H. Wieman, and W. B. Ingalls (1973), *Phys. Rev. Lett.* **30**, 823.

Burgdörfer, J. (1984), *Forward Electron Ejection in Ion Collisions, Lecture Notes in Physics,* K.-O. Groeneveld, W. Meckbach, and I. A. Sellin, Eds., Springer-Verlag, Berlin.

Burke, P. G., and D. D. McVicar (1965), *Proc. Phys. Soc. London* **86**, 989.

Burke, P. G., D. D. McVicar, and K. Smith (1963), *Phys. Rev. Lett.* **11**, 559.

Cacak, R. K., Q. C. Kessel, and M. E. Rudd (1970), *Phys. Rev. A* **2**, 1327.

Chandrasekhar, S. (1941), *Astrophys. J.* **93**, 285.

Cheng, W.-Q., M. E. Rudd, and Y.-Y. Hsu (1989a), *Phys. Rev. A* **39**, 2539.

Cheng, W.-Q., M. E. Rudd, and Y.-Y. Hsu (1989b), *Phys. Rev. A* **40**, 3599.

Cocke, C. L., and R. E. Olson (1991), *Phys. Rep.* **205**, 153.

Compton, A. H. (1923), *Phys. Rev.* **22**, 483.

Coolidge, A. S., H. M. James, and R. D. Present (1936), *J. Chem. Phys.* **4**, 193.

Crooks, J. B. (1974), "Absolute Doubly Differential Cross Sections for Producing Secondary Positive Ions from 50–200 keV Ar^+–Ar Collisions and 50–200 keV H^+ Collisions with He, Ar, H_2, and N_2," unpublished doctoral dissertation, University of Nebraska.

Crooks, G. B., and M. E. Rudd (1970), *Phys. Rev. Lett.* **25**, 1599.

Crooks, J. B., and M. E. Rudd (1971), *Phys. Rev. A1* **3**, 1628.

Dahl, P., M. Rødbro, B. Fastrup, and M. E. Rudd (1976), *J. Phys. B: At. Mol. Phys.* **9**, 1567.

Datz, S., (1977), *Ann. Israel Phys. Soc.* **1**, 196.

De Heer, F. J., J. Schutten, and H. Moustafa (1966), *Physica (Utrecht)* **32**, 1766.

Desesquelles, J., G. D. Cao, and M. C. Dufay (1966), *C.R. Acad. Sci. Ser. B* **262**, 1329.

Dettmann, K., K. G. Harrison, and M. W. Lucas (1974), *J. Phys. B* **7**, 269.

Dolder, K., and B. Peart (1985), *Rep. Prog. Phys.* **48**, 1283.

Dolder, K., and B. Peart (1986), *Adv. At. Mol. Phys.* **22**, 197.

Drepper, F., and J. S. Briggs (1976), *J. Phys. B: At. Mol. Phys.* **9**, 2063.

DuBois, R. D. (1984a), *Phys. Rev. Lett.* **52**, 2348.

DuBois, R. D. (1984b), *X84 Int. Conf. X-Ray and Inner-Shell Processes in Atoms, Molecules and Solids,* Karl-Marx Universität, Leipzig.

DuBois, R. D., and A. Kövèr (1989), *Phys. Rev. A* **40**, 3605.

DuBois, R. D., and S. T. Manson (1987), *Phys. Rev. A* **35**, 2007.

DuBois, R. D., and S. T. Manson (1990), *Phys. Rev. A* **42**, 1222.

DuBois, R. D., and L. H. Toburen (1988), *Phys. Rev. A* **38**, 3960.

DuBois, R. D., L. H. Toburen, and M. E. Rudd (1984a), *Phys. Rev. A* **29**, 70.

DuBois, R. D., J. P. Giese, and C. L. Cocke (1984b), *Phys. Rev. A* **29**, 1079.

Duncan, M. M., and M. G. Menendez (1976), *Phys. Rev. A* **13**, 566.

Duncan, M. M., M. G. Menendez, and F. L. Eisele (1977), *Phys. Rev. A* **15**, 1785.

Dunn, K. F., G. C. Angel, and H. B. Gilbody (1979), *J. Phys. B: At. Mol. Phys.* **12**, L623.

Edwards, A. K., and M. E. Rudd (1968), *Phys. Rev.* **170**, 140.

Edwards, A. K., R. M. Wood, and M. F. Steuer (1977), *Phys. Rev. A* **16**, 1385.

Elston, S. B. (1981), in D. J. Fabian, H. Kleinpoppen, and L. M. Watson, Eds., *Inner Shell and X-Ray Physics of Atoms and Solids*, Plenum Press, New York, p. 127.

Everhart, E., and Q. C. Kessel (1965), *Phys. Rev. Lett.* **14**, 247.

Fadeev, L. D. (1960), *Zh. Eksp. Teor. Fiz.* **39**, 1459. [transl. in *Sov. Phys. JETP* **12**, 1014 (1961)].

Fainstein, P. D., and R. D. Rivarola (1987), *J. Phys. B: At. Mol. Phys.* **20**, 1285.

Fainstein, P. D., V. H. Ponce, and R. D. Rivarola (1991), *J. Phys. B: At. Mol. Opt. Phys.* **24**, 3091.

Fano, U. (1954), *Phys. Rev.* **95**, 1198.

Fano, U. (1961), *Phys. Rev.* **124**, 1866.

Fano, U., and W. Lichten (1965), *Phys. Rev. Lett.* **14**, 627.

Fastrup, B. (1975), *Int. Conf. Physics of Electronic and Atomic Collisions*, invited papers.

Fastrup, B. (1977), in P. Richard, Ed., *Methods of Experimental Physics*, Vol. 14, Academic Press, New York.

Fastrup, B., G. Hermann, and K. J. Smith (1971), *Phys. Rev.* **A3**, 1591.

Fedorenko, N. V., and V. V. Afrosimov (1956), *Zh. Tekh. Fiz.* **26** [transl. in *Sov. Phys. Tech. Phys.* **1**, 1872.]

Fender, F. G., and J. P. Vinti (1934), *Phys. Rev.* **46**, 77.

Fermi, E. (1934), *Nuovo Cim.* **11**, 157.

Flannery, M. R. (1973), *Ann. Phys. (N.Y.)* **79**, 480.

Fock, V. (1935), *Z. Physik* **98**, 145.

Fontijn, A., Ed. (1985), *Gas-Phase Chemi-luminescence and Chemi-Ionization*, North-Holland, Amsterdam.

Garcia, J. D. (1970), *Phys. Rev.* **A1**, 280.

Garcia, J. D., R. J. Fortner, and T. M. Kavanagh (1973), *Rev. Mod. Phys.* **45**, 111.

Gardner, R. K., and T. G. Gray (1978), *At. Data Nucl. Data Tables* **21**, 515.

Garibotti, C. R., and J. E. Miraglia (1980), *Phys. Rev. A* **21**, 572.

Garibotti, C. R., and J. E. Miraglia (1983), *Phys. Rev. A* **25**, 1440.

Gay, T. J., H. G. Berry, E. B. Hale, V. D. Irby, and R. E. Olson (1988), *Nucl. Instrum. Methods Phys. Res.* **B31**, 336.

Gay, T. J., M. W. Gealy, and M. E. Rudd (1990), *J. Phys. B: At. Mol. Opt. Phys.* **23**, L823.

Gerjuoy, E. (1966), *Phys. Rev.* **148**, 54.

Gibson, D. K., and I. D. Reid (1984), *J. Phys. E: Sci. Instrum.* **17**, 1227.

Gilbody, H. B. (1986), *Adv. At. Mol. Phys.* **22**, 143.

Gilbody, H. B., and A. R. Lee (1963), *Proc. R. Soc. London Ser. A* **274**, 365.

Gillespie, G. H. (1982), *J. Phys. B: At. Mol. Phys.* **15**, L729.

Godunov, A. L., Sh. D. Kunikeev, V. N. Mileev, V. S. Senashenko (1983), in J. Eichler, W. Fritsch, I. V. Hertel, N. Stolterfoht, and U. Wille, Eds., *Electronic and Atomic Collisions*, Abstracts of Contributed Papers, Berlin, p. 380.

Goldman, F. (1932), *Ann. Phys. (Leipzig)* **10**, 460.

Gordeev, Yu. S., and G. N. Ogurtsov (1971), *Zh. Eksp. Teor. Fiz.* **60**, 2051 [transl. in *Sov. Phys. JETP* **33**, 1105].

Gordeev, Yu. S., and M. N. Panov (1964), *Zh. Tekh. Fiz.* **34**, 857 [transl. in *Sov. Phys. Tech. Phys.* **9**, 656].

Gordeev, Yu. S., P. H. Woerlee, H. de Waard, and F. W. Saris (1981), *J. Phys. B: At. Mol. Phys.* **14**, 513.

Green, A. E. S., and R. J. McNeal (1971), *J. Geophys. Res.* **76**, 133.

Groenveld, K.-O., W. Meckbach, I. A. Sellin, and J. Burgdörfer (1984), *Comments At. Mol. Phys.* **14**, 187.

Gryzinski, M. (1957), *Phys. Rev.* **107**, 1471.

Gryzinski, M. (1959), *Phys. Rev.* **115**, 374.

Gryzinski, M. (1965), *Phys. Rev.* **138**, A305, A322, A336.

Harrison, K. G., and M. W. Lucas (1970), *Phys. Lett.* **33A**, 142.

Hasted, J. B. (1960), *Adv. Electron. Electron Phys.* **13**, 1.

Hasted, J. B. (1972), *Physics of Atomic Collisions*, 2nd ed., Butterworth, London.

Haugen, H. K., L. H. Andersen, P. Hvelplund, and H. Knudsen (1982), *Phys. Rev. A* **26**, 1962.

Heckman, H. H., and P. J. Lindstrom (1969), *Phys. Rev. Lett.* **22**, 871.

Heil, O., R. D. DuBois, R. Maier, M. Kuzel, and K.-O. Groeneveld (1991), *Z. Phys.* **21**, 235.

Heil, O., R. D. DuBois, R. Maier, M. Kuzel, and K.-O. Groeneveld (1992), *Phys. Rev. A* **45**, 2850.

Herzberg, G. (1950), *Molecular Spectra and Molecular Structure*, Vol. 1, *Spectra of Diatomic Molecules*, 2nd ed., Van Nostrand, Toronto.

Hollricher, O. (1965), *Z. Phys.* **41**, 187.

Hooper, J. W., E. W. McDaniel, D. W. Martin, and D. S. Harmer (1961), *Phys. Rev.* **121**, 1123.

Hotop, H., and A. Niehaus (1967), *J. Chem. Phys.* **47**, 2506.

Hvelplund, P., H. K. Haugen, and H. Knudsen (1980), *Phys. Rev. A* **22**, 1930.

Inokuti, M. (1971), *Rev. Mod. Phys.* **43**, 297.

Itoh, A., and N. Stolterfoht (1985), *Nucl. Instrum. Methods* **N10/11**, 97.

Itoh, A., T. Schneider, G. Schiwietz, Z. Roller, H. Platter, G. Nolte, D. Schneider, and N. Stolterfoht (1983), *J. Phys. B: At. Mol. Phys.* **16**, 3965.

Itoh, A., D. Schneider, T. Schneider, T. J. Zouros, G. Nolte, G. Schiwietz, W. Zeitz and N. Stolterfoht (1985), *Phys. Rev. A* **31**, 684.

Jakubassa-Amundsen, D. H. (1983), *J. Phys. B: At. Mol. Phys.* **16**, 1767.

Jakubassa-Amundsen, D. H. (1989), *J. Phys. B: At. Mol. Phys.* **22**, 3989.

Jhanwar, B. L., S. P. Khare, and A. Kumar (1975), *Indian J. Pure Appl. Phys.* **4**, 33.

Kanter, E. P., P. J. Cooney, D. S. Gemmell, K.-O. Groeneveld, W. J. Pietsch, A. J. Ratkowski, Z. Vager, and B. J. Zabransky (1979), *Phys. Rev. A* **20**, 834.

Keene, J. P. (1949), *Philos. Mag.* **40**, 369.

Kessel, Q. C., A. Russek, and E. Everhart (1965), *Phys. Rev. Lett.* **14**, 484.

Kessel, Q. C. (1969), in E. W. McDaniel and M. R. C. McDowell, Eds., *Case Studies in Atomic Collision Physics*, Vol. 1, North-Holland, Amsterdam, p. 399.

Kessel, Q. C., and E. Everhart (1966), *Phys. Rev.* **146**, 16.

Kessel, Q. C., and B. Fastrup (1973), in M. R. C. McDowell and E. W. McDaniel, Eds., *Case Studies in Atomic Physics*, Vol. 3, North-Holland, Amsterdam, p. 137.

Kessel, Q. C., E. Pollack, and W. W. Smith (1978), in R. G. Cooks, Ed., *Collision Spectroscopy*, Plenum Press, New York, p. 147.

Kessel, Q. C., R. Morgenstern, B. Müller, and A. Niehaus (1979), *Phys. Rev. A* **20**, 804.

Khare, S. P., and A. Kumar (1980), *Physica* **100C**, 135.

Kim, Y.-K. (1975), *Radiat. Res.* **61**, 21.

Kim, Y.-K., and G. Magelssen, Eds. (1979), *Report on the Workshop on Atomic and Plasma Physics Requirements for Heavy Ion Fusion*, ANL-80-17, Argonne National Laboratory, Lemont, Ill.

Knudsen, H., L. H. Andersen, P. Hvelplund, G. Astner, H. Cederquist, H. Danared, L. Liljeby, and K.-G. Rensfelt (1984), *J. Phys. B* **17**, 3545.

Koschar, P., A. Clouvas, O. Heil, M. Burkhard, J. Kemmler, and K.-O. Groeneveld (1987), *Nucl. Instrum. Methods Phys. Res.* **B24/25**, 153.

Kövér, Á., D. Varga, G. Szabó, D. Berényi, I. Kádár, S. Ricz, J. Végh, and G. Hock (1983), *J. Phys. B: At. Mol. Phys.* **16**, 1017.

Kuyatt, C. E. (1968), *Methods Exp. Phys.* **7A**, 1.

Kuyatt, C. E., and T. Jorgensen, Jr. (1963), *Phys. Rev.* **130**, 1444.

Lichten, W. (1967), *Phys. Rev.* **164**, 131.

Lin, C. D. (1986), *Adv. At. Mol. Phys.* **22**, 77.

Lindsay, B. G., F. B. Yousif, F. R. Simpson, and C. J. Latimer (1987), *J. Phys. B: At. Mol. Phys.* **20**, 2759.

Lipsky, L., and A. Russek (1966), *Phys. Rev.* **142**, 59.

Lipsky, L., R. Anania, and M. J. Conneely (1977), *At. Data Nucl. Data Tables* **20**, 127.

Lotz, W. (1970), *J. Opt. Soc. Am.* **60**, 206.

Lucas, M. W., K. F. Man, and W. Steckelmacher (1984), in K.-O. Groeneveld, W. Meckbach, and I. A. Sellin, Eds., *Forward Electron Ejection in Ion Collisions*, Lecture Notes in Physics, Springer-Verlag, Berlin.

Lynch, D. J., L. H. Toburen, and W. E. Wilson (1976), *J. Chem. Phys.* **64**, 2616.

Macek, J. (1970a), *Int. Conf. Physics of Electronic and Atomic Collisions*, Abstracts of Papers, Cambridge, Mass., p. 687.

Macek, J. (1970b), *Phys. Rev. A* **1**, 235.

Madden, R. P., and K. Codling (1963), *Phys. Rev. Lett.* **10**, 516.

Madison, D. H. (1973), *Phys. Rev. A* **8**, 2449.

Madison, D. H., and S. T. Manson (1979), *Phys. Rev. A* **20**, 825.

Manson, S. T., and L. H. Toburen (1981), *Phys. Rev. Lett.* **46**, 529.

Manson, S. T., L. H. Toburen, D. H. Madison, and N. Stolterfoht (1975), *Phys. Rev. A* **12**, 60.

Manson, S. T., R. D. DuBois, and L. H. Toburen (1983), *Phys. Rev. Lett.* **51**, 1542.

Massey, H. S. W. (1971), *Electronic and Ionic Impact Phenomena*, 2nd ed., Vol. 3, Oxford University Press, Oxford.

Massey, H. S. W., E. H. S. Burhop, and H. B. Gilbody (1974), *Electronic and Ionic Impact Phenomena*, Vol. 4, 2nd ed., Clarendon Press, Oxford.

Matsuzawa, M. (1980), in N. Oda and K. Takayanagi, Eds., *Electronic and Atomic Collisions*, North-Holland, Amsterdam, p. 493.

Matsuzawa, M. (1983), in R. F. Stebbings and F. B. Dunning, Eds., *Rydberg States of Atoms and Molecules*, Cambridge University Press, New York.

McDaniel, E. W. (1964), *Collision Phenomena in Ionized Gases*, Wiley, New York.

McDaniel, E. W. and E. J. Mansky (1993), *AAMOP* **31**.

McDaniel, E. W., J. W. Hooper, D. W. Martin, and D. S. Harmer (1961), *Proc. 5th Int. Conf. Ionization Phenomena in Gases* North-Holland, Amsterdam, p. 60.

McDaniel, E. W., M. R. Flannery, H. W. Ellis, F. L. Eisele, W. Pope, and T. G. Roberts (1977–1982), *Compilation of Data Relevant to Rare Gas–Rare Gas and Rare Gas–Monohalide Excimer Lasers*, Technical Report H-78-1, High Energy Laser Laboratory, U.S. Army Missile Research and Development Command, Redstone Arsenal, Ala.

McDaniel, E. W., M. R. Flannery, E. W. Thomas, and S. T. Manson (1985), *ADNDT* **33**, 1–148.

McDowell, M. R. C., and J. P. Coleman (1970), *Introduction to the Theory of Ion–Atom Collisions*, North-Holland/American Elsevier, Amsterdam.

McGuire, J. H. (1987), *Phys. Rev. A* **36**, 1114.

McNeal, R. J. (1970), *J. Chem. Phys.* **53**, 4308.

Meckbach, W., P. J. Focke, A. R. Goñi, S. Suárez, J. Macek, and M. G. Menendez (1986), *Phys. Rev. Lett.* **57**, 1587.

Mehlhorn, W. (1969), *Lectures on the Auger Effect*, University of Nebraska, unpublished.

Mehlhorn, W. (1978), *Electron Spectroscopy of Auger and Autoionizing States: Experiment and Theory*, Institute of Physics, University of Aarhus, unpublished.

Mehlhorn, W. (1985), in Bernd Craseman, Ed., *Atomic Inner-Shell Physics*, Plenum Press, New York.

Merzbacher, E., and H. W. Lewis (1958), *Handb. Phys.* **34**, 166.

Meyerhof, W. E. (1973), *Phys. Rev. Lett.* **31**, 1341.

Miller, J. H., L. H. Toburen, and S. T. Manson (1983), *Phys. Rev. A* **27**, 1337.

Miraglia, J. E., and J. Macek (1991), *Phys. Rev. A* **43**, 5919.

Mitchell, J. B. A., K. F. Dunn, G. C. Angel, R. Browning, and H. B. Gilbody (1977), *J. Phys. B: At. Mol. Phys.* **10**, 1897.

Moe, D. E., and O. H. Petsch (1958), *Phys. Rev.* **110**, 1358.

Mott, N. F., and H. S. W. Massey (1965), *The Theory of Atomic Collisions*, 3rd ed., Clarendon Press, Oxford.

Nagy, P., A. Skutlartz, and V. Schmidt (1980), *J. Phys. B* **13**, 1249.

Neelavathi, V. N., R. H. Ritchie, and W. Brandt (1974), *Phys. Rev. Lett.* **33**, 302.

Niehaus, A. (1990), *Phys. Rep.* **186**, 149.

Ogurtsov, G. N. (1972), *Rev. Mod. Phys.* **44**, 1.

Ogurtsov, V. I., and Yu. F. Bydin (1969), *Zh. Eksp. Teor. Pis. Red.* **10**, 134 [transl. in *Sov. Phys. JETP Lett.* **10**, 85].

Oldham, W. J. B., Jr. (1967), *Phys. Rev.* **161**, 1.

Olson, R. E. (1979), *J. Phys. B: At. Mol. Phys.* **12**, 1843.

Olson, R. E. (1983), *Phys. Rev. A* **27**, 1871.

Olson, R. E. (1987), *Phys. Rev. A* **36**, 1519.

Olson, R. E., and T. J. Gay (1988), *Phys. Rev. Lett.* **61**, 302.

Olson, R. E., and A. Salop (1977), *Phys. Rev. A* **16**, 531.

Olson, R. E., T. J. Gay, H. G. Berry, E. B. Hale, and V. D. Irby, (1987), *Phys. Rev. Lett.* **59**, 36.

O'Malley, T. F., and S. Geltman (1965), *Phys. Rev. A* **137**, 1344.

Park, J. T. (1978), in R. G. Cooks, Ed., *Collision Spectroscopy*, Plenum Press, New York, p. 19.

Park, J. T., and F. D. Schowengerdt (1969), *Phys. Rev.* **185**, 152.

Park, J. T., J. E. Aldag, J. L. Peacher, and J. M. George (1978), *Phys. Rev. Lett.* **40**, 1646.

Paul, H., and J. Muhr (1986), *Phys. Rep.* **135**, 48.

Paul, H., and J. Sacher (1989), *At. Data Nucl. Data Tables* **42**, 106.

Pivovar, L. I., and Yu. Z. Levchenko (1967), *Zh. Eksp. Teor. Fiz.* **52**, 42 [transl. in *Sov. Phys. JETP* **25**, 27].

Poulaert, G., F. Brouillard, W. Claeys, J. W. McGowan, and G. Van Wassenhove (1978), *J. Phys. B: At. Mol. Phys.* **11**, L671.

Puckett, L. J., and D. W. Martin (1970), *Phys. Rev. A* **1**, 1432.

Reading, J. F., and A. L. Ford (1987), *Phys. Rev. Lett.* **58**, 543.

Reinhold, C. O., J. Burgdörfer, and J. Kemmler (1992), *Phys. Rev. A* **45**, R2655.

Robinson, B. B. (1965), *Phys. Rev.* **140**, A764.

Rødbro, M., and F. D. Anderson (1979), *J. Phys. B: At. Mol. Phys.* **12**, 2883.

Rudd, M. E. (1964), *Phys. Rev. Lett.* **13**, 503.

Rudd, M. E. (1965), *Phys. Rev. Lett.* **15**, 580.

Rudd, M. E. (1979), *Phys. Rev. A* **20**, 787.

Rudd, M. E. (1988), *Phys. Rev. A* **38**, 6129.

Rudd, M. E., and T. Jorgensen, Jr. (1963), *Phys. Rev.* **131**, 666.

Rudd, M. E., and J. H. Macek (1972), M. R. C. McDowell and E. W. McDaniel, Eds., *Case Studies in Atomic Physics*, Vol. 3, No. 2, North-Holland, Amsterdam, p. 47.

Rudd, M. E., and D. H. Madison (1976), *Phys. Rev. A* **14**, 128.

Rudd, M. E., and K. Smith (1968), *Phys. Rev.* **169**, 79.

Rudd, M. E., C. A. Sautter, and C. L. Bailey (1966a), *Phys. Rev.* **151**, 20.

Rudd, M. E., T. Jorgensen, Jr., and D. J. Volz (1966b), *Phys. Rev. Lett.* **16**, 929.

Rudd, M. E., T. Jorgensen, Jr., and D. J. Volz (1966c), *Phys. Rev.* **151**, 28.

Rudd, M. E., D. Gregoire, and J. B. Crooks (1971), *Phys. Rev. A* **3**, 1635.

Rudd, M. E., L. H. Toburen, and N. Stolterfoht (1976), *At. Data Nucl. Data Tables* **18**, 413.

Rudd, M. E., L. H. Toburen, and N. Stolterfoht (1979), *At. Data Nucl. Data Tables* **23**, 4053.

Rudd, M. E., J. S. Risley, J. Fryar, and R. G. Rolfes (1980), *Phys. Rev. A* **21**, 506.

Rudd, M. E., R. D. DuBois, L. H. Toburen, C. A. Ratcliffe, and T. V. Goffe (1983), *Phys. Rev. A* **28**, 3244.

Rudd, M. E., Y.-K. Kim, D. H. Madison, and J. W. Gallagher (1985a), *Rev. Mod. Phys.* **57**, 965.

Rudd, M. E., A. Itoh, and T. V. Goffe (1985b), *Phys. Rev. A* **32**, 2499.

Rudd, M. E., T. V. Goffe, A. Itoh, and R. D. DuBois (1985c), *Phys. Rev. A* **32**, 829.

Rudd, M. E., Y.-K. Kim, D. H. Madison, and T. J. Gay (1992), *Rev. Mod. Phys.* **64**, 441.

Rutherford, E. (1911), *Philos. Mag.* **21**, 669.

Rutledge, C. H., and R. L. Watson (1973), *At. Data Nucl. Data Tables* **12**, 195.

Salin, A. (1969a), *Int. Conf. Physics of Electronic and Atomic Collisions*, Abstracts of Papers, Cambridge, Mass., p. 684.

Salin, A. (1969b), *J. Phys. B* **2**, 631.

Saris, F. W., and D. Onderdelinden (1970), *Physica* **49**, 441.

Sarkadi, L., J. Bossler, R. Hippler, and H. O. Lutz (1983), *J. Phys. B: At. Mol. Phys.* **16**, 71.

Schiff, L. I. (1949), *Quantum Mechanics*, McGraw-Hill, New York, p. 168.

Schneider, D., M. Prost, N. Stolterfoht, G. Nolte, and R. DuBois (1983), *Phys. Rev. A* **28**, 649.

Schowengerdt, F. D., and M. E. Rudd (1972), *Phys. Rev. Lett.* **28**, 127.

Schram, B. L., A. J. H. Boerboom, and J. Kistemaker (1966), *Physica (Utrecht)* **32**, 185.

Schultz, D. R., R. E. Olson, C. O. Reinhold, G. W. Kerby III, M. W. Gealy, Y.-Y. Hsu, and M. E. Rudd (1991), *J. Phys. B: At. Mol. Phys.* **24**, L599.

Schwirzke, F. (1960), *Z. Phys.* **157**, 510.

Sellin, I. A. (1982), in S. Datz, Ed., *Physics of Electronic and Atomic Collisions*, North-Holland, Amsterdam, p. 195.

Sellin, I. A., S. D. Berry, M. Breinig, C. Bottcher, R. Latz, M. Burkhard, H. Folger, J. J. Frischkorn, K.-O. Groeneveld, D. Hofmann, and P. Koschar (1985), *Lecture Notes in Physics 213: Forward Electron Ejection in Ion Collisions*, Berlin: Springer-Verlag, p. 109.

Sevier, K. (1972), *Low Energy Electron Spectrometry*, Wiley-Interscience, New York.

Shah, M. B., and H. B. Gilbody (1982), *J. Phys. B* **15**, 3441.

Shah, M. B., and H. B. Gilbody (1985), *J. Phys. B* **18**, 899.

Shakeshaft, R., and L. Spruch (1978), *Phys. Rev. Lett.* **41**, 1037.

Shakeshaft, R., and L. Spruch (1980), *Phys. Rev. A* **21**, 1161.

Shore, B. W. (1967) *Rev. Mod. Phys.* **39**, 439.

Simpson, J. A., S. R. Mielczarek, and J. Cooper (1964), *J. Opt. Soc. Am.* **54**, 269.

Smith, P. T. (1930), *Phys. Rev.* **36**, 1293.

Solov'ev, E. S., R. N. Il'in, V. A. Oparin, and N. V. Fedorenko (1962), *Zh. Eksp. Teor. Phys.* **42**, 659 [transl. in *Sov. Phys. JETP* **15**, 459].

Stephen, K., H. Helm, and T. D. Märk (1980), *J. Chem. Phys.* **73**, 3763.

Steuer, M. F., R. M. Wood, and A. K. Edwards (1977), *Phys. Rev. A* **16**, 1873.

Stolterfoht, N. (1971), *Z. Physik* **248**, 92

Stolterfoht, N. (1978a), *Proc. 9th Summer School and Symposium on the Physics of Ionized Gases,* Invited Lectures, Dubrovnik, Yugoslavia.

Stolterfoht, N. (1978b), in I. A. Sellin, Ed., *Topics in Current Physics*, Vol. 5, Springer-Verlag, Berlin, p. 155.

Stolterfoht, N. (1987a), *J. Phys.*, suppl. 12, **48**, C9-177.

Stolterfoht, N. (1987b), *Phys. Rep.* **146**, 315.

Stolterfoht, N. (1988), in H. B. Gilbody, W. R. Newell, F. H. Read, and A. C. H. Smith, Eds., *Electronic and Atomic Collisions*, North-Holland, Amsterdam, p. 661.

Stolterfoht, N. (1990), *Phys. Scr.* **42**, 192.

Stolterfoht, N. (1991), *Nucl. Instrum. Methods Phys. Res.* **B53**, 477.

Stolterfoht, N., D. Ridder, and P. Ziem (1972a), *Phys. Lett.* **42A**, 240.

Stolterfoht, N., P. Ziem, and D. Ridder (1972b), *6th Natl. Conf. Electronic and Atomic Collisions*, Liege, Belgium.

Stolterfoht, N., H. Gabler, and U. Leithäuser (1973), *Phys. Lett.* **45A**, 351.

Stolterfoht, N., D. Schneider, D. Burch, H. Wieman, and J. S. Risley (1974), *Phys. Rev. Lett.* **33**, 59.

Stolterfoht, N., D. Schneider, J. Tanis, H. Altevogt, A. Salin, P. D. Fainstein, R. Rivarola, J. P. Grandin, J. N. Scheurer, S. Andriamonje, D. Bertault, and J. F. Chemin (1987), *Europhys. Lett.* **4**, 899.

Stolterfoht, N., K. Sommer, D. C. Griffin, C. C. Havener, M. S. Huq, R. A. Phaneuf, J. K. Swenson, and F. W. Meyer (1989), *Nucl. Instrum. Methods* **B40/41**, 28.

Tate, J. T., and P. T. Smith (1932), *Phys. Rev.* **39**, 270.

Thomas, L. H. (1927), *Proc. Cambridge Philos. Soc.* **23**, 713.

Thomson, J. J. (1912), *Philos. Mag.* **23**, 449.

Thomson, G. M., P. C. Laudieri, and E. Everhart (1970), *Phys. Rev. A* **1**, 1439.

Toburen, L. H. (1971), *Phys. Rev. A* **3**, 216.

Toburen, L. H. (1974), *Phys. Rev. A* **9**, 2505.

Toburen, L. H. (1982), in D. Berényi and G. Hock, Eds., *Nuclear Methods 2; High-Energy Ion–Atom Collisions*, Elsevier, New York, p. 53.

Toburen, L. H. (1990), *Scanning Microsc. Suppl.* **4**, 239.

Toburen, L. H., and W. E. Wilson (1972), *Phys. Rev. A* **5**, 247.

Toburen, L. H., and W. E. Wilson (1975), *Rev. Sci. Instrum.* **46**, 851.

Toburen, L. H., and W. E. Wilson (1979), *Proc. 5th Conf. Use of Small Accelerators, IEEE Trans. Nucl. Sci.* **NS-26**, 1056.

Toburen, L. H., S. T. Manson, and Y.-K. Kim (1978), *Phys. Rev. A* **17**, 148.

Toburen, L. H., W. E. Wilson, and R. J. Popowich (1980), *Radiat. Res.* **82**, 27.

Toburen, L. H., R. D. DuBois, C. O. Reinhold, D. R. Schultz, and R. E. Olson (1990), *Phys. Rev. A* **42**, 5338.

Ullrich, J., H. Schmidt-Böcking, and C. Kelbch (1988), *Nucl. Instrum. Methods Phys. Res.* **A268**, 216.

Vane, C. R., I. A. Sellin, M. Suter, G. D. Alton, S. B. Elston, P. M. Griffin, and R. S. Thoe (1978), *Phys. Rev. Lett.* **40**, 1020.

Vriens, L. (1967), *Proc. Phys. Soc. London* **90**, 935.

Wang, J., C. O. Reinhold, and J. Burgdörfer (1991), *Phys. Rev. A* **44**, 7243.

Wexler, S. (1964), *J. Chem. Phys.* **41**, 1714; **44**, 2221.

Wille, U., and R. Hippler (1986), *Phys. Rep.* **132**, 129.

Williams, E. J. (1927), *Nature* **119**, 489.

Wilson, W. E., and L. H. Toburen (1973), *Phys. Rev. A* **7**, 1535.

Wilson, W. E., and L. H. Toburen (1975), *Phys. Rev. A* **11**, 1303.

Woerlee, P. H., Yu. S. Gordeev, H. de Waard, and F. W. Saris (1981), *J. Phys. B: At. Mol. Phys.* **14**, 527.

Wood, R. M., A. K. Edwards, and M. F. Steuer (1976), *Rev. Sci. Instrum.* **47**, 1471.

Wood, R. M., A. K. Edwards, and M. F. Steuer (1977), *Phys. Rev. A* **15**, 1433.

Yagashita, A., H. Oomoto, K. Wakiya, H. Suzuki, and F. Koike (1978), *J. Phys. B: At. Mol. Phys.* **11**, L111.

Yamazaki, Y., and N. Oda (1984), *Phys. Rev. Lett.* **52**, 29.

Yu, Y. C., and G. Lapicki (1987), *Phys. Rev. A* **36**, 4710.

5

CHARGE TRANSFER

5-1. INTRODUCTION

From studies of canal rays in the first decade of this century, it was realized that an ion beam could become partly neutralized in collisions with the molecules of the gas through which it was traveling. Henderson (1922) made measurements of charge-changing collisions of alpha particles and Rutherford (1924) measured equilibrium fractions of helium ions when alpha particles passed through foils.

Besides the intrinsic interest in charge transfer as a fundamental atomic process, a knowledge of its systematics and mechanisms is a prerequisite for an understanding of radiation detectors, radiation damage in biological and other material, inertial and magnetic confinement thermonuclear fusion systems, astrophysical processes, gas discharges, mass spectrometry, and numerous other practical processes and devices. In tandem acclerators particles are accelerated once as negative ions, two (or more) electrons are removed in a charge-transfer process, and the resulting positive ions are then accelerated further by the same electric potential.

In electron capture, an incident particle picks up one or more electrons from the target, while in an electron loss or stripping reaction an incident particle loses an electron, which usually goes into the continuum. Electron loss from a negative ion is usually called *detachment*. Electron capture and loss are the two types of charge transfer. The process has also been called charge exchange, but this is not a very descriptive term since no actual exchange of electrons takes place. Electron loss and detachment should, strictly speaking, be classified as examples of ionization since the electron usually goes into the continuum. This is especially evident when viewed in the projectile's frame of reference. Quite often, however, it is only the charge state of the fast projectile that is of interest, and along with capture, electron loss is an example of such a *charge-changing* collision.

The quantitative investigation of charge transfer that began in the 1920s used naturally radioactive sources of alpha particles. After about 1950, high-energy studies of charge transfer were facilitated when the cyclotron and other high-energy accelerators superseded the lower-energy Cockroft–Walton accelerators for use in nuclear physics research, thus making several of the latter machines available for atomic physics work. The intensity of effort, both experimental and theoretical, has increased ever since. In the 1960s tandem van de Graaff accelerators began to come into use for accelerating high-energy heavy ions. More recently, new techniques have been developed to produce highly charged heavy ions all the way up to completely stripped uranium ions. This has opened up additional opportunities in the study of charge transfer, and in the past decade or two, the level of effort has increased even further, making charge transfer the most widely studied of the basic atomic collision processes.

Reviews of charge transfer have been given by Massey and Burhop (1952), Allison (1958), Allison and Garcia-Munoz (1962), Hasted (1964), Betz (1972), Tawara and Russek (1973), Massey and Gilbody (1974), Gilbody (1986), and Bransden and McDowell (1992). Linder (1988) has reviewed ion–molecule charge transfer at low energies. Janev and Presnyakov (1981) have presented a review of collision processes involving multiply charged ions. Compilations of data on charge transfer have been given by Barnett et al. (1977), McDaniel et al. (1977–1982), and by Tawara et al. (1985). Three series of reports giving bibliographies as well as extensive data compilations for various processes, including charge transfer, are issued at irregular intervals by Japanese agencies. These are the JAERI-M (1982–   ) series published by the Japan Atomic Energy Research Institute at Tokai-mura, Naka-gun, Ibaraki-ken; the NIFS-DATA (1990–   ) series put out by the National Institute for Fusion Science at Nagoya; and the IPPJ-AM (1977–   ) series from the Institute of Plasma Physics at Nagoya. These three contain up-to-date compilations of charge-transfer cross sections.

We first describe the various charge-changing processes and give equations defining and relating the cross sections. The methods of measurement of such cross sections are then taken up. Some theoretical aspects of capture are discussed next, followed by descriptions of several specialized processes.

5-2. DESCRIPTION OF CHARGE-CHANGING PROCESSES

In most of the early charge-transfer studies, the change of the charge state of the projectile was the main consideration; the accompanying production of slow ions or free electrons was of lesser interest. More recently, coincidence, mass spectrometer, and other methods have been employed to learn more about the secondary products as well. From such measurements the specific elementary processes involved in charge transfer can be determined. We concentrate on the projectile charge state studies in this and the following section and discuss some of the more detailed studies in later sections.

The charge-transfer cross section is defined similar to that for other atomic processes. If N_i particles in initial charge state i are incident on a target of number density n and linear thickness l, the number of particles exiting from the target in a given final charge state f is given by

$$N_f = N_i n l \sigma_{if} \qquad (5\text{-}2\text{-}1)$$

where σ_{if} is the cross section for that change in charge. The target is assumed to be "thin"; that is, the thickness $\pi = nl$ is such that $\pi\sigma \ll 1$, where σ is the cross section for any collision process having an effect on the charge state. This ensures that most of the reactions result from single collisions.

Since atoms of hydrogen can exist in any of three charge states, H^+, H^0, and H^-, we will take hydrogen as the prototype of a three-component system. Collisions of hydrogen atoms or ions with a target can occur in which the projectile, initially in any of the three states, can emerge from the collision either in its original charge state or in any of the others. There are, therefore, six possible charge-changing transitions, the cross sections of which are designated by σ_{if}. These cross sections have different impact energy dependences, as shown in Fig. 5-2-1 using a helium target as an example. Note that above about 200 keV (100 keV for most other targets) the cross sections for transfer into the H^- final state become very small compared to those for transitions out of that state. Therefore, in that energy range, if the initial charge state of the projectile is 0 or $+1$, it is a good approximation to consider only those two charge states. Under those conditions, hydrogen becomes a simpler two-component system and only the two cross sections σ_{10} and σ_{01} are needed to describe the transitions.

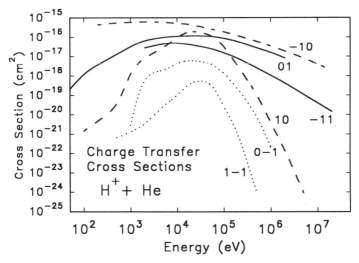

Figure 5-2-1. Cross sections for charge transfer from charge state i to charge state f (indicated by "if") for proton impact on helium atoms as a function of the projectile energy. [Data from Nakai et al. (1983b).]

A. Two-Component System

If a projectile makes multiple collisions while traversing a thick target, it may change its charge state many times. Considering a two-component system such as the one described above, we can write equations relating the fractions F_0 and F_1 of projectiles in charge states 0 and 1 after traversing a target thickness $\pi = nl$. The increase in F_0 in traversing a thickness $d\pi$ is equal to the gain of neutral particles, $F_1 \sigma_{10} \, d\pi$, minus the loss, $F_0 \sigma_{01} \, d\pi$. Thus

$$dF_0 = F_1 \sigma_{10} \, d\pi - F_0 \sigma_{01} \, d\pi. \tag{5-2-2}$$

Using the condition $F_0 + F_1 = 1$ and the initial conditions that $F_0 = F_{00}$ and $F_1 = F_{10}$ when $\pi = 0$ leads to the integral

$$\int_{F_{00}}^{F_0} \frac{dF_0}{\sigma_{10} - F_0(\sigma_{01} + \sigma_{10})} = \int_0^\pi d\pi, \tag{5-2-3}$$

which yields the general solutions

$$F_1 = \left(\frac{\sigma_{10}}{\sigma_{01} + \sigma_{10}} - F_{00} \right) e^{-\pi(\sigma_{01} + \sigma_{10})} + \frac{\sigma_{01}}{\sigma_{01} + \sigma_{10}} \tag{5-2-4}$$

and

$$F_0 = \left(\frac{\sigma_{01}}{\sigma_{01} + \sigma_{10}} - F_{10} \right) e^{-\pi(\sigma_{01} + \sigma_{10})} + \frac{\sigma_{10}}{\sigma_{01} + \sigma_{10}}. \tag{5-2-5}$$

Consider three situations which are important special cases of these solutions.

1. *Equilibrated Beam.* Regardless of the initial charge state composition of the beam, if the thickness π is great enough that $\pi(\sigma_{10} + \sigma_{01}) \gg 1$, the exponential term drops out and we have

$$F_1 \rightarrow F_{1\infty} = \frac{\sigma_{01}}{\sigma_{01} + \sigma_{10}}, \tag{5-2-6}$$

$$F_0 \rightarrow F_{0\infty} = \frac{\sigma_{10}}{\sigma_{01} + \sigma_{10}}, \tag{5-2-7}$$

and

$$\frac{F_{0\infty}}{F_{1\infty}} = \frac{\sigma_{10}}{\sigma_{01}}. \tag{5-2-8}$$

This result shows that the composition of an equilibrated beam is determined solely by the cross sections involved. This is also true for a multicomponent system.

2. *Pure Incident Beam.* If the incident beam is pure H^0, then $F_{00} = 1$, $F_{10} = 0$, and (5-2-4) and (5-2-5) reduce to

$$F_0 = \frac{\sigma_{01}e^{-\pi(\sigma_{01}+\sigma_{10})} + \sigma_{10}}{\sigma_{01} + \sigma_{10}}$$

(5-2-9)

and

$$F_1 = \frac{\sigma_{01}}{\sigma_{01} + \sigma_{10}}[1 - e^{-\pi(\sigma_{01}+\sigma_{10})}].$$

(5-2-10)

Similarly, if the incident beam is pure H^+, then $F_{10} = 1$, $F_{00} = 0$, and we obtain

$$F_0 = \frac{\sigma_{10}}{\sigma_{01} + \sigma_{10}}[1 - e^{-\pi(\sigma_{01}+\sigma_{10})}]$$

(5-2-11)

and

$$F_1 = \frac{\sigma_{10}e^{-\pi(\sigma_{01}+\sigma_{10})} + \sigma_{01}}{\sigma_{01} + \sigma_{10}}.$$

(5-2-12)

3. *Thin Target, Pure Incident Beam.* For a target thin enough that the probability of more than one collision is very small, $\pi(\sigma_{01} + \sigma_{10}) \ll 1$. Then the exponential of (5-2-10) may be expanded, yielding to second order

$$F_1 = \pi\sigma_{01} - \frac{\pi^2}{2}\sigma_{01}(\sigma_{01} + \sigma_{10}) \quad \text{and} \quad F_0 = 1 - F_1$$

(5-2-13)

for incident H^0, and similarly,

$$F_0 = \pi\sigma_{10} - \frac{\pi^2}{2}\sigma_{10}(\sigma_{01} + \sigma_{10}) \quad \text{and} \quad F_1 = 1 - F_0$$

(5-2-14)

for incident H^+. For small thicknesses the growth of a component charge state is linear.

B. Three-Component System

The corresponding equations for a three-component system are more complicated but are obtained in a similar fashion. They can be stated in a generalized form by letting i, j, and k stand for the three possible charge states of the system, (e.q., $-1, 0$, and 1 for hydrogen). The three differential equations are given by the cyclic permutations of the indices of the equation

$$\frac{dF_i}{d\pi} = F_j\sigma_{ji} + F_k\sigma_{ki} - F_i(\sigma_{ij} + \sigma_{ik}). \tag{5-2-15}$$

Only two of the three equations are independent because of the condition that

$$F_i + F_j + F_k = 1. \tag{5-2-16}$$

The general solution of the three simultaneous differential equations was given by Allison and Warshaw (1953), but it had typographical errors that were corrected by Phillips and Tuck (1956). The solution was also given by Allison (1958). It can be written in the form

$$F_i = F_{i\infty} + a_i e^{-r_1\pi} + b_i e^{-r_2\pi}, \tag{5-2-17}$$

where $F_{i\infty}$, a_i, b_i, r_1, and r_2 are functions of the six cross sections. Because the functions are rather complex, we will not give them here but instead, go directly to the simpler solutions for two important special cases.

1. *Equilibrated Beam.* The solution for $\pi \to \infty$ (i.e., for an equilibrated beam) is

$$F_{i\infty} = \frac{\sigma_{ji}(\sigma_{kj} + \sigma_{ki}) + \sigma_{ki}\sigma_{jk}}{D}, \tag{5-2-18}$$

where

$$D = \sigma_{jk}(\sigma_{ij} + \sigma_{ik} + \sigma_{ki}) + \sigma_{ji}(\sigma_{kj} + \sigma_{ik} + \sigma_{ki}) + \sigma_{kj}(\sigma_{ij} + \sigma_{ik}) + \sigma_{ij}\sigma_{ki}. \tag{5-2-19}$$

2. *Thin Target, Pure Incident Beam.* The solution for a pure incident beam on a thin target can be written in a considerably simplified form by expanding the exponentials in the general solution and neglecting powers of π higher than the second. The result is

$$F_j = \sigma_{ij}\pi + [(\sigma_{kj} - \sigma_{ij})\sigma_{ik} - (\sigma_{ij} + \sigma_{ji} + \sigma_{jk})\sigma_{ij}]\frac{\pi^2}{2}, \tag{5-2-20}$$

where i and j are the charges of the incident and observed beam particles, respectively, and k is the third of the three charge states.*

The equations for a two-component system can be obtained from the three-component equations by setting the cross sections involving the third compo-

*We wish to thank Eric Rudd for his elegant solution of the differential equations for the three-component system.

nent equal to zero. Note that there are sign errors in two of the terms of (5-2-20) in many of the earlier references [e.g., Fogel and Mitin (1955), Fogel et al. (1957), Allison (1958), Massey and Gilbody (1974), and Nakai et al. (1983a)].

C. Heavy Ion Projectiles

Hydrogen atoms have three possible states, but one is unlikely to be formed, so hydrogen is a two-component system for most purposes. Helium has four possible charge states, but again the negative charge state is unlikely to be populated, so it may be treated as a three-component system. For two- and three-component systems, (5-2-2) and (5-2-15) may be solved analytically as indicated above. However, for heavier atoms more charge states are important. The general differential equation for an n-component system is

$$\frac{dF_i}{d\pi} = \sum_{\substack{j=1 \\ j \neq i}}^{n} (\sigma_{ji} F_j - \sigma_{ij} F_i) \tag{5-2-21}$$

subject to the condition

$$\sum_{i}^{n} F_i = 1. \tag{5-2-22}$$

The analytical solutions rapidly become too complicated to use conveniently as n increases. However, the simultaneous differential equations can be solved numerically. An example of the results is shown in Fig. 5-2-2 for 15-MeV ^{127}I incident on O_2 with the iodine initially mostly in charge state $+11$ but with fractions of the initial charge states $+9$, $+10$, and $+12$ indicated by the arrows at the left of the graph. The distribution of equilibrium charge states depends on the projectile energy; the higher the energy, the higher the average charge state, as shown in Fig. 5-2-3. The distributions themselves for the case of boron projectiles on carbon foils are shown in Fig. 5-2-4. The dashed lines represent the theoretical calculation by Zaidins (1968). The distribution over charge states is usually Gaussian with a width and mean energy that are functions of the projectile atomic number Z_1 and the projectile velocity (Shima et al., 1989). The width shows an oscillatory behavior when plotted versus the nuclear charge Z_1 of the projectile. These oscillations are related to the shell structure.

Figure 5-2-5 illustrates an important difference between the charge-state distributions resulting from the passage of ions through gases and solids. In these measurements, 110-MeV ^{127}I ions are passed through two different targets, one solid and one a gas, each with a thickness sufficient to reach an equilibrium charge distribution (Datz et al., 1971). The distribution from the solid is centered at a considerably higher charge state. This "density" effect is attributed to the fact that the time between collisions is much shorter in the case of the solid. This increases the population of excited states, which tends to increase the loss cross section but decrease the capture probability.

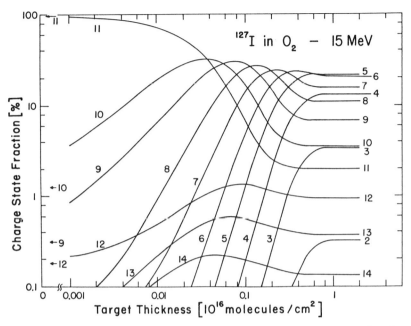

Figure 5-2-2. Charge-state distribution for 15-MeV iodine ions in oxygen as a function of target thickness. The incident distribution is shown by the arrows at the left. [From Betz (1972).]

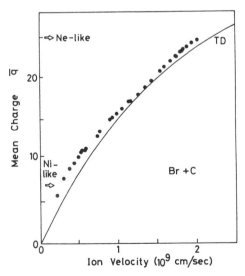

Figure 5-2-3. Equilibrium mean charges of bromine ions after passage through a carbon foil. Solid curve indicates an empirical relation given by To and Drouin (1976). [From Shima et al. (1989).]

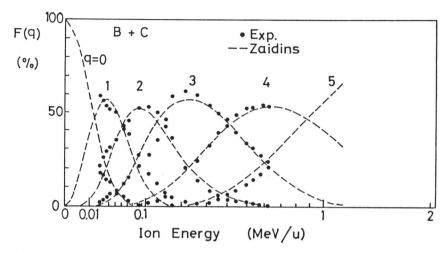

Figure 5-2-4. Equilibrium charge fractions for bromine in carbon as a function of projectile energy. Points, experimental data of Wittkower and Betz (1973) and Shima et al. (1986, 1989); dashed line, theoretical results of Zaidins (1968). [From Shima et al. (1991).]

D. Effects of Charge Transfer on Beam Experiments

In a common method of measuring collision cross sections, such as those for excitation or ionization, a beam is passed through a target and the signal from a product, such as photons, electrons, or ions, resulting from the collisions in a section of beam path are counted. The beam is typically caught in a Faraday cup or other detector and the ratio of the signal current (or counts) to the beam detector current (or counts) is used to compute the cross section. This technique requires the total thickness of the target to be small enough that the beam is not appreciably altered in energy, charge, direction, or intensity as it passes through the target. If there is a significant alteration of the beam, a correction must be made to achieve accurate cross-section data.

One of the more important alterations in the beam is a change of charge due to the capture or loss of electrons. For example, as an ion beam traverses a target gas, it is partially neutralized by capture. This will cause two errors. First, there will be a mixed beam of neutrals and ions at the target region and the neutrals will, in general, have a different cross section for producing the measured signal than ions. Second, if a Faraday cup is used, only the charged component will be read, giving a false measure of the ion current at the target and no measure of the neutral current. Corrections can be made by using the equations for charge transfer, provided that the necessary cross sections are known and the pertinent thicknesses of gas along the beam path can be measured or calculated. It is necessary, for example, to know the pressure and path length all the way back to the point where the beam is purified, usually at an analyzing magnet. Charge

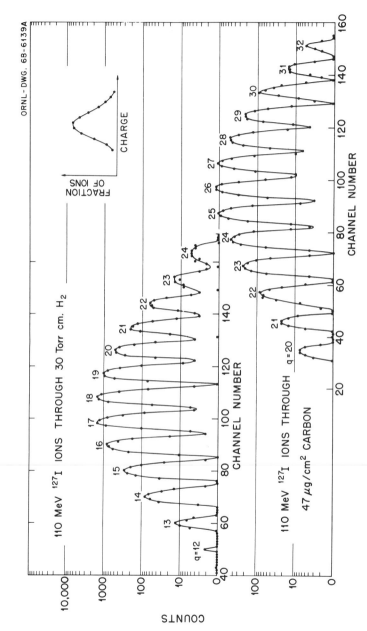

Figure 5-2-5. Charge-state distributions for 110-MeV ^{127}I ions traversing a gaseous (hydrogen) or a solid (carbon) target. [From Datz et al. (1971).]

transfer in a long beam pipe leading to the target can be important, as well as the charge transfer taking place in the target chamber itself.

In the following derivation we assume that the type of gas is the same along the entire beam path. If it is not, appropriate changes can be made by breaking up the path into segments and applying the cross sections for the particular type of gas in each. If, as is usual, the density varies from point to point, the thickness π of the gas is defined as $\int n(x)\,dx$, where $n(x)$ is the particle density as a function of distance x along the beam path. We also assume a two-component system and for concreteness, use the example of a pure incident ion beam. The results can easily be modified for any other two-component system and can serve as a basis for an approximation for a system of more components.

From (5-2-11) and (5-2-12) the current of ions and the equivalent current of neutrals are in general

$$I^+ = \frac{I_i}{\sigma}(\sigma_{10}e^{-\pi\sigma} + \sigma_{01}) \tag{5-2-23}$$

and

$$I^0 = \frac{I_i\sigma_{10}}{\sigma}(1 - e^{-\pi\sigma}), \tag{5-2-24}$$

where I_i is the initial beam current and where $\sigma = \sigma_{10} + \sigma_{01}$. Let π_1, π_2, and π_3 be the target thicknesses from the starting point (where the beam is pure) to the first point of the viewed region, to the last point of that region, and to the Faraday cup or other beam detector, respectively. Then the current to the Faraday cup is given by

$$I_F = I_i\left[1 - \frac{\sigma_{10}}{\sigma}(1 - e^{-\pi_3\sigma})\right]. \tag{5-2-25}$$

The signal viewed between π_1 and π_2 is

$$I_s = \frac{I_i}{\sigma}\left[\sigma^+\int_{\pi_1}^{\pi_2}(\sigma_{10}e^{-\pi\sigma} + \sigma_{01})\,d\pi + \sigma^0\sigma_{10}\int_{\pi_1}^{\pi_2}(1 - e^{-\pi\sigma})\,d\pi\right], \tag{5-2-26}$$

where σ^+ and σ^0 are the cross sections for producing the signal by ions and by neutrals, respectively. This yields the result

$$\frac{I_s\sigma}{I_i\sigma_{10}} = \sigma^+\left[\frac{\sigma_{01}}{\sigma_{10}}(\pi_2 - \pi_1) + \frac{e^{-\pi_1\sigma} - e^{-\pi_2\sigma}}{\sigma}\right]$$

$$+ \sigma^0\left[\pi_2 - \pi_1 - \frac{e^{-\pi_1\sigma} - e^{-\pi_2\sigma}}{\sigma}\right], \tag{5-2-27}$$

which is the exact solution. However, it is useful only when the signal current is referred to the input current I_i. Since that current is difficult to measure, it is more usual for the signal to be referred to the Faraday cup current I_F. In that case, (5-2-25) may be used to eliminate the unknown I_i.

If the thicknesses are not too great, we can simplify the equations by expanding the exponentials. To second order,

$$e^{-\pi_1\sigma} - e^{-\pi_2\sigma} = \sigma\,\Delta\pi(1 - \pi_{av}\sigma), \tag{5-2-28}$$

where $\pi_{av} = (\pi_1 + \pi_2)/2$ and $\Delta\pi = \pi_2 - \pi_1$. Using this, (5-2-27) reduces to

$$\frac{I_s}{\Delta\pi\,I_i} = \sigma^+(1 - \sigma_{10}\pi_{av}) + \sigma_{10}\pi_{av}\sigma^0. \tag{5-2-29}$$

If we now expand the exponential in (5-2-25) and use the binomial theorem,

$$\frac{I_s}{\Delta\pi\,I_F} = \sigma^+[1 + \sigma_{10}(\pi_3 - \pi_{av})] + \sigma_{10}\sigma^0\pi_{av}. \tag{5-2-30}$$

Finally, taking the uncorrected cross section to be $\sigma_u^+ = I_s/\Delta\pi\,I_F$, the corrected cross section becomes

$$\sigma^+ = \sigma_u^+\left[1 - \sigma_{10}(\pi_3 - \pi_{av}) + \pi_{av}\sigma_{10}\frac{\sigma^0}{\sigma^+}\right]. \tag{5-2-31}$$

This equation may be used to correct the measured cross section for partial neutralization of the beam.

E. Cross-Section Relations

Because of the conservation of charge there are important relationships among the various cross sections that can be used to evaluate or adjust data. If we let σ_+ and σ_- be the cross sections for gross production of positive and negative charge, respectively, σ_k^c be the cross section for capture of k electrons, and σ_j^l be the cross section for the loss of j electrons, then assuming a neutral target,

$$\sigma_+ - \sigma_- = \sum_k k\sigma_k^c - \sum_j j\sigma_j^l, \tag{5-2-32}$$

where the sums are over all possible values of j and k. For example, for incident H^+,

$$\sigma_+ - \sigma_- = \sigma_{10} + 2\sigma_{1-1}, \tag{5-2-33}$$

for incident H^0,

$$\sigma_+ - \sigma_- = \sigma_{0-1} - \sigma_{01}, \qquad (5\text{-}2\text{-}34)$$

for incident He^+,

$$\sigma_+ - \sigma_- = \sigma_{10} + 2\sigma_{1-1} - \sigma_{12}, \qquad (5\text{-}2\text{-}35)$$

and for incident He^{2+},

$$\sigma_+ - \sigma_- = \sigma_{21} + 2\sigma_{20} + 3\sigma_{2-1}. \qquad (5\text{-}2\text{-}36)$$

Some cross sections, such as σ_{2-1} and σ_{1-1}, are small enough that for many purposes they can be neglected, thus simplifying the equations. Rudd et al. (1985a,b) utilized these relations in their measurements of capture cross sections for He^+ and He^{2+}. Since they also measured σ_+ and σ_-, they had sufficient data to make a weighted least-squares adjustment of the data, which made their results consistent with the equations above. This is a well-known procedure used, for example, in adjusting interrelated values of fundamental constants. Some results of their measurements of σ_{21} are shown in Fig. 5-2-6 compared with data of other investigators.

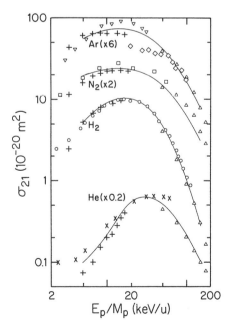

Figure 5-2-6. Values of σ_{21} for He^{2+} incident on various targets. $\diamondsuit$, Baragiola and Nemirovsky (1973); $\times$ Berkner et al. (1968); $\square$, Stearns et al. (1968); ∇, Bayfield and Khayralla (1975); $\triangle$, Pivovar et al. (1961, 1962); $\bigcirc$ and $+$, Shah and Gilbody (1974, 1978); solid lines, Rudd et al. (1985b). [From Rudd et al. (1985b).]

5-3.   MEASUREMENT OF CHARGE-TRANSFER CROSS SECTIONS

A number of different ways of measuring the charge-transfer cross sections are suggested by the equations in Section 5-2, which relate the cross sections to the fractions of various charge states under different conditions. Several of the methods used will be described followed by a discussion of the systematic dependence of the measured cross sections on various parameters.

A.   Methods of Measuring Cross Sections

1.   *Growth Method*

In the schematic diagram of the apparatus (Fig. 5-3-1), a pure incident beam in charge state i passes through the target gas at low pressure. If the length of the gas cell is l, the thickness of the target is $\pi = nl$ where n is the number density of the target. After leaving the gas cell, a magnetic or electric field deflects the components of the beam in directions determined by the final charge states. As gas is admitted to the cell, the rate of increase in the currents to the various detectors are read. If the target is thin enough, the first term of (5-2-20) suffices to determine the cross section from the slope of the growth curve. Actually, a single measurement of F_j and π would suffice provided that it is known that the operating point is on the linear part of the growth curve. The growth method has a number of advantages over most of the other methods. It is not only a relatively easy way to determine cross sections, but its sensitivity allows the measurement of very small cross sections. Furthermore, cross sections are uniquely determined by this method even if many charge states are involved. However, it requires a pure incident beam.

An example of a measurement making use of this method is Gealy and Van Zyl's (1987) determination of σ_{10} for $H^+ + H$ and $H^+ + H_2$ and of σ_{-10} for $H^- + H$ and $H^- + H_2$. The apparatus, shown in Fig. 5-3-2, provides for measurement of the beam in all three possible charge states. The projectile ion beam, at energies from 0.063 to 2.0 keV, came from a duoplasmatron source. The pair of deflection plates just before the target chamber was needed in a different

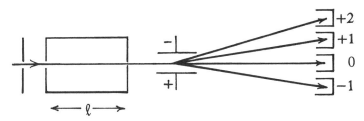

Figure 5-3-1. Schematic diagram of apparatus used in the growth method of measuring charge-transfer cross sections.

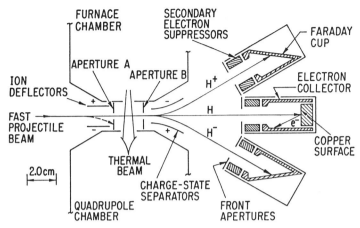

Figure 5-3-2. Apparatus used to measure capture cross sections by the growth method for $H^+ + H, H_2$ collisions at low energies. [From Gealy and Van Zyl (1987).]

experiment to remove all remaining ions when using a fast neutral beam. An oven, not shown, provided the thermal beam of atomic hydrogen used as the target. The fast neutral component of the beam was detected by a secondary emission detector while the charged components were collected in Faraday cups.

2. *Equilibrium Method*

In the equilibrium method an apparatus similar to that in Fig. 5-3-1 is used except that higher pressures and/or longer paths are needed than in the growth method. In this case, either a pure beam or one of mixed charge states may be used. The thickness of the target is increased until the composition of the beam emerging from the gas cell becomes stable (i.e., does not change with a further increase in gas pressure). The beam is then said to be equilibrated and (5-2-8) may be used to determine the ratio σ_{10}/σ_{01}. Since this method provides only ratios, it must be used with other methods to obtain the cross sections themselves. An apparatus from Allison (1958) (Fig. 5-3-3) utilizes a magnetic field to separate the beam particle charge states.

3. *Condenser Method*

Referring to Fig. 5-3-4, a pure beam enters the target gas, producing secondary products (ions and electrons) along its path. A transverse electric field supplied by pairs of plane parallel plates is small enough that it does not appreciably affect the beam trajectory but large enough to sweep out the slow ions and electrons produced along the beam path in the target. The currents I_+ and I_- to one pair of plates, suitably guarded by similar plates on either side, are measured along with the target thickness $\pi = nl$, where l is the length along the beam

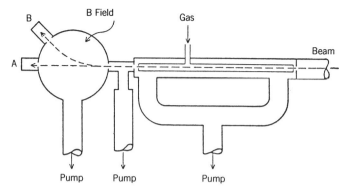

Figure 5-3-3. Apparatus for measuring charge-transfer cross sections by the equilibrium method. The neutral beam is detected at A and the charged components are directed to B by varying the magnetic field. [From Allison (1958).]

direction of the central collecting plates. The cross sections for slow positive and negative charge production are then obtained from the equations

$$I_\pm = I_0 \sigma_{\pm} \pi. \tag{5-3-1}$$

Then assuming that the contribution to σ_- by negative ions is negligible, (5-2-33) can be used to obtain σ_{10} for proton impact. Equation (5-2-32) holds in the general case, yielding a combination of cross sections.

This method is subject to numerous spurious effects that must be eliminated or corrected for. For example, electrons will be liberated from the ion collection plate when it is struck by ions or by ultraviolet photons. In either case, the resulting secondary electrons will be attracted to the electron collection plate, resulting in spurious currents to both plates. A suitably biased grid in front of the ion plate is usually used to suppress these currents, although magnetic fields

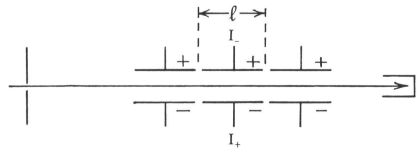

Figure 5-3-4. Schematic diagram of apparatus used in the condenser method of measuring charge-transfer cross sections.

may also be used. An extensive discussion of these and other precautions is given by Rudd et al. (1983) for their apparatus (Fig. 5-3-5).

The condenser method is used primarily at low energies, where charge transfer is often the dominant process. At high energies, ionization becomes the most likely process, and then to use (5-2-33) one must subtract two nearly equal cross sections, a practice that can lead to large errors. The condenser system cannot be used at very low energies either, since the transverse field may cause too large a deflection of the beam.

4. *Mass Spectrometer Method*

This is a variation of the condenser method that allows identification of the mass/charge ratio of the secondary products, which is particularly useful when dissociation is involved or when it is important to know the products and identify the elementary processes taking place.

The condenser apparatus in Fig. 5-3-4 is modified by making a hole in the ion collection plate to allow a sample of the secondary ions to emerge into a mass spectrometer. A difficulty arises because the ions produced by dissociation of molecular targets generally have energies of several electron volts rather than the millielectron volt recoil energies typically resulting from nondissociating collisions. The much larger fields required for high collection efficiency of these fast recoils will deflect the beam, possibly causing it to miss the Faraday cup as described in Section 4-13.

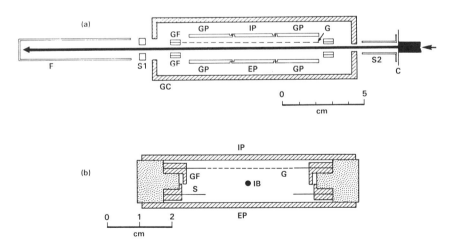

Figure 5-3-5. Condenser-type apparatus: (*a*) side view; (*b*) end view. IB, ion beam; C, collimator; S1 and S2, suppressors; IP and EP, ion and electron collecting plates; GP, guard plates; G, grid to suppress secondary electrons; GF, grid frame; S, shield. [From Rudd et al. (1983).]

5. *Beam Attenuation Method*

There are several variations of this method, each involving a measurement of the decrease of the fraction of a given charge in a beam as it passes through the target gas. Starting with no target gas (i.e., $\pi = 0$), the current to the detector is I_0. After gas is added, the target thickness π and the new current I are measured. Then

$$I = I_0 e^{-\sigma \pi} \tag{5-3-2}$$

where σ represents the sum of all cross sections that change the charge state of the beam from its original state. For example, for a He^+ beam, $\sigma = \sigma_{1-1} + \sigma_{10} + \sigma_{12}$, while for H^0, $\sigma = \sigma_{0-1} + \sigma_{01}$. This method yields only sums of cross sections rather than individual ones for a system of three or more components. Only for a two-component system is an individual cross section measured directly.

In one version, the entire target region is placed in a magnetic or electric field. The detector is positioned such that the desired incident beam strikes it with no target gas present. When gas is added to the chamber, only those beam particles that suffer no charge-changing collisions continue to reach the detector. Any change of charge along the path results in a different deflection and thus a loss of beam to the detector. Stray electric or magnetic fields have no effect on the measurement since it is only the change in the current due to the admission of the target gas that is important. If the detector is in the gas cell, care must be taken to ensure that the presence of the target gas does not change its detection efficiency. In addition, the geometry of the beam and the detector and the target density must be such that no appreciable fraction of beam particles is elastically scattered out of the beam when gas is admitted. The latter precaution is especially important for low energies.

Stier and Barnett (1956) and Barnett and Reynolds (1958) used the attenuation method combined with the equilibrium method to measure the capture and loss cross sections for hydrogen atoms in various gases. The apparatus shown in Fig. 5-3-6 was used for both measurements. A proton beam traversed a neutralizer gas and the remaining positive ions were removed from the beam by deflection plates D_1 so that a neutral beam entered the target gas cell. The transverse field provided by deflection plate system D_2 in the gas cell removed any ions formed by electron loss and the detector read the number of remaining neutrals as a function of target pressure. Making the assumption that the cross section σ_{0-1} is negligible, σ_{01} could then be determined using (5-2-13) by dropping the second-order term. Since the ions had to travel a finite distance to be sufficiently deflected by the transverse field, the effective path length was somewhat less than the geometric length. By applying the field to successive segments of the deflection system, it was possible to calculate the effective length, which was approximately 5% less than the geometric length.

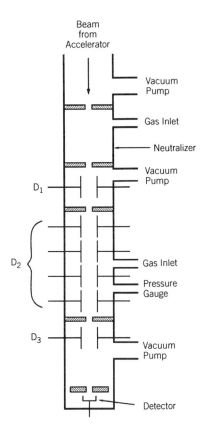

Figure 5-3-6. Schematic diagram of apparatus used in the beam attenuation and equilibrium methods of measuring charge-transfer cross sections. [Adapted from Barnett and Reynolds (1958).]

Then the deflection plates D_2 were turned off, the target gas pressure increased to obtain an equilibrated beam, and the deflection plates D_3 activated to measure the equilibrium charge distribution. This yielded the cross-section ratio by (5-2-8). Combined with the previous measurement of σ_{01}, the capture cross section σ_{10} could also be determined. Incident beams of neutrals and protons were both used to demonstrate that the equilibrium charge fraction was independent of the incident charge.

In another version of the attenuation method, the target is field free and the beam components are separated after the beam leaves the target. In this case a pure incident beam is needed. In the $H^0 + Na$ measurements of Howald et al. (1984), for example, they combined the attenuation and growth methods in one apparatus. After the charge components of the beam emerging from the target were separated by a magnetic field, the attenuation method was used for the neutral fraction of the beam and the growth method for the positive and negative components.

In a measurement of the charge-changing cross section for 400 to 2000-eV He^+ + He collisions, Nagy et al. (1969) determined the attenuated beam by first subtracting the signal due to the neutral beam component from the unattenuated beam signal. The neutral beam was found by putting a large enough retarding potential on a grid before the detector to eliminate the ion component.

6. *Retardation Method*

This method was devised to measure the cross section for double charge transfer for very low energy ions for which elastic scattering is a major problem. In particular, the method was used by Kozlov and Rozhkov (1962) to measure σ_{1-1} for protons of 0.5- to 5-keV energy.

In their apparatus shown schematically in Fig. 5-3-7, the beam of energy E enters the target gas through the small aperture A. In the path between A and the large Faraday cup F are three grids. G_1 is grounded while G_2 is held at a positive potential V_1 such that $eV_1 > E$, where e is the electronic charge. This provides a potential hill that prevents the positive component of the beam from reaching the cup but allows the negative component to pass. Secondary electrons formed at the grids are suppressed by a magnetic field, while a negative potential $-V_3$ on G_3 prevents any negative ions emitted from G_1 from entering the cup. This potential must be such that $E > eV_3 > E_{max}$, where E_{max} is the maximum energy with which a negative ion can be ejected from the grid.

7. *Merging Beam Method*

As the energy of an ion beam is reduced, the space-charge repulsion of the beam causes it to spread out, ultimately limiting the usefulness of any of the previous methods. To overcome that difficulty, a technique was developed that makes it possible to measure cross sections at energies as low as a fraction of an electron volt. This was done, for example, by Trujillo et al. (1966) and Belyaev et al. (1967), who used an arrangement in which two relatively high energy beams were merged so as to travel together in the same direction. By making the two

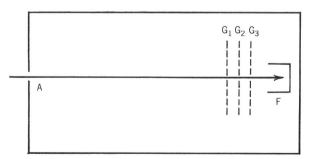

Figure 5-3-7. Schematic diagram of apparatus used in the retardation method of measuring charge-transfer cross sections.

beam energies nearly equal, the relative velocity can be made very small. The interaction energy W is related to the two energies E_1 and E_2 by the equation

$$W = \mu \left[\left(\frac{E_2}{m_2} \right)^{1/2} - \left(\frac{E_1}{m_1} \right)^{1/2} \right]^2, \qquad (5\text{-}3\text{-}3)$$

where μ is the reduced mass and m_1 and m_2 are the masses of the two interacting particles. Because of the subtraction, the uncertainty in the relative beam energy is compressed and is much smaller than the uncertainties in the energies of the individual beams. See Section 7-1-C of the companion volume.

The apparatus of Trujillo et al. (Fig. 5-3-8) was used to measure resonant charge transfer of argon. They used primary energies of 2000 and 3465 eV to obtain an interaction energy of 100 eV. A difficulty of the method, however, is that the low densities of particles in typical beams results in a very small signal. Synchronous detection may be needed to retrieve the signal from the background.

B. Systematics of Charge-Transfer Cross Sections

Figure 5-2-1 is typical of the general trend of many charge-transfer cross sections with incident particle energy. The cross sections are generally small at low energies since the electrons have time to adjust adiabatically to the changing potential as the collision proceeds and therefore the probability of a transition is small. There are, of course, exceptions to this behavior due to electron orbital promotion and to curve crossing in the temporary molecule formed in the collision. The probability of a charge transfer increases with increasing energy, but as the velocity of the projectile is increased, the time of the interaction is reduced. This eventually results in a lower probability again at high energies. Furthermore, unlike excitation, in electron capture the electron must change its momentum when picked up by the projectile. This makes capture less and less likely as the projectile velocity increases.

We have already noted from Fig. 5-2-1 that the cross sections σ_{1-1} and σ_{0-1} for capture processes in which the projectile becomes negatively charged are relatively quite small and fall off rapidly at high energies. The cross sections σ_{-10} and σ_{-11} for the loss of one or two electrons from a projectile initially charged negatively are both fairly large and are similar in their energy dependence. Similarly, the cross sections σ_{10} and σ_{1-1} for capture of one and two electrons have similar energy dependences, although the probability of capture of two electrons is about three orders of magnitude smaller than for capture of a single electron.

When two identical nuclei interact, there is no internal energy difference that depends on which one of the pair the electron is finally attached to. Since in this symmetric or resonant case the transition requires no energy to be supplied from the kinetic energy of the projectile, it can take place at very low incident energies and the cross section remains large even at low energies. Figure 5-3-9 compares

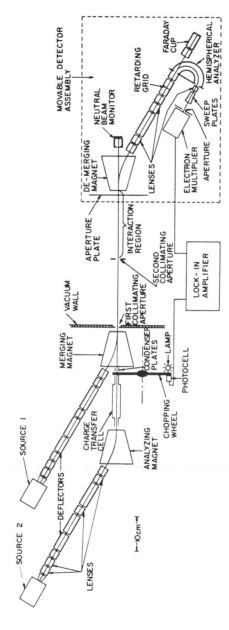

Figure 5-3-8. Diagram of a merging beam apparatus for studying ion–neutral reactions. The beam from source 2 is neutralized in the charge transfer cell before joining the ion beam from source 1 in the merging magnet. The beams travel together until they reach the demerging magnet. [From Trujillo et al. (1966).]

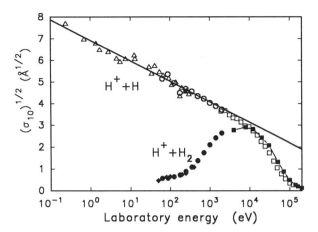

Figure 5-3-9. Capture cross sections for $H^+ + H$ (open symbols) and $H^+ + H_2$ (filled symbols). $\bigcirc$, Gealy and Van Zyl (1987); $\triangle$, Newman et al. (1982); $\blacklozenge$, Cramer (1961); $\blacksquare$, Stier and Barnett (1956); $\square$, McClure (1964); line, theory of Dalgarno and Yadav (1953). [Adapted from Gealy and Van Zyl (1987).]

the cross sections for the resonant case, $H^+ + H$, to the nonresonant case $H^+ + H_2$, where the large difference in behavior at low energies is evident.

Figure 5-3-10 shows the capture cross section σ_{10} for protons on various rare gas atoms. As expected, the heavier targets generally have larger cross sections since there are more electrons available to be captured. The high-energy behavior is similar for all four targets, except that the cross sections for the two

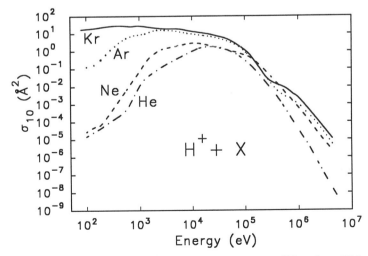

Figure 5-3-10. Capture cross sections for protons on rare gases. [Data from Nakai et al. (1983b).]

heavier atoms exhibit a high-energy shoulder attributable to additional capture from inner shells. At low energies, the progressive increase in cross section with heavier targets is at least partly attributable to the fact that the ionization potentials of He, Ne, Ar, and Kr, which are 24.6, 21.6, 15.8, and 14.0 eV, respectively, approach that of H (which is 13.6 eV) more and more closely and therefore are closer to resonance.

Figure 5-3-11 illustrates the energy dependence of the capture and loss cross sections for fast, heavy, highly charged ions. In this work by Clark et al. (1986), those cross sections were measured for 2.5- to 200-MeV S^{13+} + He collisions. While the electron-loss cross section varies only by a factor of 2 over that energy range, the capture cross section changes by nearly six orders of magnitude. The authors find a reasonably good description of the loss cross section in the plane-wave Born approximation and have used an empirical equation to fit the capture cross sections. The empirical equation, given by Schlachter et al. (1983),

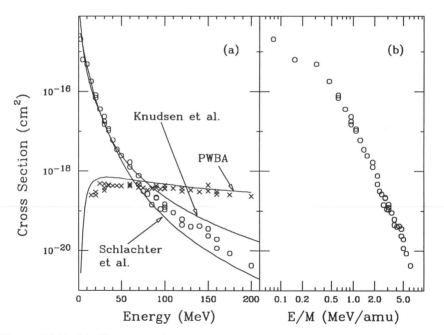

Figure 5-3-11. (*a*) Charge-transfer cross sections for 2.5- to 200-MeV S^{13+} + He collisions. ○, Electron capture; ×, electron loss; line labeled Schlachter et al., empirical scaling equation of Schlachter et al. (1983); line labeled Knudsen et al., classical scaling equation of Knudsen et al. (1981); line labeled PWBA, plane-wave Born approximation results from Choi et al. (1973). (*b*) Electron capture data of (*a*) plotted on a log-log scale. [From Clark et al. (1986).]

is a scaling rule that is useful for capture by high-energy, highly charged projectiles. The equation is given in terms of the reduced parameters

$$\tilde{\sigma} = \frac{\sigma Z_2^{1.8}}{q^{0.5}} \quad \text{and} \quad \tilde{E} = \frac{E}{Z_2^{1.25} q^{0.7}}, \tag{5-3-4}$$

where σ is the capture cross section, q the projectile charge state, E the projectile energy per nucleon in keV/u, and Z_2 the atomic number of the target. The equation is

$$\tilde{\sigma} = \frac{1.1 \times 10^{-8}}{\tilde{E}^{4.8}} [1 - \exp(-0.037\tilde{E}^{2.2})][1 - \exp(-2.44 \times 10^{-5}\tilde{E}^{2.6})] \tag{5-3-5}$$

The scaling law is found to fit data from a wide variety of targets and energies subject to the conditions $\tilde{E} \geqslant 10$ and $q \geqslant 3$. The authors note that this restriction on charge state can be removed if q is replaced by $q - 4$ in (5-3-4).

Toburen et al. (1968) and others have successfully used the *Bragg additivity* rule to equate the cross section for a molecule to the sum of the cross sections for its individual atoms. This procedure is based on the idea that at least for high-energy incident particles, the molecular forces are negligible and the collision with the molecule can be treated as a succession of collisions with the separate atoms. However, Tuan and Gerjuoy (1960) showed theoretically that such a rule cannot be exactly correct even in the simple cases of electron capture by protons from simple diatomic molecules. Moreover, Wittkower and Betz (1971) found that no such simple rule could relate charge-transfer cross sections involving 12-MeV I^{5+} incident on various targets. The rule is, of course, only an approximation that holds reasonably well in some cases, but cannot be relied upon to hold in general [see also Betz (1972)].

5-4. THEORETICAL ASPECTS OF CHARGE TRANSFER

Charge transfer is a difficult collision process to treat theoretically and despite its importance and the intensive studies made of it, reliable predictions of cross sections for each process at any given energy and pair of collision partners are not yet possible, although considerable progress has been made. Since capture of a free electron by a charged projectile colliding with it would violate the laws of conservation of energy and momentum, binary collision models are not suitable and any theoretical account of charge transfer must take account of the presence of the residual ion. Another consideration is that during the transition from one atom to the other, the electron must undergo a sudden change in momentum. In some theoretical models the lack of the proper translational factors in the electron wave function to account for this motion limits their usefulness. This change of momentum is minimized when the projectile velocity is equal to the orbital electron velocity in the target, and in most cases this is also where the maximum probability of a charge-transfer transition occurs.

We first discuss the application of the Massey adiabatic criterion to charge transfer. The quantum mechanical theories treated next are divided into high- and low-energy approximations. In the former the interaction potential is taken as the perturbation that causes the transitions, while in the latter molecular effects predominate and the relative motion of the two atomic particles is the perturbation. The section closes with a consideration of several classical treatments of electron transfer.

The theory of charge transfer has been treated in many papers. Reviews, some comprehensive and some on specific aspects, have been given by Bates (1962), Bates and McCarroll (1962), Mott and Massey (1965), Mapleton (1972), Tawara and Russek (1973), Shakeshaft and Spruch (1979), Shakeshaft (1982), Greenland (1982), Macek (1984), Janev and Winter (1985), Chibisov and Janev (1988), Kimura and Lane (1990), Briggs and Macek (1991), and Bransden and McDowell (1992).

A. Massey Adiabatic Criterion

The Massey adiabatic criterion, discussed in Section 2-9 in connection with excitation processes, can also be applied to charge transfer. The transfer of an electron from an atom 2 to an ion 1 involves an internal energy change $\Delta E = \chi_1 - \chi_2$, where χ_1 and χ_2 are the binding energies of the electron in the corresponding atoms. If $\Delta E > 0$, the reaction is exoergic and can proceed without the need for an external supply of energy. For endoergic reactions, for which $\Delta E < 0$, the needed energy usually has to be taken from the kinetic energy of the projectile, thus putting a limitation on the primary energy range for which the process can proceed. For charge transfer we can identify ΔE with the energy defect in the adiabatic criterion, which states that the projectile velocity for the maximum in the cross section is given by

$$v_1 \approx \frac{a|\Delta E|}{h}, \tag{5-4-1}$$

as in (2-9-1). For symmetric collision systems (e.g., $He^+ + He$) the value of the energy defect is zero and such a process is known as resonant charge transfer. In some nonsymmetric collision systems there are cases of accidental resonance for which ΔE is not zero but is very small. Exceptions to the Massey criterion are usually due to electron orbital promotion and curve crossing in the temporary molecule formed in the collision.

Bohr (1948) [see also Betz (1972)] described a related criterion for the average charge state of a fast, heavy ion passing through a gas. According to this criterion, electrons with smaller orbital velocities than the velocity of the ion are lost in collisions, while those electrons with larger velocities are able to adjust adiabatically and are thus retained by the ion. Arguing from this basic idea,

Bohr derived a formula for the average charge q_{av} in terms of the nuclear charge Z_1 of the ion:

$$\frac{q_{av}}{Z_1} = \frac{v_1}{v_0 Z_1^{2/3}}, \qquad 1 < \frac{v_1}{v_0} < Z_1^{2/3}, \qquad (5\text{-}4\text{-}2)$$

where v_0 is the Bohr velocity. The result is only valid if q_{av} is less than, but not much less than, Z_1.

B. High-Energy Quantum Mechanical Methods

Oppenheimer (1928) and Brinkman and Kramers (1930) were the first to adapt the Born approximation to calculations of electron capture cross sections. In this treatment the wave functions are expanded in terms of the atomic orbitals (AOs) of the two collision partners. Such an expansion cannot describe the electron density correctly when the two nuclei are close together since that requires molecular orbitals (MOs). Therefore, this method cannot be expected to be accurate except under two conditions: (1) the relative velocity must be so high that MOs do not have time to form, and (2) the impact parameter must not be too small.

Electron capture is an example of a rearrangement collision in which an electron, initially in state n in atom 2, makes a transition to state n' in atom 1. If μ is the reduced mass of the system, the Born approximation yields the result

$$\sigma = \frac{8\pi^3 \mu^2}{h^4} \int_0^\pi \left| \iint V(r_e, \rho) \phi_{n'}(r_e) \psi_n(r_e) \right.$$

$$\left. \times \exp[ik(n_0 r - n\rho)]\, dr_e\, d\rho \right|^2 \sin\theta\, d\theta, \qquad (5\text{-}4\text{-}3)$$

where $V(r_e, \rho)$ is the interaction energy between the electron and nucleus A, $\psi_n(r_e)$ is the wave function of the electron in state n in atom 1, $\phi_{n'}(r_e)$ is the wave function of the same electron in state n' in atom 2, ρ is the displacement of 2 relative to 1, and r_e is the coordinate of the electron. This equation assumes that the velocity of the projectile is not changed in the collision. In addition, the internuclear potential was omitted from the calculation, as it was not expected to contribute to the capture probability. Replacing the coordinates ρ and r_e by r_1 and r_2, the coordinates of the electron relative to nuclei 1 and 2, it is possible to do the integrals for simple wave functions. Brinkman and Kramers carried out the integration for $1s$ initial and final states, obtaining the result

$$\sigma_{OBK} = \frac{2^{18}\pi a_0^2}{5 v_1^2 (v_1^2 + 4)^5}. \qquad (5\text{-}4\text{-}4)$$

The OBK formulation, then, yields a $1/v_1^{12}$ dependence on velocity at high energies as compared to the $(1/v_1^2)\ln v_1$ dependence for ionization. This is in accord with the well-known fact that while charge transfer is an important process at low energies, it becomes relatively much less important at very high energies. McDowell and Coleman (1970) give a more general version of this equation which holds for capture from the $n = 1$ state of any atom 1 to excited states of atom 2.

The neglect of the nucleus–nucleus interaction, while it would be justified in an exact calculation, leads to an overestimation of the cross section in the approximation. Nevertheless, the OBK approximation does yield the proper dependence on the nuclear charge of the target, the incident energy, and the principal quantum number of the electron in the target and the projectile. Jackson and Schiff (1953) included the nucleus–nucleus interaction in their calculation and obtained for transfer to the ground state the equation

$$\sigma_1 = \sigma_{OBK}\left[\frac{1}{192}\left(127 + \frac{14}{E_1} + \frac{2}{E_1^2}\right) - \frac{\tan^{-1}E_1^{1/2}}{96E_1^{1/2}}\left(83 + \frac{15}{E_1} + \frac{2}{E_1^2}\right)\right.$$
$$\left. + \frac{(\tan^{-1}E_1^{1/2})^2}{96E_1}\left(31 + \frac{8}{E_1} + \frac{1}{E_1^2}\right)\right]. \tag{5-4-5}$$

The correction factor to the OBK indicated in their treatment varies from 0.117 at zero energy to 0.661 at very high energies. When they include this factor and also make allowance for capture into excited states, the comparison to data on capture by protons from hydrogen gas (see Fig. 5-4-1) shows good agreement at

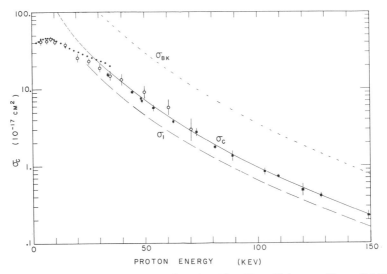

Figure 5-4-1. Electron capture cross sections for $H^+ + H_2$ collisions. $\bullet$, Keene (1949); $\blacksquare$, Ribe (1951); $\bigcirc$, Whittier (1952); dotted line, calculations of Brinkman and Kramers (1930); solid line, calculations for ground-state and excited-state capture by Jackson and Schiff (1953); dashed line, ground state only. [From Jackson and Schiff (1953).]

high energies. Below about 25 keV, however, the theoretical cross section continues to rise while the experimental data fall.

In a charge-transfer transition the electron is influenced by the potentials of both the target nucleus and the projectile. Second Born theory can be applied to the problem, but this treats the two interactions on an equal footing. However, in a highly asymmetric case the stronger potential has the greater influence on the electron's motion. This fact is taken into account in the strong potential Born (SPB) approximation developed by Macek and Alston (1982) to describe the capture of an electron from a hydrogenlike ion of large nuclear charge by a bare ion. Both discrete and continuum levels of the target (which provides the stronger potential) are represented by exact wave functions, thus incorporating the correct spectrum of intermediate states.

Both the OBK and SPB formulations have been criticized by Dewangen and Eichler (1985, 1986) on the grounds that in their rigorous form they do not satisfy the proper Coulomb boundary conditions, a situation which they correct in their work.

C.   Low-Energy Quantum Mechanical Methods

A charge-transfer process taking place at low energies generally involves a temporary molecule formed as an intermediate state. A typical process may be written

$$A^+ + B \rightarrow (AB^+)^* \rightarrow A + B^+. \tag{5-4-6}$$

A general analytical molecular orbital (MO) model based on the work of Landau (1932), Zener (1932), and Stückelberg (1932) (LZS) treats transitions in which there is a crossing of two molecular states at a finite internuclear distance R_c. Assuming that the transition takes place only at the crossing point (where the energy is H_{12}), the coupled differential equations may be solved exactly leading to a transition probability as a function of the relative velocity v given by the equation

$$P_{LZS} = 4p(1 - p)\sin^2\eta_{LZS}, \tag{5-4-7}$$

where

$$p = \exp\left(-\frac{\pi H_{12}}{v\alpha}\right) \tag{5-4-8}$$

is the crossing probability with

$$\alpha = \left|\frac{dH_{11}}{dR} - \frac{dH_{22}}{dR}\right| \tag{5-4-9}$$

evaluated at R_c, and where

$$\eta_{LZS} = \int (\Delta\varepsilon^2 + 4H_{12}^2)^{1/2} \, dt \qquad (5\text{-}4\text{-}10)$$

with $\Delta\varepsilon$ being the energy difference between the two diabatic states involved. The derivatives in (5-4-9) are the slopes of the two crossing potential energy curves as a function of internuclear separation. The LZS model may be applied to various types of transitions, including charge transfer, provided that the MO states exhibit a near crossing over a narrow range of internuclear separations.

While the LZS model assumes that the interaction energy is constant, the Demkov (1963) model takes it to be an exponential function of internuclear separation. The LZS and the Demkov models both use stationary orbitals and thus do not take account of the motion of the electron as a result of the relative motion of the nuclei. Since they do not contain electron translation factors, they are expected to work best for low collison velocities. Briggs and Taulbjerg (1975) and others have shown the importance of translational factors. Pfeifer and Garcia (1981) have improved on the Demkov model as applied to inner-shell excitation by including the necessary translational factors. This resulted in a rather simple analytic expression for the charge-transfer probability as a function of the relative velocity of the collision pair,

$$P^0 = \mathrm{sech}^2 \left[\frac{\pi}{2} \frac{\Delta E + v^2/2}{\alpha v} \right] \sin^2 \left[\frac{1}{2} \int_{-\infty}^{+\infty} \Delta E(R) \, dt \right], \qquad (5\text{-}4\text{-}11)$$

where

$$\alpha = \frac{\chi_1^{1/2} + \chi_2^{1/2}}{2^{1/2}} \qquad (5\text{-}4\text{-}12)$$

and where χ_1 and χ_2 are the ionization potentials of the projectile and target, ΔE the energy difference of the separated atom states, and $\Delta E(R)$ the interaction energy as a function of internuclear distance. Yenen et al. (1984) have successfully applied this modified Demkov model to $H^+ + H_2$ collisions.

Single electron capture from highly charged ions in collisions of complex atoms or molecules was treated by Olson and Salop (1976) using an absorbing-sphere model based on the LZS method. A large number of final product channels ensured that there are many crossings in the internuclear distance around a critical value R_c. Assuming unit probability for the reaction within that impact parameter, the cross section is simply

$$\sigma = \pi R_c^2. \qquad (5\text{-}4\text{-}13)$$

Approximate expressions were found for the adiabatic splitting and the difference in slopes of the diabatic potential energy curves at the curve crossings. The value of R_c for a given case is the solution of the equation

$$R_c^2 \exp\left(-\frac{2.648\alpha R_c}{Z_1^{1/2}}\right) = \frac{2.864 \times 10^{-4} Z_1(Z_1 - 1)v_1}{q}, \qquad (5\text{-}4\text{-}14)$$

where q is the Franck–Condon factor (for molecular targets), Z_1 the charge of the projectile, v_1 the incident velocity in atomic units, and

$$\alpha = \left(\frac{\chi_2}{13.6}\right)^{1/2}, \qquad (5\text{-}4\text{-}15)$$

where χ_2 is the ionization potential of the target atom. The authors claim an accuracy of 40% with this model for Z_1 greater than about 10.

In contrast to the classical Coulomb model discussed in the next section, this treatment yields a nearly linear dependence on Z_1, although the models agree in predicting little dependence on v_1 in the low-velocity range.

Kimura and Lin (1985a, b) have developed a unified AO–MO matching method in which they divide the varying internuclear separation into outer regions, which are appropriate before and after the collision, and an inner region that is traversed during the part of the collision when the nuclei are close enough together to justify the use of molecular orbitals to describe the wave function. Atomic orbitals are used in the outer regions and the solutions are matched at the boundaries of the regions, one on the incoming and the other on the outgoing trajectory. In the outer regions the electronic eigenstates are atomic states associated with either the stationary target or the moving projectile and therefore translational factors can be unambiguously assigned [see also Kimura and Lane (1990)].

D. Classical Methods

1. *Simple Classical Model*

Bell (1953) argued that the capture of electrons by fission fragments was primarily into the higher excited states, and for this situation a classical picture should be valid. In this model, as an ion approaches a neutral atom, an electron bound in the atom feels an increasing Coulomb attraction to the ion. When that force exceeds the force binding the electron to the atom, capture takes place. Assuming a Fermi–Thomas model of the gas atom, Bell found the probability of charge transfer by this means as a function of the velocity of the electron and the ion. Beuhler et al. (1979) used a simplified version of Bell's model in which the ion velocity was assumed to be smaller than that of the electron in the target. In this case the cross section is independent of the velocity of the ion. The critical distance of interaction for capture is given by

$$R_c = \frac{Z_2 e^2}{\chi_2}, \tag{5-4-16}$$

where Z_2 is the nuclear charge and χ_2 is the ionization potential of the target atom. The total cross section for capture is then simply

$$\sigma = \pi R_c^2 = \frac{\pi e^4 Z_2^2}{\chi_2^2}. \tag{5-4-17}$$

Beuhler et al. (1979) measured capture cross sections for $Xe^{n+} + Xe$ to various final charge states in the energy range 15 to 1600 keV and observed a relative insensitivity of the cross sections to projectile kinetic energy in accord with the model and also found that the cross sections follow the predicted Z_2^2 dependence. However, the calculated values agreed only roughly with the measured cross sections.

2. Binary Encounter Approximation

In Section 4-4 the binary encounter approximation (BEA) was discussed in its application to ionization. The basic quantity in the BEA is the cross section for energy transfer Q, which is calculated as a function of the initial velocities of the incident and the target particles. Gryzinski (1959, 1965) proposed that the total cross sections for various processes could be obtained by integrating that basic cross section over all energy transfers Q appropriate to the particular process under study. For ionization the range is $\chi_2 \leqslant Q \leqslant E_1$ while for capture the range is taken to be $T + \chi_2 - \chi_1 \leqslant Q \leqslant T + \chi_2 + \chi_1$, where $T = m_e v_1^2/2, m_e$ is the electron mass, v_1 the projectile velocity, E_1 the projectile energy, and χ_1 and χ_2 the binding energies of the electron in the projectile and target, respectively. In each case the cross section is also integrated over the speed distribution of the electron in the target atom. Gryzinski used additional approximations in his development, but Garcia et al. (1968) used the exact expression as derived by Gerjuoy (1966). Their conclusion, based on a comparison to experimental data for capture by protons on rare gases, was that the BEA does not generally provide reliable capture cross sections.

Garcia et al. (1968) go on to analyze some of the difficulties with the BEA. One problem arises when the binding energy χ_1 is less than or equal to χ_2. In that case the integral diverges at some value of incident energy. In addition, Gryzinski's prescription is unsymmetric in the way it treats the incident and target particles, leading to a violation of detailed balance. Attempts by Garcia to modify the BEA to get around these difficulties were only partially successful. However, McGuire (1973) showed that some aspects of inner-shell vacancy production could be explained with the BEA by adding the contributions due to charge transfer and ionization (see Section 5-5-D).

3. *Thomas Double-Scattering Model*

As mentioned earlier, energy and momentum conservation would be violated by a binary collision of an ion with an electron that resulted in capture. The third body that is necessary to absorb excess momentum is usually the residual target ion. Thomas (1927) suggested a classical double-collision mechanism by which the process could take place. In the first collision, the electron in the target atom recoils from the incident ion at 60°, which is the angle for which its speed is equal to v_1, the magnitude of the projectile velocity. If initially situated properly in the atom, the electron recoiling from the first collision makes a second collision with the target nucleus which scatters it without change of velocity into the same direction as the incident ion, thus putting it into a favorable position for capture (see Fig. 5-4-2).

If applied to capture in the limiting case of zero binding energy, the first Born approximation would describe the capture of a free electron, a process forbidden by the conservation laws as just mentioned. The two-collision nature of the Thomas double-collision mechanism corresponds, instead, to the second Born contribution in the quantum treatment. Calculations of this mechanism by the second Born have been reviewed by Shakeshaft and Spruch (1979) and by Briggs and Macek (1991). The latter have applied the impulse approximation to the double collision and obtained the result

$$\sigma = \sigma_{OBK} \left[0.295 + \frac{5\pi v_1}{2^{11}(Z_1 + Z_2)} \right] \tag{5-4-18}$$

for the total capture cross section where σ_{OBK} is the first Born or Oppenheimer–Brinkman–Kramers cross section given in (5-4-4). At high velocities this yields a v_1^{-11} dependence of the cross section as compared to the v_1^{-12} for the OBK. However, the second Born cross section is expected to be important only at high energies and does not become dominant until the energy is above about 40 MeV.

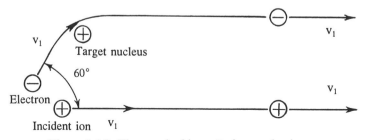

Figure 5-4-2. Thomas double-scattering mechanism.

Unfortunately, at such high energies the capture cross sections are extremely small making experimental studies difficult.

Horsdal-Pedersen et al. (1983) overcame these difficulties by utilizing the fact that the incident ion suffers a small but measurable deflection in its initial collision with the electron. They reported on measurements of capture in $H^+ + He$ collisions at 2.82, 5.42, and 7.40 MeV. The resultant neutral particle was separated magnetically from the more numerous protons and detected as a function of deflection angle. The apparatus is shown in Fig. 5-4-3. A tightly collimated beam of protons passed through the helium target after which the protons were deflected away by a magnetic field leaving the neutrals to strike a bowtie-shaped position-sensitive detector. This yielded a signal proportional to the differential cross section as a function of the deflection angle of the neutrals. The Thomas peak is expected at an angle given by

$$\theta = \frac{m_e}{m_p} \sin 60°, \qquad (5\text{-}4\text{-}19)$$

where m_e and m_p are the electron and projectile masses, respectively. For protons this angle is 0.47 mrad. The angular dependence of the differential cross sections obtained for the three energies is shown in Fig. 5-4-4, where the data points are the experimental values and the lines are theoretical calculations folded into the experimental resolution function.

The plane-wave second Born calculations for protons on helium, shown by the dashed lines, were carried out by Simony et al. (1982). The solid lines represent calculations by Macek and Alston (1982) using the strong-potential Born (SPB) theory with the peaking approximation. The SPB fits the data rather well in the region of the Thomas peak but is too low by a factor of 2 at small angles where most of the contribution to the integrated cross section lies. The second Born overestimates the central maximum. Both theory and experiment indicate a rapid weakening of the Thomas peak with decreasing energy.

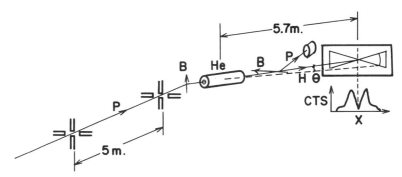

Figure 5-4-3. Apparatus used to detect electron capture by the Thomas double-scattering mechanism. [From Horsdal-Pedersen et al. (1983).]

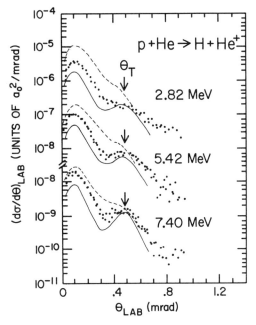

Figure 5-4-4. Angular distribution of cross sections for electron capture in H^+ + He collisions at three energies. Peak labeled θ_T is the Thomas peak. [From Horsdal-Pedersen et al. (1983).]

McGuire et al. (1989) considered the possibility of observing energy-nonconserving aspects of the intermediate states through which a system passes during such a collision.

4. *Classical-Trajectory Monte Carlo Method*

In Section 4-6 the basic approach of the classical-trajectory Monte Carlo (CTMC) method was discussed as a method of calculating ionization cross sections. It has also been applied by Olson and his collaborators at the University of Missouri at Rolla to the calculation of charge-transfer cross sections. Olson et al. (1977), for example, used that method to calculate σ_{21} for He^{2+} + H, H_2 collisions. Hamilton's equations of motion were solved numerically for a large number of trajectories for which the initial conditions, such as impact parameter, were chosen randomly. The hydrogen atom itself was represented by a microcanonical momentum distribution. Individual trajectories were followed until the collision partners were well separated. When the electron(s) were found after the collision to be bound to the He^{2+}, electron capture was deemed to have occurred. From the probabilities of such collisions, the appropriate cross sections were computed. The CTMC method has also been applied to collisions of multiply charged heavy ions [see Meng et al. (1989)].

5-5.   ONE-ELECTRON TRANSFER PROCESSES

If one is interested only in the change in the charge state of the projectile, any resulting excitation of either collision partner can be ignored. However, to obtain a more complete picture of charge transfer and to understand the fundamental physical processes taking place, one must take such excitations into account. Studies of alignment and orientation as a function of scattering angle provide further information (see, e.g., Andersen et al. (1988)]. There is, in fact, a rich variety of possible transfer processes involving various initial and final charge states and/or excited states of both the projectile and target. This has produced a large and rapidly growing literature on this subject, and in this section it will only be possible to make a sampling of the progress that has been made during the past few decades.

Only a single electron takes part in some processes; in others, two or more electrons are involved. Although according to conventional wisdom, the simpler processes are the more likely ones, the rule is often broken and the probability of some of the more complex reactions is actually greater than that for simple transfer. In this section we take up some of the single-electron processes, and in the following section more complex processes involving more than one electron will be treated.

A.   Electron Capture to an Excited State

An incident particle that finds itself in an excited state after capturing an electron will shed its excess energy, making a transition to the ground state by emitting a photon, or if its excitation energy is greater than the ionization potential, by the ejection of an electron through autoionization. In the case of an inner-shell excitation, the emission is in the form of an X-ray photon or an Auger electron.

The search for a method of producing lasers in the extreme ultraviolet or X-ray regions is one of several motivations for work in this area. It was suggested, for example, by Louisell et al. (1975), that since an inverted population could be obtained by electron capture to He^{2+}, which favors the $2p$ over the $1s$ state at moderate primary energies, this process could form the basis for such a laser.

In some early studies, excitation to highly excited states was detected by field ionization. For example, Riviere and Sweetman (1963) used the technique to study electron capture into excited states of hydrogen, and Il'in et al. (1965) and Oparin et al. (1967) studied the production of highly excited hydrogen atoms during electron transfer in proton–metal vapor collisions from 10 to 180 keV. For some thermonuclear fusion work hydrogen atoms in highly excited states are important. Metal vapors, especially magnesium, were found to be much more effective than rare gas atoms in producing H* with principal quantum numbers ranging from 9 to 16. At low energies the main contribution to the process was found to be from the weakly bound outer-shell electrons of the target while inner-shell electrons dominated at higher energies. Berkner et al. (1969) used an optical technique to show that the capture reaction

$H^+ + Mg \rightarrow H\,(n = 6) + Mg^+$ at 15 keV had a cross section over 60 times the corresponding cross section using neon as a target.

An optical method was also used by Pretzer et al. (1963) to measure cross sections for capture into the $2p$ state of hydrogen in 1.5- to 23-keV proton and deuteron collisions with rare gas atoms. The ensuing Lyman-alpha radiation was detected with an oxygen-filtered iodine-filled Geiger counter. Later Jaecks et al. (1965) measured similar cross sections for the production of the $2s$ state of hydrogen by observing the Lyman-alpha radiation from the $2s$ atoms in a quenching field where they were mixed with the $2p$ state. It was believed that only a small fraction of the apparent $2s$ and $2p$ cross sections resulted from cascades from higher excited states, and therefore the emission cross section was essentially equal to the excitation cross section. Figure 5-5-1 shows the results of the $2s$ and $2p$ measurements for argon targets compared to the total capture cross section. The difference in the position of the main maximum for the two cases is consistent with the prediction of the Massey adiabatic criterion. The

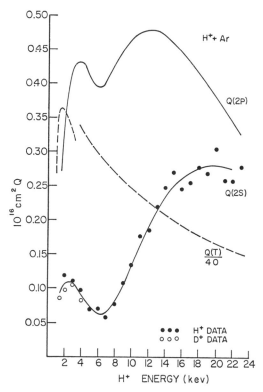

Figure 5-5-1. Capture cross sections for H^+, $D^+ + Ar$ collisions. $Q(T)$, total charge-transfer cross section from Stedeford and Hasted (1955); $Q(2P)$, capture cross section to the $2p$ state from Pretzer et al. (1963); $Q(2S)$, capture cross section to the $2s$ state from Jaecks et al. (1965). [From Jaecks et al. (1965).]

small, low-energy maxima in both the $2s$ and $2p$ curves come at the same energy as the maximum in the total capture cross section and were explained as being due to a direct coupling between the incoming $H^+ + He$ state and the $H + He^+$ final state.

Donnally et al. (1964) and Peterson and Lorents (1969) showed the importance of the accidental near-resonance of excited states in enhancing some cross sections for charge transfer. If the binding energy of an electron in an excited state of the projectile is nearly equal to the ionization potential of the target atom, the energy defect ΔE is very small, making a transition more likely. For example, Peterson and Lorents (1969) found cross sections in excess of $1.5 \times 10^{-14} \, \text{cm}^2$ at 1.5-keV primary energy for the capture of an electron from cesium by a helium ion and attributed the size to near-resonant reactions such as

$$He^+ + Ce \rightarrow He(2^1S) + Ce^+ + 0.08 \, eV$$

and

$$He^+ + Ce \rightarrow He(2^3P) + Ce^+ - 0.27 \, eV.$$

The case of charge transfer when the projectile is in a metastable excited state was studied at a low energy (6 eV center-of-mass energy) by Rothwell et al. (1978). They produced Ar^+ in an electron impact ion source and measured the production of light from two excited $3p^54s$ states of argon and the $2p$ state of hydrogen in $Ar^+ + H_2$ collisions as a function of the energy of the electron in the ion source. The production of the three spectral lines increased roughly in proportion to the number of metastable ions in the beam.

Hippler et al. (1988) measured partial cross sections for capture into states of specific principal quantum number n for $I^{q+} + H_2$ collisions where $q = 12$ to 18 in the range 0.1 to 0.25 MeV/u. Excitation for which $n = 9$ to 13 was measured by detection of light in a vacuum-UV spectrometer corrected by subtracting the contribution due to cascades from higher states. Their results and the earlier measurements by Hvelplund et al. (1983) using Au^{14+} showed large discrepancies with several theoretical models as illustrated in Fig. 5-5-2.

B. Electron Capture to the Continuum

In addition to capture into the ground state and capture into excited states, there is the third possibility, electron capture into the continuum (ECC). This process was long overlooked, perhaps because it was considered to be indistinguishable from direct ionization. However, ECC has its own unique signature: a sharply peaked forward electron emission in the laboratory frame which yields a cusp-shaped peak in the electron energy spectrum at a velocity equal to the projectile velocity. As pointed out in Section 4-8, where it is discussed more thoroughly, ECC is at the same time true ionization and true charge transfer and

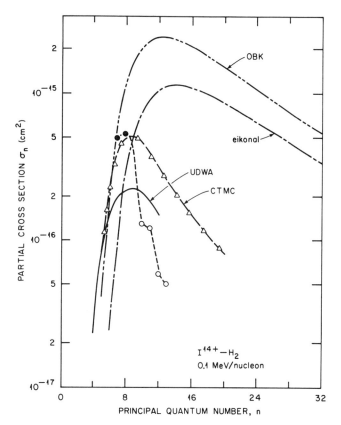

Figure 5-5-2. Cross sections for electron capture to specific states of principal quantum number n for 0.1-MeV/u collisions. ●, $I^{14+} + H_2$ (Hippler et al., 1988); ○, $Au^{14+} + H_2$ (Hvelplund et al., 1983); solid line, unitarized distorted wave calculations (Ruyufuku and Watanabe, 1978); dash-dot line, eikonal calculation (Eichler and Chan, 1979); dash-double dot line, Oppenheimer, Brinkman, Kramers calculation (Eichler and Chan, 1979); △, classical-trajectory Monte Carlo calculation (Olson, 1981). Theoretical calculations made for $Si^{14+} + H$. [From Hippler et al. (1988).]

therefore forms an important link between the two elementary processes. It was first identified as a charge-transfer process by Macek (1970), who described it by making a Neumann expansion of the Fadeev equations. ECC has been the subject of intensive experimental and theoretical work since that time, and along with the related subject of electron loss to the continuum (ELC), it was the subject of an international symposium [see Groeneveld et al. (1984a)].

C. Electron Loss to the Continuum

If incident particles carry electrons into a collision, one or more of those electrons may be released into the continuum. As mentioned in the introduction to the chapter, this process is actually ionization taking place in the frame of reference of the projectile and is also discussed in Chapter 4. But since it entails a change of charge of the incident particle, it is often also considered under charge transfer. Of course, it is also possible for the electron lost from the incident particle to be captured by the target instead of going into the continuum. When it goes into an unbound state it is usually labeled ELC (electron loss to the continuum) in analogy to ECC (electron capture to the continuum.) Reviews that deal with ELC have been written by Breinig et al. (1982), Sellin (1982), Lucas et al. (1984), and Groeneveld et al. (1984b).

ELC was identified independently by two groups, in both cases through studies of electron spectra. Wilson and Toburen (1973) measured spectra of electrons from 0.6- to 1.5-MeV $H_2^+ + H_2$ collisions which were ejected at 20° to 125° and found a prominent peak at an electron energy corresponding to the velocity of the incident ion. The peak was seen at all angles but was strongest in the forward and backward directions. Similar equivelocity peaks were seen by Burch et al. (1973) at 90° in 17- to 41-MeV $O^{n+} + Ar$ collisions. Not only did the position of their peak vary as expected with beam energy, but also the height of the peak depended on the charge state of the O^{n+} projectile. Figure 5-5-3 shows that the size of the peak decreased as n was increased from 3 to 8 since the projectile carried fewer and fewer electrons which could be detached in the collision. When only the more tightly bound K electrons remained, there was little evidence of a peak.

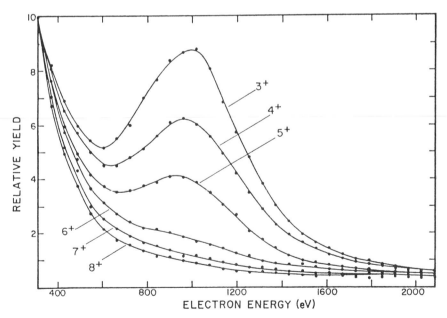

Figure 5-5-3. Electron spectra at 90° from 30-MeV $O^{n+} + Ar$ collisions. The yields are normalized to the same value at 310 eV. [From Burch et al (1973).]

As with the ECC process, the ELC spectrum is sharply peaked, resulting in a cusp-shaped peak in the energy spectrum of electrons seen at $0°$. In fact, it was found that the ELC peak was narrower in both angle (Duncan and Menendez, 1976) and energy (Rudd and Macek, 1972; Breinig et al., 1982). The ELC peak was also found to be larger and more symmetric.

Drepper and Briggs (1976) obtained a good account of ELC from a transformation to the laboratory frame of the Born cross section for ionization in the projectile frame. This treatment agreed with the experimental finding that while the equivelocity peak at zero angle is sharp and cusp-shaped, at angles greater than about $25°$ it is broader due to the momentum distribution of the projectile bound state. The incident particle with its retinue of electrons interacting with the target atom can be pictured as a group of free electrons with the proper distribution of momenta scattering elastically from the target. For a further elaboration of this idea, see Section 4-3.

D.   Electron Capture from Inner Shells

Capture is an important mechanism for producing inner-shell vacancies and can contribute a significant fraction of the cross section. This is especially the case when the nuclear charge of the projectile approaches that of the target. The theory of this near-resonant charge transfer between inner shells has been thoroughly reviewed by Stolterfoht (1987), who also discussed the experimental aspects of the subject. Wille and Hippler (1986) have reviewed mechanisms for producing inner-shell vacancies.

Vinogradov and Shevel'ko (1970) derived the Brinkman–Kramers equation in a form that they could apply to the capture of electrons from shells of different principal quantum number n by an incident proton. Using hydrogenlike radial wave functions, they showed that the ratio of the cross sections for capture of electrons from two different shells, 1 and 2, is

$$\frac{\sigma_1}{\sigma_2} = \left(\frac{n_1}{n_2}\right)^2 \alpha^{-15/2}\left(\frac{1+x}{1+x/\alpha}\right)^{10}, \tag{5-5-1}$$

where $x = v^2/\chi_2$ and $\alpha = \chi_1/\chi_2 > 1$ and where v is the projectile velocity and χ the ionization potential. From this they conclude that while the contribution from the outer-shell electrons predominates at low energy, that of the more tightly bound inner shells is most important at high energies. Taking account of the inner shells yields results in much better agreement with measurements involving heavier targets, where the falloff deviates from the E_1^{-6} or E_1^{-7} behavior at high energies.

As examples of the importance of inner-shell electrons to the capture process, Macdonald et al. (1974) used coincidence techniques to show that for protons on argon, the K-shell contribution to the total capture cross section increased from 0.5% at 2.5 MeV to 47% at 12 MeV. Conversely, the contribution of capture to the total K-shell vacancy production cross section went from 0.4% to 0.1% over the same energy range.

Some systematics of capture from the K shell of argon were investigated experimentally by Macdonald et al. (1972, 1973). In one experiment the X-ray yield was measured as a function of the nuclear charge of incident ions at 1.88 MeV/u. The dependence is shown in Fig. 5-5-4, where the cross section is divided by the corresponding cross section for proton impact (taken as $3.3 \times 10^{-21} \, \text{cm}^2$) and by Z_1^2. This is compared to the sum of the ionization and charge-transfer cross sections calculated by McGuire (1973) using the binary encounter approximation (BEA) and the agreement is seen to be satisfactory. Note that the proportionality of the cross section to Z_1^2 is neither predicted nor observed except at very low Z_1 values. In the other measurement the cross section was determined for K-shell production in argon by 35.7-MeV F^{q+} as a function of the charge state q. Reasonably good agreement was obtained with the BEA calculations when the ionic charge q was used rather than the nuclear charge Z_1.

The relation between the K-shell cross sections by ionization and charge

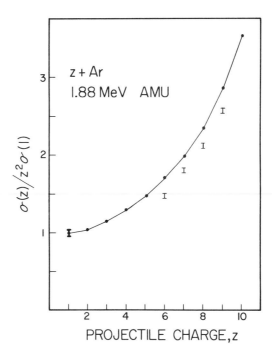

Figure 5-5-4. K-vacancy production cross sections for bare nuclei of charge Z_1 incident on argon, shown as the ratio of the cross section for projectile charge Z_1 to that for protons multiplied by Z_1^2. Error bars, experimental values from Macdonald et al. (1973); line and dots, values calculated by the binary encounter approximation. In each case the sum of the ionization and charge-transfer cross sections is used. [From McGuire (1973).]

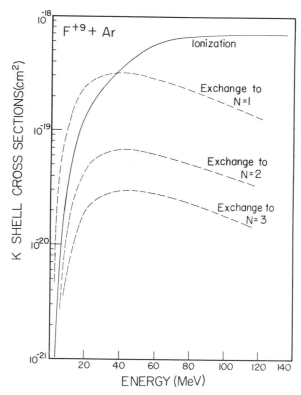

Figure 5-5-5. K-vacancy production cross sections by ionization and by charge transfer to excited states indicated by the principal quantum number N, for F^{9+} + Ar collisions. [From McGuire (1973).]

transfer can be gauged from the BEA calculations for F^{9+} + Ar as shown in Fig. 5-5-5. While ionization dominates at high energies, charge transfer becomes more important at low energies, although the BEA may overestimate the latter.

Lapicki and McDaniel (1980) have developed formulas for capture from inner shells by fully stripped ions which is analogous to the treatment of inner-shell ionization by Brandt and Lapicki (1974, 1979). Based on the OBK approximation at low energies and on the second Born approximation at high energies, their treatment takes account of such effects as Coulomb deflection and increased electron binding. With these corrections, they obtain good agreement with a large amount of data when $Z_1/Z_2 < 1/2$.

E. Differential Probabilities for Resonant Capture

In the early 1960s, an experimental group led by Edgar Everhart at the University of Connecticut discovered a remarkable behavior of the capture probability in small-impact-parameter ion–atom collisions. Lockwood and

Everhart (1962) measured probabilities for electron capture in very close $H^+ + H$ collisions as a function of energy from 0.7 to 50 keV. They restricted their study to violent or close collisions by selecting beam particles that had been scattered through an angle of 3°. (Later work was done at different angles.) In their apparatus (Fig. 5-5-6) the proton beam passed through a tungsten furnace into which hydrogen gas was admitted. A temperature of 2400 K was sufficient to thermally dissociate more than 90% of the molecular hydrogen into atomic hydrogen which formed the target. A collimator system selected only those incident particles that had been scattered through an angle of 3°, and an electrostatic analyzer separated the ones that had captured an electron from those that had not. If the numbers of neutrals and charged particles detected are denoted by N_0 and N_+, respectively, the capture probability is given by

$$P_0 = \frac{N_0}{N_0 + N_+}. \tag{5-5-2}$$

This measurement can be made without the necessity of determining target gas densities or geometric quantities.

In earlier work (Ziemba and Everhart, 1959; Ziemba et al., 1960) this experimental group had discovered that a particularly useful way to display results of this type was on a plot of capture probability versus reciprocal velocity. Their $H^+ + H$ data plotted in this way are shown in Fig. 5-5-7. Three features are evident: (1) there is a strong oscillatory structure with probabilities varying between 0.1 and 0.9; (2) plotted versus reciprocal velocity, there is a uniform spacing of the maxima and minima; and (3) there is a decrease in amplitude with an increase in $1/v$, indicating damping.

The oscillatory structure can be understood by the following argument. Since the collision is a close one, the electron orbits around both nuclei, thus oscillating back and forth in its proximity to one or the other. The position of

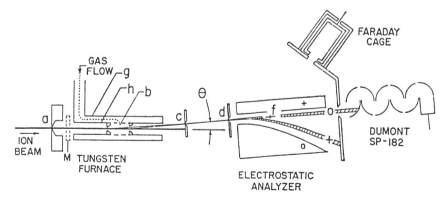

Figure 5-5-6. Schematic diagram of apparatus used to measure resonant electron capture. [From Lockwood and Everhart (1962).]

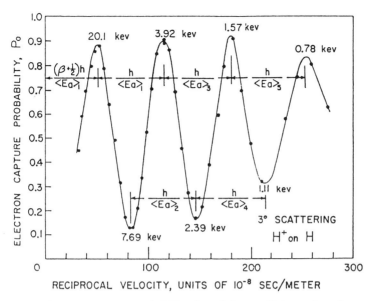

Figure 5-5-7. Electron capture probability for violent collisions plotted versus the reciprocal of the projectile velocity. [From Lockwood and Everhart (1962).]

the electron at the time of separation determines which nucleus it stays with. The time of the collision is proportional to $1/v$, and as that quantity varies, the probability of the electron ending up on the projectile oscillates. If the period of one oscillation is set equal to Planck's constant h divided by the interaction energy E associated with the oscillation, then

$$\langle Ea \rangle_n = \frac{h}{(1/v_{n+2}) - (1/v_n)}, \qquad (5\text{-}5\text{-}3)$$

where a is the path length over which the collision takes place. The experimental value of $\langle Ea \rangle$ obtained from the period of the oscillation was 63.7 ± 1 eV-Å.

We can also look at the collision event from a somewhat different point of view. As a ground-state hydrogen atom and a proton are brought together, two different states of the H_2^+ molecular ion can form, the $1s\sigma_g$ and the $2p\sigma_u$. The oscillatory structure can be interpreted as an interference between the waves representing these two states. Lichten (1963) showed that summing the amplitudes along the particle trajectory leads to a charge-transfer probability given by

$$P_0 \propto \sin^2 \left[\frac{1}{v} \int_{r_0}^{\infty} (E_{2p\sigma_u} - E_{1s\sigma_g}) \, dr \right], \qquad (5\text{-}5\text{-}4)$$

where r_0 is the distance of closest approach of the two collision partners. In the

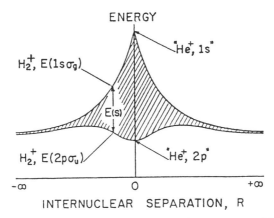

Figure 5-5-8. Energy-level diagram showing $1s\sigma_g$ and $2p\sigma_u$ states of H_2^+. The capture probability is related to the area between the two curves, indicated by the crosshatching. [From Lockwood and Everhart (1962).]

energy-level diagram of Fig. 5-5-8 showing the two lowest states of H_2^+, it is seen that this integral is just the area between the two states, indicated by the cross-hatching. Yenen et al. (1984) have developed a molecular two-state model for dealing with the similar data on $H^+ + H_2$ presented by Lockwood and Everhart (1962).

Bates and Williams (1964) showed that additional factors, such as $2p\sigma_u - 2p\pi_u$ coupling near the united atom limit and the incorporation of translational factors into the electron wave function, accounted for the damping of the probability function. Similar measurements on the $He^+ + He$ system were also carried out at the same laboratory (Lockwood et al., 1963) with similar oscillatory results. An important finding to come out of these data was the observation by Lichten (1963) that it was necessary to invoke diabatic molecular orbital states rather than adiabatic ones (see Section 2-9-B) in the interpretation. The initial assumption that the collisions were adiabatic with its implication that crossings of energy levels must be avoided led to transitions to states for which the calculated integral did not agree with the experimental oscillation periods. However, if the states were taken as diabatic, curve crossings were allowed, and the states then available gave areas in agreement with experiment.

5-6.   MULTIELECTRON TRANSFER PROCESSES

A.   Capture of Two or More Electrons

Measurements were made of the cross sections for double capture as early as 1955. Fogel and Mitin (1955) determined values of σ_{1-1} for $H^+ + H_2$ from 9.5 to 29 keV, finding that the cross section for double capture was 1 to 2% of the value for the single-capture cross section. In addition, they measured several other targets, and in the case of helium the ratio of single to double capture ranged from 120 to 250. Since double capture by a singly ionized projectile results in a negative ion, it can take place only when the projectile has a positive electron affinity in its neutral state.

Electron correlation (i.e., the interaction of electrons in an atom with each other) is difficult to identify experimentally. Tanis et al. (1989) used 3.5-MeV/u Ne^{10+} + Ne collisions in an ingenious experiment to expose the presence of electron correlation. In their apparatus the incident beam passed successively through the target gas, an electrostatic electron energy analyzer, and a magnetic beam charge-state analyzer. Electrons in the ECC peak ejected at zero angle were detected in coincidence with the projectile. Electrons in coincidence with Ne^{9+} came from double capture, with one electron going to the continuum and the other to a bound state. Events in coincidence with Ne^{10+} resulted from single electron capture to the continuum. The ratio of coincidence double- to single-capture events was calculated and compared with the ratio of total double- to single-capture cross sections. At the high energy used, electrons other than K electrons were not involved to any appreciable extent. Therefore, it was assumed that only two electrons were available in the target for capture. Then a simple application of the independent particle model (IPM) showed that the two ratios should differ by a factor of 2. Instead, the two ratios differed by a factor of about 8, thus indicating the influence of electron correlation. The use of ratios eliminates any dependence on the specific assumptions used in any particular IPM.

The experiment was also done using capture to a Rydberg state instead of to a continuum state. In this case the Rydberg electron became detached in the fringing field of the electrostatic analyzer and was detected in coincidence as for the ECC electrons. Helium and argon were also used as targets in the experiment.

An example of multielectron capture to excited states was investigated by Zouros et al. (1987). The reaction studied was

$$He^{2+} + He \rightarrow He(2l\,nl') + He^{2+} \tag{5-6-1}$$

in which the fast doubly charged helium ion captures two electrons. Autoionization of the doubly excited projectile was detected by observing the electron spectrum (see Fig. 5-6-1) at zero angle. While the $2s^2(^1S)$, $2p^2(^1D)$, and $2s2p(^1P)$ states result from a double electron capture, the $2s2p(^3P)$ state cannot be formed in this way since this would require a spin flip of one of the electrons. Since the spin-orbit interaction is small in such low-Z systems, it is more likely that the state is formed by two successive one-electron capture events, one taking place in the beam line before the scattering chamber. This interpretation is borne out by the quadratic pressure dependence for the formation of the triplet state as compared to the linear dependence for the singlets. Cross sections for the ^1S state extracted from the data were 6 to 20 times larger than those for the other singlet states.

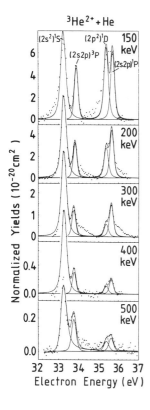

$^3\mathrm{He}^{2+} + \mathrm{He}$

Figure 5-6-1. Electron spectra from He^{2+} + He collisions at various energies after subtraction of the background continuum and transforming to the projectile rest frame. [From Zouros et al. (1987).]

An example of the rich Auger spectra obtained from double capture by highly charged ions in collisions is shown in Fig. 5-6-2 for 40-keV C^{4+} + He by Stolterfoht et al. (1987). Electrons ejected at zero angle by Coster-Kronig and L-Auger processes dominate the spectrum. The mechanism proposed by the authors to account for the data is a correlated double capture at an internuclear distance of about 3 or 4 a.u. At this separation one of the electrons in the target is transferred to the $2p$ state of the projectile and the other to a Rydberg state.

B. Transfer Excitation

Transfer excitation (TE) is a two-electron process involving the transfer of one electron to the projectile and the simultaneous excitation of another projectile electron. If the process is due to the interaction of the two electrons, the process is known as resonant transfer excitation (RTE), but if the two events are independent, it is nonresonant transfer excitation (NTE). The RTE process is closely related to dielectronic recombination (DR, about which see Section 7-4 of the companion volume), which, in turn, bears an inverse relation to

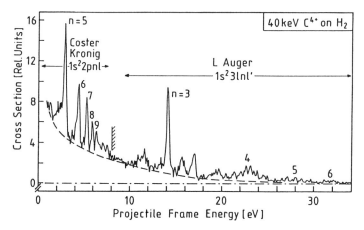

Figure 5-6-2. Spectrum of $L_1L_{2,3}X$ Coster–Kronig electrons and L-Auger electrons from 40-keV C^{4+} + He collisions. [From Stolterfoht et al. (1987).]

autoionization (AI) or Auger electron emission. The three reactions, in their simplest forms, are:

$$\text{RTE} \qquad A^+ + B \rightarrow A^* + B^+ \qquad (5\text{-}6\text{-}2)$$

$$\text{DR} \qquad e^- + A^+ \rightarrow A^* \qquad (5\text{-}6\text{-}3)$$

$$\text{AI or Auger} \qquad A^* \rightarrow A^+ + e^-. \qquad (5\text{-}6\text{-}4)$$

The main difference between RTE and DR is that the active electron is provided by an atom in RTE but by electron impact in DR. Processes that involve an inner-shell vacancy result in an Auger electron or an X-ray photon, while those exciting two electrons yield autoionization electrons. The capture processes dealing with inner-shell vacancies, where most of the attention has been directed, are sometimes labeled RTEX and RTEA to distinguish between cases involving the emission of X-rays and Auger electrons, respectively. Transfer excitation can be "resonant" since a peak is found in the measured RTEA cross section at a projectile velocity equal to the ejected Auger velocity. The width of the resonance results from the momentum distribution of the electrons in the target (i.e., the Compton profile).

Among the first to study RTE were Tanis et al. (1981, 1982), who observed sulfur K X-ray emission in coincidence with electron capture events from 70-MeV S^{q+} ($q = 13$ to 16) ions incident on argon. Both high-resolution X-ray spectroscopy [e.g., Pepmiller and Richard (1982)] and high-resolution zero-angle Auger spectroscopy [e.g., Swenson et al. (1986)] have been employed to select particular excited states. (See Section 4-11 for an explanation of zero-angle Auger spectroscopy.)

The strongly resonant character of some reactions is shown in the data of Zouros et al. (1989) displayed in Fig. 5-6-3, where the cross section for electron ejection from $1s2s2p^2(^{1,3}D)$ states was measured as a function of projectile energy for F^{6+} collisions with helium or hydrogen. The RTEA cross sections, calculated in the impulse approximation, yielded the same energy dependence of the cross sections as experiment but were found to be smaller by factors of 2.2 to 3.2.

The dependence of the energy of resonance on the charge of the projectile is shown in the data of Tanis et al. (1985) in which highly charged projectiles of sulfur ($Z_1 = 16$), calcium ($Z_1 = 20$), and vanadium ($Z_1 = 23$) were incident on helium at energies up to 460 MeV. Cross sections were measured for K X-ray

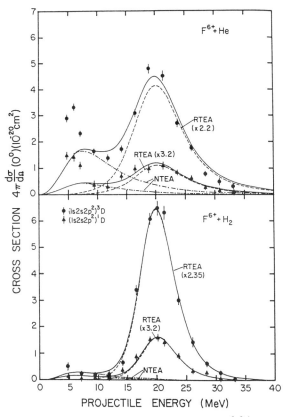

Figure 5-6-3. Electron production cross sections for $1s2s2p^2(^{3,1}D)$ states produced by transfer excitation (TE) in F^{6+} $(1s^2 2s)$ + He and H_2 collisions as functions of the projectile energy. Solid line, scaled sum $\sigma_{TE} = a_R \sigma_{RTEA} + a_N \sigma_{NTE}$; dashed line, RTEA impulse approximation of Brandt (1983a) and Swenson et al. (1986); dash-dotted line, NTEA calculation of Brandt (1983b). The values of the scaling coefficient a_R are shown in parentheses. [From Zouros et al. (1989).]

production in coincidence with projectile ions that had captured one electron. The results of these measurements are summarized in the calculations shown in Fig. 5-6-4 for lithiumlike projectile ions from Z_1 ranging from 14 to 26. These calculations were made using the method of Brandt (1983a) based on dielectronic recombination cross sections. The double peak which is resolved in the curves for the heavier ions is due to intermediate states involving different principal quantum numbers.

Hahn (1989) used the impulse approximation to formulate a unified description of RTE, NTE, and a new mechanism which he proposed called *uncorrelated transfer excitation* (UTE). As with the other two mechanisms, UTE involves two electrons in the target. One electron excites the projectile but remains unbound while the second is captured, forming a doubly excited state. His treatment also encompases dielectronic recombination. A review of transfer excitation has been written by Tanis (1987).

C.   Transfer Ionization

A process in which the number of electrons lost from the target exceeds the number of electrons captured by the projectile is called *transfer ionization* (TI). As its name implies, it is a combination of an electron transfer and an ionization. At least two electrons are involved, which distinguishes it from electron capture

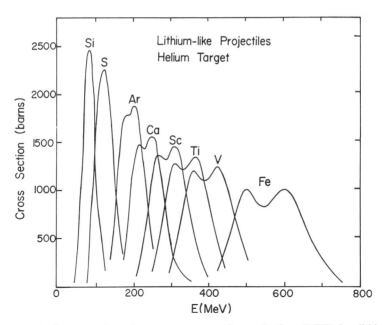

Figure 5-6-4. Cross sections for resonant transfer excitation (RTE) by lithiumlike projectiles on helium calculated by method of Brandt (1983a) using dielectronic recombination data from McLaughlin and Hahn (1982). [From Tanis et al. (1985).]

to the continuum (ECC), for which only a single electron is needed. A transfer ionization process can be written

$$A^{q+} + B \rightarrow A^{(q-k)+} + B^{i+} + (i - k)e, \qquad (5\text{-}6\text{-}5)$$

where A^{q+} is the charged projectile striking the neutral target B from which i electrons are removed, k of which are transferred to the projectile and the rest of which go into the continuum. In the simplest case when the target is neutral, $i = 2$ and $k = 1$, so one electron is transferred and one liberated. In this case if the ion A^{q+} has a recombination energy I_A and B has first and second ionization potentials χ_{B1} and χ_{B2}, the net potential energy available in the process is $\Delta E = I_A - (\chi_{B1} + \chi_{B2})$. TI is also possible in interactions between two ions.

Originally, TI referred only to processes for which the energy ΔE was positive. Such an exoergic process can take place even at vanishingly small relative energies and is therefore an important process occurring in hot plasmas. However, the term has come to mean any process in which more electrons are removed from the target than are captured by the projectile. If the reaction is endoergic, of course, it can proceed only at the expense of the kinetic energy of the projectile. Cocke et al. (1981) have shown that for low-energy collisions of highly charged argon ions with rare gases, TI generally has a much larger cross section than either electron capture or ionization alone. This result was also noted in Section 4-2.

Niehaus (1980) has given a description of the process in terms of an electronic-coupling matrix element $M_{i\varepsilon}(r)$ between a discrete initial state i and a state in which the electron has an energy ε in the ionization continuum. The description uses the Born–Oppenheimer approximation with a complex potential

$$\tilde{V}(r) = V_i(r) = \frac{i}{2} \Gamma_{i\varepsilon}(r), \qquad (5\text{-}6\text{-}6)$$

where $\Gamma_{i\varepsilon}(r) = 2\pi |M_{i\varepsilon}(r)|^2 \rho(E)$ is the reciprocal of the lifetime of the system for the particular value of internuclear separation and where $\rho(E)$ is the density of final states. This leads to an expression for the probability of a TI process for a given trajectory T:

$$P_T(v_1) = \int_T \frac{\Gamma(r)}{hv(r)} \exp\left[-\int_T \frac{\Gamma(r')\,dr'}{hv(r')} \right] dr, \qquad (5\text{-}6\text{-}7)$$

where v_1 is the relative collision velocity and $v(R)$ is the instantaneous radial velocity of relative motion. The equation yields a E_1^2 dependence of the cross section on collision energy.

Transfer ionization can take place through a number of different mechanisms. An autoionization can occur from the quasi-molecule formed during the collision, or double capture can be followed by autoionization of the projectile after separation (Cocke et al, 1981). Transfer of two electrons, one to a bound state and one to a continuum state of the projectile, is also a possibility.

If highly charged ions interact with high Z atoms, the multiple ionization and multiple electron transfer that result can be explained in terms of a statistical sharing of the available energy. As an example of this, Groh et al. (1983) and Müller et al. (1983) studied the production of Xe^{i+} in collisions Xe^{q+} with He. They were able to show that the probability of various final charge states of the projectile and target ions could be calculated from a statistical model originally derived by Russek (1963) and Russek and Meli (1970). If B^{k+} is a target that has lost k of its outer-shell electrons to the projectile, the probability that of its remaining N electrons, n additional ones escape to the continuum is given by

$$P_n(N, \Delta E_m) = \binom{N}{n} \sum_{j=0}^{l} (-1)^j \binom{N-n}{j} \left(1 - \frac{n+j}{\Delta E_m/\langle \chi_B \rangle}\right)^{N-1}, \quad (5\text{-}6\text{-}8)$$

where l is defined by $n + l \leqslant \Delta E_m/\langle \chi_B \rangle \leqslant n + l + 1$ and $\langle \chi_B \rangle$ is the averaged ionization potential for electrons "boiling off" the ions B^{k+}. ΔE_m is the available energy. The value of $\langle \chi_B \rangle$ is approximated for the case of xenon by

$$\langle \chi_B \rangle = 13.3 \, \text{eV} + 2.4 \, \text{eV} \left(\frac{\Delta E_m}{1 \, \text{eV}}\right)^{1/2}. \quad (5\text{-}6\text{-}9)$$

Figure 5-6-5 shows the measured fractions as a function of the available energy compared to those calculated from (5-6-8) and (5-6-9).

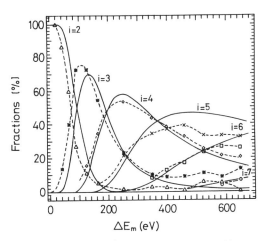

Figure 5-6-5. Charge-state fractions of Xe^{i+} target ions produced in two-electron capture by Xe^{q+} incident on Xe. Data points and dashed lines, experimental results of Müller et al. (1983); solid lines, calculations from Eq. (5-6-8). [From Müller et al. (1983).]

5-7. ION–MOLECULE TRANSFER PROCESSES

A. Transfer with Vibrational and Rotational Excitation

If at least one of the collision partners is a molecule, additional dynamic effects may come into play during charge-transfer events, namely vibrational and rotational excitation and dissociation. Cross sections for such processes of 10^{-15} to 10^{-14} cm^2 at energies from thermal energies up to a few keV are not uncommon. Reviews of low-energy ion–molecule charge transfer have been given by Koyano et al. (1986) and by Linder (1988). Charge-transfer reactions with molecules generally have large cross sections at low energies, especially if the reaction is exoergic.

To understand the transfer of energy in the collision, investigations have been made which involve a precise knowledge of either the initial or final vibrational state or both. Several ways of preparing molecules in selected states and of detecting such states have been developed. Although lasers were first used to excite molecules to specified vibrational states [see, e.g., Zare and Bernstein (1980)], a Japanese group [see Koyano et al. (1986)] developed a better technique which they called TESICO (threshold electron–secondary ion coincidence). In this method, which was also described briefly in Section 3-8-D, an electron–ion pair is formed by photoionization. Electron spectroscopy is used to determine the state of the ion, which is then accelerated to the collision region. The product ions are detected in coincidence with the photoelectron, thus selecting only the desired vibrational state.

Using this technique, a group at Orsay (Guyon et al., 1986; Govers et al., 1983) built the double time-of-flight (TOF) spectrometer shown in Fig. 5-7-1 to study $N_2 + Ar$ collisions. Pulsed synchrotron radiation from the Orsay storage

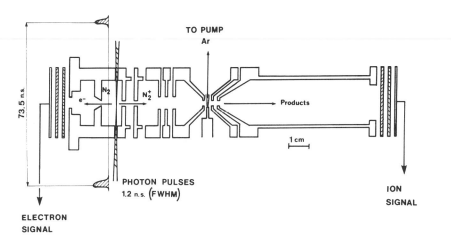

Figure 5-7-1. Double electron-ion time-of-flight spectrometer used to measure vibration state-selected molecules for charge-transfer measurements. [From Guyon et al. (1986).]

ring was monochromatized before being directed into the N_2 gas, where it produced photoions and photoelectrons. Threshold electrons were used by selecting only those electrons with zero energy within the 20-meV passband of the TOF analyzer. The numbers of such threshold electrons are shown in Fig. 5-7-2 plotted against the wavelength of the radiation selected. The levels $v = 0$ to 4 of the $X^2\Sigma_g^+$ state of N_2^+ and $v = 0$ to 6 of the $A^2\pi_u$ state are well resolved. The N_2^+ ions are accelerated to the desired energy, in this case 8 eV, and pass through the target gas effusing from a jet where they may make a charge-changing collision. Product ions from the collision are detected in coincidence with the threshold photoelectrons. The coincidence spectrum for the

$$N_2^+(X, v = 2) + Ar \rightarrow N_2 + Ar^+$$

process, shown in Fig. 5-7-3, has one peak for Ar^+ and a larger N_2^+ peak from the unreacted ions. By measuring the relative areas of the two peaks, the cross sections can be obtained. The resolution obtainable by this method may even be good enough to separate rotational states.

To resolve final vibrational and rotational states, the group at Kaiserslautern [see, e.g., Hermann et al. (1978)] have used a high-resolution electrostatic analyzer. In their apparatus a mass-analyzed and energy-selected ion beam is crossed at right angles with a target beam from a supersonic nozzle. The ions scattered at a given angle were energy analyzed with a double 180° hemispherical energy analyzer. Spectra at two combinations of incident energy and scattering angle are shown in Fig. 5-7-4 for $H^+ + H_2$ collisions. The individual vibrational transitions are clearly resolved in the 20-eV spectrum. In the 6-eV run several rotational states are also resolved.

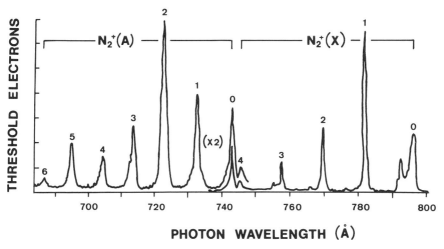

Figure 5-7-2. Threshold photoelectrons from N_2 as a function of the wavelength of the exciting photons. [From Guyon et al. (1986).]

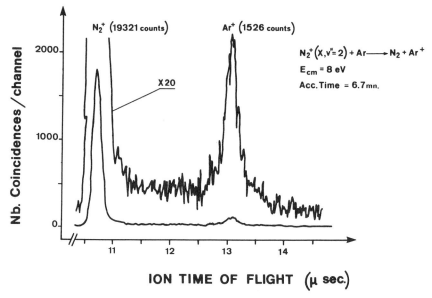

Figure 5-7-3. Time-of-flight spectrum from $N_2^+ + Ar$. Left-hand peak, unreacted nitrogen ions; right-hand peak, argon ions resulting from a capture reaction. [From Guyon et al. (1986).]

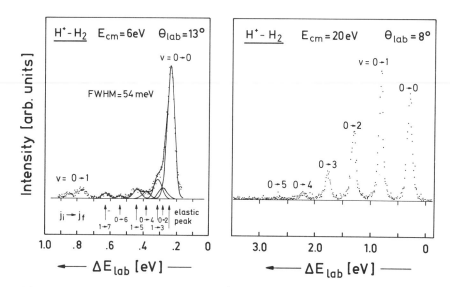

Figure 5-7-4. Energy-loss spectra from $H^+ + H_2$ collisions. [From Linder (1979).]

Time-of-flight analysis was also used by the group at Göttingen (Noll and Toennies, 1986) to study the products of collisions of the $H^+ + O_2$ system. A mass- and energy-selected ion beam was chopped before it intersected a target beam. A detector 3 m from the interaction region detected the scattered projectiles. To study the charge-transfer reaction $H^+ + O_2 \rightarrow H + O_2^+$ a repeller at a positive potential placed before the detector allowed only the neutrals to pass. In the TOF spectrum in Fig. 5-7-5 the large central peak is due to elastically scattered H^+. The peaks on the right are from inelastically scattered H^+, and those on the left were obtained with the repeller potential and are due to charge transfer in which the oxygen molecule is left in various vibrational states. The times of flight for the charge-transfer process are negative since the process is exoergic.

A semiclassical model for vibrational excitation in molecular collisions has been proposed by Giese and Gentry (1974). On the basis of their experimental measurements of transition probabilities and cross sections for nonreactive collisions of H^+ with H_2, HD, and D_2, they concluded that the major mechanism for vibrational excitation is what they term *bond dilution*. As the incident proton approaches the molecule, some of the electronic charge density holding the molecule together is attracted toward the proton, causing an

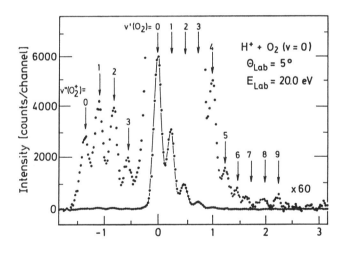

Flight time delay [μs]

Figure 5-7-5. Time-of-flight spectrum of ions from $H^+ + O_2$ collisions. Lower spectrum and magnified spectrum to the right, elastic transition and inelastically scattered H^+ leaving the O_2 in various vibrational states; spectrum on the left, neutral H atoms from charge transfer leaving the O_2^+ in various states. The time of flight, relative to the elastic peak, is negative for the latter because the charge transfer is an exoergic reaction. [From Noll and Toennies (1986).]

increase in the internuclear separation. After the proton has left, the resulting bond stretching leaves the molecule in a higher vibrational state. To calculate transition probabilities, a Monte Carlo method is adopted. For each set of initial conditions, an exact trajectory and the energy transfer to vibration are calculated classically. A semiclassical prescription then determines the final quantum vibrational state into which the molecule goes. The method is the DECENT model, an acronym for "distribution (among quantum states) of exact classical energy transfer". This model was also discussed briefly in Note 2-7.

The amount of vibrational energy transferred to the molecule is dependent on the relation between the collision time t_{coll} and the period of vibration t_{vib}. In a slow collision, the molecule adjusts quasistatically, and a transition causing an energy transfer is unlikely to take place. During a fast collision, the perturbation acts for too short a time to be effective in causing a transition. The cross section is expected, therefore, to reach a maximum for energies for which $t_{coll} \approx t_{vib}$. This is also what the results of calculations on the DECENT model indicate as shown in Fig. 5-7-6 for the $H^+ + H_2$ collision. The value of t_{vib} is 8×10^{-15} s. Taking an interaction length of $8a_0$, the energy at the maximum should be about 35 eV, which is approximately where the maxima in the figure come.

B.   Transfer with Dissociation

When the target is a molecule, the possibility of its breakup during or immediately following the capture of one of its electrons is another complication that occurs frequently. If BC is a molecule and A^+ the incident ion, the process is represented by the reaction

$$A^+ + BC \rightarrow A + [BC]^+ \tag{5-7-1}$$

followed by either

$$[BC]^+ \rightarrow B + C^+ \tag{5-7-2}$$

or

$$[BC]^+ \rightarrow B^+ + C \tag{5-7-3}$$

or a combination of dissociative charge transfer and ionization,

$$[BC]^+ \rightarrow B^+ + C^+ + e^-. \tag{5-7-4}$$

These processes compete with the simple dissociative ionization process:

$$A^+ + BC \rightarrow A^+ + B^+ + C + e^-. \tag{5-7-5}$$

The simple dissociative charge-transfer process can proceed if the recom-

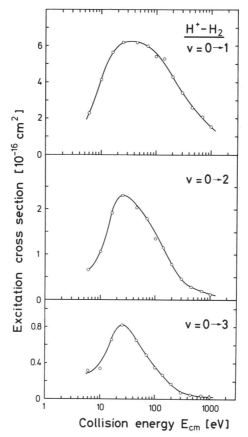

Figure 5-7-6. Cross sections for vibrational excitation to three different states in $H^+ + H_2$ collisions. ○, Experimental data; lines, calculations using the DECENT model by Gentry and Giese (1975). [From Linder (1984).]

bination energy of the incident positive ion exceeds the sum of the ionization potential and dissociative energy of the molecule. Browning et al. (1969) showed that such a process can become dominant at energies below a few keV where ionization is unlikely, but it also takes place at higher energies.

Frequently, the focus of studies of these processes is on the dissociation itself without regard to whether ionization or charge transfer is the mechanism responsible. An important application of the dissociation process is in producing hot, dense plasmas by injection of high-energy hydrogen molecules. To trap and hold them in the magnetic mirror field, their charge-to-mass ratio must be changed, and one way to effect this is by dissociating the molecular ions through collisions.

Figure 5-7-7, showing the potential energy curves for selected states of H_2, H_2^+, and H_2^{2+}, can be used to understand the basic mechanism for dissociation.

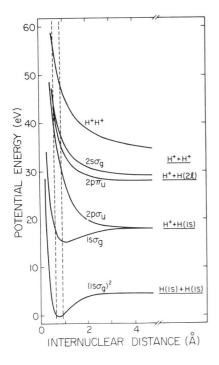

Figure 5-7-7. Potential curves for some states of H_2, H_2^+, and H_2^{2+} as a function of internuclear separation. [From Wood et al. (1977).]

The vertical dashed lines encompass the range of internuclear distances for the H_2 molecule in its ground or lowest vibrational state. The Franck–Condon principle asserts that most electronic transitions take place too rapidly for the internuclear separation to change appreciably. Therefore, capture and/or ionization transitions taking place during collisions raise the molecule from the ground state to one of the higher repulsive states within the dashed vertical lines. If the molecule is left in the $1s\sigma_g$ state, it does not have enough energy to dissociate and the atoms remain bound as a H_2^+ molecular ion in that state. However, if it makes a transition to any of the higher states, such as the $2p\sigma_u$, it immediately follows that potential energy curve to a lower energy at a large internuclear distance. The loss of potential energy, in this case approximately $35 - 18 = 17\,eV$, becomes kinetic energy, which is divided equally between the H^+ ion and the $H(1s)$ atom coming from the dissociation of the molecule.

Some early measurements of capture from molecular targets used the condenser method and therefore collected ions from dissociative capture as well as from nondissociative capture. Since the former results in energetic recoil ions, a large transverse field is needed to ensure complete collection. But since the size of the field is limited by the resulting deflection of the projectile beam, some recoil ions may be lost. For example, Koopman (1968) could not be certain if his system collected all of the products of the 40- to 1500-eV $He^+ + O_2$ collisions since the graph of ion current versus transverse field intensity did not saturate.

Bischof et al. (1983) measured the energy spectrum of O^+ ions from the reaction $He^+ + O_2 \rightarrow He + O + O^+$ and found that even at an impact energy of only 5 eV the recoil ion spectrum had peaks at energies ranging from under 1 eV to above 3 eV. At a much higher impact energy (1 MeV), Steuer et al. (1977) found for the same reactants that the O^+ fragment spectrum extended to 20 eV and the O^{3+} spectrum to about 55 eV. Bischof's ion spectrum is shown in Fig. 5-7-8.

Using the apparatus shown in Fig. 4-2-3, the Leningrad group headed by V. V. Afrosimov (Afrosimov et al., 1969) made coincidence measurements of slow ions and neutrals resulting from 5- to 50-keV $H^+ + H_2$ and $H^0 + H_2$ and $H^+ + CO$ collisions. By measuring the kinetic energies of the fragments they were able to identify not only the final charge states of the products, but also the electronic states to which the molecule was excited. This enabled them to determine cross sections for transitions to specific excited states in the particular processes of capture, ionization, and stripping. The work was summarized by Afrosimov (1972).

In an interesting application of charge transfer, deBruijn and Los (1982) and deBruijn and Helm (1986) developed a new method of studying photodissociation of metastable molecules. By charge transfer in cesium H_2^+ was formed in their experiment into a neutral H_2 beam in its metastable $c^3\Pi_u^-$ state, after which it was photodissociated by a laser. A position-sensitive detector a distance downstream recorded the arrival of the photofragments. Their distance of separation at the detector yielded information about their kinetic energies, and

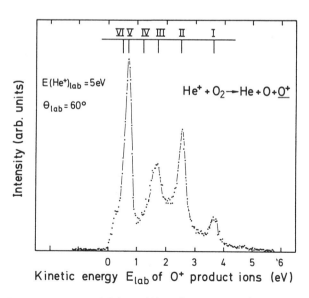

Figure 5-7-8. Energy spectrum of O^+ recoil ions from 5-eV $He^+ + O_2$ collisions. [From Bischof et al. (1983).]

from the difference in the time of arrival the orientation of the molecule at the time of the dissociation could be determined. Thus by a combination of translational spectroscopy and time-of-flight analysis of the photofragments, a great deal of new information about the decay mechanisms, level positions, and lifetimes of the ro-vibrational states was obtained. For further information about dissociation, see Part $\mathscr{B}$ of Chapter 2 and also Section 5-5 of the companion volume.

5-8.   ION–ION CHARGE TRANSFER

Because of the need for data on collisions between ions in the high-temperature plasmas of interest in thermonuclear fusion research, a few experimental and theoretical studies have been made of ion–ion charge transfer. The difficult experimental problem of providing target ions of sufficient density has been solved by the use of intersecting beam technology.

The first to measure cross sections for such collisions was the Belfast group headed by H. B. Gilbody. Mitchell et al. (1977) obtained cross sections for the production of He^{2+} from the two processes

$$H^+ + He^+ \rightarrow H^+ + He^{2+} + e \qquad \text{(ionization)} \qquad (5\text{-}8\text{-}1)$$

$$\rightarrow H + He^{2+} \qquad \text{(charge transfer)} \qquad (5\text{-}8\text{-}2)$$

at center-of-mass energies of 72 to 402 keV. Their apparatus is shown in Fig. 5-8-1. A 10-keV beam of He^+ from one positive ion accelerator crosses a higher energy beam of H^+ at right angles. Both beams were chopped at a frequency of 93 Hz. The high vacuum needed to minimize the interaction of the beams with the background gas was obtained by using high-speed, high-vacuum pumps which yielded a pressure in the collision chamber of 4×10^{-10} torr when the beams were off. After crossing the proton beam at T, the helium beam entered a parallel-plate electrostatic analyzer A_1 which separated the three charge states. The background count rate was reduced by baffles in the analyzer and further reduced for the much smaller He^{2+} component by sending it through a second analyzer, not shown. A particle detector counted the He^{2+} and a Faraday cup registered the current of He^+ ions. The neutral component could also be recorded by a channeltron detector. Provision was made to measure the beam profiles of both beams so that overlap integrals could be calculated.

The two beams were chopped $90°$ out of phase. One scaler counted when either beam was on alone and a second scaler counted when both beams were on or off together. By proper subtraction, background counts and electronic noise counts could be corrected for. Later measurements by Peart et al. (1983) and by Rinn et al. (1985) brought to light a 25% arithmetical error in the earlier calculation of the cross sections. Subsequently, this was confirmed by a redetermination by Watts et al. (1986).

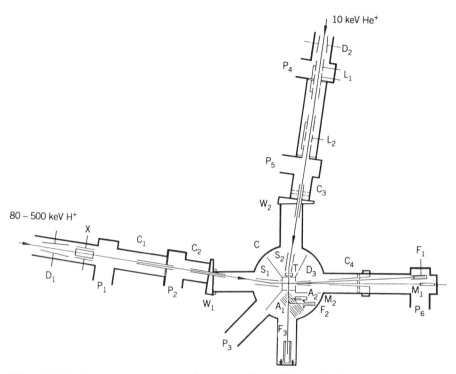

Figure 5-8-1. Apparatus to measure charge transfer and ionization in ion–ion collisions. [From Mitchell et al. (1977).]

Angel et al. (1978) used the apparatus described above to count He^{2+} products in coincidence with the fast H atoms to determine separately the charge-transfer cross section σ_c. Since

$$\sigma(He^{2+}) = \sigma_c + \sigma_i \qquad (5\text{-}8\text{-}3)$$

where σ_i is the cross section for ionization, all three cross sections could be determined.

A novel folded-beam ion–ion collider was devised at Oak Ridge National Laboratory (Kim and Janev, 1987) to study symmetric electron-loss collisions between multicharged heavy ions. Cross sections for loss of one electron were measured for the $Ar^{3+} + Ar^{3+}$ and $Kr^{3+} + Kr^{3+}$ systems at 60-keV CM energy. In the apparatus, shown in Fig. 5-8-2, the incident beam enters a magnet from the upper left and emerges into an ultrahigh-vacuum region (5×10^{-10} torr pressure). It then enters a three-element retarding-lens-type electrostatic mirror that stops the beam and redirects it in the opposite direction with the same energy and charge. It then interacts with the beam going in the original direction before reentering the magnet, where those ions with the original charge are

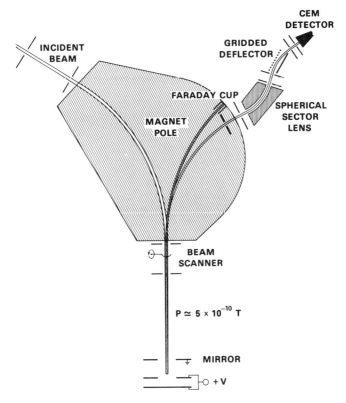

Figure 5-8-2. Schematic diagram of the folded-beam ion–ion collider. [From Kim and Janev (1987).]

deflected into a Faraday cup while those that have lost 1 unit of charge are directed to a particle detector. The ratio of the two currents is used to calculate cross sections. This clever arrangement has the advantage of requiring only a single accelerated beam of heavy ions.

5-9. PROBLEMS

5-1. Derive (5-2-3).

5-2. Derive (5-2-4) and (5-2-5).

5-3. Derive (5-2-6) and (5-2-7).

5-4. Derive (5-2-13) and (5-2-14).

5-5. Show that (5-2-15) reduces to (5-2-2) for a two-component system.

5-6. Derive (5-2-13) and (5-2-14) from (5-2-9) and (5-2-10).

5-7. Show that (5-2-18) reduces to (5-2-6) when $\sigma_{ik} = \sigma_{jk} = 0$.

5-8. Show that (5-2-20) reduces to (5-2-13) for the two-component case.

5-9. Read off approximate values of the cross sections at 10 keV from Fig. 5-2-1 and calculate the equilibrium charge fractions for the three charge states of hydrogen.

5-10. From Fig. 5-2-4 read off (from the dashed lines) enough information to calculate the mean charge states at 0.01, 0.1, and 1.0 MeV/u.

5-11. In (5-2-31), take $\sigma^0/\sigma^+ = 1.6$ and $\sigma_{10} = 2 \times 10^{-16}$ cm^2. The distances corresponding to π_3 and π_{av} are 8 and 2 cm, respectively. At room temperature, what is the gas pressure in millitorr (assumed the same all along the beam path) that makes the beam neutralization correction equal to 10%?

5-12. Summarize the advantages and disadvantages of the various methods of measuring charge-transfer cross sections.

5-13. In the condenser method, suppose that the length of the collecting plates is 5 cm, the primary beam current 1 μA, and the pressure 1 mT. If a current of 10^{-11} A is the smallest that one can reliably read on the collecting plate, what is the smallest cross section that can be measured?

5-14. In the reaction He$^+$ + Ar $\rightarrow$ He + Ar$^+$, find the impact energy at which the Massey criterion predicts a maximum cross section. Use the quantities $a = 7$ Å, χ(He) = 24.6 eV, χ(Ar) = 15.8 eV.

5-15. From (5-4-13), find the dependence of σ on Z_1. Use a root-finder computer program to solve (5-4-14) for R_c at several values of Z_1 from 5 to 50. Take $\alpha = 1$, $q = 1$, and $v = 7 \times 10^5$ m/s. Compare with the graphs in Olson and Salop (1976).

5-16. Prove (5-4-19).

5-17. Using (5-6-8) with (5-6-9), calculate the fractions in Fig. 5-6-5 for $i = 2$ and 3. Take $N = 6$ and $n = i - 2$.

5-18. Suppose that a collision causes a Franck–Condon transition to the $2p\sigma_u$ state of H$_2$. From Fig. 5-7-7 estimate the kinetic energy of the ion resulting from the dissociation. Do the same for the H$^+$ + H$^+$ state.

REFERENCES

Afrosimov, V. V. (1972), *ICPEAC VII*, Amsterdam, p. 84.

Afrosimov, V. V., G. A. Leiko, Yu. A. Mamaev, and M. N. Panov (1969), *Zh. Eksp. Teor. Fiz.* **56**, 1204 [transl. in *Soviet Phys. JETP* **28**, 648].

Allison, S. K. (1958), *Rev. Mod. Phys.* **30**, 1137.

Allison, S. K., and M. Garcia-Munoz (1962), in D. R. Bates, Ed., *Atomic and Molecular Processes*, Academic Press, New York.

Allison, S. K., and S. D. Warshaw (1953), *Rev. Mod. Phys.* **25**, 779.

Andersen, N., J. W. Gallagher, and I. V. Hertel (1988), *Phys. Rep.* **165**, 1.

Angel, G. C., E. C. Sewell, K. F. Dunn, and H. B. Gilbody (1978), *J. Phys. B: At. Mol. Phys.* **11**, L297.

Baragiola, R. A., and I. B. Nemirovsky (1973), *Nucl. Instrum. Methods* **110**, 511.

Barnett, C. F., and H. K. Reynolds (1958), *Phys. Rev.* **109**, 355.

Barnett, C. F., J. A. Ray, E. Ricci, M. I. Wilker, E. W. McDaniel, E. W. Thomas, and H. B. Gilbody (1977), *Atomic Data for Controlled Fusion Research*, ORNL-5206, Oak Ridge National Laboratory, Oak Ridge, Tenn.

Bates, D. R. (1962), Chap. 14 in D. R. Bates, Ed., *Atomic and Molecular Processes*, Academic Press, New York.

Bates, D. R., and R. McCarroll (1962), *Adv. Phys.* **11**, 79.

Bates, D. R., and D. A. Williams (1964), *Proc. Phys. Soc.* **83**, 425.

Bayfield, J. E., and G. A. Khayrallah (1975), *Phys. Rev. A* **11**, 920.

Bell, G. I. (1953), *Phys. Rev.* **90**, 548.

Belyaev, V. A., B. G. Brezhnev, and E. M. Erastov (1967), *Sov. Phys. JETP* **25**, 777.

Berkner, K. H., R. V. Pyle, J. W. Stearns, and J. C. Warren (1968), *Phys. Rev.* **166**, 44.

Berkner, K. H., W. S. Cooper III, S. N. Kaplan, and R. V. Pyle (1969), *Phys. Rev.* **182**, 103.

Betz, H.-D. (1972), *Rev. Mod. Phys.* **44**, 465.

Beuhler, R. J., L. Friedman, and R. F. Porter (1979), *Phys. Rev. A* **19**, 486.

Bischof, G., P. Reinig, and F. Linder (1983), *ICPEAC XIII*, Berlin, Abstracts of Contributed Papers, J. Eichler, W. Fritsch, I. V. Hertel, N. Stolterfoht, and U. Wille, Eds. p. 628.

Bohr, N. (1984), *K. Dan. Vidensk. Selsk. Mat. Fys. Medd.* **18**(8).

Brandt, D. (1983a), *Phys. Rev. A* **27**, 1314.

Brandt, D. (1983b), *Nucl. Instrum. Methods* **214**, 93.

Brandt, W., and G. Lapicki (1974), *Phys. Rev. A* **10**, 474.

Brandt, W., and G. Lapicki (1979), *Phys. Rev. A* **20**, 465.

Bransden, B. H. and M. R. C. McDowell (1992), *Charge Exchange and the Theory of Ion–Atom Collisions*, Oxford University Press, New York.

Breinig, M., S. B. Elston, S. Huldt, L. Liljcby, C. R. Vane, S. D. Berry, G. A. Glass, M. Schauer, I. A. Sellin, G. D. Alton, S. Datz, S. Overbury, R. Laubert, and M. Suter (1982), *Phys. Rev. A* **25**, 3015.

Briggs, J. S., and J. H. Macek (1991), *Adv. At. Mol. Opt. Phys.* **28**, 1.

Briggs, J. S., and K. Taulbjerg (1975), *J. Phys. B* **8**, 1909.

Brinkman, H. C., and H. A. Kramers (1930), *Proc. Acad. Sci. Amsterdam* **33**, 973.

Browning, R., C. J. Latimer, and H. B. Gilbody (1969), *J. Phys. B: At. Mol. Phys.* **2**, 534.

Burch, D., H. Wieman, and W. B. Ingalls (1973), *Phys. Rev. Lett.* **30**, 823.

Chibisov, M. I., and R. K. Janev (1988), *Phys. Rep.* **166**, 1.

Choi, B.-H., E. Merzbacher, and G. S. Khandelwal (1973), *At. Data* **5**, 291.

Clark, M. W., E. M. Bernstein, J. A. Tanis, W. G. Graham, R. H. McFarland, T. J. Morgan, B. M. Johnson, K. W. Jones, and M. Meron (1986), *Phys. Rev. A* **33**, 762.

Cocke, C. L., R. DuBois, T. J. Gray, E. Justiniano, and C. Can (1981), *Phys. Rev. Lett.* **46**, 1671.

Cramer, W. H. (1961), *J. Chem. Phys.* **35**, 836.

Dalgarno, A., and H. N. Yadav (1953), *Proc. Phys. Soc. London* A **66**, 173.

Datz, S., C. D. Moak, H. O. Lutz, L. C. Northcliffe, and L. B. Bridwell (1971), *At. Data* **2**, 273.

DeBruijn, D. P., and H. Helm (1986), *Phys. Rev. A* **34**, 3855.

DeBruijn, D. P., and J. Los (1982), *Rev. Sci. Instrum.* **53**, 1020.

Demkov, Yu. N. (1963), Zh. Eksp. Teor. Fiz. **45**, 195 [transl. in Sov. Phys. JETP **18**, 138 (1964)].

Dewangen, D. P., and J. Eichler (1985), *J. Phys. B: At. Mol. Phys.* **18**, L65.

Dewangen, D. P., and J. Eichler (1986), *J. Phys. B: At. Mol. Phys.* **19**, 2939.

Donnally, B. L., T. Clapp, W. Sawyer, and M. Schultz (1964), *Phys. Rev. Lett.* **12**, 502.

Drepper, F., and J. S. Briggs (1976), *J. Phys. B: At. Mol. Phys.* **9**, 2063.

Duncan, M. M., and M. G. Menendez (1976), *Phys. Rev. A* **13**, 566.

Eichler, J., and F. T. Chan (1979), *Phys. Rev. A* **20**, 104.

Fogel, Ia. M., and R. V. Mitin (1955), *J. Exp. Theor. Phys.* **30**, 450 [transl. in *Sov. Phys. JETP* **3**, 334].

Fogel, Ia. M., V. A. Ankudinov, D. V. Pilipenko, and N. V. Topolia (1957), *J. Exp. Theor. Phys.* **34**, 579 [transl. in *Sov. Phys. JETP* **34**, 400].

Garcia, J. D., E. Gerjuoy, and J. E. Welker (1968), *Phys. Rev.* **165**, 72.

Gealy, M. W., and B. Van Zyl (1987), *Phys. Rev. A* **36**, 3091.

Gentry, W. R. and C. F. Giese (1975), *Phys. Rev. A* **11**, 90.

Gerjuoy, E. (1966), *Phys. Rev.* **159**, 39.

Giese, C. F., and W. R. Gentry (1974), *Phys. Rev. A* **10**, 2156.

Gilbody, H. B. (1986), *Adv. At. Mol. Phys.* **22**, 143.

Govers, T. R., P. M. Guyon, T. Baer, K. Cole, H. Frohlich, and M. Lavollée (1983), *Chem. Phys.* **87**, 373.

Greenland, P. T. (1982), *Phys. Rep.* **81**, 132.

Groeneveld, K.-O., W. Meckbach, and I. A. Sellin, Eds. (1984a), *Forward Electron Ejection in Ion Collisions*, *Proceedings*, Aarhus, Denmark, Springer-Verlag, Berlin.

Groeneveld, K.-O., W. Meckbach, I. A. Sellin, and J. Burgdörfer (1984b), *Comments At. Phys.* **14**, 187.

Groh, W., A. Müller, B. Schuch, A. S. Schlachter, and E. Salzborn (1983), *ICPEAC XIII*, Berlin, Abstracts of Papers, p. 576.

Gryzinski, M. (1959), *Phys. Rev.* **115**, 374.

Gryzinski, M. (1965), *Phys. Rev.* **138**, A336.

Guyon, P. M., T. R. Govers, and T. Baer (1986), *Z. Phys. D At., Mol. Clusters* **4**, 89.

Hahn, Y. (1989), *Phys. Rev. A* **40**, 2950.

Hasted, J. B. (1964), *Physics of Atomic Collisions*, Butterworth, London, p. 421.

Henderson, G. H. (1922), *Proc. R. Soc. London* **A102**, 496.

Hermann, V., H. Schmidt, and F. Linder (1978), *J. Phys. B* **11**, 493.

Hippler, R., S. Datz, H. F. Krause, P. D. Miller, P. L. Pepmiller, and P. F. Dittner (1988), *Phys. Rev. A* **37**, 3201.

Horsdal-Pedersen, E., C. L. Cocke, and M. Stockli (1983), *Phys. Rev. Lett.* **50**, 1910.

Howald, A. M., R. E. Miers, J. S. Allen, L. W. Anderson, and C. C. Lin (1984), *Phys. Rev. A* **29**, 1083.

Hvelplund, P., E. Samsøe, L. H. Andersen, H. K. Haugen, and H. Knudsen (1983), *Phys. Scr.* **T3**, 176.

Il'in, R. N., V. A. Oparin, E. S. Solov'ev, and N. V. Fedorenko (1965), *JETP Lett.* **2**, 310 [transl. in *Sov. Phys. JETP Lett.* **2**, 197].

IPPJ-AM (1977–   ), *Reports of the Institute of Plasma Physics*, Nagoya University, Chikusa-ku, Nagoya, Japan.

Jackson, J. D., and H. Schiff (1953), *Phys. Rev.* **89**, 359.

Jaecks, D., B. Van Zyl, and R. Geballe (1965), *Phys. Rev.* **137**, A340.

JAERI-M (1982–   ), *Reports of the Japan Atomic Energy Research Institute*, Tokai-mura, Naka-gun, Ibaraki-ken, Japan.

Janev, R. K., and L. P. Presnyakov (1981), *Phys. Rep.* **70**, 1.

Janev, R. K., and H. Winter (1985), *Phys. Rep.* **117**, 266, 2463.

Keene, J. P. (1949), *Philos. Mag.* **40**, 369.

Kim, H. J., and R. K. Janev (1987), *Phys. Rev. Lett.* **58**, 1837.

Kimura, M. and N. F. Lane (1990), *Adv. At. Mol. Opt. Phys.* **26**, 50.

Kimura, M., and C. D. Lin (1985a), Phys. Rev. A **31**, 590.

Kimura, M., and C. D. Lin (1985b), Phys. Rev. A **32**, 1357.

Knudsen, H., H. K. Haugen, and P. Hvelplund (1981), *Phys. Rev. A* **23**, 597.

Koopman, D. W. (1968), *Phys. Rev.* **166**, 57.

Koyano, I., K. Tanaka, and T. Kato (1986), *ICPEAC XIV*, Stanford, Calif., p. 529.

Kozlov, V. F., and A. M. Rozhkov (1962), *Zh. Tekh. Fiz.* **32**, 719 [transl. in *Sov. Phys. Tech. Phys.* **7**, 524].

Landau, L. D. (1932), *Phys. Z. Sowjetunion* **1**, 46.

Lapicki, G., and F. D. McDaniel (1980), *Phys. Rev. A* **22**, 1896.

Lichten, W. (1963), *Phys. Rev.* **131**, 229.

Linder, F. (1979), *ICPEAC XI*, Kyoto, Japan, p. 535.

Linder, F. (1984), Invited lecture at the XII SPIG '84, Šibenik, Yugoslavia, Sept. 3–7.

Linder, F. (1988), *ICPEAC XV*, Brighton, England, p. 287.

Lockwood, G. J., and E. Everhart (1962), *Phys. Rev.* **125**, 567.

Lockwood, G. J., H. F. Helbig, and E. Everhart (1963), *Phys. Rev.* **132**, 2078.

Louisell, W. H., M. O. Scully, and W. B. McKnight (1975), *Phys. Rev. A* **11**, 989.

Lucas, M. W., K. F. Man, and W. Steckelmacher (1984), *Forward Electron Ejection in Ion Collisions*, Springer-Verlag, Berlin, p. 1.

Macdonald, J. R., L. Winters, M. D. Brown, T. Chiao, and L. D. Ellsworth (1972), *Phys. Rev. Lett.* **29**, 1291.

Macdonald, J. R., L. Winters, M. D. Brown, L. D. Ellsworth, T. Chiao, and E. W. Pettus (1973), *Phys. Rev. Lett.* **30**, 251.

Macdonald, J. R., C. L. Cocke, and W. W. Eidson (1974), *Phys. Rev. Lett.* **32**, 648.

Macek, J. (1970), *Phys. Rev. A* **1**, 235.

Macek, J. (1984), *ICPEAC XIII*, Berlin, p. 317.

Macek, J., and S. Alston (1982), *Phys. Rev. A* **26**, 250.

Mapleton, R. A. (1972), *Theory of Charge Exchange*, Wiley-Interscience, New York.

Massey, H. S. W., and E. H. S. Burhop (1952), *Electronic and Ionic Impact Phenomena*, Clarendon Press, Oxford.

Massey, H. S. W., and H. B. Gilbody (1974), *Electronic and Ionic Impact Phenomena, Vol.* 4, Clarendon Press, Oxford.

McClure, G. W. (1964), *Phys. Rev.* **132**, 1636.

McDaniel, E. W., M. R. Flannery, H. W. Ellis, F. L. Eisele, W. Pope, and T. G. Roberts (1977–1982), *Compilation of Data Relevant to Rare Gas–Rare Gas and Rare Gas–Monohalide Excimer Lasers*, Tech. Rep. H-78-1, U.S. Army Missile Research and Development Command, Redstone Arsenal, Ala.

McDowell, M. R. C., and J. P. Coleman (1970), *Introduction to the Theory of Ion–Atom Collisions*, North-Holland, Amsterdam, p. 379.

McGuire, J. H. (1973), *Phys. Rev. A* **8**, 2760.

McGuire, J. H., J. C. Straton, W. J. Axmann, T. Ishihara, and E. Horsdal (1989), *Phys. Rev. Lett.* **62**, 2933.

McLaughlin, D. J., and Y. Hahn (1982), *Phys. Lett.* **88A**, 394.

Meng, L., C. O. Reinhold, and R. E. Olson (1989), *Phys. Rev. A* **40**, 3637.

Mitchell, J. B. A., K. F. Dunn, G. C. Angel, R. Browning, and H. B. Gilbody (1977), *J. Phys. B: At. Mol. Phys.* **10**, 1897.

Mott, N. F., and H. S. W. Massey (1965), *Theory of Atomic Collisions*, 3rd ed., Oxford University Press, Oxford.

Müller, A., W. Groh, and E. Salzborn (1983), *Phys. Rev. Lett.* **51**, 107.

Nagy, S. W., W. J. Savola, Jr., and E. Pollack (1969), *Phys. Rev.* **177**, 71.

Nakai, Y., A. Kikuchi, T. Shirai, and M. Sataka (1983a), *Data on Collisions of Hydrogen Atoms and Ions with Atoms and Molecules (I)*, JAERI-M, 83-013, Japan Atomic Energy Research Institute, Tokai-mura, Naka-gun, Ibaraki-ken.

Nakai, Y., A. Kikuchi, T. Shirai, and M. Sataka (1983b), *Data on Collisions of Hydrogen Atoms and Ions with Atoms and Molecules (II)*, JAERI-M, 83-143, Japan Atomic Energy Research Institute, Tokai-mura, Naka-gun, Ibaraki-ken.

Newman, J. H., J. D. Cogan, D. L. Ziegler, D. E. Nitz, R. D. Rundel, K. A. Smith, and R. F. Stebbings (1982), *Phys. Rev. A* **25**, 2976.

Niehaus, A. (1980), *Comments At. Mol. Phys.* **9**, 153.

NIFS-DATA (1990–), *Reports of the National Institute for Fusion Science*, Chikusa-Ku, Nagoya, Japan.

Noll, M., and J. P. Teonnies (1986), *J. Chem. Phys.* **85**, 3313.

Olson, R. E. (1981), *Phys. Rev. A* **24**, 1726.

Olson, R. E., and A. Salop (1976), *Phys. Rev. A* **14**, 579.

Olson, R. E., A. Salop, R. A. Phaneuf, and F. W. Meyer (1977), *Phys. Rev. A* **16**, 1867.

Oparin, V. A., R. N. Il'in, and E. S. Solov'ev (1967), *J. Exp. Theor. Phys. (USSR)* **52**, 369 [transl. in *Sov. Phys. JETP* **25**, 240].

Oppenheimer, J. R. (1928), *Phys. Rev.* **31**, 349.

Peart, B., K. Rinn, and K. Dolder (1983), *J. Phys. B: At. Mol. Phys.* **16**, 1461.

Pepmiller, P. L., and P. Richard (1982), *Phys. Rev. A* **26**, 786.

Peterson, J. R., and D. C. Lorents (1969), *Phys. Rev.* **182**, 152.

Pfeifer, S. J., and J. D. Garcia (1981), *Phys. Rev. A* **23**, 2267.

Phillips, J. A., and J. L. Tuck (1956), *Rev. Sci. Instrum.* **27**, 97.

Pivovar, L. T., V. M. Tubaev, and M. T. Novikov (1961), *Zh. Eksp. Teor. Fiz.* **41**, 26 [transl. in *Sov. Phys. JETP* **14**, 20 (1962)].

Pivovar, L. T., M. T. Novikov, and V. M. Tubaev (1962), *Zh. Eksp. Teor. Fiz.* **42**, 1490 [transl. in *Sov. Phys. JETP* **15**, 1035 (1962)].

Pretzer, D., B. Van Zyl, and R. Geballe (1963), *Phys. Rev. Lett.* **10**, 340.

Ribe, F. (1951), *Phys. Rev.* **83**, 1217.

Rinn, K., F. Melchert, and E. Salzborn (1985), *J. Phys. B: At. Mol. Phys.* **18**, 3783.

Riviere, A. C., and D. R. Sweetman (1963), *Atomic Collision Processes, Proc. ICPEAC III*, North-Holland, Amsterdam.

Rothwell, H. L., Jr., R. C. Amme, and B. Van Zyl (1978), *J. Chem. Phys.* **68**, 4326.

Rudd, M. E., and J. H. Macek (1972), in M. R. C. McDowell and E. W. McDaniel, Eds., *Case Studies in Atomic Physics*, Vol. 3, No. 2, North-Holland, Amsterdam, p. 47.

Rudd, M. E., R. D. DuBois, L. H. Toburen, C. A. Ratcliffe, and T. V. Goffe (1983), *Phys. Rev. A* **28**, 3244.

Rudd, M. E., T. V. Goffe, A. Itoh, and R. D. DuBois (1985a), *Phys. Rev. A* **32**, 829.

Rudd, M. E., T. V. Goffe, A. Itoh, and R. D. DuBois (1985b), *Phys. Rev. A* **32**, 2128.

Russek, A. (1963), *Phys. Rev.* **132**, 246.

Russek, A., and J. Meli (1970), *Physica (Utrecht)* **46**, 222.

Rutherford, E. (1924), *Philos. Mag.* **47**, 277.

Ruyufuku, H., and T. Watanabe (1978), *Phys. Rev. A* **18**, 2005.

Schlachter, A. S., J. W. Stearns, W. G. Graham, K. H. Berkner, R. V. Pyle, and J. A. Tanis (1983), *Phys. Rev. A* **27**, 3372.

Sellin, I. A. (1982), *ICPEAC XII*, Gatlinburg, Tenn., p. 195.

Shah, M. B., and H. B. Gilbody (1974), *J. Phys. B* **7**, 256.

Shah, M. B., and H. B. Gilbody (1978), *J. Phys. B* **11**, 121.

Shakeshaft, R. (1982), *ICPEAC XII*, Gatlinburg, Tenn.

Shakeshaft, R., and L. Spruch (1979), *Rev. Mod. Phys.* **51**, 369.

Shima, K., T. Mikumo, and H. Tawara (1986), *At. Data Nucl. Data Tables* **34**, 357.

Shima, K., N. Kuno, and M. Yamanouchi (1989), *Phys. Rev. A* **40**, 3557.

Shima, K., N. Kuno, M. Yamanouchi, and H. Tawara (1991), *Equilibrium Charge Fraction of Ions of $Z = 4 - 92$ (0.02 $- 6$ MeV/u) and $Z = 4 - 20$ (up to 40 MeV/u) Emerging from a Carbon Foil*, NIFS-DATA-10, ISSN 0915-6364, National Institute for Fusion Science, Nagoya, Japan.

Simony, P. R., J. H. McGuire, and J. Eichler (1982), *Phys. Rev. A* **26**, 1337.

Stearns, J. W., K. H. Berkner, V. J. Honey, and R. V. Pyle (1968), *Phys. Rev.* **166**, 40.

Stedeford, J. B. H., and J. B. Hasted (1955), *Proc. R. Soc. London* **A227**, 466.

Steuer, M. F., R. M. Wood, and A. K. Edwards (1977), *Phys. Rev. A* **16**, 1873.

Stier, P. M., and C. F. Barnett (1956), *Phys. Rev.* **103**, 896.

Stolterfoht, N. (1987), in H. Kleinpoppen, Ed., *Progress in Atomic Spectroscopy*, Part D, Plenum Press, New York, p. 415.

Stolterfoht, N., C. C. Havener, R. A. Phaneuf, J. K. Swenson, S. Shafroth, and F. W. Meyer (1987), *Nucl. Instrum. Methods Phys. Res.* **B27**, 584.

Stückelberg, E. C. G. (1932), *Helv. Phys. Acta* **5**, 369.

Swenson, J. K., Y. Yamazaki, P. D. Miller, H. F. Krause, P. F. Dittner, P. L. Pepmiller, S. Datz, and N. Stolterfoht (1986), *Phys. Rev. Lett.* **57**, 3042.

Tanis, J. A. (1987), *Nucl. Instrum. Methods Phys. Res.* **A262**, 52.

Tanis, J. A., S. M. Shafroth, J. E. Willis, M. Clark, J. Swenson, E. N. Strait, and J. R. Mowat (1981), *Phys. Rev. Lett.* **47**, 828.

Tanis, J. A., E. M. Bernstein, W. G. Graham, M. Clark, S. M. Shafroth, B. M. Johnson, K. W. Jones, and M. Meron (1982), *Phys. Rev. Lett.* **49**, 1325.

Tanis, J. A., E. M. Bernstein, C. S. Oglesby, W. G. Graham, M. Clark, R. H. McFarland, T. J. Morgan, M. P. Stockli, K. H. Berkner, A. S. Schlachter, J. W. Stearns, B. M. Johnson, K. W. Jones, and M. Meron (1985), *Nucl. Instrum. Methods Phys. Res.* **B10/11**, 128.

Tanis, J. A., G. Schiwietz, D. Schneider, N. Stolterfoht, W. G. Graham, H. Altevogt, R. Kowallik, A. Mattis, B. Skogvall, T. Schneider, and E. Szmola (1989), *Phys. Rev. A* **39**, 1571.

Tawara, H., and A. Russek (1973), *Rev. Mod. Phys.* **45**, 178.

Tawara, H., T. Kato, and Y. Nakai (1985), *At. Data Nucl. Data Tables* **32**, 235.

Thomas, L. H. (1927), *Proc. R. Soc.* **114**, 561.

To, K. X., and R. Drouin (1976), *Phys. Scr.* **14**, 277.

Toburen, L. H., M. Y. Nakai, and R. A. Langley (1968), *Phys. Rev.* **171**, 114.

Trujillo, S. M., R. H. Neynaber, and E. W. Rothe (1966), *Rev. Sci. Instrum.* **37**, 1655.

Tuan, T. F., and E. Gerjuoy (1960), *Phys. Rev.* **117**, 756.

Vinogradov, A. V., and V. P. Shevel'ko (1970), *Zh. Eksp. Teor. Fiz.* **59**, 593 [trans. in *Sov. Phys. JETP* **32**, 323 (1971)].

Watts, M. F., K. F. Dunn, and H. B. Gilbody (1986), *J. Phys. B: At. Mol. Phys.* **19**, L35.

Whittier, A. C. (1952), Ph. D. thesis, McGill University.

Wille, U., and R. Hippler (1986), *Phys. Rep.* **132**, 129.

Wilson, W. E., and L. H. Toburen (1973), *Phys. Rev. A* **7**, 1535.

Wittkower, A. B., and H. D. Betz (1971), *J. Phys. B* **4**, 1173.

Wittkower, A. B., and H. D. Betz (1973), *At. Data Nucl. Data Tables* **5**, 113.

Wood, R. M., A. K. Edwards, and M. F. Steuer (1977), *Phys. Rev. A* **15**, 1433.

Yenen, O., D. H. Jaecks, and J. Macek (1984), *Phys. Rev. A* **30**, 597.

Zaidins, C. (1968), in J. B. Marion and F. C. Young, Eds., *Nuclear Reaction Analysis: Graphs and Tables*, North-Holland, Amsterdam, p. 34.

Zare, R. N., and R. B. Bernstein (1980), *Phys. Today* **33**(11), 43.

Zener, C. (1932), *Proc. R. Soc.* **A137**, 696.

Ziemba, F. P., and E. Everhart (1959), *Phys. Rev. Lett.* **2**, 299.

Ziemba, F. P., G. J. Lockwood, G. H. Morgan, and E. Everhart (1960), *Phys. Rev.* **118**, 1552.

Zouros, T. J. M., D. Schneider, and N. Stolterfoht (1987), *Phys. Rev. A* **35**, 1963.

Zouros, T. J. M., D. H. Lee, P. Richard, J. M. Sanders, J. L. Shinpaugh, S. L. Varghese, K. R. Karim, and C. P. Bhalla (1989), *Phys. Rev. A* **40**, 6246.

6

NEGATIVE IONS

6-1. INTRODUCTION

A large proportion of this volume is concerned with interactions involving positively charged ions, which are usually formed through violent electron ejection events. There is, however, another class of ion. Negative ions arise when a neutral atom or molecule attaches an extra electron and hence acquires a negative charge. They are created and destroyed by processes that are typically more gentle than their positively charged counterparts, and in many cases they can form at thermal energies. Because of this fact, they have many important practical applications, particularly in the area of electrostatics, atmospheric phenomena, electrical technology, and accelerator physics. Negative ions are formed naturally in the atmosphere and play important roles in the nucleation of aerosols and in lightning production. In an ionized gas, their formation leads to a decrease in charged particle mobility since they are much heavier than electrons. Compounds that readily form negative ions are, therefore, often used in high-voltage insulation applications. Since electrons are readily detached from negative ions, they are used in conjunction with tandem accelerators to produce very high-energy positive-ion beams. In recent years a great deal of effort has gone into understanding the mechanism for negative-ion formation in high current ion sources, designed for use in neutral beam injection applications for fusion plasma heating as well as for military applications.

Detailed reviews of recent negative-ion physics have been given by Massey (1976), Smirnov (1982), Janousek and Brauman (1979), and Christophorou (1984), while early studies are described in McDaniel (1964). Applications where negative ions play an important role are discussed in Massey et al. (1982–1984).

6-2. STRUCTURE AND SPECTRA OF NEGATIVE IONS

A. Negative Atomic Ions

A neutral atom may be regarded as consisting of a point-charge nucleus surrounded by a cloud of atomic electrons. If an additional electron is brought near the atom, it finds itself in an attractive force field that falls off much more rapidly with distance than that acting on one of the atomic electrons in the neutral atom. In the neutral atom, the asymptotic behavior of the field is coulombic, whereas an additional electron in the field of a hydrogen atom, for instance, feels something like an inverse-fourth-power attractive potential. A negative ion may thus be considered in terms of the stationary states of an electron in an attractive field which falls off very rapidly with distance. The number of stationary states is therefore likely to be finite, as contrasted with the infinite number in a Coulomb field. This fact, together with the Pauli exclusion principle, imposes a severe restriction on the number of elements that can form a negative ion.

In the H atom there is a vacancy in the lowest ($n = 1$) level and the possibility exists for capture of a second electron into the 1s subshell, close to the nucleus of the atom. The shielding of the nuclear field by the 1s electron already present might then be thought to be weak enough to permit the second electron to be bound by the nuclear field, and the existence of the H^- ion, identified experimentally many years ago, is not at all surprising.

In He, however, the $n = 1$ level is fully occupied and an additional electron can be bound to the atom only in the $n = 2$ state or one of higher quantum number. The nuclear field in the He atom is strongly attenuated by the atomic electrons at distances corresponding to shells with $n > 1$, and at one time it was thought that He^- could not exist at all. The ion was, however, observed experimentally [first in the 1930s by Hilby (1939)], and a variational calculation by Holøien and Midtdal (1955) showed that the energy of the $(1s2s2p)^4P$ state is less than the 2^3S state of He so that it is metastable against autoionization. In general, those atoms with completely filled outer shells are unlikely to form negative ions. The additional electron would have to be attached in a state of higher total quantum number than that of the outermost atomic electrons, and usually the attractive force is too weak at distances corresponding to these states. Those atoms with single vacancies in their outer shells should be the most likely of all to attach electrons, since in these structures, the outer atomic electrons have very little effect in shielding the added electron from the nucleus. Thus it is to be expected that the noble gases should have difficulty in forming stable negative ions but that the halogens should prove to be highly electronegative; experiment confirms this expectation.

Although the foregoing considerations help us to understand why some atoms form negative ions and others do not, it is necessary to make detailed calculations to obtain quantitative information about an individual atomic species. A necessary condition is supplied by energy considerations, for if a given species of negative ion is to be stable, its binding energy must be greater than that of the corresponding atom. The criterion for stability of a negative ion formed by an atom of Z electrons may therefore be expressed as

$$E_1 + \sum_{i=1}^{z} E_i^- > \sum_{i=1}^{z} E_i^0 \qquad (6\text{-}2\text{-}1)$$

or

$$E_1 - \sum_{i=1}^{z} (E_i^0 - E_i^-) > 0, \qquad (6\text{-}2\text{-}2)$$

where E_i^0 is the binding energy of the ith atomic electron before the attachment takes place, E_i^- the binding energy of that electron after the attachment, and E_1 the binding energy of the attached electron. The quantity on the left-hand side of (6-2-1) is called the *electron affinity* of the atom and represents the difference in total energy of the normal states of the atom and the negative ion. This energy, which is denoted by EA, also equals that required to detach the least tightly bound electron from the ion (the detachment energy of the ion). Since binding energies are negative quantities, a positive electron affinity indicates the stability of the negative ion. It should be emphasized that the binding energies E_i^- are those in the atomic field as modified by the presence of the additional electron. The change in the field resulting from the attachment of the electron, usually referred to as electron correlation, can be decisive in determining whether the ion is stable.

There is an enormous body of work concerning electron affinities. For example, a recent compilation by Christodoulides et al. (1984) lists 819 entries covering the measurement and calculation of electron affinities for 71 elements. Comprehensive reviews of the methods used to obtain electron affinities have been given by those authors and by Massey (1976) and Smirnov (1982). Negative-ion research has been revolutionized by the advent of high-intensity tunable visible lasers and by the advances made in the measurement of photoelectron spectra using high-resolution electron analyzers. This technology has allowed the accuracy of measured electron affinities to be increased by more than four orders of magnitude over studies described in McDaniel (1964). Indeed, the values for sodium, rubidium, cesium, and gold are now known to five decimal places:

$$EA(Na) = 0.547930 \pm 0.000025 \, eV$$

$$EA(Rb) = 0.48592 \pm 0.00002 \, eV$$

$$EA(Cs) = 0.471630 \pm 0.000025 \, eV$$

$$EA(Au) = 2.30863 \pm 0.00003 \, eV$$

while those for hydrogen, deuterium, oxygen, and sulfur are known to six places:

$$EA(H) = 0.754195 \pm 0.000019 \, eV$$

$$EA(D) = 0.754593 \pm 0.000074 \, eV$$

$$EA(O) = 1.461125 \pm 0.000001 \, eV$$

$$EA(S) = 2.077120 \pm 0.000001 \, eV.$$

A much more extensive list of electron affinities for atomic species appears in Table 6-2-1. (A similar list for molecular species is given in Table 6-2-2.) Most of the values given here have been derived from experimental measurements. Where this information is not available, data taken from quantum mechanical or semiempirical calculations are shown. The elements have been listed according to their groups in the periodic table, along with the electronic configurations of the negative ions.

It can be seen that, indeed, the noble gases, with filled electron shells, do not form negative ions except in excited states. Furthermore, the group II elements, Be, Mg, Ca, and so on, which have filled s subshells, also have difficulty binding an extra electron. The group VII elements, the halogens, have large electron affinities, as the addition of the extra electron leads to the complete filling of electron shells. Most of the other elements exhibit positive electron affinities ranging from a few tenths to 1 or 2 eV. Semiempirical methods have been developed to try to predict relationships between neighboring members of element periods, groups, and isoelectronic series, and a fuller discussion of these

Table 6-2-1. Electron Affinities of the Elements

At. No.	Element	EA	Method[a]	At. No.	Element	EA	Method[a]
	Group 0	$ns^2 S_{1/2}$			Group IIA	$ns^2 n'l \, {}^2L$	
2	He	<0	T	4	Be	<0	T
10	Ne	<0	T	12	Mg	<0	EI
18	Ar	<0	T	20	Ca	0.043	PES
36	Kr	<0	T	38	Sr	0.11	T
54	Xe	<0	T	56	Ba	0.15	T
	Group IA	$ns^2 \, {}^1S_0$					
1	H	0.754195	LPT		Group IIIA	$nl^3 \, {}^3L_0$	
3	Li	0.6180	LPT	5	B	0.277	PES
11	Na	0.547926	LPT	13	Al	0.441	PES
19	K	0.50147	LPT	31	Ga	0.3	PT
37	Rb	0.48592	LPT	49	In	0.3	PT
55	Cs	0.471626	LPT	81	Tl	0.2	PT

Table 6-2-1 (*Continued*)

At. No.	Element	EA	Method[a]	At. No.	Element	EA	Method[A]
	Group IVA	$np^3\ ^4S_{3/2}$			*Group IIIB*	$nd^2(n+1)s^2\ ^3F$	
6	C	1.2629	LPT	21	Sc	0.188	PES
14	Si	1.385	PES	39	Y	0.307	PES
32	Ge	1.233	PES	57	La	0.5	T
50	Sn	1.112	PES				
82	Pb	0.364	PES		*Group IVB*	$nd^3(n+1)s^2\ ^4F$	
				22	Ti	0.079	PES
	Group VA	$np^4\ ^3P_2$		40	Zr	0.426	PES
				72	Hf	0.04	T
7	N	−0.07	DA				
15	P	0.7565	LPT		*Group VB*	$nd^4(n+1)s^2\ ^5D$	
33	As	0.810000	PT				
51	Sb	1.07	PT	23	V	0.525	PES
83	Bi	0.946	PES	41	Nb	0.893	PES
				73	Ta	0.322	PES
	Group VIA	$np^5\ ^2P_{3/2}$			*Group VIB*	$nd^5(n+1)s^2\ ^6S_{1/2}$	
8	O	1.4611103	LPT	24	Cr	0.666	PES
16	S	2.077104	LPT	42	Mo	0.746	PES
34	Se	2.020670	LPT	74	W	0.815	PES
52	Te	1.9708	LPT				
84	Po	1.9	T		*Group VIIB*	$nd^6(n+1)s^2\ ^5D$	
				25	Mn	<0	T
	Group VIIA	$np^6\ ^1S_0$		43	Tc	0.55	T
				75	Re	0.15	T
9	F	3.40119	LPT				
17	Cl	3.61269	LPT		*Group VIII*	$nd^7(n+1)s^2\ ^4F$	
35	Br	3.36359	LPT				
53	I	3.0591	LOGS	26	Fe	0.151	PES
				44	Ru	1.05	T
				76	Os	1.1	T
	Group IB	$nd^{10}(n+1)s^2\ ^1S_0$					
					Group VIII	$nd^8(n+1)s^2\ ^3F$	
29	Cu	1.235	PES				
47	Ag	1.302	PES	27	Co	0.662	PES
79	Au	2.30863	LPT	45	Rh	1.137	PES
				77	Ir	1.565	PES
	Group IIB	$ns^2(n+1)s\ ^2S_{1/2}$			*Group VIII*	$nd^9(n+1)s^2\ ^2D_{3/2}$	
30	Zn	<0	ETS	28	Ni	1.156	PES
48	Cd	<0	ETS	46	Pd	0.557	PES
80	Hg	<0	ETS	78	Pt	2.128	PES

Source: Miller (1991).

[a]T, theory; LPT, laser photodetachment threshold; PT, photodetachment threshold; EI, electron impact; PES, photoelectron spectroscopy; DA, dissociative attachment; ETS, electron transmission spectroscopy; LOGS, laser optogalvanic spectroscopy.

Table 6-2-2. Electron Affinities of Molecules

Molecule	EA	Method[a]	Molecule	EA	Method[a]
			PtF_4	5.25	CTR
	Diatomics		PtF_5	6.5	CTR
OH	1.82767	LPT	PtF_6	>5.14	CITS
SH	2.314344	LPT	ReF_6	5.1	CITS
O_2	0.451	PES	SF_3	3.07	CITS
SO	1.125	PES	SF_4	1.5	CTR
S_2	1.67	PES	SF_5	>3.7	CTR
NO	0.026	PES	SF_6	1.05	CITS
N_2	−1.9	AI	SiF_3	<2.95	PT
CO	−1.3	AI	SiF_5	6.3, 6.4	T
C_2	3.269	PES	TeF_4	2.2	DEAR
CN	3.821	LOGS	TeF_5	4.2	DEAR
F_2	3.08	CTR	TeF_6	3.3	CITS
Cl_2	2.38	CTR	UF_5	3.3	CTR
Br_2	2.55	CTR	UF_6	>5.1	CITS
I_2	2.55	CTR			
IBr	2.55	CTR			
				Polyatomic Organic Molecules	
	Triatomics		$CClF_2$	1.6	CITS
H_3	<0	T	$CClF_3$	−0.4	T
O_3	2.1028	LPT	CCl_2F	1.1	CITS
NO_2	2.275	PES	CCl_2F_2	0.4	CITS
N_2O	0.22	EI	CCl_3	1.3	CITS
N_3	2.69	PT	CCl_3F	1.1	CITS
CO_2	−0.6	CTR	CCl_4	2	CITS
C_3	1.981	PES	PAHs		
AlO_2	4.11	ETR	$C_{14}H_{10}$	0.48, 0.52	EAR
BO_2	3.57	CTR	Anthracene		
Cl_3	4.6, 5.1	CITS	$C_{16}H_{10}$	0.5	EAR
			Pyrene		
			$C_{18}H_{12}$	0.397, 0.419	EAR
	Polyatomic Inorganic Molecules		Chrysene		
AlF_4	5.4, 5.5	T	$C_{20}H_{12}$	0.68	EAR
CO_3	2.69	PES	Benzopyrene		
CeF_4	3.6	CTR	$C_{24}H_{12}$	0.32	S
FeF_4	5.4	CTR	Coronene		
IrF_6	>5.14	CITS	Clusters		
MgF_3	3.2, 3.8	T	$(Cl_2)_2$	2.2, 2.8	CITS
MnF_3	4.36	CTR	Cl_5	4.7, 5.2	CITS
MnF_4	5.27	CTR	$(Cl_2)_3$	2.3, 2.8	CITS
MoF_6	3.6	CTR	Cl_7	4.7, 5.2	CITS
$MoOF_4$	4	CTR	$(Cl_2)_4$	2.3, 2.8	CITS
OsF_6	5.97, 6.17, 7.3	S			

Source: Christodoulides et al. (1984).

[a] T, theory; PT, photodetachment threshold; LPT, laser photodetachment threshold; AI, atom impact; LOGS, laser optogalvanic spectroscopy; EI, electron impact; CTR, charge transfer; CITS, collisional ionization threshold spectra; ETR, electron transfer reaction; DEAR, dissociative electron attachment reaction; EAR, electron attachment reaction.

methods is given in Massey (1976). There are a few notable oddities throughout the table, including N^-, which appears to be unstable, although other members of group VA have small but significant EA values. A similar situation arises for Mn^-. Pegg et al. (1987) have discovered that Ca^-, a group IIA element, is in fact stable, with a binding energy of 43 meV.* This experimental finding has been confirmed by the quantum mechanical calculations of Froese-Fischer et al. (1987). The reason for this stability is that Ca^- assumes the configuration $4s^24p(^2P)$ rather than $3d4s^2(^2D)$ as assumed previously. The calculations imply that the extra electron experiences less Coulomb repulsion from the other electrons in the former state than in the latter, and it is this fact that brings the energy of the state below the binding energy of the $4s^2(^1S)$ ground state of the neutral calcium atom. A similar situation occurs with the neighboring scandium negative ion, where again capture occurs into the $3d4s^24p(^1D$ or $^3D)$ configuration rather than the $3d^24s^2(^3F)$ configuration, because of the reduced repulsion experienced by the added electron in the former state.

B. Excited States of Negative Atomic Ions

Table 6-2-1 refers to the stable ground states of negative ions. Short-lived autodetaching negative-ion states are often formed temporarily during collision processes; these are discussed in Chapter 5 of the companion volume and so are not dealt with here. Some possible candidates for having bound excited states, however, are C^-, Si^-, Ge^-, Sn^-, Ni^-, Pd^-, and Pt^-. In each case, calculation of the energies of the first excited states indicate that these states may have binding energies of a few meV. In addition to stable excited states there are also negative ions that exist in metastable autodetaching states which are sufficiently long-lived to be seen in ion beams. Most notable of these is helium, which forms a metastable negative ion in the $1s2s2p(^4P)$ configuration even though the ground-state $1s^22s(^2S_{1/2})$ configuration is unstable. Autodetachment from this state is spin forbidden. The $^4P_{1/2}$ and $^4P_{3/2}$ states can decay via spin-orbit and spin-spin interactions, while the $^4P_{5/2}$ state can decay only via the latter and so is considerably longer lived. Calculations of Manson (1971) indicate the lifetime for the $\frac{3}{2}$ level to be 33 μs, compared to 1000 μs for the $\frac{5}{2}$ level. The lifetime for the $\frac{1}{2}$ level could not be calculated. Experimental determination of the He^- lifetimes by Blau et al. (1970) and Novick and Weinflash (1970) estimate values of 500 ± 200 μs for the $\frac{5}{2}$ level, 10 ± 2 μs for the $\frac{3}{2}$ level, and 16 ± 4 μs for the $\frac{1}{2}$ level. The lifetime of the $\frac{5}{2}$ level is sufficiently long that He^- ions can be used routinely for the production of high-energy He^+ ions in tandem accelerators.

A number of metastable negative ions have been identified by Peterson and coworkers (Hanstorp et al., 1989) using laser detachment from negative ion beams formed via near-resonant charge exchange between excited atoms and cesium vapor targets. To date, metastable excited states of the ions Ar^-, Be^-, and Ca^- have been seen as well as the molecular ion He_2^-. However, with the

*A recent study by Walter et al. (1991) has indicated an EA value of 9 meV with respect to the ^1S state of Ca.

exception of Ca$^-$, which is weakly bound, these ions do not support stable ground states. Searches made for metastable states of otherwise unbound negative ions Mg$^-$, Ne$^-$, Kr$^-$, and Xe$^-$ have failed to show evidence of their existence.

To date, there is no evidence for the existence of stable doubly charged atomic negative ions, although short-lived species have been observed in a number of beam experiments [see Massey (1976)].

C. Negative Molecular Ions

To appreciate the structure of negative molecular ions and the various collision processes involving them, a familiarity with molecular potential energy curves is necessary. This subject is sketched in Chapter 5 together with a discussion of dissociation processes resulting from transitions between the various molecular states. As pointed out there, electronic transitions between pairs of molecular states occur on a time scale that is generally much shorter than characteristic vibrational periods of the nuclear motion, so the nuclei can be considered to be at rest during these transitions. This simplification is known as the *Born–Oppenheimer approximation*, and within this approximation, electronic transitions are considered to be vertical on a potential energy plot (i.e., the nuclei do not have time to change their separation during the transition). This behavior, governed by the *Franck–Condon principle*, introduces a complication compared to the case for atomic transitions, for now the energy required to drive the transition usually differs substantially from the asymptotic energy separations of the initial and final states. The *electron affinity* is defined here as the difference in the energies of the neutral molecule and its molecular ion when both structures are in their normal electronic and nuclear states. Thus if the molecular negative-ion state has a minimum at an internuclear separation different from that of the neutral molecule, the electron affinity will differ from the energy required to remove the electron from the ion by, for example, electron or photon impact. The latter quantity is known as the *vertical detachment energy*. This situation is illustrated in Fig. 6-2-1.

As in the case of atoms, extensive studies have been performed to determine the electron affinities of a wide variety of molecular species. Table 6-2-2 gives a selection of electron affinities for diatomic, triatomic, and polyatomic molecules and for radicals. Much more extensive listings are given by Christodoulides et al. (1984) and Miller (1991).

6-3. METHODS OF MEASUREMENT OF ELECTRON AFFINITIES

As mentioned in Section 6-2-A, the laser threshold photodetachment technique has dominated the measurement of electron affinities for atomic ions in recent years. It is, however, much more limited when applied to molecules. The reason

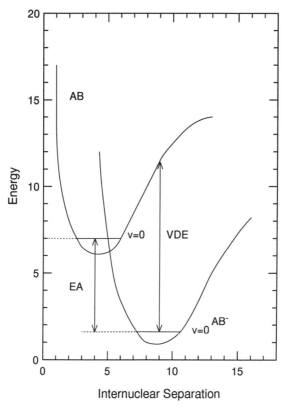

Figure 6-2-1. Potential energy curves of a diatomic molecule AB and its negative ion AB⁻. EA is the *electron affinity* of AB and VDE the *vertical detachment energy* of AB⁻.

for this is that when the minima of the ground electronic states of the neutral molecule and negative ion are offset from each other, the neutral molecule following photodetachment can be left in a vibrationally excited state. There are instances, however, when conditions are appropriate for its implementation, and it has been used for the determination of very accurate electron affinities for OH⁻ and SH⁻ (Schulz et al., 1982; Breyer et al., 1981).

A more widely applicable technique is that of photoelectron spectroscopy (Note 6-2), where the measured energy of the photoelectrons ejected yields information on the various excitation states produced by the photodetachment process. Provided that the spectrum is not too complicated, unambiguous electron affinities can be obtained.

A third optical absorption technique involves measuring the threshold for a polar dissociation process, such as

$$hv + HCN \rightarrow H^+ + CN^-.$$

Since the ionization potential of atomic hydrogen is known very accurately, a precise determination of the electron affinity for the negative ion can be made (Berkowitz et al., 1969).

Electron impact studies involving dissociative attachment,

$$e + AB \to A + B^-,$$

or nondissociative attachment,

$$e + AB \to (AB^-)^*,$$

have also been used (Burrow et al., 1976) to yield electron affinities, although the energy resolution of these techniques is inferior to that for photodetachment.

A technique that has seen widespread use (Chupka et al., 1971) is that of measuring the energy thresholds for endothermic charge exchange processes, such as

$$Cl_2 + I^- \to Cl_2^- + I.$$

In this particular case, the electron affinity for I^- is known (3.08 eV) and the threshold of the reaction was measured to be 0.66 eV. Hence the electron affinity of Cl_2^- is determined to be 2.4 eV. Due to the difficulty of making a precise determination of the position of the threshold, which is smeared out by the energy spread of the reactants, the accuracy is limited to about 0.2 eV. In some circumstances the electron affinity can be determined by bracketing a series of charge-transfer reactions (Viggiano et al., 1991).

A related technique (Compton et al., 1978) involves the measurement of thresholds for collisional ionization reactions involving atom impact, such as

$$Cs + SF_6 \to Cs^+ + SF_6^-.$$

In a few cases where negative ions are unbound, the electron affinity can be determined by measuring the translational energy spectra of the products.

Electron affinities have also been determined by measuring the equilibrium constants for reactions such as the electron transfer process,

$$X^- + AlO_2 \to X + AlO_2^-$$

(Srivastava et al., 1972).

6-4.   MECHANISMS FOR FORMATION OF NEGATIVE IONS

It can be seen from Tables 6-2-1 and 6-2-2 that even for the most electronegative atoms and compounds, the electron affinities are relatively small. Negative ions are rather fragile entities and hence their formation generally arises from subtle collision processes whose cross sections often fall off sharply above threshold. In several examples, the threshold for electron attachment peaks at zero electron energy. The following specific processes are the primary formation mechanisms for negative ions.

1. The *radiative capture* of a free electron by an atom:

$$e + A \rightarrow A^- + hv.$$

2. The capture of a free electron by an atom with a third body taking up the excess energy:

$$e + A + B \rightarrow A^- + B + \text{kinetic energy}.$$

3. The capture of a free electron by a molecule, accompanied by vibrational excitation of the molecular ion with subsequent stabilization in a collision with another molecule:

$$e + XY \rightarrow [XY^-]^*$$

$$[XY^-]^* + A \rightarrow [XY^-] + A + \text{kinetic energy} + \text{potential energy}.$$

4. *Dissociative attachment*; the capture of a free electron by a molecule, the excess energy going into dissociation of the molecule:

$$e + XY \rightarrow [XY^-]^* \rightarrow X + Y^- + \text{kinetic energy}$$

$$e + XY^+ \rightarrow X^+ + Y^-.$$

5. *Ion-pair production*; the noncapture dissociation of a neutral molecule or molecular ion into positive and negative ions by electron or heavy particle impact:

$$e + XY \rightarrow e + X^+ + Y^-$$

$$A^+ + XY \rightarrow A^+ + X^+ + Y^-$$

$$A^* + XY \rightarrow A^+ + XY^-$$

$$XYZ^+ + A \rightarrow X^+ + Y^- + Z^+ + A.$$

These mechanisms are often referred to as *polar dissociation processes*.

6. The transfer of one or more electrons to a neutral structure or a positive ion in a collision with another particle:

$$A + B \rightarrow A^- + B^+; \text{ or } C^+ + D \rightarrow C^- + D^{++}$$

$$A^- + B \rightarrow A + B^-.$$

Under ordinary laboratory conditions and excluding charge-changing collisions, processes 2, 3, and 4 are most probable. For electrons with energies in excess of 20 eV, process 5 will also become important. At very low pressures, process 1 must be considered while processes 2 and 3 become unimportant compared to process 4. It should be noted that the processes involving capture of free electrons are exactly equivalent to electron–ion recombination processes (O'Malley, 1971). The latter are, however, generally more complex, due to the interaction of neutral Rydberg states and are discussed in Chapter 9. The reactions listed above are discussed individually in greater detail below.

A.   Attachment of Free Electrons to Neutral Atoms

The problems associated with this process are similar to those involved with the radiative recombination of electrons with atomic ions discussed in Chapter 9. For the system to undergo a transition from the neutral atom to the negative ion, energy must be removed, equal to the sum of the electron affinity plus the kinetic energy of the electron. For a binary electron–atom collision, the only mechanism available for the removal of this energy is radiation. Since the typical time for a thermal electron to traverse an atom is 10^{-15} s and that for radiation to occur is 10^{-9} s, radiative attachment processes occur with a very small rate, which decreases rapidly with increasing electron energy. Calculated radiative attachment rates for atomic hydrogen (Massey, 1976) are shown in Table 6-4-1. As with electron–ion recombination, this rate can be greatly enhanced if a third body such as an electron, an atom, or a molecule is available to remove the excess energy.

In McDaniel (1964), a third process, referred to as *dielectronic attachment,*

Table 6-4-1. Calculated Radiative Attachment Cross Sections for Atomic Hydrogen

Electron Energy (eV)	Cross Section (10^{-22} cm^2)
0.135	0.456
0.406	0.570
0.677	0.583
1.354	0.547
2.708	0.455
5.416	0.315
8.124	0.300
10.832	0.270

Source: Massey (1976).

was described. This is the neutral atom complement to dielectronic recombination, which has risen to prominence in the last decade, with a number of experimental rate and cross-section measurements being published, along with accurate theoretical studies. (A discussion of these is given in Section 7-4 of the companion volume and in Chapter 9.) Dielectronic recombination of ions is a very important process since it occurs at the elevated temperatures encountered in stars and in fusion reactors. Dielectronic attachment of electrons to neutral atoms, however, is not likely to be an important process in equilibrium systems, for at the elevated temperatures required for its rate to be appreciable, the system is likely to have been converted to plasma form. Then any negative ion thus formed will probably suffer rapid electron detachment by energetic atom or ion collisions. Dielectronic attachment has never been observed in the laboratory.

B.   Negative-Ion Formation in Collisions with Molecules

As the list of formation mechanisms at the beginning of this section shows, negative ions may be produced in impacts of electrons with molecules in reactions of types 3 and 4, which involve capture of the electron, or in process 5, in which the electron merely causes the molecule to break up into charged fragments. The transfer of the excess energy in the capture of an electron is greatly facilitated when the electron is attached to a molecule because of the possibility of increasing the kinetic energy of relative motion of the nuclei in the molecule. In fact, the attachment may be thought of as a three-body process in which the third body is actually bound to the capturing atom or radical. The molecular ion produced in this fashion is left in an excited electronic or vibrational state. The attachment process can be stabilized by either a transfer of the excitation energy to a third body as in processes 2 and 3 or via dissociation, where the potential energy of the excited state is converted to kinetic energy of the dissociation products as in process 4, known as *dissociative attachment*. If stabilization does not occur, the electron will undergo autodetachment on a time-scale of 10^{-11} s if the molecule is small. For larger molecules, internal conversion of energy into vibrational modes can lead to longer-lived intermediate negative-ion states.

The range of electron energy over which processes 3 and 4 occur is quite narrow, typically from 0 to about 15 eV. The rate of the three-body processes 2 and 3 is also determined by the number density of the third body available for subsequent collision.

In process 5 the electron acts simply as a source of energy and excites the molecule into an unstable state that dissociates into a negative and a positive ion. Of course, there is a certain excitation energy below which this process of ion-pair production cannot occur. The probability of its taking place depends on the energy of the electron and the availability of suitable molecular states that decay asymptotically to ion pairs.

6-5.   MECHANISMS FOR DESTRUCTION OF NEGATIVE IONS

Processes leading to detachment of electrons from negative ions include:

1. Collisions with electrons or fast ions or fast molecules. The detachment cross sections can be larger than gas kinetic cross sections:

$$X^- + Y \to X + Y + e.$$

2. Absorption of radiation by the ions (photodetachment):

$$hv + X^- \to X + e.$$

3. Collisions of negative ions with excited atoms:

$$X^- + Y^* \to X + Y + e.$$

4. Collisions with low-energy ions or molecules. These processes include charge exchange reactions:

$$X^- + Y \to X + Y^- \qquad \text{or} \qquad X^- + Y^+ \to X^+ + Y^-$$

and reactive collisions such as

$$X^- + YZ \to XY + Z + e.$$

5. Collisions with neutral atoms leading to molecule formation:

$$X^- + Y \to XY + e.$$

This process is the inverse of dissociative attachment and is similar to the process of associative ionization where two atoms combine to form a molecular ion and a free electron.

6. Recombination processes with positive ions, including three-body recombination:

$$X^- + Y^+ + Z \to XY + Z$$

and mutual neutralization:

$$X^- + Y^+ \to X + Y.$$

7. Field detachment. This process is of particular relevance in the design of cyclotrons and synchro-cyclotrons, where the Lorentz electric field in-

duced by the ions motion through the strong magnetic fields can strip off the rather weakly attached electron.

8. Collision of negative ions with surfaces provide the most effective means of all of electron detachment if the work function of the surface exceeds the electron affinity.

Processes 1 to 5 are discussed in more detail in the following sections of this chapter, and ion–ion recombination processes in Chapter 9. Field detachment has been studied theoretically by Hiskes (1962) and Mullen and Vogt (1968).

6-6. EXPERIMENTAL METHODS FOR STUDY OF NEGATIVE-ION FORMATION

In the following a summary examination of techniques used for the study of dissociative and nondissociative electron attachment will be given. A discussion of charge-transfer studies is given in Chapter 5. The experiments described here are aimed at examining the formation of stable negative ions. Electron scattering experiments where temporary negative-ion resonance states are formed have already been described in Section 5-4 of the companion volume.

There are two primary thrusts in the experimental study of electron attachment processes. The first is to determine the absolute attachment rates to learn which compounds are particularly effective at forming negative ions. The motivation for this work is practical in nature, although through these experiments, a broad picture of the attachment process in complex species has been developed. The main tool in this field of study is the *drift tube*, in which the interaction of electron swarms with attaching gases is examined.

The second thrust is toward a detailed understanding of the mechanisms underlying attachment processes, the states involved, the dependence of the attachment cross section on both the kinetic energy of the electron and the internal energy of the targets, and the branching ratios for competing dissociation channels. Here the high-energy resolution of electron beam experiments is a great advantage. Developments in both these areas will be discussed.

A. Electron Beam Methods

The basic procedure behind these methods is to prepare an electron beam of known energy, allow it to interact with the gas whose attachment cross section is to be determined, and collect the negative ions thus formed as a function of electron energy. In early studies such as those of Lozier (1930) and Tate and Lozier (1935), an effort was made to ensure that complete collection of the ions was accomplished so that accurate absolute cross sections could be obtained. However, this condition was not always met (Note 6-1). Emphasis on complete collection has been continued in a series of apparatus constructed over the years that have employed updated vacuum and data collection techniques to improve the accuracy of the measurements (Schulz, 1959; Asundi et al., 1963; Buchel'nikova, 1958; Rapp and Briglia, 1965). The latter apparatus has been described in Section 6-3-A of the companion volume. Variations on this basic technique have addressed the problems of improving energy resolution, identifying the ions, changing the internal energy of the attaching molecules, and measuring angular distributions of the products.

Increased energy resolution is important in determining the details of the process and for the cases where the cross sections peak at or near zero electron energy. A major advance in the study of electron attachment came with the introduction by Stamatovic and Schulz (1970) of the trochoidal electron analyzer for preselecting the energy spread of the electrons prior to their interaction with the gas under study. The trochoidal analyzer consists of a pair of parallel metal plates across which is applied an electric field E. Electrons from an indirectly heated cathode or thoriated filament are accelerated by and pass between the plates. An axial magnetic field is applied and the electrons are deflected into a helical path as they pass through the plates. As they emerge from the analyzer, their path has been laterally shifted by an amount d. This shift may be determined from the equation

$$E(\text{V/cm}) = 12.56\left(\frac{V_0}{l^2}\right)nd, \qquad (6\text{-}6\text{-}1)$$

where n is the number of complete orbits performed by the electrons, V_0 the acceleration voltage of the electrons, E the electric field, and l the length of the plates in centimeters.

For a given V_0 and l, n is determined by the applied magnetic field strength, thus:

$$B(\text{gauss}) = 21.16\left(\frac{V_0^{1/2}}{l}\right)n. \qquad (6\text{-}6\text{-}2)$$

By restricting the size of the entrance and exit apertures, the energy resolution of the emerging beam can be controlled and is given by

$$\frac{\Delta E}{E} = \frac{2\Delta D}{d}, \qquad (6\text{-}6\text{-}3)$$

while ΔD is the sum of the entrance and exit aperture diameters. The best energy resolution is obtained for low-energy electrons, and the axial magnetic field serves to confine the beam, making the use of low-energy beams feasible. Strictly speaking, the trochoidal analyzer selects electrons according to their axial velocity and does not produce an improvement of the transverse energy spread of the electrons, induced by the thermal energy imparted to them by their thermionic origin. Stamatovic and Schulz (1970) pointed out that the trochoidal analyzer can reject electrons with a large energy spread if it is operated so that the electrons are retarded prior to entering the deflection region. Their studies of the shape of the attachment cross sections for electrons on SF_6 indicate that electron beams with an energy distribution half-width of 20 meV could be produced using this device. Schulz's attachment apparatus also had the facility of being able to determine the identity of the negative ions thus formed by using a quadrupole mass spectrometer to mass analyze the ions prior to detection.

In a subsequent study of electron excitation and attachment processes using trochoidal analyzer–based apparatus, Verhaart and Brongersma (1980) found that the energy resolution obtained in their instrument was poorer than that obtained by Schulz. Retarding potential analysis of the beam showed that the axial spread of the beam was low, so they attributed the overall spread to electrons with significant thermally induced transverse velocities passing unhindered through the spectrometer. They circumvented this problem by using thick slits in front of the cathode emitter to remove electrons with large transverse velocities. They point out that thin slits are insufficient for this purpose because such electrons move with a spiraling motion due to the magnetic field and so are able to worm their way through these slits. They are, however, stopped by slits whose thickness is several times the length of the pitch of the electrons' helical motion. This length is

$$p = \frac{2\pi R v_{\parallel}}{v_t}, \tag{6-6-4}$$

where $v_{\parallel}$ and v_t are the velocities of the electrons parallel to and perpendicular to the direction of motion, and R is the radius of the helix given by

$$R = \frac{(2mE_t)^{1/2}}{eB}, \tag{6-6-5}$$

E_t being the transverse kinetic energy associated with the transverse velocity v_t, that is,

$$E_t = \tfrac{1}{2}mv_t^2. \tag{6-6-6}$$

Using this modification, Verhaart et al. were able to achieve energy resolutions of 76 meV with usable beam currents of 10 nA while using a directly heated tungsten filament. This filament was necessary because indirectly heated oxide cathodes, which produce smaller transverse energy spreads, are quickly poisoned by the types of gases (chlorofluorocarbons) studied by these workers. Without the modification it was found that full-width half-maximum spreads of more than 100 meV were obtained even with an indirectly heated cathode. In these studies, the axial energy distributions, measured using the retarding potential method, were found to be 25 to 30 meV with or without the limiting slit.

The incorporation of mass analysis of the ions allows relative cross sections for competing dissociative attachment processes to be measured. This technique was first adopted by Stamatovic and Schulz (1970) and by Chantry (1968). Theoretical models of the dissociative attachment process indicate that the cross section should be a sensitive function of the internal energy of the target molecules in some cases, so methods have been incorporated that allow the cross section to be measured as a function of varying target temperature [Chantry 1969, Henderson et al. (1969), Allan and Wong (1978)].

An apparatus that allows the energy and angular distribution of the negative ions, formed by electron attachment, to be measured is shown in Fig. 6-6-1. This instrument uses 127° analyzers for selection of the incident electron energy and for determining the energy of the scattered electrons or of the negative ions. The electrons and ions are distinguished by incorporating a magnetic momentum filter into the second analyzer which, when operating, prevents electrons from reaching the detector. The energy resolution of the incident electron beam is typically 100 meV with a beam current of 150 nA. The resolution of the ion energy analysis is degraded, however, to about 160 meV for 1-eV ions, due to the thermal motion of the target molecules, which amplifies the energy spread of the fragment ions because of momentum conservation (Chantry and Schulz, 1967).

The 127° analyzer is less well suited to very low energy collision processes, so this technique is generally used to study the details of attachment processes at energies of a few eV.

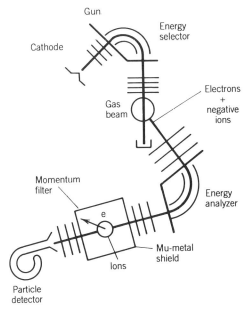

Figure 6-6-1. Hall's crossed electron-molecule beams apparatus for dissociative attachment studies. [Adapted from Hall et al. (1977).]

B.   Threshold Photoelectron Attachment

For large molecules, electron attachment occurs predominantly via the trapping of very low energy electrons into excited states. Conventional electron beam methods have difficulty in reaching such low energies. A method that overcomes this problem is that of Chutjian (1981) and Ajello and Chutjian (1979), in which very low energy photoelectrons are produced by shining monochromatized light onto a suitable buffer gas such as xenon or krypton. These electrons then attach to the gas under study. The ions so formed are extracted from the collision region, mass analyzed, and detected. This technique is capable of measuring relative attachment cross sections in the energy range 0 to 200 meV with a resolution of about 5 meV.

C.   Electron Attachment via Rydberg Atom Collisions

An even more novel technique for producing very slow collisions between molecules and electrons is that in which the electron is not in fact strictly free but, rather, is itself bound in a high principal quantum number, "Rydberg" orbit of an atom such as Ar, Kr, *or* Xe (see Note 6-1 in the companion volume). Since the binding energy of the electron in these high orbits is so weak, the electron can be considered to be quasi-free from the core ion in an *independent-particle approximation*. The electrons are so much lighter than the nuclei that their effective collision energy can be very small, being determined primarily by the relative velocity of the heavy collidants. The electron energy can be represented by the reduced energy

$$\varepsilon = \frac{E m_e}{M},\qquad(6\text{-}6\text{-}7)$$

where E and M are the energy and mass of the projectile atom and m_e is the electron mass. In this way it is possible to measure attachment processes at near-zero energies. This technique is particularly useful for studying attachment to cluster molecules, where the electron affinities are often small but greater than the cluster binding energies. In fact, for clusters of increasing size, the trend is typically for the electron affinities to increase while the binding energies decrease.

1.   *Dunning's Apparatus*

This apparatus is illustrated schematically in Fig. 6-6-2, and the technique has been described in detail by Dunning (1987). A beam of alkali metal atoms is directed through the collision region as shown. A pulse of Rydberg atoms is formed within the beam by laser excitation from the alkali metal ground state. The Rydberg atoms thus formed react with the attaching gas under study for a

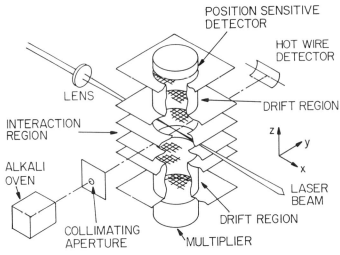

Figure 6-6-2. Dunning's Rydberg atom attachment apparatus. [From Dunning (1987).]

period of a few microseconds, whereupon the negative and positive ions created through the electron transfer reaction

$$X^* + AB \rightarrow X^+ + AB^- \tag{I}$$

are expelled from the interaction region by a weak, pulsed electric field and are detected in coincidence by the pair of opposing detectors shown in the figure. Time-of-flight techniques can be used to determine the mass of the negative ions, using the expelling pulse to start the measurement and the particle detection to stop it.

The density of Rydberg atoms in the interaction region can be determined by applying a sufficiently strong electric field to ionize the atoms and then counting the ions thus formed. The strength of the field required to ionize an atom in a Rydberg state n is given by classical theory as

$$E_c = \frac{3.2 \times 10^8 \text{ V/cm}}{n^4}. \tag{6-6-8}$$

Since the Rydberg atoms have finite lifetimes and are very sensitive to excitation and deexcitation in collisions, the gas pressure and the time spent by the atoms in the interaction region must be kept to minimum values so that state mixing does not occur. Rate constants for the electron attachment reaction (I) are determined by monitoring the growth of the ion signals to the detector and the decay of the Rydberg atom concentration in the interaction region following their formation by the laser pulse. By assuming that reaction (I) is equivalent to a free-electron attachment process, the cross section as a function of electron velocity $\sigma(v_e)$ for this process is related to the rate coefficient k by the expression

$$k = \int v_e \sigma(v_e) f(v_e) \, dv_e, \qquad (6\text{-}6\text{-}9)$$

where $f(v_e)$ is the velocity distribution of the electron. For orbits with principal quantum number n, the root-mean-square orbital velocity of the Rydberg electron is given by $2.2 \times 10^{-8}/n$ cm/s, and one can define the velocity-averaged attachment cross section

$$\bar{\sigma} = \frac{k}{v_{\text{rms}}}. \qquad (6\text{-}6\text{-}10)$$

Dunning (1987) has gone to some lengths to test the correctness of the hypothesis that the Rydberg atom electron transfer reaction is in fact equivalent to a free-electron attachment process, and Fig. 6-6-3 shows rate coefficients for reaction (I) measured for a variety of gases as a function of principal quantum number. It can be seen that for $n > 40$, the process is independent of quantum number, suggesting that in fact the atomic core plays no part in the reaction for such states, so the electron can be treated as being free.

A similar technique using high-energy (100-keV) Rydberg atoms has been used by Wang et al. (1986) to study the interaction of quasi-free electrons with N_2, CO_2, and SF_6 molecules. They have observed the same negative-ion resonant structure as seen in conventional electron scattering experiments.

2. Kondow's Technique

An alternative method of studying Rydberg atom–molecule collisions is that of Kondow (1987). He forms a static target of Rydberg atoms via electron impact excitation using the apparatus shown in Fig. 6-6-4 and passes a beam of the attaching gas through the collision cell containing these Rydberg atoms. Negative ions thus formed are mass analyzed using a quadrupole mass spectrometer and subsequently detected. The advantage of this method is that it can be used for studying attachment to clusters that are generally formed in molecular beams. The technique is not as selective as Dunning's with regard to the identity of the particular Rydberg level involved in a given collision. However, by adjusting the potentials on the various grids in the generator, high-lying levels are field ionized, while lower states decay to the ground state before reaching the collision center. There will certainly be low-lying metastable states present in the collision region, but the cross section for negative-ion formation via electron capture from such a state will be very small and can be ignored. Typical Rydberg states surviving in the collision region are in the range $n = 25$ to 35. Kondow has used this technique to study the attachment of electrons to SF_6, and as can be seen from Fig. 6-6-3, the free-electron approximation seems

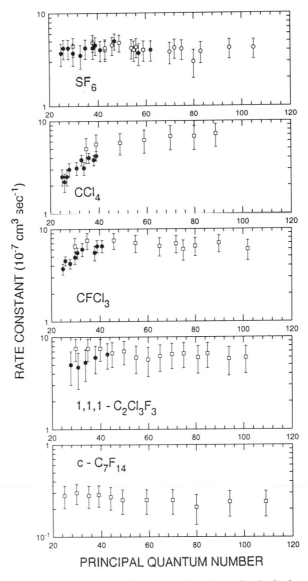

Figure 6-6-3. Attachment rate coefficient as a function of principal quantum number. [From Dunning (1987).]

to be valid for n values above 20 for this molecule. However, for other molecules such as $CFCl_3$ and CCl_4, Dunning's findings suggest that a value of $n > 40$ should be used, so Kondow's technique might be suspect in these cases. Kraft et al. (1989) have studied the attachment to clusters from state-selected Rydberg atoms using a crossed cluster beam–Rydberg atom beam apparatus and laser

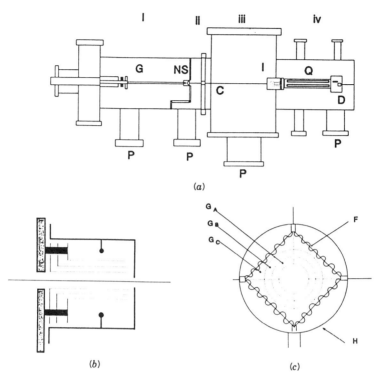

Figure 6-6-4. (*a*) Schematic diagram of Kondow's apparatus. G, inlet of source gas; NS, nozzle/skimmer assembly; C, collimator; I, ion source; Q, quadrupole mass spectrometer; D, detector; P, diffusion pump. (*b*) Side view of ion source and (*c*) front view of ion source. Krypton gas is excited by impact of electrons emitted from four filaments (F). Three concentric grids, G_A, G_B, and G_C, are installed in a housing (H) of the ion source to prevent ionic species and electrons from entering the central region surrounded by G_A. A cluster beam passes along the axis of the ion source and collides with high-Rydberg krypton atoms in the central region. [Adapted from Kondow (1987).]

excitation to select individual Rydberg levels. This study has revealed that in fact the n independence is not achieved for large clusters even for high n values. This result is illustrated in Fig. 6-6-5. The absolute rate for forming Ar^+ is independent of n but that for forming a particular negative ion is not. For low n values, the electrostatic interaction between the negative and positive ions leads to the formation of a complex that stays together for sufficient time for the electron transferred to the cluster to be localized and for the cluster to achieve a stable configuration, thus enhancing the attachment process. Hence there is an important core effect for this type of attachment. As the principal quantum number increases, the influence of the core decreases and the enhancement falls off, leading to a decreased attachment rate. There appears to be no such n dependence for small clusters, suggesting that a different mechanism is operating

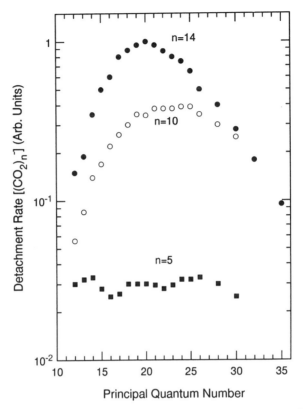

Figure 6-6-5. n dependence of attachment rate for cluster ions. [From Kraft et al. (1989).]

here, possibly a direct charge-transfer process involving crossings of potential energy curves.

3. Hotop's Technique

This is a crossed-beam photoionization method (Klar et al., 1991) in which a beam of metastable $Ar^*(4s^3P_2)$ gas atoms is illuminated first by a single-mode dye laser to produce $Ar^*(4p^3D_3)$ and then by a tunable multimode intracavity dye laser to produce either high-Rydberg argon atoms or low-energy free electrons (and argon ions) as the photon frequency is increased. The illumination region is also intersected by a beam of the target molecules to which electron attachment will occur. Negative ions so formed are extracted into a quadrupole mass spectrometer using a pulsed electric field. The use of the laser provides high photon energy resolution, resulting in very low free-electron energy spreads. Energy resolutions of less than 1 meV have been reported. Figure 6-6-6 shows the ion yield spectrum for SF_6^- ions in the energy range -40 to $+24$ meV. The negative value refers to electron attachment of bound

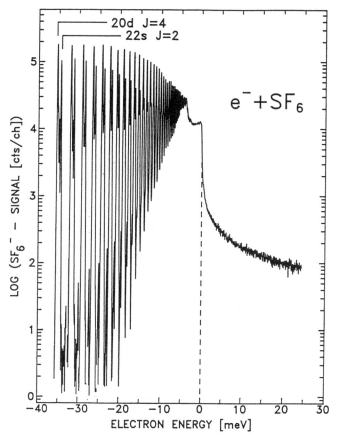

Figure 6-6-6. Results from Hotop's experiments on electron attachment from laser-excited Rydberg atoms and photodetached threshold electrons. [From Klar et al. (1991).]

Rydberg electrons, as occurs in Dunning's method. The step at $-4\,\text{meV}$ is due to field ionization of Rydberg levels with $n > 60$ in the ion extraction field, resulting in a decrease of SF_6^- ions. The vertical dashed line indicates the ionization limit. Above this limit the electrons are free to attach as in the threshold photoelectron attachment method. It can be seen that the attachment signal falls rapidly as the electron energy increases. This method is particularly suitable for the study of energy dependencies of attachment cross sections in the region close to zero energy.

D. Swarm Techniques

The principle behind the swarm experiments is to generate electrons in a gas and to allow them to diffuse through the gas. If an electric field is applied, a drift motion, in the direction opposite to the field, can be imparted to the electrons.

Swarm methods have been used successfully for many years in the study of a wide variety of collision phenomena (see Chapters 7 and 8). They have the advantage that very low electron energies can be achieved and for many years were the chief source of thermal energy electron data. In particular, they have been a major source of information concerning electron attachment processes, although in recent years, they have had strong competition from the threshold photoelectron spectroscopy technique and the Rydberg atom collision method.

The principal difference between the swarm and the electron beam techniques is that in the former, the electron energy distribution is broad, so attachment rate coefficients k_a are measured rather than cross sections $\sigma(\varepsilon)$ as in the latter methods. If $f(\varepsilon)$ is the electron energy distribution function, k_a and $\sigma(\varepsilon)$ are related by

$$k_a = \left(\frac{2}{m}\right)^{1/2} \int \sigma(\varepsilon)\varepsilon^{1/2} f(\varepsilon)\, d\varepsilon. \qquad (6\text{-}6\text{-}11)$$

Table 6-6-1. Mean Electron Energies as a Function of E/N

N_2		Ar	
E/N 10^{-19} (V/cm^2)	Energy (eV)	E/N 10^{-19} (V/cm^2)	Energy (eV)
3	0.040	1	0.305
6	0.046	2	0.398
9	0.054	3	0.466
12	0.064	4.5	0.551
15	0.074	6	0.624
20	0.093	9	0.751
25	0.112	12	0.862
30	0.130	15	0.960
40	0.164	18	1.050
50	0.195	22	1.160
60	0.226	25	1.230
90	0.319	30	1.350
120	0.408	45	1.640
180	0.547	60	1.890
240	0.638	75	2.100
300	0.703	90	2.290
350	0.744	120	2.640
400	0.778	150	2.950
500	0.829	220	3.570
600	0.866	300	4.180

Source: Christophorou (1984).

Christophorou (1981) and McCorkle et al. (1980) have developed methods for unfolding attachment cross sections from the rate coefficients, provided that the electron energy distribution function is accurately known. For thermal energy electrons, where there is no applied field, a Maxwellian distribution function is a good approximation to $f(\varepsilon)$. Whenever there is an applied field, the electron drift velocity is a function of the density-reduced electric field E/N, which must be incorporated into the electron energy distribution function. $f(\varepsilon, E/N)$ can be derived from a solution to the Boltzmann equation provided that accurate elastic and inelastic cross-section data are available. Generally, the gas whose attachment coefficient is to be determined is added at low concentration to a buffer gas such as N_2 or Ar so that the drift velocity of the electrons is determined by the collisions with the buffer gas atoms or molecules. The mean electron energies as a function of E/N for N_2 and Ar are listed in Table 6-6-1, and they are used together with an iterative procedure to determine σ.

The main swarm methods currently in use for the study of electron attachment processes are the pulse-shape analysis technique, used extensively in recent years by Christophorou and coworkers (Hunter et al., 1986), the flowing afterglow (FA) technique of Ferguson et al. (1969), the flowing afterglow Langmuir probe (FALP) method of Smith and Adams (1983), and the Cavalleri method (Cavalleri 1969, Gibson et al., 1973). Hatano and coworkers (Shimamori and Hatano, 1976; Toriumi and Hatano, 1985) have developed the microwave-cavity/pulsed-radiolysis (MCPR) technique for the measurement of attachment rates in high-pressure gases. The FA technique has been used extensively for studies of ion–molecule reactions and is described in Chapter 8. The FALP technique is discussed in detail in Chapter 9.

1. *Pulsed Drift-Tube Technique*

This method was first used for electron attachment in the 1950s by Doehring (1952) and developed further by Chanin et al. (1959, 1962). A more modern apparatus used at Oak Ridge National Laboratory by Christophorou and coworkers (Hunter et al., 1986) is shown in Fig. 6-6-7. The most significant innovations are in the pulse production and data acquisition methods. The principle behind the technique is as follows.

A burst of photons liberates an electron pulse from the photocathode, and this drifts under the influence of an applied electric field toward the anode. As the electrons pass through the gas, between the electrodes, some undergo attachment, forming negative ions which also drift toward the anode but with a drift velocity v_i about 1000 times less than the electron drift velocity v_e. A prompt electron pulse will therefore be detected at the anode, followed by a delayed negative-ion pulse. Ions arriving at time t after the electron pulse will have been formed at a distance $v_i t$ from the anode. If $x = d - v_i t$, d being the separation of the electrodes, the electron pulse intensity will be reduced by a factor of $\exp(-k_a x)$, where k_a is the attachment rate coefficient. The negative-ion pulse height will therefore vary as $\exp(k_a v_i t)$, cutting off abruptly at a time t_c,

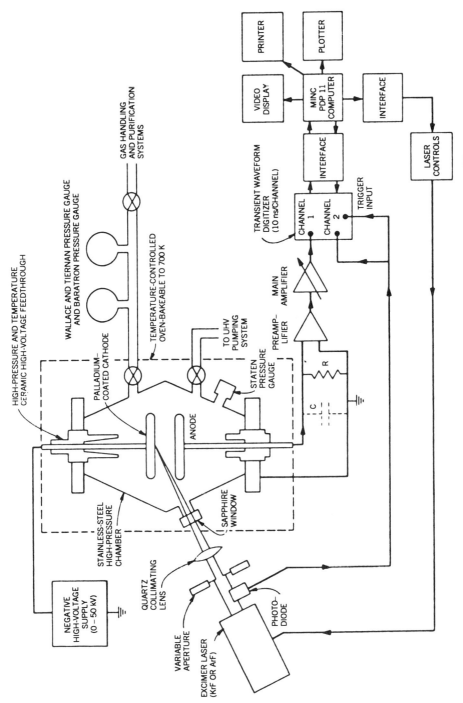

Figure 6-6-7. Pulsed drift-tube apparatus at Oak Ridge National Laboratory. [From Hunter et al. (1986).]

where $v_i = d/t_c$. The shape of the measured negative-ion pulse is determined using a transient digitizer, and hence the rate coefficient is determined. The attachment cross section can then be evaluated using the procedure described above.

2. Cavalleri Method

This method, used extensively by Crompton and coworkers (Gibson et al., 1973; Petrovic and Crompton, 1985 and references therein), allows very accurate attachment rate coefficients to be determined for electronegative, permanent gases such as $CFCl_3$ and SF_6. A cylindrical Pyrex glass cell is filled with the gas to be studied. Electrons are produced in the cell by a short burst of X-rays. After a delay time of S microseconds, a high-voltage RF pulse is applied to electrodes, external to the cell. This pulse generates an electron avalalanche, resulting in the emission of light. The intensity of the light is proportional to the number of electrons initially present in the cell. About 1 s later a second X-ray pulse is applied, but this time there is an additional delay time D before the RF sampling pulse is applied. The ratio of the light intensities measured at time S and $S + D$ is proportional to the decrease in the electron density in the cell, and from this the electron density decay constant, and hence the attachment rate, can be determined. Assuming an exponential decay, the time constant τ is given in terms of the electron densities $N_e(S)$ and $N_e(S + D)$ at times S and $S + D$:

$$\tau = D \left[\ln \frac{N_e(S)}{N_e(S + D)} \right]^{-1}. \tag{6-6-12}$$

The temperature of the cell can be varied, so the temperature dependence of the attachment rate can be obtained.

3. Microwave-Cavity/Pulsed-Radiolysis Technique

In the MCPR technique, a high-energy electron beam from a Febotron accelerator strikes a thin tungsten foil, generating a pulse of X-rays that illuminate a high-pressure (10 to 100 kPa) gas mixture within a microwave cavity. The gas mixture consists of the gas to be studied, AB, and a buffer gas, M. The electrons thus generated in the gas produce change in the resonant frequency of the cavity. The magnitude of the change is proportional to the electron density, and thus the electron density decay constant can be determined. At high pressures, both two- and three-body attachment can occur, and the product of the decay constant and the gas density, $\tau[AB]$, is given by

$$\tau[AB] = \frac{1}{k_i} + \frac{1}{k_M[M]}, \tag{6-6-13}$$

where k_i is the two-body attachment rate and k_M is the three-body rate. By plotting $\tau[AB]$ versus $[M]^{-1}$, one obtains k_M from the slope of the graph and k_i from the intercept. Unlike the other swarm methods discussed above this method produces electron heating while leaving the temperature of the target unaffected. It is therefore possible to study the temperature dependence of the attachment to van der Waals molecules, which would be destroyed by heating the gas.

6-7. DISSOCIATIVE ATTACHMENT

To appreciate the significance of the results obtained from measurements of free-electron attachment to molecules, one should first understand the mechanisms and subtleties underlying the dissociative attachment process. A detailed theoretical analysis of this fascinating reaction was first provided by Bardsley et al. (1966). A more general review of this subject, which places dissociative attachment within the context of molecular ion resonance theory and shows its link to the dissociative recombination of electrons and molecular ions, was later published by Bardsley and Mandl (1968). The theoretical framework of Bardsley et al. was subsequently elaborated upon by O'Malley (1966) and by O'Malley and Taylor (1968), who developed a method for calculating differential cross sections for dissociative attachment as a function of angle of emission of the fragments.

As shown earlier in this chapter, many molecules have positive electron affinities and can therefore form negative ions. Once formed, energy is required to remove the electron, either in the form of photon absorption or via the transfer of kinetic energy from an impacting particle. During the negative-ion formation process, therefore, this attachment energy must be removed. For atomic species, the only way in which this can be accomplished is via photon emission (radiative attachment) or via transfer to a third particle (charge-transfer or three-body attachment). For molecules there is often an alternative method, and that is through dissociation of the temporarily formed negative-ion complex into charged and uncharged fragments. For this to happen, the attachment of the electron to the parent molecule must result in promotion of the system to an alternative electronic state which subsequently dissociates. Such a process is illustrated in Fig. 6-7-1. In this example transition to the intermediate state requires the input of energy, so the process is endothermic (i.e., it will display an energy threshold). Note that the potential energy of the fragments $A^- + B$ is less than for the neutral fragments $A + B$, where A and B can be atoms, molecules, or radicals. The dissociation process allows this energy to be released in the form of kinetic energy of the fragments. There are variations on this basic mechanism, depending on the exact form of the intermediate state, and they will be discussed with regard to specific examples in the next section.

Bardsley et al. (1966) showed that this intermediate state is in fact a resonance

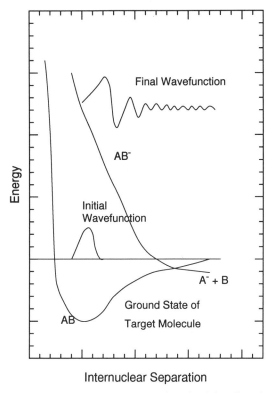

Internuclear Separation

Figure 6-7-1. Schematic diagram of the states involved in dissociative attachment. [From Bardsley and Mandl (1968).]

that decays naturally via autoionization or via dissociation. Thus the process can be represented by

$$e + AB \rightleftharpoons AB^{*-} \rightarrow A^{-} + B. \tag{II}$$

By detailed balance, for the initial capture process to have a large cross section, the autoionization width must be large (i.e., the autoionization lifetime of AB^{*-} should be small). Of course, for the stabilized negative-ion formation process to be effective, the time for dissociation must also be fast, so that it can compete effectively with the autoionization process, which will destroy AB^{*-}. In Fig. 6-7-1 the ground-state molecule will spend most of its time at the equilibrium internuclear separation, R_0, and the intermediate state is formed if the incoming electron has sufficient energy to excite a Franck–Condon (vertical) transition to the upper state. AB^{*-} has a decaying wave function. The system will in time either revert back to the molecule's electronic ground state with the emission of the electron, or it will dissociate; that is, the internuclear separation will increase and the potential energy will fall, being converted into kinetic energy of the individual fragments A$^-$ and B. If the system arrives at R_s without autoionizing, the negative-ion formation will be stabilized since the state AB* no longer has enough potential energy to convert to electron kinetic energy. R_s is known as the stabilization point and the likelihood of the system reaching R_s is given by the survival factor,

$$S_F = \exp\left[\int_{R_e}^{R_s} \frac{\Gamma_n(R)}{\hbar} \frac{dR}{v(R)}\right], \tag{6-7-1}$$

where R_e is the internuclear separation at the capture point, $T_n(R)$ is the autoionization width, and $v(R)$ is the relative velocity of nuclear motion. The dissociative attachment cross section σ_{DA} is then given by

$$\sigma_{DA} = \sigma_C S_F, \tag{6-7-2}$$

where

$$\sigma_c = \frac{4\pi^{3/2}}{k_\mu^2} \frac{2S_n + 1}{2(2S_\mu + 1)} \frac{\Gamma_c(R_e)}{2W_1 a}$$

$$\times \exp\left\{-\frac{[E - E_n(R_0)]^2 - \Gamma_n^2(R_0)/4}{W_1^2 a^2}\right\}, \tag{6-7-3}$$

$T_c(R_e)$ is the capture width at R_e, S_μ and S_n the spins of the initial and resonant states, k_μ the wave number of the incident electron (i.e., its velocity is given by $\hbar k_\mu/m_e$), W_1 the slope of the resonance potential at the turning point, that is,

$$-W_1 = \frac{\partial E(R_n)}{\partial R}, \tag{6-7-4}$$

and a the vibrational amplitude of the initial nuclear wave function ζ_μ. The exponential term, often known as the Franck–Condon factor, represents the overlap of the initial and final nuclear wave functions. $E_0(R_0)$ is the real part of the resonance energy at the equilibrium position, and E is the energy of the incoming electron. It should be noted that σ_c will be a maximum when there is optimum overlap between the two states, and this factor will determine the energy dependence for the overall dissociative attachment cross section. It should be noted, however, that σ_c also decreases with increasing k_μ (i.e., with electron velocity). For situations in which there is optimum overlap over small energy intervals, where the Franck–Condon factor changes slowly, the variation of σ_c with electron velocity can have a significant effect. This is the case for the dissociative attachment of H_2 molecules in vibrationally excited states and for dissociative recombination processes involving molecular ions.

The survival factor represents the probability of the negative-ion resonant state living long enough to reach the stabilization position R_s, and is given by $dR/v(R)$, while $\Gamma_n(R)/\hbar$ is the decay rate at R. One can therefore write σ_{DA} in the form

$$\sigma_{DA} = \sigma_c \exp\left(\frac{-\bar{\Gamma}_n \tau}{\hbar}\right), \qquad (6\text{-}7\text{-}5)$$

$\bar{\Gamma}$ is the mean autoionization width between R_e and R_s, and τ is the time taken to move between these points.

Now the survival factor is a function of the relative velocity of the nuclei, which in turn is proportional to $M^{-1/2}$ (and so will differ from one isotopic species to another). Hence S_F decreases as the molecular weight increases. σ_c is also mass dependent since a, the amplitude of the nuclear wave function, is proportional to $M^{-1/2}$. The overall effect is to enhance the capture cross section for heavier-molecular-weight isotopic species. If the resonant state is located, with respect to the initial state, so that the capture occurs far from R_s, the influence of the survival factor on M is likely to dominate. Otherwise, the M dependence of σ_c may take prominence.

Of more significance to the absolute value of the dissociative attachment cross section is the location of the resonant state with respect not only to the initial state, but also to other states into which it can decay. Calculations of the process must therefore begin with very accurate determinations of these states. It also means that there can be an almost incredible sensitivity to the initial internal energy of the molecule undergoing attachment. This sensitivity is illustrated vividly in the case of H^- formation, discussed in the next section.

6-8. DATA ON ELECTRON ATTACHMENT

A. Molecular Hydrogen

Total cross sections for the dissociative attachment of electrons to H_2 have been measured by Khvostenko and Dukel'skii (1958), Schulz (1959), Schulz and Asundi (1967), and Rapp et al. (1965). The latter authors also studied attachment to HD and to D_2, and their results are shown in Fig. 6-8-1. The threshold for the process lies at 3.75 eV and there is a sharp resonance there, first observed by Schulz and Asundi (1967) and illustrated in the figure.

The first thing to notice is that the cross section is very small and displays a striking dependence on isotopic composition. Figure 6-8-2 shows the potential energy curves for the $X^1\Sigma_g^+$ and the $b^3\Sigma_u^+$ states of H_2 along with the real part of the energies for two negative-ion resonant states that participate in this reaction. The resonance structure in the cross section between 3.75 and 4 eV is due to capture into the $^2\Sigma_u^+$ resonance, which can decay back to a vibrationally excited

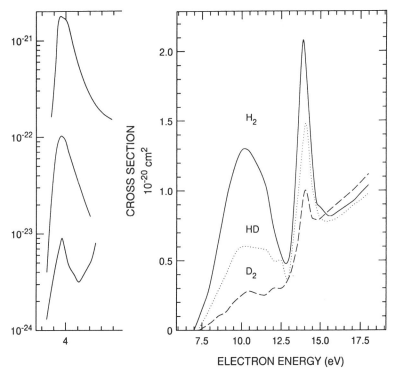

Figure 6-8-1. Dissociative attachment cross sections for H_2, HD, and D_2. Due to the small cross sections, the low energy resonances are shown separately. [From Rapp et al. (1965).]

level of the ground state. This is a shape resonance (see Chapter 4 of the companion volume) with a very short lifetime and so is likely to autoionize before there is time to stabilize the negative-ion formation. For dissociation to be energetically possible, the capture must occur to the left of the equilibrium position, leading not only to a long stabilization time but also a nonideal overlap between the wave nuclear functions of the $v = 0$ level H_2 and of the resonant state. The resultant attachment cross section is very small, decreasing dramatically in going from H_2 to HD to D_2, the peak cross sections being 1.6×10^{-21}, 1×10^{-22}, and $8 \times 10^{-24}\,cm^2$, respectively. The broad feature between 8 and 13 eV is associated with the repulsive $^2\Sigma_g^+$ state of H_2^-. In this instance the isotope effect is dominated by the survival factor's $M^{-1/2}$ dependence, and from this it can be determined that the mean autoionization width is $\bar{\Gamma} = 0.8\,eV$. Using this value in turn with the measured dissociative attachment cross section allows σ_c to be determined, and it is found that $\Gamma_c = 0.004\,eV$. In the other words, the probability of capture into the resonant state from the ground state of H_2 is small, and once formed, the probability of the resonant state autoionizing before being stabilized is large. As mentioned

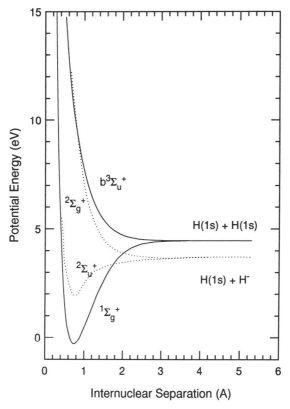

Figure 6-8-2. Potential energy curves of H_2 and H_2^-. [Adapted from Sharp (1971).]

earlier, the capture width from the initial state is proportional to the auto-ionization width back to that state. In this instance, however, the resonant state can autoionize to the $^3\Sigma_u^+$ repulsive state of H_2, also shown in the figure.

The third large feature in the total dissociative attachment cross section is between 13 and 15 eV, and it arises via the capture into the repulsive part of the bound $^2\Sigma_g^+$ resonance, also shown in the figure. The products arising from this attachment are H^- and $H(n = 2)$. This resonance also gives rise to vibrational structure superimposed on the 12-eV structure. High-resolution studies of the 14-eV feature by Tronc et al. (1977) indicate the presence of vibrational structure superimposed on it, and they attribute this structure to the presence of an additional $^2\Delta_g$ resonance. This conclusion is supported by variational cal-culations of resonance curves by Bardsley and Cohen (1978).

Given the small size of the cross sections measured for H_2 attachment, these studies might have passed for an interesting academic study of the mechanisms underlying a somewhat exotic atomic reaction with little practical application. In fact, however, a revival of interest in this reaction has come from an application in nuclear engineering. Large negative-ion sources are required for the production of fast neutral beams needed to heat fusion plasmas in Tokamak confinement devices. Negative hydrogen ions formed in these sources can be electrostatically accelerated to the desired energy and then stripped in a gas cell or by laser detachment to yield fast neutral atoms which can penetrate the magnetic confining fields in the plasma apparatus. They then make energy-transferring collisions with the plasma ions, thus heating the plasma. Studies of the negative-ion flux from volume sources (i.e., sources where the ions are produced in the gas phase rather than through wall reactions) indicate that there must be an efficient two-body reaction that produces copious amounts of H^- ions. It is generally believed that this reaction involves the attachment of thermal electrons to vibrationally excited hydrogen molecules. Measurements have been performed by Allan and Wong (1978) on the dissociative attachment of electrons to hydrogen molecules contained in an oven whose temperature could be varied from 300 to 1600 K. Their data indicate that the cross section for H^- attachment increases by four orders of magnitude on going from $v = 0$ to $v = 4$ and that for D^- production by five orders of magnitude. The effect of rotational excitation is less dramatic, being about a factor of 5 between $J = 0$ and $J = 7$. Wadehra and Bardsley (1978) have modeled this behavior and explain the increase in σ_{DA} as being due to improvement of the overlap between the initial- and final-state nuclear wave functions as the vibrational energy of the initial state is increased. Cross sections calculated by these authors for different vibrational levels are shown in Fig. 6-8-3.

B. Carbon Monoxide

The dissociative attachment of electrons to carbon monoxide has received considerable attention and has been extremely well characterized. Hypothetical potential energy curves involved in this process are shown in Fig. 6-8-4, and it can be seen that from the ground state of the molecule, two intermediate resonances are easily accessible, one decaying to $O^-(^2P) + C(^3P)$, the other to $O(^3P) + C^-(^4S)$. For molecules in the ground vibrational state, thresholds for these channels are 9.62 and 10.2 eV, respectively. While the maximum value of the cross section for the former channel is 2×10^{-19} cm^2, that for the latter is only 6×10^{-23} cm^2. There is also a third channel observed leading to $O^-(^2P) + C(^1D)$, with a threshold at 10.88 eV and a maximum cross section of 9.5×10^{-21} cm^2. The cross sections for O^- formation peak at threshold in both cases and fall off slowly to the baseline again as illustrated in Fig. 6-8-5. The great differences in the measured cross sections may perhaps be attributable to differences in the capture widths, but this question can only be answered by calculating these widths, and this has not been done.

Hall et al. (1977) have studied the angular distributions of the fragments, and using the theory of O'Malley and Taylor (1968), have identified the symmetries of the resonances leading to O^- channels. The ground state of CO has Σ symmetry and this means that the $O^-(^2P) + C(^3P)$ channel can arise from a Σ, Π, or Δ symmetry resonance. O'Malley and Taylor have shown that the

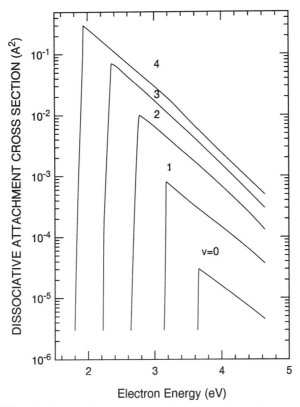

Figure 6-8-3. Dissociative attachment cross section for various vibrational states of H_2, each in the rotational state, $J = 0$. [From Wadehra and Bardsley (1978).]

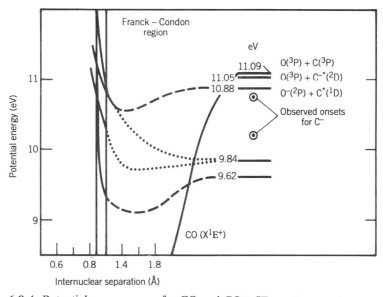

Figure 6-8-4. Potential energy curves for CO and CO^-. [From Stamatovic and Schulz (1970); see also Morgan (1991).]

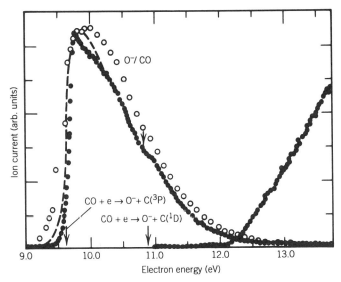

Figure 6-8-5. Cross sections for electron attachment to CO. [From Stamatovic and Schulz (1970).]

differential cross sections for dissociative recombination take the following forms for the transitions indicated:

$$\Sigma \to \Sigma: \quad \sigma(\theta) \sim (1 + 3.46 A_1 A_2 \cos\theta + 3A_1^2 \cos^2\theta) \tag{6-8-1}$$

$$\Sigma \to \Pi: \quad \sigma(\theta) \sim \sin^2\theta(1 + 4.47 A_1 A_2 \cos\theta + 5A_1^2 \cos^2\theta) \tag{6-8-2}$$

$$\Sigma \to \Delta: \quad \sigma(\theta) \sim \sin^4\theta(1 + 5.29 A_1 A_2 \cos\theta + 7A_1^2 \cos^2\theta) \tag{6-8-3}$$

By fitting the measured angular distributions to these theoretical expressions, it was found that the $O^-(^2P) + C(^3P)$ channel arises from a resonance with Π symmetry.

A similar procedure was applied to the measured angular distributions for the $O^-(^2P) + C(^1D)$ channel, and in this case the possibility of a $\Sigma \to \Phi$ transition with the form

$$\sigma(\theta) \sim \sin^6\theta(1 + 6.00 A_1 A_2 \cos\theta + 9A_1^2 \cos^2\theta) \tag{6-8-4}$$

was included in the fitting procedure. Again, however, it was found that a $\Sigma \to \Pi$ transition gave the best fit to the data.

In a later study, Abouaf et al. (1981) used a deconvolution technique to analyze the spectrum of signals obtained from the $C^-(^4S) + O(^3P)$ channel and showed that it arises due to the predissociation of the $^2\Sigma^+$ CO^- resonance by a second resonant state, also possibly with $^2\Sigma^+$ symmetry.

C. Other Diatomics

Extensive studies have been performed on other diatomic systems, such as O_2, NO, HF, HCl, HBr, and so on, and this work has been reviewed extensively by Christophorou (1984) and by Massey (1976). Generally, similar results are found for these molecules; that is, the cross sections are small (although widely varying in magnitude) and display thresholds as available resonant states open up. Angular scattering measurements have allowed the identity of the resonant states to be determined for these molecules.

D. Halogens

The diatomic halogens deserve special mention, because the form of the cross sections for these species is considerably different from that for other diatomics. The dependence of the attachment cross section for F_2, Cl_2, and Br_2 is illustrated in Fig. 6-8-6a. It can be seen that while, as for other diatomics, these molecules display broad structures at energies of several eV, in each case they have their maximum cross sections at zero energy. Furthermore, the cross sections are substantially larger than for other diatomics. This difference is due to the reciprocal energy dependence of the capture cross section alluded to earlier.

In each case the dissociation limit is $X^-(^1S) + X(^2P)$, and according to the Wigner–Wittmer rules, this limit correlates to four compound states with symmetries $^2\Sigma_u^+$, $^2\Pi_g$, $^2\Pi_u$, and $^2\Sigma_g^+$. The ground state of the halogen molecules has $^1\Sigma_g^+$ symmetry. In this experiment the negative ions are detected at right angles to the direction of motion of the electrons, and according to the selection rules laid down by Dunn (1962) (see p. 416 in the companion volume), $\Sigma_g^+ \rightarrow \Sigma_u^+$ and $\Sigma_u^+ \rightarrow \Pi_u$ transitions are forbidden for dissociations perpendicular to the electron momentum vector. Because of the finite acceptance angle of the detector, however, such transitions could in fact be observed in this apparatus. An estimate of the energies of these resonant states can be obtained from a comparison with photoelectron spectra, and approximate curves for these states are shown for the case of Cl_2 in Fig. 6-8-6b. The zero-energy peak for these molecules is ascribed to the $^2\Pi_g$ state, which is believed to pass through the minimum of the ground-state potential curve. Even though a transition to this state is unfavored on symmetry considerations, the strong influence of the reciprocal energy dependence compensates for this, yielding a large cross section.

For fluorine the situation is somewhat different, for in this case the $^2\Pi_g$ state lies higher in energy with respect to the ground state, so the zero-energy peak must arise from the $^2\Sigma_u^+$ state, again symmetry forbidden. Hazi et al. (1981) have performed calculations of the resonant states for F_2^- and have used them to calculate the dissociative attachment cross section for F_2. Although above a few tenths of an electron volt, agreement with experiment is quite good, their calculation fails to display the observed zero-energy resonance. This failure illustrates the sensitivity of dissociative attachment to the location of

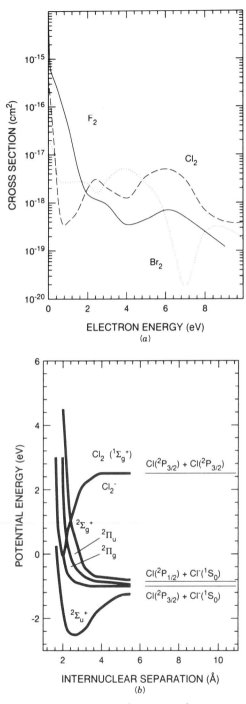

Figure 6-8-6. (*a*) Total attachment cross section versus electron energy for the halogens. Energy scale is expanded below 2 eV. (b) Approximate potential energy curves for the ground state of Cl_2 and for the four states of Cl_2^-, correlating to the dissociation limit of $Cl(^2P) + Cl^-(^1S)$. [Adapted from Hunter et al. (1984).]

the resonant states and the corresponding difficulty in producing reliable calculations for the process. This sensitivity is shared in the related process of dissociative recombination of electrons and molecular ions discussed in Chapter 9.

E.   Triatomics

Extensive studies have been performed on a number of triatomic molecules, such as H_2O, CO_2, and N_2O. Generally, the cross sections exhibit features similar to those found for diatomics, although the greater complexity of these species leads to additional decay channels.

F.   Halocarbons

Halocarbons comprise a class of molecules that have received much attention, for some members of this family of compounds have very large attachment rates, making them attractive for high-voltage insulation applications. On the negative side, the attachment of electrons to these otherwise very stable compounds yields reactive radical fragments which play an important role in the depletion of the ozone layer high in the earth's atmosphere.

Figure 6-8-7 shows how the attachment rates and cross sections vary widely depending on the number of halogen atoms in the molecule and the isomeric form. This topic has been discussed extensively by Christophorou (1984).

G.   Sulfur Hexafluoride

The attachment of electrons to SF_6 has been studied extensively since, as illustrated in Fig. 6-8-8, the cross section for the formation of SF_6^- is extremely large at zero energy. SF_6 is widely used in high-voltage devices, where it helps to prevent discharges from occurring by soaking up electrons through the formation of less mobile negative ions. The SF_6^- ion is actually in a relatively long-lived (6 μs) resonant state, which subsequently decays by reverting back to the neutral SF_6 plus a free electron. The potential energy curves for the SF_6 and SF_6^- states are shown in Fig. 6-8-9a. It can be seen from this figure that the formation of SF_5^- occurs via the dissociation of a higher-lying SF_6^- resonance, and this fact explains the much smaller cross section for this process. Chen and Chantry (1979), however, have shown that this cross section displays a strong dependence on the initial rotational and vibrational state of the SF_6 molecule (Fig. 6-8-9b). This observation—that a large attachment cross section arises at zero energy for a complex molecule, and that it is due to the formation of a relatively long-lived resonant state, which subsequently decays back to the neutral state—is common for other complex species. The long lifetime of these states is attributable to rapid redistribution of the attachment energy among the various degrees of freedom of the negative-ion complex, which delays the autodetachment process.

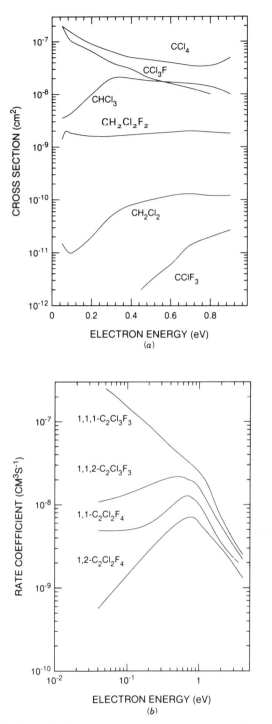

Figure 6-8-7. (*a*) Electron attachment rate constant versus mean electron energy: CCl_4, $CHCl_3$, CH_2Cl_2, CCl_3F, CCl_2F_2, and $CClF_3$. (*b*) Total electron attachment-rate constants as a function of mean electron energy for chlorofluoroethanes. [Adapted from Hunter et al. (1984).]

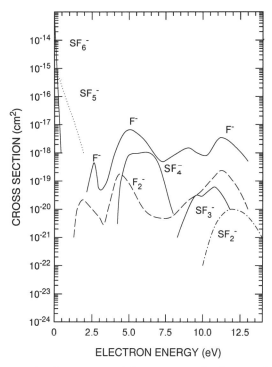

Figure 6-8-8. Cross section for parent and fragment negative ions produced by electron impact on SF_6. [From Kline et al. (1979).]

6-9.   THREE-BODY ATTACHMENT

Approximate potential energy curves for O_2 and O_2^- are shown in Fig. 6-9-1, and it can be seen that in this case there is a striking resonant curve that is not repulsive but in fact is strongly bound, supporting many vibrational levels. Measurements of the dissociative attachment cross section for O_2 display a single maximum at 6.7 eV with a peak value of $1.4 \times 10^{-18}\, cm^2$, but this maximum is due to a higher-lying resonance that decays to $O^-(^2P) + O(^3P)$. Attachment to O_2 at low energies and at elevated gas densities leads to the formation of a long-lived O_2^- molecule. The mechanism for the formation of this molecule was first identified by Bloch and Bradbury (1935) and later quantified by Herzenberg (1969). If the incoming electron has sufficient energy (0.082 eV), it can be captured into the $v = 4$ level of O_2^-, which can decay back to the $v = 0$ level of O_2 by casting off the electron again. If, however, the O_2^- collides with another molecule, it can be deexcited to a lower vibrational level that is non-autodetaching, that is,

$$e + O_2(X^3\Sigma_g^-, v = 0) \rightarrow O_2^-(X^2\Pi_g, v > 3)$$

$$O_2^-(X^2\Pi_g, v > 3) + M \rightarrow O_2^-(X^2\Pi_g, v \leqslant 3) + M$$

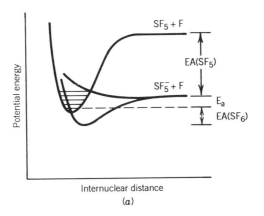

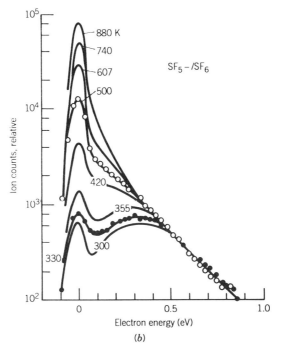

Figure 6-8-9. (*a*) Schematic potential energy curves for SF_6 and SF_6^-. (*b*) Electron attachment to form SF_5^- from SF_6 versus electron energy, for various gas temperatures. [From Chen and Chantry (1979).]

A wide variation in attachment rate is observed for different collision partners, M, as illustrated in Table 6-9-1. Rare gases are least efficient because the attachment energy can only be transferred into translational motion, and the observed attachment rate is small. On the other hand, if the excited O_2^- collides with a polyatomic molecule, the excitation energy can be efficiently transferred

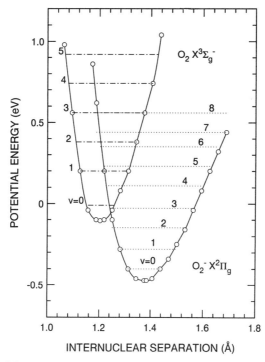

Figure 6-9-1. Approximate potential energy curves for O_2 and O_2^-.

Table 6-9-1. Three-Body Attachment Rates for O_2

Collision Partner M	Attachment Rate k $(10^{-30}\,cm^6/s)$
He	0.033
Ne	0.023
Ar	0.050
Kr	0.050
Xe	0.085
H_2	0.480
D_2	0.140
N_2	0.090
CH_4	0.340
C_2H_4	2.000
$n\text{-}C_5H_{12}$	8.000
H_2O	14.000
C_2H_5OH	18.000
CH_3COCH_3	27.000

Source: Christophorou (1984).

to its internal vibrational and rotational modes, thus producing a large attachment rate. At higher pressures, the Bloch–Bradbury model is insufficient to explain adequately the observed attachment characteristics as $O_2^- - M$ collision complexes are formed and they can be stabilized by yet another collision partner.

6-10.   DETACHMENT PROCESSES

A.   Photodetachment of Negative Ions

The detachment of electrons from negative ions by photon impact has been the subject of intensive investigation since the pioneering work of Branscomb and collaborators in the 1950s and 1960s (Branscomb, 1962). Recent reviews of progress in this area have been given by Miller (1981), Drzaic et al. (1984), Mead et al. (1984), and Ervin and Lineberger (1991). By measuring the location of photodetachment thresholds, it is possible to obtain extremely accurate values of electron affinities for atomic and some molecular negative ions, as discussed in Section 6-2. The high photon flux available from laser light sources has led to improvements in accuracy of many orders of magnitude for such measurements.

A typical apparatus used for photodetachment studies is that shown in Fig. 6-10-1. Negative ions are produced in a hot-cathode arc discharge or cold-

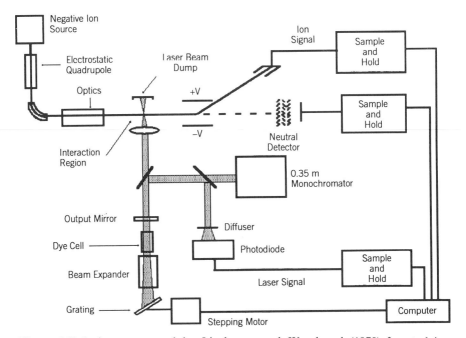

Figure 6-10-1. Apparatus used by Lineberger and Woodward (1970) for studying photodetachment with a tunable dye laser as a light source.

cathode discharge sputter ion source, accelerated and mass analyzed, before being intersected by an intense pulsed beam from a tunable dye laser. Typical wavelength resolutions of 1 Å in the range 4500 to 6000 Å are obtainable with energies of 1 to 5 mJ per 300-ns pulse. Following the detachment, the fast neutral products thus produced are detected using a photomultiplier. The counting system is gated in and out of sequence with the laser pulse to discriminate against neutrals formed in collision of the negative ions with the background gas.

The central feature of any photodetachment apparatus is the method of producing the negative ions themselves. Apart from ion sources, other techniques, such as ion traps (Smyth and Brauman, 1972; McMahon and Beauchamp, 1972; Eyler, 1974) and drift tubes (Moseley et al., 1974) have been used. The latest version of Lineberger's apparatus at JILA employs a flowing afterglow source to produce vibrationally and rotationally relaxed ions of a wide variety of species (Ervin and Lineberger, 1991).

An apparatus that deserves special mention is that of Bryant et al. (1977). A schematic of the experimental setup is shown in Fig. 6-10-2. The 800-MeV H^- beam from the Los Alamos Meson Factory (LAMPF) linear accelerator is crossed at a variable angle by a laser beam, and the photodetached electrons are deflected magnetically into a silicon detector. The photon energy in the collision frame is Doppler tuned by rotating the mirror assembly shown in the figure so as to alter the angle of intersection with the ion beam. If α is the intersection angle, the Doppler-shifted photon energy is given by

$$E = \gamma E_L(1 + \beta \cos \alpha), \tag{6-10-1}$$

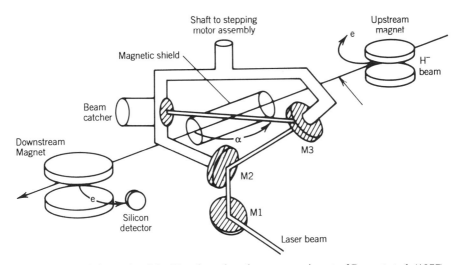

Figure 6-10-2. Schematic of the H^- photodetachment experiment of Bryant et al. (1977).

where $\beta = v/c$, v being the ion velocity, c the speed of light, E_L the laboratory photon energy, and

$$\gamma = \left(1 - \frac{v^2}{c^2}\right)^{-1/2}. \tag{6-10-2}$$

Here $\alpha = 0$ means "head-on." By using an argon fluoride 6.42-eV laser, the photon energy can be tuned from 2.5 to 21.2 eV with an energy resolution of 1 meV! Lower energies can be achieved by using lower photon energy lasers. This unique apparatus* has been used for high-resolution studies of the photodetachment of H^-.

B. Threshold Studies

Applying the Wigner threshold law (Wigner, 1948) leads to the prediction that the cross section for photodetachment should have the general form

$$\sigma \sim (E - E_{thr})^{(2L+1)/2}, \tag{6-10-3}$$

where L is the orbital angular momentum of the detached electron with respect to the atom, E_{thr} the threshold energy, and E the energy of the photon producing the detachment. The detached electron carries away an excess energy of $E - E_{thr}$. Wigner's law applies to processes in which the final-state interaction falls off faster than $1/r^2$.

For the case of Se^-, the extra electron is in a p level and upon detachment can be either an s or a d orbital, so Wigner's law predicts a $\frac{5}{2}$ and $\frac{1}{2}$ excess energy dependence, respectively, for these two cases. Since, near threshold, the excess energy is small, detachment to the d state will increase much more slowly than to the s state and so can be neglected. Experimental values for the Se^- photodetachment cross section measured by Hotop et al. (1973) in the vicinity of the threshold are illustrated in Fig. 6-10-3a and confirm the $(E - E_{thr})^{1/2}$ dependence. The transition thresholds between the fine-structure levels of Se and Se^- (see Fig. 6-10-3b) are indicated as shown. Deviations from the Wigner threshold law are, however, seen for photon energies greater than 5 meV above threshold.

For the case of Au^-, where the extra electron resides in an s state, it will detach into a p orbital, so Wigner's law predicts an $(E - E_{thr})^{3/2}$ energy dependence. This energy dependence can be seen in the results of Hotop et al. (1973) shown in Fig. 6-10-4. In this case, Wigner's law is followed up to excess energies of 60 meV.

For molecular ions, the situation is complicated by the many transitions between rotational and vibrational levels of the ion and the neutral, making

---

*Bryant (private communication, 1991) has expressed interest in constructing a similar apparatus at the proposed Superconducting Supercollider (SSC) in Texas. This machine will have a 1-GeV linear accelerator and would have the advantage of greater access to beam time than the LAMPF machine.

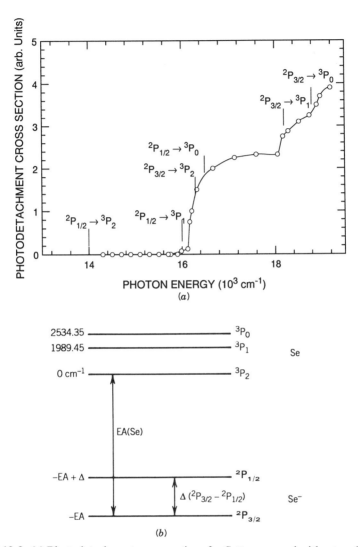

Figure 6-10-3. (*a*) Photodetachment cross sections for Se⁻, measured with a tunable dye laser as a light source, by Lineberger and Woodward (1970) and by Hotop et al. (1973), respectively. (Not all measured data points are shown. (*b*) Energy levels of Se and Se⁻.

interpretation of the process difficult. In recent years, however, a better understanding of photodetachment from negative ions is being achieved, and in particular, the role of autodetachment is becoming clear. In this process, the incoming photon induces a transition to a vibrationally or rotationally excited state of the negative ion. This state subsequently decays into the neutral plus an electron. An example of this phenomenon is seen in Fig. 6-10-5, which shows the threshold region for the photodetachment of CH_2CN^-. Superimposed on the

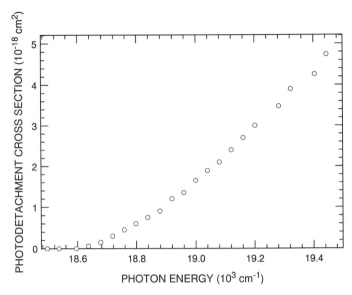

Figure 6-10-4. Photodetachment cross section for Au⁻, observed by Hotop et al. (1973). (Not all measured data points are shown.)

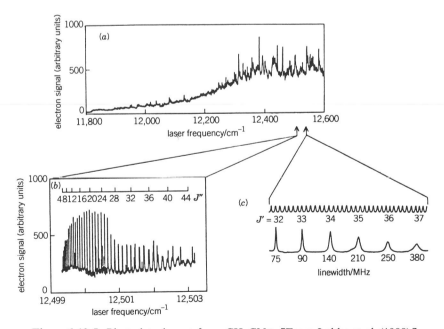

Figure 6-10-5. Photodetachment from CH_2CN^-. [From Lykke et al. (1988).]

direct photodetachment background are sharp peaks arising from autodetachment. Studies of this type are very useful for determining the structure of the negative ion and the interactions between the electron and the neutral molecule. A review of this field has been given by Lykke et al. (1988).

C. Photodetachment Cross Sections

Despite the extensive studies of photodetachment threshold phenomena that have been conducted over the last two decades using the laser photodetachment technique, information concerning the behavior of the cross section over a wide range of photon energies is surprisingly scarce. Perhaps even more surprising is that with the exception of H^-, all recent photodetachment cross-section studies have been normalized to the early results of Branscomb et al. (1958), who made an absolute measurement of O^- photodetachment. A check of the Branscomb et al. calibration was performed by Lee and Smith (1979) at SRI International, and their data for O^- are shown in Fig. 6-10-6. The agreement with the Branscomb et al. data is excellent.

Figure 6-10-7 shows very accurate theoretical photodetachment cross sections for H^- (Broad and Reinhardt, 1976) as a function of photon energy. Figure 6-10-8 shows the resonances, between 10 and 11 eV, predicted by this calculation together with experimental results taken at Los Alamos (Gram et al., 1978). The sharp feature is a Feshbach resonance just below the threshold for the $n = 2$

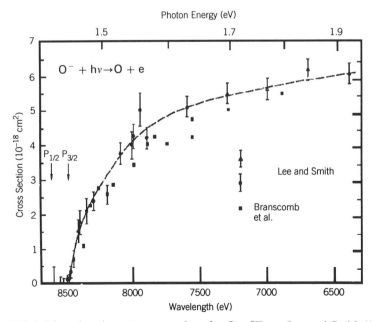

Figure 6-10-6. Photodetachment cross sections for O^-. [From Lee and Smith (1979).]

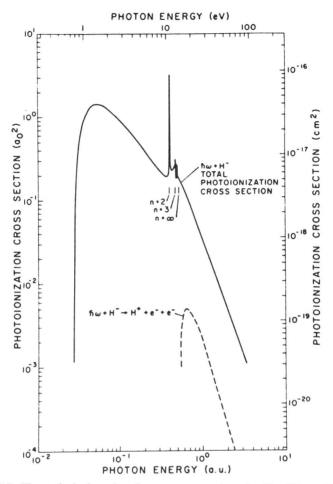

Figure 6-10-7. Theoretical photodetachment cross sections for H^-. [From Broad and Reinhardt (1976).]

state of atomic hydrogen, while the broader structure is a shape resonance lying just above the threshold. Bryant's group has also observed resonances near the $n = 3$ thresholds (Hamm et al., 1979) and has studied the effects of electric fields on the resonant structures.

Calculations and experiments have been performed on negative alkali ions (Kaiser et al., 1974) for which the maximum cross sections rise to 1 to 3×10^{-16} cm^2 and fall again to zero within a few eV. With the widespread availability of synchrotron radiation sources offering fairly intense beams of tunable radiation over a wide energy range, it is possible to extend these studies to many other species, and this will undoubtedly be an area of great interest in the next few years.

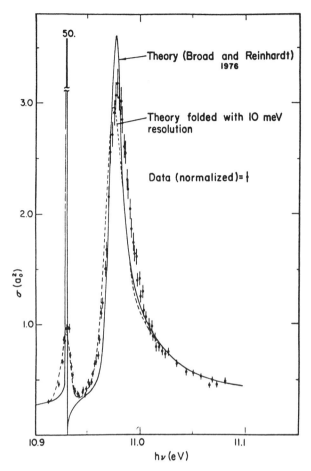

Figure 6-10-8. Comparison of calculated photodetachment cross section for H^- (Broad and Reinhardt, 1976) with experimental results measured using the LAMPF accelerator. [From Gram et al. (1978).]

D. Collisional Detachment by Atom Impact

When a negative ion, X^-, undergoes a low-energy collision with an atom Y, it can lose its attached electron via a transition between the $X-Y^-$ and $X-Y$ states if these states intersect. For example, the collision between H^- and He can be treated by considering the molecular states HeH^- and HeH. These states have been calculated by Olson and Liu (1980) and are shown in Fig. 6-10-9a. If the reactants make a sufficiently close collision so that their internuclear separation R is less than R_x, the system can cross to the HeH state, ejecting the attached electron. For D^- + He collisions, the D^-, being heavier, spends more time in the vicinity of the curve-crossing region, so the detachment cross section is

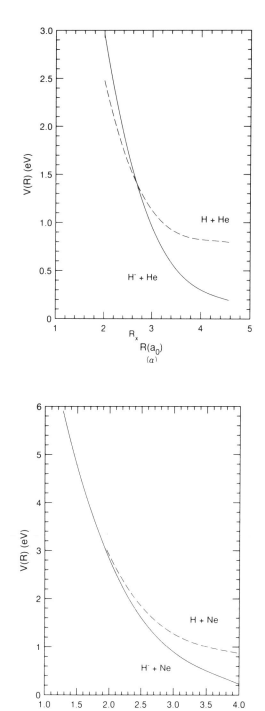

Figure 6-10-9. Internuclear potentials for the ground states of (a) HeH and HeH$^-$ and (b) NeH and NeH$^-$. [From Olson and Liu (1980).]

slightly larger than for the H^- + He case. This simple model gives good agreement with measured detachment cross sections in H^- + He collisions.

For H^- + Ne collisions, however, the situation is somewhat different, for a calculation of the NeH$^-$ and NeH states reveals that these states do not intersect but, in fact, merge at low R (Fig. 6-10-9b). As a result, the detachment cross section is smaller by about a factor of $\frac{2}{3}$ from that for H^- + He. The threshold for the H^- + Ne reaction is less well defined than for H^- + He and the isotope effect is reversed, the D^- + Ne cross section being smaller. This reaction is better characterized by a dynamic mechanism in which the collision dislodges the electron into the continuum. At energies greater than several hundred kilovolts, other detachment processes involving excitation of the reactants become important and are discussed in Chapter 5.

E.   Collisional Detachment by Molecule Impact

The greater complexity of the molecular target offers the possibility for many more reactions to occur, including charge transfers to molecular shape resonances which subsequently autoionize, and reactive collisions in which new molecular species can be formed. Of particular note is the process of associative detachment

$$A^- + BC \to ABC + e,$$

which is important at energies less than 1 eV and has been studied extensively using flow-tube techniques (Albritton, 1978). An extensive review of collisional detachment involving both atomic and molecular targets has been given by Champion and Doverspike (1984).

6-11.   NOTES

6-1. Some Sources of Error in the Study of Dissociative Processes. Ions produced in dissociative processes may have appreciable kinetic energy (as much as several eV), whereas ions formed in nondissociative reactions do not. A given apparatus may discriminate in favor of or against ions produced with initial kinetic energy. Furthermore, ions resulting from dissociative processes may have angular distributions that are far from isotropic, and angular anisotropies can strongly influence experiments in which ions are collected through a small acceptance angle in a fixed direction (Section 6-10 of the companion volume). The anisotropies can cause very serious errors, not only in magnitude of measured attachment cross sections, but also in their shapes. For these reasons, some of the early attachment data obtained with "Lozier tubes" and mass spectrometers are faulty. The Lozier apparatus is discussed on pp. 393–395 of the companion volume and alluded to in Section 6-6-A.

6-2. *Negative-Ion Photoelectron Spectroscopy: Comparison with Photodetachment Threshold Spectroscopy.* The spectroscopy of negative ions involves the study of bound-free transitions (i.e., photodetachment) because negative ions generally have only one, or a few, bound states, as opposed to neutral molecules, which have an infinite number. The two main experimental techniques used in such studies are (1) photodetachment threshold spectroscopy and (2) photoelectron spectroscopy of negative ions.

Photodetachment threshold spectroscopy (PDTS) was discussed in Sections 6-10-A through 6-10-C. It is the negative-ion counterpart of studies of threshold ionization of neutral molecules by electron impact (Section 6-8 of the companion volume) and by photon impact (Section 8-16-A-7 of the companion volume). PDTS can determine electron affinities with an accuracy set mainly by the wavelength resolution of the light source, but it requires an intense light source tunable in the wavelength region of the threshold.

Negative-ion photoelectron spectroscopy (NIPS) involves crossing a mass-analyzed negative-ion beam with a laser beam of fixed frequency. Electrons photodetached into a small solid angle are energy analyzed and counted. We see that NIPS is the counterpart of the study of ionization of neutrals by photons of fixed energy (see Section 8-16-A-2 of the companion volume on photoelectron spectroscopy of neutrals). Its utility for the determination of electron affinities is degraded by the relatively poor energy resolution of the electron energy analyzers used to measure the energy of the ejected electrons. However, here, the incident photon beam energy can be set at any energy above the photodetachment threshold, and the requirement of an intense light source can easily be met.

As Lineberger (1982) points out, three kinds of information can be obtained from the photoelectron spectrum of negative ions: (1) electron affinities and energy levels of negative ions, (2) dynamical information, especially electron–molecule scattering cross sections, and (3) information on the excited states of neutral species produced by photodetachment. The easiest negative target ions to prepare are closed-shell ions, which yield neutral free radicals. By contrast, the easiest target molecules to prepare in neutral studies are closed-shell neutrals, which yield radical positive ions. We see that negative-ion photoelectron spectroscopy provides relatively easy access to these neutral species, which are very difficult to study by conventional methods, and NIPS has become an important and exciting branch of spectroscopy. Interesting case studies appear in the review by Lineberger (1982).

6-12. PROBLEMS

6-1. Calculate the potential energy of an electron in the field of a ground-state hydrogen atom. Neglect polarization effects, although they are important. [See Massey (1976, pp. 1–3).]

6-2. The average total collision cross section, $\bar{q}_t$, for electrons and the average cross section for electron attachment (or capture) to form negative ions, $\bar{q}_a$, are frequently used today in discussions of negative ions. Here we also wish to use the average electron mean free path λ, the collision frequency v, and the attachment probability h, which is the probability that an electron will be attached by a given neutral species of molecule on a single impact. The attachment probability h is given in terms of the cross sections by the equation

$$\bar{q}_a = h\bar{q}_t. \qquad (6\text{-}12\text{-}1)$$

Now consider a swarm of electrons with average speed $\bar{v}$ that is drifting with a constant average velocity v_d through a gas under the influence of a constant electric field. Show that if n_0 electrons are present in the swarm at $x = 0$, the number surviving attachment over a drift distance d is

$$n = n_0 \exp\left(\frac{-h\bar{v}d}{\lambda v_d}\right). \qquad (6\text{-}12\text{-}2)$$

The drift velocity v_d of ions and electrons is discussed in Chapters 7 and 8.

6-3. Study the experimental methods for measuring electron affinities that are discussed in Massey (1976, Chap. 3).

6-4. The effects that free electrons in the earth's ionosphere have on the propagation of electromagnetic waves are well known and well understood (McDaniel, 1964, pp. 421–422). Less well known are the effects of negative ions on electromagnetic radiation of extremely low frequency. Read about this subject in McDaniel and Viehland (1984).

REFERENCES

Aboauf, R., D. Teillet-Billy, and S. Goursaud (1981), *J. Phys. B* **14**, 3517.

Ajello, J. M., and A. Chutjian (1979), *J. Chem. Phys.* **71**, 1079.

Albritton, D. L. (1978), *ADNDT* **22**, 1.

Allan, M., and S. F. Wong (1978), *Phys. Rev. Lett.* **41**, 1791.

Asundi, R. K., J. D. Craggs, and M. V. Kurepa (1963), *Proc. Phys. Soc. London* **82**, 967.

Bardsley, J. N., and J. S. Cohen (1978), *J. Phys. B* **11**, 3645.

Bardsley, J. N., and F. Mandl (1968), *Rep. Prog. Phys.* **31**, 471.

Bardsley, J. N., A. Herzenberg, and F. Mandl (1966), *Proc. Phys. Soc.* **89**, 305, 321.

Berkowitz, J., W. A. Chupka, and T. A. Walter (1969), *J. Chem. Phys.* **50**, 1497.

Blau, L. M., R. Novick, and D. Weinflash (1970), *Phys. Rev. Lett.* **24**, 1268.

Bloch, F., and N. E. Bradbury (1935), *Phys. Rev.* **48**, 689.

Branscomb, L. M. (1962), in D. R. Bates, Ed., *Atomic and Molecular Processes*, Academic Press, New York, p. 100.

Branscomb, L. M., D. S. Burch, S. J. Smith, and S. Geltman (1958), *Phys. Rev.*, **111**, 504.

Breyer, F., P. Frey, and H. Hotop (1981), *Z. Phys. A* **300**, 7.

Broad, J. T., and W. P. Reinhardt (1976), *Phys. Rev. A* **14**, 2159.

Bryant, H. C., B. D. Dieterle, J. Donahue, H. Sharifian, H. Tootoonchi, D. M. Wolfe, P. A. M. Gram, and M. A. Yates-Williams (1977), *Phys. Rev. Lett.* **38**, 228.

Buchel'nikova, I. S. (1958), *Sov. Phys. JETP* **8**, 783.

Burrow, P. D., J. A. Michejda, and J. Comer (1976), *J. Phys. B* **9**, 3225.

Cavalleri, G. (1969), *Phys. Rev.* **179**, 86.

Champion, R., and L. D. Doverspike (1984), in L. G. Christophorou, Ed., *Electron–Molecule Interactions and Their Applications*, Vol. 2, Academic Press, New York, p. 619.

Chanin, L. M., A. V. Phelps, and M. A. Biondi (1959), *Phys. Rev. Lett.* **2**, 344.

Chanin, L. M., A. V. Phelps, and M. A. Biondi (1962), *Phys. Rev.* **128**, 219.

Chantry, P. J. (1968) *Phys. Rev.* **172**, 125.

Chantry, P. J. (1969) *J. Chem Phys.* **51**, 3369.

Chantry, P. J., and G. J. Schulz (1967), *Phys. Rev.* **156**, 134.

Chen, C. L., and P. J. Chantry (1979), *J. Chem. Phys.* **71**, 3897.

Christodoulides, A. A., D. L. McCorkle, and L. G. Christophorou (1984), in L. G. Christophorou, Ed., *Electron–Molecule Interactions and Their Applications*, Vol. 1, Academic Press, New York, p. 423.

Christophorou, L. G. (1981), *Electron and Ion Swarms*, Pergamon Press, Elmsford, N.Y.

Christophorou, L. G. (1984), *Electron–Molecule Interactions and Their Applications*, Vols. 1 and 2, Academic Press, New York.

Chupka, W. A., J. Berkowitz, and D. Gutman (1971), *J. Chem. Phys.* **55**, 2724.

Chutjian, A. (1981) *Phys. Rev. Lett.* **46**, 1511.

Compton, R. N., P. W. Reinhardt, and C. D. Cooper (1978), *J. Chem. Phys.* **68**, 2023.

Doehring, A. (1952), *Z. Naturforsch.* **7A**, 253.

Drzaic, P. S., J. Marks, and J. I. Brauman (1984), *Gas Phase Ion Chem.* **3**, 167.

Dunn, G. H. (1962), *Phys. Lett.* **8**, 62.

Dunning, F. B. (1987), *J. Phys. Chem.* **91**, 2244.

Ervin, K. M., and W. C. Lineberger (1991), in N. G. Adams and L. M. Babcock, Eds., *Advances in Gas Phase Ion Chemistry*, Vol. 1, JAI Press, Greenwich, Conn.

Eyler, J. R. (1974), *Rev. Sci. Instrum.* **45**, 1154.

Ferguson, E. E., F. C. Fehsenfeld, and A. L. Schmeltekopf (1969), *AAMP* **5**, 1.

Froese-Fischer, C., J. Lagowski, and S. Vosko (1987), *Phys. Rev.* **59**, 2263.

Gibson, D. K., R. W. Crompton, and G. Cavalleri (1973), *J. Phys. B* **6**, 1118.

Gram, P. A. M., J. C. Pratt, M. A. Yates-Williams, H. C. Bryant, J. Donahue, H. Sharifian, and H. Tootoonchi (1978) *Phys. Rev. Lett.* **40**, 107.

Hall, R. I., I. Cadez, C. Schermann, and M. Tronc (1977), *Phys. Rev.* **A15**, 599.

Hamm, M. E., R. W. Hamm, J. Donahue, P. A. M. Gram, J. C. Pratt, M. A. Yates, R. D. Bolton, D. A. Clark, H. C. Bryant, C. A. Frost, and W. W. Smith (1979), *Phys. Rev. Lett.* **43**, 1715.

Hanstorp, D., P. Devynck, W. G. Graham, and J. R. Peterson (1989), *Phys. Rev. Lett.* **63**, 368.

Hazi, A. U., A. E. Orel, and T. N. Rescigno (1981), *Phys. Rev. Lett.* **46**, 918.

Henderson, W. R., W. L. Fite, and R. T. Brackmann (1969), *Phys. Rev.* **183**, 157.

Herzenberg, A. (1969), *J. Chem. Phys.* **51**, 4942.

Hilby, J. W. (1939), *Ann. Phys.* **34**, 473.

Hiskes, J. R. (1962), Quoted in D. L. Judd, *Nucl. Instrum. Methods* **18**, 70, 1962.

Holøien, E., and J. Midtdal (1955), *Proc. Phys. Soc.* **A68**, 815.

Hotop, H., and W. C. Lineberger (1975), *J. Phys. Chem. Ref. Data* **4**, 539.

Hotop, H., R. A. Bennett, and W. C. Lineberger (1973), *J. Chem. Phys.* **58**, 2373.

Hunter, S. R., J. G. Carter, and L. G. Christophorou (1986), *J. Appl. Phys.* **60**, 24.

Janousek, B. K., and J. I. Brauman (1979), in M. T. Bowers, Ed., *Gas Phase Ion Chemistry*, Vol. 2, Academic Press, New York, p. 53.

Kaiser, H. J., E. Heinicke, R. Rackwitz, and D. Feldmann (1974), *Z. Phys.* **270**, 259.

Khvostenko, N. I., and V. M. Dukel'skii (1958), *Sov. Phys. JETP* **6**, 657.

Klar, D., M.-W. Ruf, and H. Hotop (1991), in *Proc. Joint Symp. Electronic Ion Swarms and Low Energy Electronic Scattering*, Bond University, Gold Coast, Australia, p. 2.

Kline, L. E., D. K. Davies, and C. L. Chen (1979), *J. Appl. Phys.* **50**, 6789.

Kondow, T. (1987), *J. Phys. Chem.* **91**, 1307.

Kraft, T., M. W. Ruf, and H. Hotop (1989), *Z. Phys. D* **14**, 179.

Lee, L. C., and G. P. Smith (1979), *J. Chem. Phys.* **70**, 1727.

Lineberger, W. C. (1982), "Negative Ion Photoelectron Spectroscopy," in Vol. 5 in H. S. W. Massey, E. W. McDaniel, and B. Bederson, Series Eds. (1982–1984), *Applied Atomic Collision Physics*, 5 vols., Academic Press, New York.

Lineberger, W. C., and B. W. Woodward (1970), *Phys. Rev. Lett.* **25**, 424.

Lozier, W. W. (1930), *Phys. Rev.* **36**, 1285.

Lykke, K. R., K. K. Murray, D. M. Neumark, and W. C. Lineberger (1988), *Philos. Trans. R. Soc.* **A324**, 179.

Manson, S. T. (1971), *Phys. Rev.* **A3**, 147.

Massey, H. S. W. (1976), *Negative Ions*, 3rd ed., Cambridge University Press, Cambridge.

Massey, H. S. W., E. W. McDaniel, and B. Bederson (1982–1984), *Applied Atomic Collision Physics*, 5 vols., Academic Press, New York.

McCorkle, D. L., A. A. Christodoulides, L. G. Christophorou, and I. Szamrej (1980), *J. Chem. Phys.* **72**, 4049.

McDaniel, E. W. (1964), *Collision Phenomena in Ionized Gases*, Wiley, New York.

McDaniel, E. W., and L. A. Viehland (1984), *Phys. Rep.* **110**, 363–365.

McMahon, T. B., and J. L. Beauchamp (1972), *Rev. Sci. Instrum.* **43**, 509.

Mead, R. D., A. E. Stevens, and W. C. Lineberger (1984), in M. T. Bowers, Ed., *Gas Phase Ion Chemistry*, Vol. 3, Academic Press, New York, p. 213.

Miller, T. M. (1981), *Adv. Electron. Electron Phys.* **55**, 119.

Miller, T. M. (1991), in D. R. Lide, Ed., *Handbook of Chemistry and Physics*, CRC Press, Boca Raton, Fla., pp. 10–180.

Morgan, L. A. (1991), *J. Phys. B* **24**, 4649.

Moseley, J. T., R. A. Bennett, and J. R. Peterson (1974), *Chem. Phys. Lett.* **26**, 288.

Mullen, B., and E. W. Vogt (1968), in Stinson, G. M. et al. (1969) *Nucl. Instrum. Methods* **74**, 333.

Novick, R., and D. Weinflash (1970), in D. N. Langenberg, and B. N. Taylor, Eds., NBS Special Publication, *Proc. Int. Conf. Precision Measurement* NBS343, p. 403.

Olson, R. E., and B. Liu (1980), *Phys. Rev. A* **22**, 1389.

O'Malley, T. F. (1966), *Phys. Rev.* **150**, 14.

O'Malley, T. F. (1971), *AAMP* **7**, 223.

O'Malley, T. F., and H. S. Taylor (1968), *Phys. Rev.* **176**, 207.

Pegg, D., J. Thompson, R. N. Compton, and G. Alton (1987), *Phys. Rev. Lett.* **59**, 2267.

Petrovic, Z. Lj., and R. W. Crompton (1985), *J. Phys. B* **17**, 2777.

Rapp, D., and D. D. Briglia (1965), *J. Chem. Phys.* **43**, 1480.

Rapp, D., T. E. Sharp, and D. D. Briglia (1965), *Phys. Rev. Lett.* **14**, 533.

Schulz, G. J. (1959), *Phys. Rev.* **113**, 816.

Schulz, G. J., and R. K. Asundi (1967), *Phys. Rev.* **158**, 25.

Schulz, P. A., R. D. Mead, P. L. Jones, and W. C. Lineberger (1982), *J. Chem. Phys.* **77**, 1153.

Sharp, T. E. (1971), *AD* **2**, 119.

Shimamori, H., and Y. Hatano (1976), *Chem. Phys. Lett.* **38**, 242.

Smirnov, B. M. (1982), *Negative Ions*, McGraw-Hill, New York.

Smith, D., and N. G. Adams (1983), *J. Phys. B* **17**, 461.

Smyth, K. C., and J. I. Brauman (1972), *J. Chem. Phys.* **56**, 1132.

Srivastava, R. D., O. M. Uy, and M. Farber (1972), *J. Chem. Soc. Faraday Trans.* **68**, 1388.

Stamatovic, A., and G. J. Schulz (1970), *J. Chem. Phys.* **53**, 2663.

Tate, J. T., and W. W. Lozier (1935), *Phys. Rev.* **39**, 254.

Toriumi, M., and Y. Hatano (1985), *J. Chem. Phys.* **82**, 254.

Tronc, M., F. Fiquet-Fayard, C. Schermann, and R. I. Hall (1977), *J. Phys. B* **10**, 305.

Verhaart, G. J., and H. H. Brongersma (1980), *Chem. Phys.* **52**, 431.

Viggiano, A. A., T. M. Miller, A. E. Stevens Miller, R. A. Morris, J. M. Van Doren, and J. F. Paulson (1991), *Int. J. Mass Spectrom. Ion Process* **109**, 327.

Wadehra, J. M., and J. N. Bardsley (1978), *Phys. Rev. Lett.* **41**, 1795.

Walter, C. W., Ch. F. Hertzler, and J. R. Peterson (1991), *Abstracts of ICPEAC XVII*, Brisbane, I. E. McCarthy, W. R. MacGillivray, and M. C. Standage, Eds., p. 66.

Wang, L. J., M. King, and T. J. Morgan (1986), *J. Phys. B* **19**, L623.

Wigner, E. (1948) *Phys. Rev.* **73**, 1002.

7

ION TRANSPORT IN GASES;
ION–MOLECULE REACTIONS
AT LOW ENERGIES

7-1. INTRODUCTION

In Part $\mathcal{A}$ we are concerned with the drift and diffusion of slow gaseous ions in an externally applied electric field. We also discuss drift-tube experiments on ion swarms that yield ion–molecule reaction rates. The energy range of interest extends from thermal values at low temperatures up to about 6 eV. The transport phenomena are first discussed in physical terms; experimental techniques for measuring ionic drift velocities, diffusion coefficients, and reaction rates with drift tubes are then described. Ionic transport theory is sketched, with emphasis placed on the aspects of modern theory that permit the determination of ion–neutral interaction potentials from experimental transport data. Except where indicated (see Section 7-5), we take the interaction potentials to be spherically symmetric and the ion–molecule collisions to be elastic. Finally, in Part $\mathcal{B}$, we discuss several versions of flow-drift apparatus and ion traps that are used for studies of ion–molecule reactions.

PART $\mathcal{A}$. ION TRANSPORT

7-2. GENERAL CONSIDERATIONS

A. Basic Phenomenology; Drift and Diffusion of Gaseous Ions

For simplicity we assume that only a single species of ion is present in a single-component gas unless otherwise specified. Also, only steady-state behavior is discussed.

When a localized, tenuous cloud of ions exists in a neutral gas of uniform temperature and pressure, diffusive flow will occur, driven by the gradient in the

485

relative concentration of the ions. If we impose the constraint that the number density of the ions, n, be sufficiently low, the ionic flux density $\mathbf{J}$ will obey Fick's law of diffusion;

$$\mathbf{J} = -D\nabla n. \tag{7-2-1}$$

The diffusion coefficient is denoted by the positive scalar D, and the negative sign indicates that the flow occurs in the direction of decreasing ionic concentration. If a weak, uniform electric field $\mathbf{E}$ is now applied throughout the gas, the ion cloud, or "swarm," will acquire a steady-state drift velocity, $\mathbf{v}_d$, in addition to the greater, random velocity. As long as the field intensity is weak, the drift velocity (which is also the average velocity of the ions) is proportional to the field strength:

$$\mathbf{v}_d = K\mathbf{E}, \tag{7-2-2}$$

where K is the (scalar) mobility of the ions in the neutral gas. Both D and K are joint properties of the ions and the gas, ultimately depending on the ion–neutral interaction potential, the energy of the particles, and the density of the gas. For weak electric field strength, the proportionality (7-2-2) is valid, the diffusion is isotropic, and D is a scalar. A simple equation known as the Nernst–Townsend–Einstein relation, or more often, the Einstein relation, then exists between the mobility and the diffusion coefficient:

$$K = \frac{e}{kT} D, \tag{7-2-3}$$

where e is the ionic charge, k is Boltzmann's constant, and T is the gas temperature.

It is important to point out that (7-2-3) is valid only when the ions are close to being in thermal equilibrium with the gas molecules, that is, when "low-field" conditions obtain. Under these conditions the ionic velocity distribution is very nearly Maxwellian. The ionic motion is largely the thermal random motion produced by the heat energy of the gas, with a small drift component superimposed in the direction of the applied field. Equation (7-2-3) is exact in the limit of vanishing electric field and ion concentration.

The constant, steady-state drift velocity of the ions given by (7-2-2) is achieved as a balance between the accelerations in the field direction between collisions with gas molecules and the decelerations that occur between collisions. Since the ionic mass is usually comparable to the molecular mass, only a few collisions are normally required for the ions to attain a steady-state condition after the electric field is applied.

If the electric field strength becomes so large that the ions acquire an average energy appreciably greater than the thermal energy, then (7-2-1) and (7-2-3) are no longer valid. The thermal energy becomes less important, but two large components of motion are produced by the drift field: a directed component along the field lines and a random component representing energy acquired from the drift field but converted into random form by collisions with molecules. The mobility K appearing in (7-2-2) is no longer a constant in general but will usually depend on the ratio of the electric field intensity to the gas number density E/N, which is the parameter that determines the average ionic energy gained from the field in the steady-state drift, above the energy associated with the thermal motion. In addition, the energy distribution of the ions becomes distinctly non-Maxwellian, and the diffusion now takes place transverse to the field direction at a rate different from that of the diffusion in the direction of the electric field. The diffusion coefficient thus becomes a tensor rather than a scalar, and has the form

$$\mathbf{D} = \begin{vmatrix} D_T & 0 & 0 \\ 0 & D_T & 0 \\ 0 & 0 & D_L \end{vmatrix}, \tag{7-2-4}$$

where D_T is the (scalar) transverse diffusion coefficient that describes the rate of diffusion in directions perpendicular to E, and D_L is the (scalar) longitudinal diffusion coefficient characterizing diffusion in the field direction (Wannier, 1953). It should be noted that the symbols $D_\parallel$ and $D_\perp$ are as often used as D_L and D_T to refer to diffusion parallel and perpendicular, respectively, to the drift field E.

The mobility of a given ionic species in a given gas is inversely proportional to the number density of the molecules but relatively insensitive to small changes (a few kelvin at room temperature) in the gas temperature if the number density is held constant. To facilitate the comparison and use of data, a measured mobility K is usually converted to a "standard," or "reduced," mobility K_0 defined by

$$K_0 = \frac{p}{760.00} \frac{273.15}{T} K, \tag{7-2-5}$$

where p is the gas pressure in torr and T is the gas temperature in kelvin at which the mobility K was obtained. Under the standard conditions of pressure and temperature (760 torr and 0°C) the gas number density $N_0 = 2.6868 \times 10^{19} \, cm^{-3}$. It must be emphasized that the use of (7-2-5) merely provides a standardization or normalization with respect to the molecular number density; the temperature to which the standard mobility actually refers is the temperature of the gas during the measurement. For ions of atmospheric interest in atmospheric gases the standard mobility is on the order of several cm^2/V-s. In the modern literature, when a single value is quoted as the "mobility" of an ion in a gas, the value cited is the standard mobility extrapolated to zero field strength.

Ions of the atmospheric gases in their parent gases have diffusion coefficients on the order of $50 \, cm^2/s$ at 1 torr pressure and low E/N. The diffusion coefficients usually increase dramatically as E/N is raised above the low-field region. As expected theoretically, $D_{L,T}$ is found to vary inversely with the number density of the gas, and diffusion coefficient data are usually presented in the form of the product $ND_{L,T}$ as a function of E/N or T.

Reliable mobility data are now available for about 300 ion–gas combinations, but diffusion coefficients for a much smaller number of systems [see tabulations by Ellis et al. (1976, 1978, 1984)]. Most of these data were obtained as a function of the energy parameter E/N. Incidentally, the unit for E/N that is commonly used today is the *Townsend*, where $1 \, Td = 10^{-17} \, V\text{-}cm^2$.

B.   Ion–Ion Interactions and Effects of Space Charge

Under the usual conditions more than one ionic species is present at a given time in an ionized gas. If the density of ionization is low, each species of ion may be considered separately, and the ions of each type drift and diffuse independently through the gas without interacting appreciably with one another or with members of the other species. This condition must obtain in ion drift experiments if reliable measurements are to be made. Space-charge blowup of a drifting ion cloud mimics diffusion of the ions and has a disastrous effect on diffusion measurements; it can even falsify measurements of the drift velocity. Transport studies should be performed under conditions where space-charge effects are not in evidence. These matters were first considered in the context of ion swarm experiments by Wannier (1953); his work has been summarized by Mason and McDaniel (1988).

C.   Ambipolar Diffusion

If we consider a gas-filled cavity in which both electrons and positive ions are diffusing toward the walls, usually one can neglect the interaction between the negative and positive particles below ionization densities of about 10^7 to $10^8 \, cm^{-3}$. Above this level, space-charge effects produced by Coulomb forces between the electrons and the positive ions must be taken into account.

It may be shown that the number density of electrons in a highly ionized gas will approximately equal the number density of positive ions at each point, provided that we are not within about 1 Debye length* of a boundary. Any deviation from charge equality produces electrical forces that oppose charge separation and tend to restore the balance. Because their diffusion coefficient is much higher than that of the ions, the electrons attempt to diffuse more rapidly than the ions toward regions of lower concentration, but their motion is

---

*The Debye length is a measure of the distance over which deviations from charge neutrality can occur in an ionized gas. It is directly proportional to the square root of the energy and inversely proportional to the square root of the number density of the charged particles in the gas.

impeded by the restraining space-charge field thus created. This field has the opposite effect on the ions and causes them to diffuse at a faster rate than they would in the absence of electrons. Both species of charged particle therefore diffuse with the same velocity, and since there is now no difference in the flow of particles of opposite sign, the diffusion is called *ambipolar*.

We now derive an expression for the coefficient of ambipolar diffusion. Let n represent the common number density of the electrons and positive ions, and v_a the velocity of ambipolar diffusion. We assume that the gas pressure is high enough for the particles to make frequent collisions. The mobility concept will then be assumed to apply not only for the ions but for the electrons as well. Let **E** denote the intensity of the electric field established by the space-charge separation. Since the velocity of diffusion is the same for both species, we have

$$v_a = -\frac{D^+}{n}\frac{dn}{dx} + K^+ E \tag{7-2-6}$$

and

$$v_a = -\frac{D^-}{n}\frac{dn}{dx} - K^- E, \tag{7-2-7}$$

where K^+ and K^- are the mobilities of the ions and electrons, respectively, and D^+ and D^- are their ordinary or "free" diffusion coefficients. All four coefficients are positive numbers. By eliminating E we obtain

$$v_a = -D_a\frac{1}{n}\frac{dn}{dx}, \tag{7-2-8}$$

where D_a is the coefficient of ambipolar diffusion defined by the equation

$$D_a = \frac{D^+ K^- + D^- K^+}{K^+ + K^-}. \tag{7-2-9}$$

D_a characterizes the diffusive motion of both species.

If we assume that $K^- \gg K^+$ and $T^- \gg T^+$ and use the Einstein relation

$$\frac{D}{K} = \frac{kT}{e}, \tag{7-2-10}$$

we find that

$$D_a \approx D^- \frac{K^+}{K^-} = \frac{kT^-}{e} K^+. \tag{7-2-11}$$

When $T^+ = T^- = T$, on the other hand,

$$D_a \approx 2D^+ = \frac{2kT}{e} K^+. \tag{7-2-12}$$

The zero-field reduced mobility K_0 is related to the ambipolar diffusion coefficient D_a by the equation

$$K_0 = \frac{D_a p}{T^2} 2.086 \times 10^3. \tag{7-2-13}$$

D.   Importance of Ionic Mobility and Diffusion Data

Data on ionic mobilities and diffusion coefficients are of both theoretical and practical interest. First, experimental values of these quantities, particularly their dependence on E/N and T, can provide information about ion–molecule interaction potentials at greater separation distances than are generally accessible in beam scattering experiments (see Chapter 1). Second, mobilities are required for the calculation of ion–ion recombination coefficients (Flannery, 1982) and the rate of dispersion of ions in a gas due to mutual repulsion (McDaniel, 1964). Ion transport data are also required for the proper analysis of various kinds of experiments on chemical reactions between ions and molecules (Mason and McDaniel, 1988). In addition, knowledge of the mobility of an ionic species in a given gas as a function of E/N permits the estimation of the average ion energy as a function of this parameter (Section 7-4-C). Finally, information on both mobilities and diffusion is required for a quantitative understanding of electrical discharges in gases and various atmospheric phenomena (McDaniel and Viehland, 1984; Massey et al., 1982–1984).

Accurate elastic momentum transfer cross sections can be obtained for atomic ion–atomic gas combinations as a function of impact energy provided that the requisite transport properties are accurately known over a wide range of E/N. It is also possible to extract cross sections for symmetric charge transfer from ion mobility measurements on such systems as Ne^+ in neon gas.

7-3.   MEASUREMENT OF IONIC TRANSPORT COEFFICIENTS

A.   Basic Aspects

The transport properties of ions in gases have been studied experimentally since shortly after the discovery of X-rays in 1895 and theoretically since 1903. The first measurements were performed by Thomson, Rutherford, and Townsend at the Cavendish Laboratory of Cambridge University late in the nineteenth century. Surveys of the history of experimentation in this field appear in Loeb (1960) and Massey (1971) and are not repeated here.

Meaningful direct measurements of ionic free diffusion were first made only in the 1960s, and comparatively few data are available. On the other hand, some reliable mobility measurements were made as long ago as the 1930s and a large number of good data are now in hand. Recent work has indicated, however, that most of the old data and even some of the newer results are either incorrect or refer to ions whose identities were not known. The main reason for this is that in most cases the drifting ions can undergo chemical reactions* with the molecules of the gases through which they are moving and thereby change their identities. Techniques for obtaining reliable results, some of them developed only recently, are discussed in Mason and McDaniel (1988) and, more briefly, here.

Ionic (and electronic) transport properties are measured with apparatus called *drift tubes.* "Conventional" ionic drift tubes usually consist of an enclosure containing gas, an ion source positioned on the axis of the enclosure, a set of electrodes that establishes a uniform axial electrostatic field along which the ions drift, and a current-measuring collector in the gas at the end of the ionic drift path. The drift field E causes the ions of any given molecular composition to "swarm" through the gas with a drift velocity and diffusion rate determined by the nature of the ions and gas molecules, the field strength, and the gas pressure and temperature. For the determination of drift velocities and longitudinal diffusion coefficients the ion source is operated in a repetitive, pulsed mode, and the spectrum of arrival times of the ions at the collector is measured electronically. (Other measurement techniques are required for the determination of transverse diffusion coefficients.)

Usually more than one kind of ion is present in the gas, and the ions can undergo chemical reactions with the gas molecules. Hence most data obtained with "conventional" drift tubes are extremely difficult to interpret and have little fundamental importance. This situation led to the development of the *drift-tube mass spectrometer* in the 1960s. This type of instrument, in various forms, has been used in about 20 laboratories and has produced a wealth of transport data as well as reaction rate coefficients for about 100 ion–molecule reactions as a function of E/N. A drift-tube mass spectrometer differs from a conventional drift tube by having the ion collector in the gas replaced by a sampling orifice located in the wall at the end of the drift tube, usually on the axis. Ions arriving at the end plate close to the axis pass through the orifice, out of the drift tube, and into an evacuated region that contains a mass spectrometer and an ion detector (usually a pulse-counting electron multiplier). The mass spectrometer can be set to transmit any one of the various ionic species entering it, the other species of ions being rejected in the mass selection process. Thus the arrival-time spectrum can be mapped separately for each kind of ion arriving at the end of the drift tube.

The pressures at which ionic drift-tube experiments have been performed

---

*We use the term *ion–molecule reaction* to refer to a heavy particle rearrangement reaction such as $A^+ + BC \rightarrow AB^+ + C$ or $A^- + B + C \rightarrow AB^- + C$. By a *charge-transfer reaction* we mean a reaction in which an electron is transferred between the colliding structures, as in $A^+ + B \rightarrow A + B^+$ or $A^- + BC \rightarrow A + BC^-$. Both kinds of reactions are called "chemical."

have ranged from about 2.5×10^{-2} torr to above 1 atm. Most measurements are now made in the pressure range 7.5×10^{-2} to 1.0 torr. At pressures lower than the minimum quoted here, the ions would make too few collisions for steady-state conditions to be achieved except in apparatus of uncommonly great length. Ion sampling difficulties may appear in drift-tube mass spectrometer measurements at pressures greater than about 1.0 torr.

The product of the gas pressure p and the drift distance d should be large enough that the ions will travel a negligible fraction of the total distance before energy equilibrium in the drift field is achieved. Drift distances of 0.5 to 44 cm have been employed. Because of the large pd product used, each ion makes many collisions with molecules as it drifts the distance d; the average number of collisions per ion usually lies somewhere between 10^2 and 10^7, depending on the value of pd.

Drift-tube measurements on ions are made at ratios of drift-field intensity to gas number density (E/N) as low as about 0.3 Td, and for systems with resonant charge transfer, as high as about 5000. In a given experiment the lowest E/N value to which measurements may be extended will depend on intensity considerations, that is, on the ion current reaching the detector and the sensitivity of the detector. The measured ion signal eventually becomes too small for accurate measurements to be made. In many experiments it has proved impossible even to approach the low-field region, where the transport and reaction coefficients become independent of E/N. The high E/N limit in drift-tube experiments is usually imposed by electrical breakdown within the apparatus. At very high E/N the average ionic energy may attain values of about 6 eV, and inelastic collisions may be important even for atomic systems.

B. Georgia Tech Drift-Tube Mass Spectrometer; Representative Data

We now return to a specific apparatus, the drift-tube mass spectrometer constructed at the Georgia Institute of Technology for the study of ionic transport and reactions at low pressures and room temperature (Albritton et al., 1968; Miller et al., 1968). An overall view is shown in Fig. 7-3-1. The entire apparatus is of ultrahigh-vacuum construction and is unusual in its large size.

The instrument consists of a 51-cm-diameter stainless steel chamber enclosing a drift tube with a movable ion source and an ion sampling and detection region. Ions which we shall initially assume to be incapable of chemical reaction are created in short (1 μs), periodic bursts at a selected position on the axis. (The ionic number density is deliberately held at such a low value that space-charge effects are completely negligible.) The uniform, axial electrostatic field that produces the drift is maintained by electrodes within the drift tube.

After their creation in the source, the ions in each successive burst are gated into the drift region, where they quickly achieve a steady-state condition in which each species of ion moves with its characteristic, constant drift velocity v_d in the axial direction.

Each ion cloud drifts the same known distance before it encounters the end

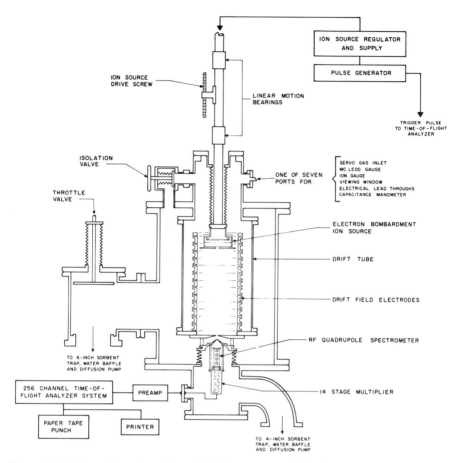

Figure 7-3-1. Overall sectioned view of the low-pressure drift-tube mass spectrometer at the Georgia Institute of Technology. The time-of-flight analyzer shown has been replaced by one having 1024 channels. [From Mason and McDaniel (1988).]

plate of the drift tube. There the axial core of each cloud passes through an exit aperture on the axis of the tube and enters a differentially pumped regon to be mass analyzed in an RF quadrupole mass filter. Ions of the mass selected for transmission are then detected with an electron multiplier operated in the pulse counting mode. The transit time of each ion detected is measured and stored by a multichannel analyzer, and repetitive pulsing of the ion source builds up an arrival-time distribution of the ionic species selected.

A differencing technique is used in the determination of the drift velocity in order to eliminate end effects. At each combination of E/N and gas pressure, the average transit time of the ions is evaluated for several different ion source positions, and the drift velocity is computed from the slope of the plot of drift distance versus average arrival time. The mobility K is then computed from the defining equation $K = v_d/E$, and the standard mobility from (7-2-5).

An analytical solution of the transport equation for the motion of nonreacting ions in the drift tube has been worked out by Gatland (1975) [see also Mason and McDaniel (1988)]. This solution provides an expression for the flux of ions on axis at any drift distance z and any time t after the ions are created. The flux as a function of time gives a theoretical arrival time spectrum that can be fitted to the experimental data for the purpose of determining D_L (see Fig. 7-3-2).

In the foregoing discussion, we assumed that no reactions occurred between the ions and molecules of the gas filling the drift tube, but in fact, ion–molecule reactions take place in most drift-tube experiments. However, the techniques outlined above may still be used to obtain accurate values of v_d and D_L for primary ions (i.e., ions produced by thermionic emission or electron impact in the ion source) provided that any reactions occurring merely deplete the population of the ions under consideration and provided that the reactions are not too fast [see, e.g., Thomson et al. (1973)]. We may also make measurements on secondary ions formed by reactions of the primaries with gas molecules by operating at relatively high gas pressures and long drift distances. This type of

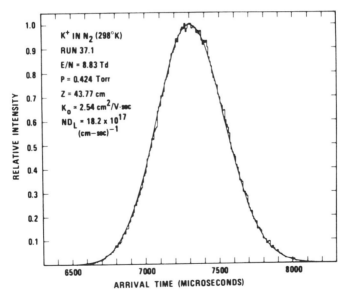

Figure 7-3-2. Arrival-time spectrum of "nonreacting" K^+ ions in nitrogen. The histogram represents experimental data; the smooth curve was calculated using Gatland's analysis. [From Thomson et al. (1973).]

operation ensures that most of these ions are formed inside or near the ion source and hence travel almost all of the way down the drift tube in this form.

Gatland (1975) has also analyzed ionic motion in the apparatus for a variety of ion–molecule reaction schemes, including sequential and reversible reactions, and has obtained solutions that may be utilized to deduce rate coefficients for the reactions from the experimental data. To date, rate coefficients have been obtained by two completely independent methods. The first involves the analysis of the shape of the arrival-time spectrum for secondary ions and is the basis for determinations of the rate coefficients for reactions in oxygen [see Fig. 7-3-3 and Snuggs et al. (1971)], carbon monoxide, hydrogen, deuterium, and

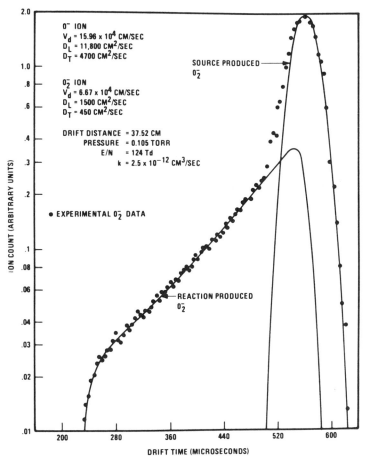

Figure 7-3-3. O_2^- arrival-time spectrum used to determine the rate coefficient for the reaction $O^- + O_2 \rightarrow O_2^- + O$. [From Snuggs et al. (1971).]

carbon dioxide. The second method involves measurement of the attenuation of the detector count rate for the primary (reactant) ions as the ion source is moved progressively farther from the drift-tube exit aperture. This second technique has been used for the study of reactions of ions in hydrogen, deuterium, nitrogen, and oxygen. The attenuation method may also be employed for the measurement of transverse diffusion coefficients [see e.g., Moseley et al. (1969)]. Comprehensive discussions of various methods of obtaining ionic transport and reaction data appear in Mason and McDaniel (1988), where special attention is paid to the techniques used for reactive systems.

Typical plots of v_d, K_0, and D_L versus E/N are shown in Figs. 7-3-4, 7-3-5, and 7-3-6. The variation of D_T with E/N is similar to what is shown in Fig. 7-3-6 for D_L, except in those cases where resonant charge transfer can take place between the ions and gas molecules. In such cases the variation of D_T with E/N is very weak. Except in ion–molecule combinations that exhibit resonant charge transfer, $D_T < D_L$ outside the low-field region.

The book by Mason and McDaniel (1988) contains descriptions of other drift-tube mass spectrometers. The three instruments built by R. Johnsen and M. A. Biondi at the University of Pittsburgh have been used for measurements of both ion mobilities and ion–molecule reaction rates.

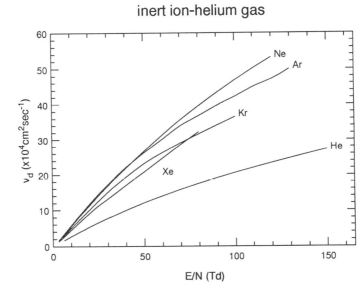

Figure 7-3-4. Experimental drift velocities of rare-gas ions in He gas at room temperature. [From Ellis et al. (1976, 1978, 1984).]

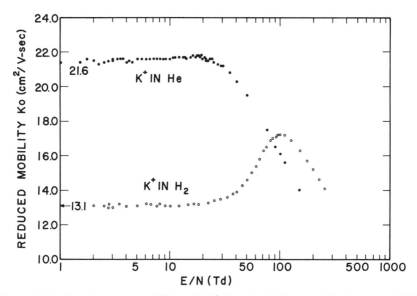

Figure 7-3-5. Experimental mobilities of K^+ ions in helium and hydrogen at 295 K. [From Thomson et al. (1973).]

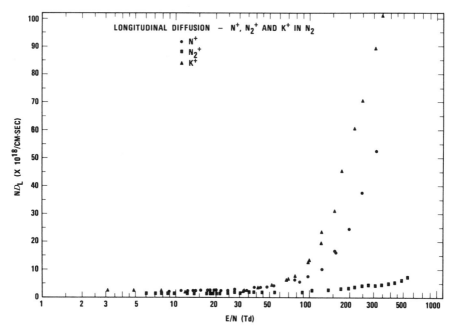

Figure 7-3-6. Longitudinal diffusion coefficients of N^+, N_2^+, and K^+ ions in nitrogen at room temperature. [From Moseley et al. (1969).]

7-4. THEORY OF ION TRANSPORT IN GASES; SPHERICALLY SYMMETRIC POTENTIALS AND ELASTIC COLLISIONS*

A.   Theories of Langevin

In 1903, soon after the first measurements of the mobility of gaseous ions, Langevin published a theory of ionic mobility based on the kinetic theory of gases, which was just becoming widely accepted. He considered the ions and molecules to be rigid spheres, the ions differing from the molecules only by possession of an electric charge. Only repulsive forces acting at the instant of impact were taken into account, and E/N was assumed to be small, so that the field energy would be negligible compared with the thermal energy. The ion density was taken to be low in order that ion–ion interactions could be ignored. A mean-free-path approach was followed.

The equation obtained by Langevin was

$$K = \frac{e\lambda}{m\bar{v}}, \tag{7-4-1}$$

where λ is the common mean free path for the molecules and ions, m is the common molecular and ionic mass, and $\bar{v}$ is the mean thermal velocity. Not surprisingly, this simple theory and its extension to unequal ion and molecule masses give only order-of-magnitude agreement with experiment.

In 1905, Langevin published an elegant and elaborate theory of ionic mobility based on the momentum transfer method that Maxwell had developed for the investigation of transport problems.† This theory applies only at low E/N values, but does take into account not only hard-core repulsion but also the inverse-fifth-power attractive force that dominates the long-range interaction between an ion and a nonpolar, but polarizable neutral molecule. The potential for the long-range interaction, the *polarization potential*, is

$$V(r) = -\frac{\alpha e^2}{2r^4}, \tag{7-4-2}$$

where α is the polarizability of the gas and r is the separation distance between the ion and molecule. In the *polarization limit*, where the other components of the interaction potential may be considered to have a negligible effect, Langevin's theory gives

$$K_{\text{pol}} = 35.9(\alpha\mu)^{-1/2} \, \text{cm}^2/V\text{-s} \tag{7-4-3}$$

---

*Sections 7-4 and 7-5 are largely taken from L. A. Viehland's discussion in McDaniel and Viehland (1984).

†A French-to-English translation of Langevin's classic 1905 paper appears in McDaniel (1964, pp. 701–726).

for the standard mobility, where α is measured in atomic units (a_0^3, with a_0 being the radius of the first Bohr orbit) and the reduced mass μ is measured in units of the proton mass. Although (7-4-3) applies rigorously for real ion–neutral systems only in the double limit of low T and low E/N, it predicts the low-field mobility at room temperature with moderate accuracy for a large number of systems.

B. Chapman–Enskog Theory

About 10 years after the publication of Langevin's theories, Chapman and Enskog developed a rigorous kinetic theory for gases composed of spherically symmetric particles with no internal degrees of freedom (Ferziger and Kaper, 1972). They applied their results to transport problems involving only neutral gases, but their expression for the mutual diffusion coefficient D_{12} can be evaluated for atomic ion–atom as well as atom–atom interaction potentials and then used for the calculation of low-field mobilities. This procedure is possible because the mobility of an ion at low E/N is related to the diffusion coefficient by the Einstein relation (7-2-3).

According to the Chapman–Enskog theory, in which the diffusion coefficient is calculated to second order, the ion mobility is given by the equation

$$K_0 = \frac{3}{8} \frac{e}{N_0} \left(\frac{\pi}{2\mu k T} \right)^{1/2} \frac{1 + \alpha_0}{\bar{\Omega}^{(1,1)}(T)}, \qquad (7\text{-}4\text{-}4)$$

where μ is the reduced mass of the ion–atom system and α_0 is a higher-order correction that is usually less than experimental error (α_0 equals zero for an inverse-fourth-power potential and has what appears to be a maximum value of 0.136 for hard-sphere interaction). If we neglect α_0, (7-4-4) shows that all of the influence of the atomic ion–atom interaction potential upon gaseous ion mobility arises through the momentum-transfer collision integral $\bar{\Omega}^{(1,1)}(T)$. Since quantum mechanical effects have been shown [see Gatland et al. (1977) and McDaniel (1964, pp. 442–447)] to become important only at temperatures far below those usually used for ion mobility measurements, this collision integral is given by the classical-mechanical equations

$$\bar{\Omega}^{(1,1)}(T) \equiv \frac{1}{2} (kT)^{-3} \int_0^\infty \varepsilon^2 Q^{(1)}(\varepsilon) \exp\left(-\frac{\varepsilon}{kT} \right) d\varepsilon, \qquad (7\text{-}4\text{-}5)$$

$$Q^{(1)}(\varepsilon) = 2\pi \int_0^\infty [1 - \cos \theta(b, \varepsilon)] b \, db, \qquad (7\text{-}4\text{-}6)$$

and

$$\theta(b, \varepsilon) = \pi - 2b \int_{r_0}^\infty \left[1 - \frac{b^2}{r^2} - \frac{V(r)}{\varepsilon} \right]^{-1/2} \frac{dr}{r^2}. \qquad (7\text{-}4\text{-}7)$$

Here ε is the relative kinetic energy of an ion–atom collision, $Q^{(1)}$ the momentum-transfer cross section, b the impact parameter of a collision, θ the scattering angle, r the ion–atom separation distance, V the interaction potential, and r_0 the turning point [the largest positive root of the bracketed quantity in (7-4-7)]. Equations (7-4-4) through (7-4-7) therefore represent the relationships needed to calculate gaseous ion mobilities in the limit of low E/N, given knowledge of the ion–atom interaction potential.

Dimensional considerations show that for a potential of the form $V(r) \sim r^{-n}$, $Q^{(1)}(\varepsilon)$ varies as $\varepsilon^{-2/n}$. It follows then from (7-4-5) that $K \sim T^{2/n} T^{-1/2}$ for potentials of this type. Hence it is clear that observations on the temperature variation of mobilities can lead to considerable information on ion–atom interactions. Note that the mobility should be temperature independent for a pure r^{-4} potential. The scattering of an ion at low temperatures is determined mainly by the long-range attractive polarization potential, which varies as r^{-4}. Thus we should expect all mobilities to become essentially independent of the temperature, provided that sufficiently low temperatures can be reached before quantum effects set in. At high temperatures, on the other hand, the scattering is determined principally by the short-range forces. If we represent these forces by a repulsive r^{-12} potential, we would predict the mobility to vary approximately as $T^{-1/3}$ at high temperatures. At intermediate temperatures there will be some cancellation of the short- and long-range forces, and the mobility should pass through a maximum as the temperature is varied. A similar explanation for a mobility maximum as the value of E/N is varied has been provided by Wannier (1970).

In summary, the Chapman–Enskog theory provides an excellent description of low-field gaseous ion transport if the atomic ion–atom interaction is accurately known. Modeling the potential with a particular functional form is usually of little use, however, because so many adjustable parameters must be used that the data cannot determine them uniquely. This problem is compounded by the fact that it is very difficult to make accurate low-field measurements of gaseous ion transport over a wide range of gas temperature; under the best conditions only temperatures from about 80 to 900 K can be sampled, with an accuracy of 1 to 2% for mobilities and 5 to 10% for diffusion coefficients. To fully connect the interaction potential with the transport coefficients, it is necessary to have a theory that applies at nonvanishing values of E/N.

C.   Theory of Wannier

The Chapman–Enskog theory makes use of the fact that the steady-state, position-independent gaseous transport coefficients can all be calculated from a knowledge of the velocity distribution functions, which are assumed to be described in terms of the cross sections governing the molecular collisions by the appropriate Boltzmann kinetic equation [see (2-12-2) of the companion volume and Section 8-1]. For gaseous ion transport coefficients, several simplifications can be built into this equation. For example, the dilute neutral gas is not disturbed by the electric field or the trace amounts of ions, and it must therefore have a Maxwell–Boltzmann velocity distribution function characterized by the temperature T. Similarly, ion–ion interactions can be neglected in the collision term of the Boltzmann equation, thereby making it a linear equation to be solved for the ion-velocity distribution function. Nevertheless, an exact solution of this linear integrodifferential equation is not possible even for the simple case of atomic ions in atomic gases.

In 1951–1952, Wannier (1953) developed a theory of gaseous ion transport, based on the Boltzmann equation, which is applicable to the high-E/N region. He applied this theory to the mean-free-time model in which the angular distribution of scattering in the center-of-mass system between colliding particles is independent of the relative speed of the particles, and the collision cross sections vary inversely with the speed. Although this model was known even to Maxwell, Wannier's approach showed that the Langevin equation (7-4-3) is exact not only at low E/N, but at high E/N as well for this model. Wannier's approach also provided an expression for the total energy of an ion at high E/N:

$$E_K = \frac{mv_d^2}{2} + \frac{Mv_d^2}{2} + \frac{3kT}{2}. \tag{7-4-8}$$

The first term on the right side is the field energy associated with the drift motion of the ion (mass m), whereas the second term is the random part of the field energy, expressed in terms of the molecule mass, M. The last term represents the thermal energy. It is apparent that

$$\frac{\text{random field energy}}{\text{drift energy}} = \frac{M}{m}. \tag{7-4-9}$$

Equation (7-4-9) illustrates the capacity that light ions in a heavy gas have for storing energy in the form of random motion. For ions traveling in the parent gas the ordered and random field energies are equal. For heavy ions in a light gas the random field energy is negligible.

In principle, the *Wannier expression* (7-4-8) provides an answer to the old question about the relationship between T and E/N dependences of the measured transport coefficients. It shows that the transport coefficients should be the same at any combination of T and E/N that leads to the same relative kinetic energy E_{rel} in the center-of-mass frame, which is given by the equation

$$E_{rel} = \left(mE_K + \frac{3MkT}{2} \right)(m + M) = \frac{3kT}{2} + \frac{Mv_d^2}{2}. \tag{7-4-10}$$

Theoretical and experimental tests of this prediction show that it is accurate to within about 10% for the indicated atomic ion–atom systems examined by McDaniel and Viehland (1984). However, little use of this prediction was made until the middle 1970s because other predictions of Wannier's theory are only semiquantitative and because there does not seem to be any systematic way of improving the accuracy of results calculated from the theory. Another reason for such hesitancy is that the prediction does not hold for molecular ion–neutral systems.

The theory of Wannier also leads to generalization to high E/N of the Einstein relations, with the results usually being given in the forms

$$\frac{eD_{\parallel}}{K} = kT_{\parallel}\left[1 + \frac{d\ln K}{d\ln(E/N)}\right] \tag{7-4-11}$$

and

$$\frac{eD_{\perp}}{K} = kT_{\perp}. \tag{7-4-12}$$

The form of these equations has been verified in many different ways. We note, however, that the expressions derived by Wannier for the parallel and perpendicular temperatures, $T_{\parallel}$ and $T_{\perp}$, are accurate only for the special cases for which they were derived. More general expressions are discussed in Section 7-4-F and in Mason and McDaniel (1988).

D.   One-Temperature Kinetic Theories

Much of Wannier's work on gaseous ion transport involved solutions of the Boltzmann equation which are solutions only in the sense that all of the experimentally important velocity averages of the ion distribution function could be obtained directly, without explicit knowledge of the distribution function.[*] At about the same time Kihara (1953) introduced a general procedure of this type, in which the Boltzmann equation is converted into an infinite set of coupled moment equations (which are also known as transport-relaxation equations or Maxwell's equations of change). These equations must be solved for the few moments of physical interest (the transport coefficients) by a successive approximation scheme. The crucial step for obtaining rapid convergence to accurate results is the choice of basis functions used to form the moment equations.

One-temperature kinetic theories (Mason and McDaniel, 1988) of gaseous ion transport use the Burnett functions to form moment equations from the Boltzmann equation. These functions are eigenfunctions of the collision operator in the Boltzmann equation for the special case where the ion–neutral potential varies as the inverse fourth power of the separation distance.

---

[*]Recent work on the determination of velocity distributions by solution of the Boltzmann equation is discussed by Ness and Viehland (1990).

Moreover, these functions are orthogonal in velocity space with respect to the weighting function

$$g_1(v) = \left(\frac{m}{2\pi kT}\right)^{3/2} \exp\left(-\frac{mv^2}{2kT}\right),$$ (7-4-13)

which therefore represents a zeroth approximation to the true ion distribution function. It should be noted that the description of these theories as "one-temperature" arises from the use of the gas temperature T to characterize the ion velocity distribution at all E/N.

Since the long-range interaction potential between an ion and a nonpolar, but polarizable, molecule always varies as r^{-4}, results calculated from the one-temperature theories must be exact in the limit of low T and low E/N. This means, for example, that they must yield the correct expression (7-4-3) for the polarization limit of the standard mobility. However, all one-temperature kinetic theories to date have yielded expressions for the transport coefficients that are power series or ratios of polynomials of E/N. Although these expressions are very accurate in the limit of low E/N, where they essentially reduce to the results of the Chapman–Enskog theory, they diverge at higher E/N. The general usefulness of one-temperature theories is therefore quite limited.

E. Two-Temperature Kinetic Theories

The essential feature of a two-temperature kinetic theory is the explicit recognition that at high E/N the trace ions in a drift tube can have a temperature appreciably greater than the neutral gas temperature. Accordingly, the basis functions used by Viehland and Mason (1975) to form moment equations from the Boltzmann equation were chosen so as to be orthogonal with respect to the weighting function

$$g_2(v) = \left(\frac{m}{2\pi kT_K}\right)^{3/2} \exp\left(-\frac{mv^2}{2kT_K}\right).$$ (7-4-14)

This zero-order distribution function differs from (7-4-13) by the use of an ion kinetic temperature T_K. Any convenient choice can be made for T_K without affecting the ultimate description of the gaseous ion transport coefficients, but different choices lead to drastically different results when the infinite set of coupled moment equations are solved at some intermediate level of approximation. Since such approximations are absolutely essential except for special cases, some care must be devoted to the selection of the parameter T_K. It has been established (Viehland and Mason, 1975) that a reasonable choice to make is to require that T_K satisfy the equation

$$\frac{3kT_K}{2} = \frac{m\langle v^2 \rangle}{2},$$ (7-4-15)

where the angular brackets indicate a velocity average or moment, taken with respect to the true velocity distribution function of the ions. The use of (7-4-15) guarantees that the final expression for the gaseous ion transport coefficients be well behaved at all E/N, a notable failure of earlier kinetic theories. Equation (7-4-15) does not imply that the true velocity distribution function of the ions must be known, or that any numerical value of T_K must be known before the two-temperature theory can be used at a particular value of T and E/N. Indeed, it simply implies that the Boltzmann equation will be solved subject to an initial constraint, which reflects our physical intuition that the true ion distribution function may be approximated, at least crudely, by (7-4-14) and (7-4-15).

In addition to using (7-4-14) and (7-4-15), Viehland and Mason (1975) adopted a particular procedure for systematically truncating the moment equations in a series of approximations. They showed that the standard mobility of a swarm of atomic ions moving in a single-component atomic gas under the influence of an electrostatic field of arbitrary strength may be written at all T and all E/N as

$$K_0 = \frac{3}{8} \frac{e}{N_0} \left(\frac{\pi}{2\mu k T_{\text{eff}}} \right)^{1/2} \frac{1 + \alpha}{\bar{\Omega}^{(1,1)}(T_{\text{eff}})}, \tag{7-4-16}$$

where the effective ion temperature T_{eff} is given by

$$k T_{\text{eff}} = kT + \frac{m v_d^2 (1 + \beta)}{3}. \tag{7-4-17}$$

If for the moment we neglect the correction terms α and β, (7-4-16) is seen to be a simple extension of the Chapman–Enskog result (7-4-4) to finite values of E/N by means of the Wannier expression (7-4-10)!

The corrections terms α and β in (7-4-16) and (7-4-17) depend in a complicated way on T, E/N, m, M, and the ion–atom interaction potential. Numerical calculations (Viehland and Mason, 1978) have shown that in most cases α and β are substantially less than 0.1, so the accuracy obtained by setting them equal to zero (which is simply the first approximation to the two-temperature theory) is generally good. Nevertheless, it is often necessary to go to the sixth approximation or beyond to ensure a converged result that is as accurate as the measured mobility. The reason for going to this large computational effort is simple to understand. Mobility data of excellent accuracy are now available for about 300 ion–neutral systems over a wide range of E/N, corresponding to T_{eff} values ranging from about 80 or 300 K to as high as 10,000 K, and the data can be used to probe details of the ion–neutral interaction potential by means of the two-temperature theory.

It might seem surprising that there is any direct method for inferring the interaction potential from measurements of K_0 as a function of E/N, even if the effects of the correction terms α and β are ignored. Equations (7-4-16) and

(7-4-17) allow values of the momentum transfer collision integral $\bar{\Omega}^{(1,1)}$ to be determined as a function of T_{eff}, but this collision integral is separated from the interaction potential by three layers of integration, given by (7-4-5) to (7-4-7), and these might plausibly be expected to wash out most details of the potential. It is nevertheless true that at a given T_{eff}, the value of $\bar{\Omega}^{(1,1)}$ is determined essentially by values of the potential over only a small range of separation distance. This idea was first incorporated into a simple computational scheme for the direct determination of interaction potentials from transport data by E. B. Smith and coworkers (Gough et al., 1972). The scheme was extended to gaseous ion mobility data by Viehland et al. (1976) and has been placed on a more secure foundation by Maitland et al. (1978).

There is one major difficulty with the two-temperature theory; it is not nearly as good for gaseous ion diffusion as for mobility. When the ion–atom mass ratio is less than 1, the convergence of successive approximations to the coefficients describing the different diffusion rates parallel and perpendicular to the electric field is nearly as good as for the mobility. When the mass ratio is between 1 and 4, the convergence is good only at low and intermediate E/N. Finally, when the mass ratio is greater than about 4, the two-temperature theory is essentially useless for diffusion (Viehland and Mason, 1978).

F.   Three-Temperature Theory and Direct Determination of Atomic Ion–Atom Interaction Potentials from Transport Data

The inability of two-temperature theory to describe gaseous ion diffusion adequately stems from two facts. First, diffusion coefficients are, compared to mobility, more intimately connected with the anisotropic nature of the ion motion that becomes more increasingly manifest as E/N and the ion–neutral mass ratio increase. Second, the two-temperature theory makes use from the start of a single ion temperature to characterize the energy distribution, thereby assuming essentially isotropic conditions. Accordingly, a *three-temperature kinetic theory* of gaseous ion transport was developed (Lin et al., 1979; Viehland and Lin, 1979). This theory was based on the use of basis functions for solving the Boltzmann equation which are orthogonal with respect to a modified Maxwellian distribution in which the ions are allowed to have different temperatures parallel $(T_{\parallel})$ and perpendicular $(T_{\perp})$ to the electric fields and to be displaced in velocity by an amount v_{dis} along the field direction. Values for $T_{\parallel}$, $T_{\perp}$, and v_{dis} are obtained from the physically appealing definitions

$$v_{dis} = \langle v_z \rangle, \tag{7-4-18}$$

$$kT_{\perp} = m\langle v_x^2 \rangle = m\langle v_y^2 \rangle, \tag{7-4-19}$$

and

$$kT_{\parallel} = m\langle (v_z - v_{dis})^2 \rangle. \tag{7-4-20}$$

The three-temperature theory is more difficult to apply than the two-temperature theory. However, these difficulties are only computational, and they have been overcome with the development of the computer programs MOBILDIF (Viehland, 1982) and MOBDIF (Viehland and Kumar, 1989). The advantage of the three-temperature theory is that it is able to describe gaseous ion diffusion as well as mobility. It leads (Waldman and Mason, 1981) to the generalized Einstein relations (7-4-11) and (7-4-12) in first approximation, but in general it provides expressions for $T_\parallel$ and $T_\perp$ that can only be evaluated numerically. Waldman and Mason (1981) have provided parameterized expressions which can provide nearly the same accuracy with considerably less effort. When resonant charge transfer does not occur, they recommend the equations

$$\frac{eD_\parallel}{K} = kT_\parallel[1 + \Delta_\parallel)K'] \qquad \left[K' = \frac{d\ln K}{d\ln(E/N)}\right], \qquad (7\text{-}4\text{-}21)$$

$$\frac{eD_\perp}{K} = kT_\perp[1 + \Delta_\perp K'(2 + K')^{-1}], \qquad (7\text{-}4\text{-}22)$$

and

$$kT_\parallel = kT + \frac{4m - (2m - M)\bar{A}}{4m + 3M\bar{A}} \, Mv_d^2(1 + \beta_\parallel K'). \qquad (7\text{-}4\text{-}23)$$

$$kT_\perp = kT + \frac{(m + M)\bar{A}}{4m + 3M\bar{A}} \, Mv_d^2, \qquad (7\text{-}4\text{-}24)$$

Here the parameters $\Delta_\parallel$, $\Delta_\perp$, and $\beta_\parallel$ are represented as a function of the ion–atom reduced mass μ in Table 3.1 of McDaniel and Viehland (1984), and the parameter $\bar{A}$ is represented in Table 3.2 as a function of the dimensionless effective temperature $\bar{T}_{\text{eff}}$, which is the ratio of the quantity $(kT + mv_d^2/3)$ to the value of this same quantity at the mobility maximum.

Recent applications of the three-temperature theory to the direct determination of atomic ion–atom interactions from transport data are described in Mason and McDaniel (1988), Kirkpatrick and Viehland (1985, 1988), and Larsen et al., (1988). Koutselos and Mason (1991) have obtained generalized Einstein relations for ions in molecular gases. They solved the Boltzman equation by means of a four-temperature theory that includes as parameters the gas temperature, two translational ion temperatures, and one internal temperature.

G. Systems with Resonant Charge Transfer

Ordinarily, charge transfer can be detected easily in a drift-tube mass spectrometer because of the change in mass of the charged particle. It can then be treated in the same manner as any other type of chemical reaction. The result is that the reported mobility and diffusion coefficients are not ordinarily "contaminated" by charge-transfer effects. However, when the ions move in their parent gas, an ion and a neutral can interchange roles by the resonant transfer of an electron. This type of charge transfer cannot be detected or compensated for when gaseous ion transport coefficients are measured in a drift tube. Therefore, the kinetic theories mentioned above must be modified to take resonant charge transfer into account when it occurs.

Resonant charge transfer is usually so probable that it dominates all other elastic scattering processes except at very low energies where the long-range ion-induced dipole scattering eventually dominates. The result is that the collision dynamics are very different, and the T and E/N dependencies of the mobilities and diffusion coefficients are greatly changed from what they would have been in the absence of charge transfer. Only minor changes are necessary, however, in the two- and three-temperature theories. Most of the theoretical effort comes, therefore, in evaluating the transport cross sections from the two or more potential energy curves that must be considered.

Although quantum mechanical calculations of the transport cross sections can be performed (Sinha et al., 1979), such effort is probably warranted only for systems involving hydrogen and helium. For heavier systems, semiclassical methods (Mason and McDaniel, 1988) should suffice because the quantum mechanical oscillations in the transport cross sections are small in the energy ranges of interest in drift tubes. Moreover, the kinetic theories only make use of energy averages of the cross sections, so the effects of quantum oscillations are reduced still further.

7-5. SYSTEMS WITH ASYMMETRIC INTERACTION POTENTIALS; SYSTEMS IN WHICH INELASTIC COLLISIONS OCCUR

The theories discussed thus far apply only to systems with symmetric interaction potentials undergoing elastic collisions. Complications arise when these restrictions are removed. The kinetic theory for molecular systems that is analogous to the two-temperature kinetic theory for atomic systems leads to the equations

$$K_0 = \frac{3}{8} \frac{e}{N_0} \left(\frac{\pi}{2\mu k T'_{\text{eff}}} \right)^{1/2} \frac{1 + \alpha'}{\Omega(T'_{\text{eff}})} \tag{7-5-1}$$

$$\frac{3}{2} k T'_{\text{eff}} = \left(1 + \frac{M\xi}{m} \right)^{-1} \left[\frac{3}{2} kT + \frac{1}{2} M v_d^2 (1 + \beta') \right]. \tag{7-5-2}$$

The collision integral Ω in (7-5-1) differs only slightly from that defined by

(7-4-5) through (7-4-7) for elastic collisions and spherical potentials, in the same manner that the collision integrals for interdiffusion of atomic and molecular gases differ. This minor difference means that there is nothing remarkable about the self-diffusion coefficients of molecular gases compared with atomic gases, and that these coefficients contain little information about the anisotropic interaction potentials and the inelastic collisions that occur when internal states are present. However, this is not true of gaseous ion transport coefficients. In neutral gases, all species are at the same temperature regardless of whether or not inelastic collisions occur, while in drift-tube experiments the value of the effective temperature in (7-5-2) can depend strongly on such collisions.

Equation (7-5-2) is a generalization to molecular systems of the Wannier expression (7-4-10) for the relative kinetic energy in the center-of-mass frame between the ions and the single-component neutral molecules. The new quantity ξ is a dimensionless ratio of the collision integral for inelastic energy loss to that for momentum transfer. In combination with the factor of M/m in the equation, we can expect the presence of anisotropic potentials and inelastic collisions in molecular systems to have the largest impact on ion mobility in the case of light ions in heavy neutral gases.

For molecular ions, the distribution of energy among the internal states must be known to fully interpret the experimental results. This distribution is characterized by the "internal temperature" that results when the difference between pre- and postcollision ion internal energies averages to zero. In general, this average energy balance can only be determined by a brute-force calculation, since it depends on the cross section for inelastic collisions, especially in comparison with those for elastic collisions. A special case that is of great practical importance is that of molecular ions in a pure atomic gas, where in steady state the internal temperature of the ions must be equal to the effective temperature given by (7-5-2). A simple physical explanation has been given for this equality (Viehland et al., 1981). Energy is fed into the internal degrees of freedom of the ions by collisions with the structureless neutrals; the source of the ion internal energy is thus its translational motion. Energy leaks out of both the internal and translational degrees of freedom of the ions only through the translational motion of the neutrals. Since the leak is the same for both forms of energy, and since the internal energy is fed by the translational, it is not surprising that the internal and effective translational temperatures must be equal in steady state.

Quantitative application of the two- and three-temperature theories is impeded by the difficulty in calculating molecular cross sections from knowledge of the ion–neutral potential energy surfaces, but this situation is improving rapidly. Considerable effort is being focused on the problem of extracting microscopic information about molecular ion–neutral interactions from ion mobility measurements. Such studies are discussed in Mason and McDaniel (1988), Viehland and Fahey (1983), Viehland (1986, 1988), and Viehland and Kumar (1989).

7-6.   LASER PROBING OF DRIFTING IONS

Leone and his colleagues in Boulder have used laser probing techniques in innovative studies of ions in flowing afterglow drift tubes (Leone, 1989). Examples of their experiments are briefly described below.

A.   Laser-Induced Fluorescence Measurements of Ionic Velocity Distributions

Penn et al. (1990) have measured velocity distributions of Ba^+ ions drifting in argon under the influence of an external electric field. Their studies involved the use of single-frequency LIF techniques (see Section 2-3). Ion mobilities, ion temperatures, and skewness parameters of the ion velocity distributions were derived from a moment analysis of the data. Doppler profiles of the drifting ions were obtained as a function of E/N with the laser beam propagating first parallel and then perpendicular to the applied E field. The velocity-component distributions perpendicular to E are nearly Maxwellian at all of the values of E/N studied, but the parallel distributions are skewed toward high velocities as the highest value reached for E/N (148 Td) is approached.

Lauenstein et al. (1991) have used similar techniques to measure rotationally resolved velocity distributions for CO^+ drifted in helium.

B.   Studies of the Excitation of Ions in Drift Fields

The increased translational energy of ions drifting in a buffer gas may be high enough to excite both the ions and the buffer molecules to higher rotational and vibrational states. Duncan et al. (1983) have made LIF studies of the rotational excitation of N_2^+ ions drifting in helium, using the transition $B^2\Sigma_u^+ \rightarrow X^2\Sigma_g^+$ at 391.4 nm. They applied the drift field continuously in some experiments, but also carried out pulsed experiments to obtain information about the time scales for rotational equilibration. The number of collisions required to achieve full rotational equilibration was estimated to be 10 or fewer. The method employed by Duncan et al. (1983) is discussed in the review by Leone (1989), which also treats "monitor ion" techniques, which probe vibrationally excited ions by selective chemical reactions.

C.   Measurements of the Alignment of Molecular Ions in Drift Fields

The alignment and orientation of particles has been discussed in Section 5-10 of the companion volume, in Sections 2-8 and 3-6, and in Note 3-4. Here we concentrate on drift-tube measurements of alignment. As Leone (1989) points out, a nonspherical potential energy surface, combined with an anisotropic spatial distribution of relative velocity vectors, may produce alignment of molecules. An external electric field affects rotational alignment by causing the angular momentum vector to precess and thereby diminish the alignment produced by collisions. As Fig. 5-10-2 of the companion volume shows, alignment is a symmetric, anisotropic distribution of the m_J components of the **J** vectors, whereas orientation is an asymmetric, anisotropic distribution of these components. The cylindrical symmetry of the impact parameters in collisions produces alignment rather than orientation. Dressler et al. (1987) have used a flow-drift apparatus for LIF probing of the rotational alignment of N_2^+ ions drifting in helium. It differs in two respects from the apparatus of Duncan et al. (1983) that was used for measurements of rotational populations: (1) a polarization rotator is added for selection of the direction of linear polarization of the exciting laser beam, and (2) a polarization analyzer is employed for determination of the polarization direction of the laser-induced fluorescence. The linearly polarized laser preferentially excites molecules whose plane of rotation lies along the laser polarization direction. The resulting fluorescence radiation is parallel to the internuclear axis of the molecule. The result is that some planes of rotation are equally excited with any laser polarization, whereas others are preferentially excited by one polarization or the other. Furthermore, the fluorescence from certain planes of rotation is transmitted with equal efficiency through the polarization analyzer, whereas the radiation from other planes is strongly attenuated or transmitted depending on the direction of the polarizer.

In their studies of N_2^+ ions drifting in helium, Dressler et al. (1987) found the angular momentum vectors of the ions to be preferentially aligned perpendicular to the electric field vector. At $E/N = 14$ Td, which corresponds to a center-of-mass energy of 52 meV, they determined that the $N = 10$ rotational state has a quadrupole moment of $A_0^{(2)} = -0.11 \pm 0.03$. (State multipoles are discussed in Section 5-10-F of the companion volume.) The population ratio of $2 : 3$ is established for finding molecules with rotational angular momentum vectors parallel and perpendicular to the **E** field, respectively.

PART $\mathscr{B}$.   STUDIES OF ION–MOLECULE REACTIONS BY FLOW-DRIFT AND ION TRAP TECHNIQUES*

7-7.   FLOWING AFTERGLOW TECHNIQUES

The flowing afterglow (FA) technique was developed in the 1960s for ion–molecule reaction studies by Ferguson et al. (1969). Their research was conducted in a laboratory that is now part of the U.S. National Oceanic and Atmospheric Administration (NOAA). By itself, the FA technique cannot be used for the study of mobility and diffusion coefficients, but it was the forerunner of techniques that have been used very successfully for such measurements, so we shall discuss the FA method in this section.

*Beam studies of ion–molecule reactions at higher energies than covered here were discussed in Chapter 3.

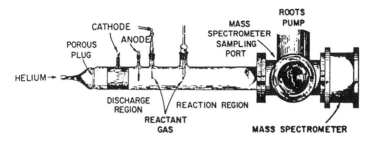

Figure 7-7-1. Flowing afterglow tube used for ionic reaction studies by Ferguson et al. (1969).

In this technique a localized electrical discharge creates a plasma at one end of a tube, through which gas flows rapidly under the action of powerful pumping. Downstream from the discharge, the afterglow* can be chemically modified and controlled in space and time. Various diagnostic tools, such as optical spectrometers, microwave interferometers, Langmuir probes, and mass spectrometers, can be used to monitor the state of the afterglow and the ions within it.

Figure 7-7-1 shows one type of flowing afterglow tube that has been used at NOAA, this one being of Pyrex glass. Here the discharge electrodes are a large cylindrical cathode and a small wire anode and the tube is operated in the dc mode. Other tubes have been made of quartz and stainless steel, all of the tubes being about 1 m in length and 8 cm in diameter, wide enough that wall effects can be ignored. In typical operation, helium is pumped through the tube at a pressure of about 0.4 torr, a flow rate of 13.0 atm-cm^3 s and a flow velocity of 10^4 cm/s. The helium is exhausted by a large Roots-type blower backed by a large mechanical pump that attains a 500 L/s pumping speed at pressures between 10 and 10^{-4} torr. The discharge produces about 10^{11} He$^+$ ions per cubic centimeter and a comparable concentration of triplet helium metastable atoms.

The gas temperature is not raised significantly by the discharge, and electrons produced in the discharge have thermalized by the time they have traveled a few centimeters downstream. The electron density can be measured with spectroscopic techniques, microwave absorption techniques, or Langmuir probes. The neutral gas temperature can be determined spectroscopically either by Doppler line width or rotational intensity distribution measurements, although it is usually taken to be ambient.

The gas jets shown in Fig. 7-7-1 permit the addition of neutral gases into the flowing helium afterglow plasma, either for the purpose of studying their reactions with the primary discharge ions or for the purpose of producing secondary ions for further reaction studies. This is a titration procedure, and

---

*The *afterglow period* is the time regime following the removal of the ionizing source, during which the gas remains ionized to an appreciable extent.

generally the character of the plasma is altered completely by the introduction of the reactant gas. The ion composition of the afterglow is recorded at the end of the reaction region by means of a quadrupole mass spectrometer, located in a differentially pumped chamber connected to the flow tube by a small aperture. The decrease in the reactant ion current and the increase in product ion current, as a function of concentration of the added neutral reactant gas, constitute the principal measurables of the system.

Reactions of atomic helium ions can be studied by introducing the neutral reactant into the helium afterglow through one of the jets and observing the loss of helium ions as a function of reactant concentration. The product ions can also be detected, and in addition, ion products of the helium metastable reactions appear. The large gas flow precludes the use of other rare gases as primary ion sources in the tube because of their cost.

More often the helium or argon afterglow plasma is used to produce a foreign ion whose reactions with neutrals are to be studied. The O^+, N^+, or C^+ ions can be produced by introducing O_2, N_2, or CO downstream from the discharge in a helium afterglow since the cross section for dissociative ionization of these gases by thermal He^+ ions is very large, over $100 \, Å^2$. Helium triplet metastables react rapidly at thermal energies (cross sections of about $10 \, Å^2$) with almost every gas to ionize it, so that positive ions of almost all stable molecules can easily be produced by this mechanism. The neon atom is an exception since its ionization energy is greater than the excitation energy of $He(2^3S)$, and not enough internal energy can be transferred to form the Ne^+ ion. No other neutral species are known to be incapable of ionization by helium triplet metastables.

The FA offers an important advantage in the knowledge and control of the quantum states of the ionic reactants. These ions will almost certainly be in the ground electronic state because of the energetics of their formation (e.g., N^+) or more commonly because ions (such as O^+) originally produced in excited states are deexcited in superelastic collisions with electrons and collisions with parent gas neutrals before they reach the position where the neutral reactant is added. The situation regarding the vibrational state of molecular ions is much less certain. The feature of reactant gas addition downstream from the site of primary ionization (the discharge) permits the study of many ion–neutral combinations with the flowing afterglow that would be inaccessible with other techniques.

Flowing afterglow ion–molecule reaction data can be interpreted in terms of rate coefficients as follows. For a reaction

$$A^+ + B \rightarrow products$$

the rate coefficient k is defined by

$$-\frac{\partial[A^+]}{\partial t} = k[A^+][B], \qquad (7\text{-}7\text{-}1)$$

where the brackets denote concentrations. Over the range of observable A^+ decrease, $[B] \gg [A^+]$, so that $[B]$ can be assumed constant from the location of the addition jet down to the mass spectrometer port. Equation (7-7-1) has the solution

$$\ln \frac{[A^+]}{[A^+]_0} = -k\tau[B], \qquad (7\text{-}7\text{-}2)$$

where $[A^+]_0$ is proportional to the A^+ signal before the B addition. Thus the slope of a logarithmic plot of primary ion current against added reactant concentration, when divided by τ, gives k. Here τ is the ion flow time. Since the absolute neutral reactant concentration must be known in order to obtain k, the neutral flow rate is measured by flowmeters. Note that only the change in the detected current is required, so possible variation in ion sampling efficiency with ion mass cannot falsify the results.

If corrections are made for radial diffusion, nonuniform ion velocity profiles, axial diffusion, axial velocity gradient, slip flow, and inlet effects, the flowing afterglow appears capable of achieving results with 5% accuracy (Ferguson et al., 1969). For most of the data, however, 20 to 30% accuracy is claimed. Flowing afterglow studies have yielded rate coefficients for several thousand ion–molecule reactions! Most of the measurements are performed at room temperature, although variable-temperature apparatus has permitted coverage of the gas temperature range extending from 8 to 900 K.

7-8. FLOW-DRIFT TUBE OF ALBRITTON AND McFARLAND

Albritton and McFarland at NOAA have extended the FA technique to allow the measurement of reaction rates and ion mobilities as well as functions of E/N and hence average ionic translational energy. The NOAA flow-drift tube (FDT) (McFarland et al., 1973) employs a four-section stainless steel tube 8 cm in diameter and 125 cm long. Ions are produced by electron impact in the first section, positive ions are separated from negative ions and electrons in the second section, ions of the chosen polarity are periodically bunched by a two-grid shutter in the third section, and then narrow pulses of ions are released at specified times into the drift-reaction section. The latter section contains drift field electrodes similar to those in the Georgia Tech drift tube. A buffer gas flow of 80 to 180 atm-cm^3/s (usually helium) is maintained through the tube, where the pressure is usually 0.19 to 1.25 torr. The apparatus is pumped and the ions detected as in the FA apparatus. The FDT is run primarily in the dc mode by leaving the shutters on. The pulsed mode is used for mobility measurements and some diagnostic studies.

7-9. SELECTED ION FLOW TUBE

One of the main reasons for the power of the FA and FDT techniques is the chemical versatility that they afford, that is, the vast number of ion–gas combinations made available for study. However, Adams and Smith (1976) at the University of Birmingham have taken the matter of primary ion formation in flow tubes yet one step further with their selected ion flow tube (SIFT). A recent version of the part of their apparatus on which we wish to focus is shown in Fig. 7-9-1. The remainder of the apparatus is similar to the reaction and detection regions of the flowing afterglow. Ions are produced in an electrical discharge or electron impact ion source remote from the flow tube. After selection in a quadrupole mass filter, Q, they are injected at low energy through an orifice O into the flow tube. The ions are rapidly thermalized in collisions with molecules of the carrier gas, which is introduced into an annular space around the orifice and enters the flow tube through twelve 1-mm-diameter apertures arranged in a circle about the ion beam. These holes direct gas at high velocity (close to the speed of sound) parallel to the ion "beam" and then down the flow tube. The rapid gas flow causes the pressure in the vicinity of O to be much lower than that downstream in the flow tube. A low pressure can be maintained in the mass filter even for a relatively large connecting orifice

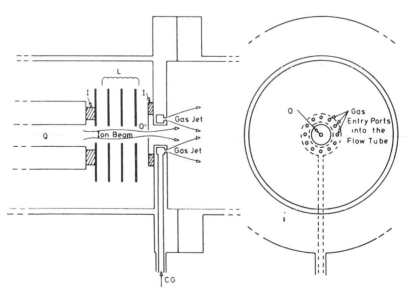

Figure 7-9-1. Elevation and end views of the SIFT ion injector. Q, quadrupole mass filter; I, insulators; L, ion lens; O, orifice; CG, carrier gas. The dashed lines indicate the tube and cavity through which the carrier gas passes before entry into the flow tube. [From Adams and Smith (1976).]

(typically, 1 mm in diameter). The venturi-aspirator action provided by the substantial and rapid gas flow allows useful ion currents to be injected at a lab frame energy of about 15 eV, typically; the center-of-mass energy, of course, can be lower. A low injection energy is essential if excitation and dissociation are to be avoided.

It is important to note that only a single species of ion is injected during a given measurement with the SIFT apparatus. The presence of more than one kind of primary ion in the FA and FD apparatus can greatly complicate the analysis of data in certain situations, such as branching ratio measurements. Of course, the fact that charge carriers of only one sign are injected in the SIFT means that no plasma is present, in contrast with the FA and FD.

7-10.   SELECTED ION FLOW-DRIFT TUBE

The selected ion flow-drift tube (SIFDT) technique was developed at NOAA as a combination of the techniques discussed above (Howorka et al., 1979; Rowe et al., 1980). SIFT furnished the method of producing, selecting, and injecting any single species of ion into a flow tube, the general configuration of which is that of the FA. The technique of incorporating a drift field in the flow tube was provided by the FD. The apparatus is illustrated in Fig. 7-10-1. Similar apparatus has been constructed at the University College of Wales (Jones et al., 1981) and at the University of Innsbruck (Villinger et al., 1984).

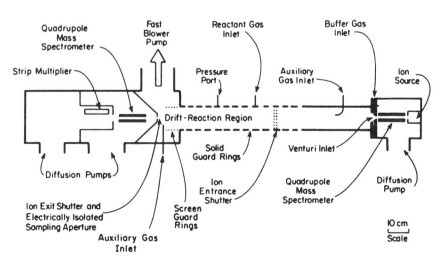

Figure 7-10-1. Selected ion flow drift-tube mass spectrometer (SIFDT) at NOAA. [From Rowe et al. (1980).]

7-11. LOW-TEMPERATURE STUDIES

One of the scientific areas that has the greatest need for data on ion–molecule reactions is interstellar chemistry. Ion–molecule and electron–ion recombination reactions (see Chapter 9) are responsible for producing a wide variety of molecular species that have been detected using astronomical spectroscopic observations (Duley and Williams, 1984). The interstellar clouds where such molecules are usually produced are typically very cold, low-density environments, so measurements designed to simulate the reactions that take place in these clouds should reflect these conditions. This means in effect that only binary collisions should be considered, the reactants should be in their ground electronic and vibrational states, and the temperatures at which the measurements are made should be in the range 10 to 100 K. While the first two conditions can usually be satisfied, few techniques in use today can reproduce the low-temperature condition for ion–molecule studies. Generally, it is assumed that ion–molecule reactions have a small temperature dependence, and measurements performed at room temperature are used directly in interstellar chemical models. The low-temperature region is beginning to be explored using ion beam methods (see Chapters 1 through 3), ion trap techniques, and an ususual type of apparatus known as a CRESU.

A. Cooled Penning Trap

The cooled Penning trap, developed at the Joint Institute for Laboratory Astrophysics (JILA) in Boulder, Colorado, allows rate coefficients for ion–molecule reactions to be measured at temperatures down to 4 K (Barlow et al., 1986). The apparatus was developed at JILA by Walls and Dunn (1974) for measurement of the dissociative recombination of molecular ions (see Chapter 9). All collision studies involving molecular ions must address the problem of internal excitation of the reactants, whether vibrational, rotational, or electronic. In high-pressure experiments such as flowing afterglows, the excited states in which the ions are formed are often deexcited in collisions. [This is not always so, and particular attention must be given to individual cases.] In the case of the ion trap technique, the long lifetime of the ions in the trap allows excited states to decay naturally via radiative transitions to the ground state. The apparatus used for ion–molecule reactions is illustrated in Fig. 7-11-1. The walls of the trap are thermally anchored to a flange that can be cooled to 4 K using liquid helium. The reaction temperature can, however, be selected by using a heater to change the wall temperature in the range 9 to 70 K. The gas from which the ions are to be formed is introduced into the trap through a tube that can be heated if condensable gases are used. The ions are then created by electron impact. Once enough time has elapsed for the ions to relax, the reactant gas is introduced and the ion–molecule collisions follow. The ions in the trap oscillate at the frequency of the radio-frequency (RF) potential applied to the electrodes. The presence of ions in the trap is determined by measuring the

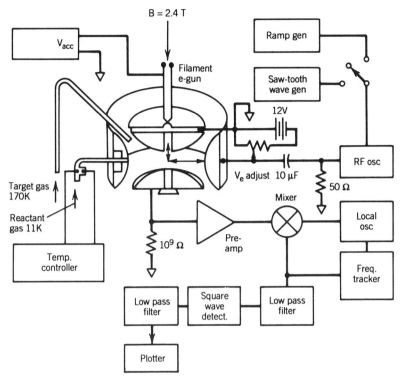

Figure 7-11-1. Schematic view of the cooled Penning ion trap. [From Barlow et al. (1986).]

image currents in the end caps induced by the oscillating charges. Mass detection can be performed selectively using cyclotron resonance heating, which results in an increase in the noise power signal proportional to the number of ion absorbers, the RF power applied, and the time the RF power is on. This method is extremely sensitive, and rate coefficients down to 10^{-15} cm³/s can be measured. Neutral reagents are, however, limited to H_2 and He, which are not condensed on the walls of the trap. Nonetheless, the cooled Penning trap has been invaluable for measurements of small rate reactions such as radiative association (Barlow et al., 1984).

B. CRESU Technique

The CRESU technique (cinétique de réactions en écoulement supersonique uniforme) developed by Rowe et al. (1984a, b) permits binary and ternary ion molecule reactions to be measured at temperatures in the range 8 to 160 K even for gases that would normally condense at these temperatures. It is therefore a very versatile technique and although as yet used only for a few cases, has great potential for the future.

The principle behind the technique is to create a supersonic expansion of a carrier gas seeded with the reactant ion parent gas. This is accomplished by passing the relatively high pressure (<0.1 torr) gas mixture through a Laval nozzle into a large vessel evacuated to ($<10^{-4}$ torr) by fast Roots blower pumps (see Fig. 7-11-2). Ionization is created downstream of the nozzle by means of electron impact. In the supersonic expansion, provided that flow conditions are uniformly maintained, the random motion of the gas molecules is directed into essentially a one-dimensional flow, so that the relative collision energies of the reactants are low. In addition, the internal rotational, and to some extent the vibrational, motions of molecular species are cooled to very low temperatures during expansion, as these degrees of freedom are converted into forward-directed translational motion. Although it may at first sight appear very similar in nature to conventional supersonic nozzle molecular beam apparatus, the CRESU is in fact a very different concept, actually closely resembling a wind tunnel into which ions are injected. The particles in the flow are in a condition in which local thermodynamic equilibrium (LTE) of the translation–rotation motion is superimposed on their mean flow velocity. The flow is isentropic, and thermalization of the motion occurs within a few microseconds. The flow temperature T at which the reactions take place is related to the stagnation temperature T_0 in the gas reservoir behind the nozzle by the expression

$$\frac{T_0}{T} = 1 + \frac{\gamma + 1}{2} M^2, \tag{7-11-1}$$

where γ is the ratio of the specific heats at constant pressure and at constant

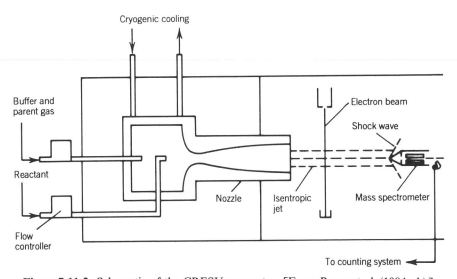

Figure 7-11-2. Schematic of the CRESU apparatus. [From Rowe et al. (1984a, b).]

volume for the buffer gas, and M is the Mach number for the flow. T_0 is measured using a thermocouple, and M using an impact pressure probe on the axis of the flow.

To date, this apparatus has been used for the study of three-body association reactions, and reactions involving atomic ions such as He^+, N^+, and C^+ with H_2, O_2, N_2, H_2O, NH_3, and CH_4. Application of SIFT injection techniques will greatly expand the versatility of this method, and plans are in place to implement this at the University of Rennes.

7-12. PROBLEMS

7-1. Before about 1970, E/p was generally used as the energy parameter instead of E/N. Show that the conversion relation is

$$\frac{E}{N} = \frac{(1.0354 \times T \times 10^{-2})E}{p} \tag{7-12-1}$$

where E/N is in townsend, T is the gas temperature in kelvin, and E/p is in V/cm-torr.

7-2. Investigate the acceleration of ions to their terminal drift velocity. [See Mason and McDaniel (1988, p. 3).]

7-3. Analyze the space charge spreading of a cloud of gaseous ions. [See Mason and McDaniel (1988, p. 7).]

7-4. Analyze the spreading of a cloud of ions through an unbounded gas. [See Mason and McDaniel (1988, pp. 11–12).]

7-5. Study the time-independent diffusion of gaseous ions in containers of the following shapes: infinite parallel plates, rectangular parallelepiped, spherical cavity, and cylindrical cavity. [See Mason and McDaniel (1988, pp. 16–25).]

7-6. Show how the ambipolar diffusion coefficient may be determined from the rate of decay of the charged particle density in a microwave cavity after the ionization source has been turned off. [See Mason and McDaniel (1988, pp. 25–27).]

7-7. Discuss the extraction of the ion mobilities, ion temperatures, and skewness parameters from the velocity distributions measured in the experiment by Penn et al. (1990).

7-8. Explain the shape of the arrival-time spectrum in Fig. 7-3-3.

7-9. A narrow pulse of O_2^+ ions is injected into a drift tube containing nitrogen at 300 K and $E/N = 5$ Td.

(a) Assume that the entering ions are all excited to a particular rotational level, and draw a rough plot of the number of rotationally excited ions present as a function of distance down the drift tube.

(b) Repeat the problem for the case of vibrationally excited ions.

7-10. Consult the literature on molecular energy levels and verify that the wavelength of the $B^2\Sigma_u^+ \rightarrow X^2\Sigma_g^+$ transition in N_2^+ is 390 nm.

7-11. Discuss the "monitor ion techniques" that probe vibrationally excited ions by selective chemical reactions.

7-12. Study the theory of alignment and multipole determinations by the LIF techniques outlined by Dressler et al. (1987).

REFERENCES

Adams, N. G., and D. Smith (1976), *Int. J. Mass Spectrom. Ion Phys.* **21**, 349.

Albritton, D. L., T. M. Miller, D. W. Martin, and E. W. McDaniel (1968), *Phys. Rev.* **171**, 94.

Barlow, S. E., G. H. Dunn, and M. Schauer (1984), *Phys. Rev. Lett.* **52**, 902.

Barlow, S. E., J. A. Luine, and G. H. Dunn (1986), *Int. J. Mass Spectrom. Ion Process.* **74**, 97.

Dressler, R. A., H. Meyer, and S. R. Leone (1987), *J. Chem. Phys.* **87**, 6029.

Duley, W. W., and D. A. Williams, (1984), *Interstellar Chemistry*, Academic Press, New York.

Duncan, M. A., V. M. Bierbaum, G. B. Ellison, and S. R. Leone (1983), *J. Chem. Phys.* **79**, 5448.

Dupeyrat, G., J. B. Marquette, and B. R. Rowe (1985), *Phys. Fluids* **28**, 1273.

Ellis, H. W., R. Y. Pai, E. W. McDaniel, E. A. Mason, and L. A. Viehland (1976), *ADNDT* **17**, 177.

Ellis, H. W., E. W. McDaniel, D. L. Albritton, L. A. Viehland, S. L. Lin, and E. A. Mason (1978), *ADNDT* **22**, 179.

Ellis, H. W., M. G. Thackston, E. W. McDaniel, and E. A. Mason (1984), *ADNDT* **31**, 113.

Ferguson, E. E., F. C. Fehsenfeld, and A. L. Schmeltekopf (1969), "Flowing Afterglow Measurements of Ion-Neutral Reactions," *AAMP*, **5**, 1.

Ferziger, J. H., and H. G. Kaper (1972), *Mathematical Theory of Transport Processes in Gases*, North-Holland, Amsterdam.

Flannery, M. R. (1982), *Philos. Trans. R. Soc.* **A-304**, 447.

Gatland, I. R. (1975), "Analysis for Ion Drift Tube Experiments," in E. W. McDaniel and M. R. C. McDowell, Eds., *Case Studies in Atomic Collision Physics*, Vol. 4, North-Holland, Amsterdam, pp. 369–437.

Gatland, I. R., W. F. Morrison, H. W. Ellis, M. G. Thackston, E. W. McDaniel, M. H. Alexander, L. A. Viehland, and E. A. Mason (1977), *J. Chem. Phys.* **66**, 5121.

Gough, D. W., G. C. Maitland, and E. B. Smith (1972), *Mol. Phys.* **24**, 151.

Heimerl, J. M., R. Johnsen, and M. A. Biondi (1969), *J. Chem. Phys.* **51**, 5041.

Howorka, F., F. C. Fehsenfeld, and D. L. Albritton (1979), *J. Phys. B* **12**, 4189.

Jones, T. T. C., J. D. C. Jones, and K. Birkinshaw (1981), *Chem. Phys. Lett.* **82**, 377.

Kihara, T. (1953), *Rev. Mod. Phys.* **25**, 844.

Kirkpatrick, C. C., and L. A. Viehland (1985), *Chem. Phys.* **98**, 221.

Kirkpatrick, C. C., and L. A. Viehland (1988), *Chem. Phys.* **120**, 235.

Koutselos, A. D., and E. A. Mason (1991) *Chem. Phys.* **153**, 351.

Larsen, P. H., H. R. Skullerud, T. H. Lovans, and Th. Stefansson (1988), *J. Phys. B* **21**, 2519.

Lauenstein, C. P., M. J. Bastian, V. M. Bierbaum, S. M. Penn, and S. R. Leone (1991), *J. Chem. Phys.* **94**, 7810.

Leone, S. R. (1989), "Laser Probing of Ion Collisions in Drift Fields: State Excitation, Velocity Distributions, and Alignment Effects," in M. N. R. Ashfold and J. E. Baggott, eds., *Gas Phase Bimolecular Collisions*, Royal Society of Chemistry, London, pp. 377–416.

Lin, S. L., L. A. Viehland, and E. A. Mason (1979), *Chem. Phys.* **37**, 411.

Loeb, L. B. (1960), *Basic Processes of Gaseous Electronics*, 2nd ed., University of California Press, Berkeley.

Maitland, G. C., E. A. Mason, L. A. Viehland, and W. A. Wakeham (1978), *Mol. Phys.* **36**, 797.

Mason, E. A., and E. W. McDaniel (1988), *Transport Properties of Ions in Gases*, Wiley, New York.

Massey, H. S. W. (1971), *Electronic and Ionic Impact Phenomena*, Vol. 3, 2nd ed., Clarendon Press, Oxford.

Massey, H. S. W., E. W. McDaniel, and B. Bederson, Eds. (1982–1984), *Applied Atomic Collision Physics*, 5 vols., Academic Press, New York.

McDaniel, E. W. (1964), *Collision Phenomena in Ionized Gases*, Wiley, New York.

McDaniel, E. W., and L. A. Viehland (1984), "The Transport of Slow Ions in Gases: Experiment, Theory and Applications," *Phys. Rep.* **110**, 333.

McFarland, M., D. L. Albritton, F. C. Fehsenfeld, E. E. Ferguson, and A. L. Schmeltekopf (1973), *J. Chem. Phys.* **59**, 6610.

Miller, T. M., J. T. Moseley, D. W. Martin, and E. W. McDaniel (1968), *Phys. Rev.* **173**, 115.

Moseley, J. T., I. R. Gatland, D. W. Martin, and E. W. McDaniel (1969), *Phys. Rev.* **178**, 234.

Ness, K. F., and L. A. Viehland (1990), *Chem. Phys.* **148**, 255.

Penn, S. M., J. P. M., Beijers, R. A. Dressler, V. M. Bierbaum, and S. R. Leone, (1990) *J. Chem. Phys.* **93**, 5118.

Rowe, B. R., D. W. Fahey, F. C. Fehsenfeld, and D. L. Albritton (1980), *J. Chem. Phys.* **73**, 194.

Rowe, B. R., G. Dupeyrat, J. B. Marquette, D. Smith, N. G. Adams, and E. E. Ferguson (1984a), *J. Chem. Phys.* **80**, 241.

Rowe, B. R., G. Dupeyrat, J. B. Marquette, and P. Gaucherel (1984b), *J. Chem. Phys.* **80**, 4915.

Sinha, S., S. L. Lin, and J. N. Bardsley (1979), *J. Phys. B* **12**, 1613.

Snuggs, R. M., D. J. Volz, I. R. Gatland, J. H. Schummers, D. W. Martin, and E. W. McDaniel (1971), *Phys. Rev.* **A-3**, 487.

Thomson, G. M., J. H. Schummers, D. R. James, E. Graham, I. R. Gatland, M. R. Flannery, and E. W. McDaniel (1973), *J. Chem. Phys.* **58**, 2402.

Viehland, L. A. (1982), *Chem. Phys.* **70**, 149.

Viehland, L. A. (1986), *Chem. Phys.* **101**, 1.

Viehland, L. A. (1988), *Chem. Phys. Lett.* **144**, 552.

Viehland, L. A., and S. L. Lin (1979) *Chem. Phys.* **43**, 135.

Viehland, L. A., and D. W. Fahey (1983), *J. Chem. Phys.* **78**, 435.

Viehland, L. A., and K. Kumar (1989), *Chem. Phys.* **131**, 295.

Viehland, L. A., and E. A. Mason (1975), *Ann. Phys. (N.Y.)* **91**, 499.

Viehland, L. A., and E. A. Mason (1978), *Ann. Phys. (N.Y.)* **110**, 287.

Viehland, L. A., M. M. Harrington, and E. A. Mason (1976), *Chem. Phys.* **17**, 433.

Viehland, L. A., S. L. Lin, and E. A. Mason (1981), *Chem. Phys. Lett.* **54**, 341.

Villinger, H., J. H. Futrell, A. Saxer, R. Richter, and W. Lindinger (1984), *J. Chem. Phys.* **80**, 2543.

Waldman, M., and E. A. Mason (1981), *Chem. Phys.* **58**, 121.

Walls, F. L., and G. H. Dunn (1974), *J. Geophys. Res.* **79**, 1911.

Wannier, G. H. (1953), "Motion of Gaseous Ions in Strong Electric Fields," *Bell Syst. Tech. J.* **32**, 170.

Wannier, G. H. (1970), *Bell Syst. Tech. J.* **49**, 343.

8

ELECTRON SWARMS AND TRANSPORT

8-1. INTRODUCTION*

Much of atomic and molecular collisions research is directed towards the study of collision processes between electrons and atoms or molecules. These processes include elastic scattering and inelastic collisions such as excitation, ionization, dissociation, recombination, and attachment. Detailed discussions of these processes are given elsewhere in this volume and in our companion volume.

In this chapter we consider electron swarms drifting and diffusing through gases at low energies (usually below a few eV), such that only elastic scattering, and excitation in molecular gases, need be considered in most cases. Here, as in Chapter 7, the emphasis is on the measurement and calculation of the drift velocity (v_D), the mobility (K), and the diffusion coefficients (D_L and D_T), which were defined in Section 7-2-A. Because of the small mass of the electrons, their transport properties have values larger by orders of magnitude than those of ions under similar conditions (see Problems 8-2 through 8-4). However, in the electron case, all of the transport coefficients are still determined by the energy parameter E/N, along with the composition and temperature (T) of the drift gas. Here N is the gas number density and E is the intensity of the applied electric field. Other points of similarity and dissimilarity are discussed in Mason and McDaniel (1988, pp. 9–10).

After our treatment of the transport coefficients, we shall discuss the mean electron thermalization times and the extraction of collision cross sections from experimental data. Before we proceed, however, it is appropriate to cite some comprehensive references. Reviews of early work on electron swarms have been given by McDaniel (1964) and Loeb (1961). More recent reviews are those of

---

*Many aspects of the discussion of ion swarms in Chapter 7 carry over here, at least qualitatively, and hence are not repeated in this chapter.

Huxley and Crompton (1974), Christophorou (1981), Hunter and Christophorou (1984), and Kumar (1984). Compilations of electron transport data have been provided by Dutton (1975), Beaty and Pitchford (1979) and Gallagher et al. (1983).

A. Transport Coefficients

As in the case of ion swarms, the usual starting point for relating the macroscopic transport coefficients (i.e., the electron drift velocity and diffusion coefficients) to the microscopic quantities characterizing individual collisions (i.e., the collision cross sections, energy transfer in the collision, etc.) is the Boltzmann equation* [see pp. 63–69 of the companion volume and Mason and McDaniel (1988, pp. 193–196)]. For electrons present in trace quantities and undergoing only elastic collisions in a mixture of gases, this equation has the form

$$\frac{\partial f}{\partial t} + \mathbf{v} \cdot \nabla_r f + \mathbf{a} \cdot \nabla_v f \equiv \frac{Df}{Dt} = \sum_j \int\int [f(\mathbf{v}', \mathbf{r}, t)F_j(\mathbf{V}'_j, \mathbf{r}, t)$$
$$-f(\mathbf{v}, \mathbf{r}, t)F_j(\mathbf{V}_j, \mathbf{r}, t)] \times v_{rj}\sigma_j(\theta, v_{rj}) \, d\Omega_j d\mathbf{V}_j. \quad (8\text{-}1\text{-}1)$$

Here $f(\mathbf{v}, \mathbf{r}, t)$ and $F_j(\mathbf{V}_j, \mathbf{r}, t)$ are the distribution functions of the electrons and the neutrals of type j, respectively, before collisions. The electron position in configuration space is denoted by $\mathbf{r}$, its velocity by $\mathbf{v}$, and its acceleration by $\mathbf{a}$. The velocity of the neutral of type j is denoted by $\mathbf{v}_j$. Primed quantities indicate the quantities evaluated after collisions. In addition,

$$\mathbf{a} = \frac{e\mathbf{E}}{m}, \quad (8\text{-}1\text{-}2)$$

$$v_{rj} = |\mathbf{v} - \mathbf{V}_j|, \quad (8\text{-}1\text{-}3)$$

$$d\Omega = \sin\theta \, d\theta \, d\phi, \quad (8\text{-}1\text{-}4)$$

and $\sigma_j(\theta, v_{rj})$ is the differential cross section for elastic scattering of electrons by gas molecules of type j. The significance of the terms in (8-1-1) is discussed by Mason and McDaniel (1988, 194) and on p. 66 of the companion volume.

The electron/neutral mass ratio is small ($m/M \sim 1/2000$), and the fractional energy transfer in an elastic collision is of the order $2m/M$. Elastic scattering can change the direction of the electron velocity by a large amount but can alter its magnitude only slightly. These facts mean that the electron distribution function will be nearly spherical in velocity space. The behavior of ions, for which the mass ratio is typically of order unity, is quite different. The considerable

---

*The moment method of solving the Boltzmann equation to obtain ionic transport coefficients has also been applied to electrons, but with much less success (Mason and McDaniel, 1988).

forward-directed momentum is difficult to disperse, so the velocity distribution function tends to be elongated in the direction of the applied electric field.

B. Two-Term Approximation

In the following we assume initially that electron scattering is isotropic and later examine the situation where this assumption breaks down. Isotropic scattering makes it possible to obtain a solution to (8-1-1) by expanding $f(v, r, t)$ in terms of Legendre polynomials:

$$f(\mathbf{v}, \mathbf{r}, t) = n(\mathbf{r}, t) \sum_{l=0} P_1(\cos \theta), \tag{8-1-5}$$

where $n(\mathbf{r}, t)$ is the number density of electrons in configuration space. It is customary to use only the first two terms of the series, corresponding to $P_0(\cos \theta) = 1$ and $P_1(\cos \theta) = \cos \theta$; this procedure is known as the *two-term approximation* (Mason and McDaniel, 1988, pp. 205–206).

We continue to consider the case where all electron–neutral collisions are elastic. We also assume that σ_m, the electron momentum transfer cross section, is independent of v, that there are no spatial gradients, and that the electron motion has reached equilibrium (so that there will be no time dependence). Then the electron velocity distribution function can be shown to be

$$f_0(v) = \frac{1}{\zeta^3 \pi \Gamma(\frac{3}{4})} \exp\left[-\left(\frac{v}{\zeta}\right)^4 \right] \tag{8-1-6}$$

with

$$\zeta^4 = \frac{4M}{3m} \left(\frac{e}{m} \frac{E}{N\sigma_m}\right)^2, \tag{8-1-7}$$

$\Gamma(\frac{3}{4}) = 1.2254$. This form of $f(v)$ is known as the *Druyvesteyn distribution*.

The more commonly used form of $f(v)$ is the *Maxwellian distribution*, which is obtained if one takes $\sigma_m \propto v^{-1}$ and is given by

$$f_0(v) = \left(\frac{1}{\zeta \pi^{1/2}}\right)^3 \exp\left[-\left(\frac{v}{\zeta}\right)^2 \right], \tag{8-1-8}$$

where $\zeta = (2kT/m)^{1/2}$ is the most probable electron speed, T the gas temperature, m the electron mass, and k is Boltzmann's constant.

When $\sigma_m \propto v^{-1}$, the distribution is Maxwellian for *all* E/N, provided that there are no inelastic collisions; for molecular gases, with inelastic collisions, the distribution is approximately Maxwellian provided that the mean electron energy is close to thermal.

The following expressions for the drift velocity v_d, the transverse diffusion coefficient D_T, and the longitudinal diffusion coefficient D_L can be derived from (8-1-1) (Huxley and Crompton, 1974):

$$v_D = -\frac{4\pi}{3}\frac{eE}{m}\int_0^\infty \frac{v^3}{v}\frac{df_0}{dv}\,dv, \tag{8-1-9}$$

$$D_T = \frac{4\pi}{3}\int_0^\infty \frac{v^4}{v}f_0\,dv, \tag{8-1-10}$$

$$D_L = D_T - \frac{4\pi}{3}\frac{eE}{m}\int_0^\infty b_1 f_0 \frac{d}{dv}\frac{v^3}{v}\,dv, \tag{8-1-11}$$

where $v = Nv\sigma_m(v)$ is the collision frequency for momentum transfer. The quantity $b_1 = b_1(v)$ depends on the form of $\sigma_m(v)$ and the spatial derivative of $f(\mathbf{v}, \mathbf{r}, t)$. The values for these transport coefficients are determined by the collision cross section, so, in principle, it is possible to obtain information concerning the cross sections by measuring the transport coefficients. The basic procedure involves taking a trial cross section and using the equations above to calculate the values for the transport coefficients. These values are then compared with measured values, and the cross sections are then adjusted to obtain a good fit. The primary weaknesses of this procedure show up when (1) the two-term approximation used to obtain (8-1-9) through (8-1-11) is itself invalid, and (2) there are too many competing collision processes contributing to the electron transport. The latter is usually referred to as the *uniqueness problem.*

The validity of the two-term approximation has been examined by a number of workers [see, e.g., Robson and Kumar (1971)] and found to be generally satisfactory for cases involving elastic scattering. One of the most closely examined cases is that of the elastic scattering of electrons in argon, since it exhibits a very deep Ramsauer minimum (see Chapter 4 of the companion volume) at low energies. One might expect that electrons with energies close to this minimum would have very long mean free paths and thus would have relatively large energy gains and direction changes between collisions. This behavior would tend to skew the electron velocity distribution. Even in this case, however, the two-term approximation is found to yield results that agree closely with results obtained using much more sophisticated solutions to the Boltzmann equation.

The two-term approximation does, however, begin to break down when the sum of inelastic cross sections becomes a significant fraction of the total scattering cross section and when there is considerable anisotropy in the electron scattering. Such situations commonly arise in molecular gases where rotational and vibrational excitation processes can contribute strongly, even at low collision energies.

When the two-term approximation fails, one must resort to more sophisticated methods for the solution of the Boltzmann equation. One might ask, if these are available, why not use them anyway? The answer to this is that they involve orders of magnitude more computing time and therefore have only been applied to a few cases. Most of the data available concerning low-energy electron collision cross sections in molecular gases is still derived from calculations performed using the two-term approximation. Christophorou and Hunter (1984) have warned that for this reason, some of these data must be regarded as suspect.

C. Higher-Order Approximations and Other Methods of Solution

Mathematical details of these methods are contained in the papers referenced below; only an overview of available techniques and their relative strengths is given here.

The most natural extension of the two-term approximation is to employ more terms in the Legendre polynomial expansion used in modeling of the electron velocity distribution function (8-1-5). Three-term expansion calculations have been performed by a number of workers, including Wilhelm and Winkler (1969), Braglia and Caraffini (1974), Ferrari (1975, 1977), Braglia et al. (1977), Makabe and Mori (1978, 1980, 1982), and Cavalleri (1981). Inclusion of the third term in the expansion greatly increases the complexity involved in the solution, but such calculations have been used to highlight the failure of the two-term approximation in a number of cases. New, more tractable mathematical techniques for dealing with multiterm approximations have been developed by Lin et al. (1979) and Pitchford et al. (1981). These methods are especially interesting, for one can see directly to what degree of approximation one must go to obtain convergence of the approximation. Pitchford (1981) has shown that generally four to six terms in the multiterm expansion are sufficient to obtain solutions of the Boltzmann equation that are of the accuracy required for confident prediction of collision cross sections over a wide range of E/N values. Their results are illustrated in Fig. 8-1-1.

Numerical methods for direct evaluation of the Boltzmann equation have been used by Kleban and coworkers (1977, 1978, 1980), Tagashira and coworkers (1978), and Kitamori et al. (1978, 1980). The accuracy of these techniques has, however, been questioned (Lin et al., 1979; Pitchford, 1981). The multiterm methods discussed above involve the conversion of the Boltzmann partial differential equation into a set of coupled ordinary differential equations. Numerical methods for the direct solution of partial differential equations have not reached the same level of sophistication as those for ordinary differential equations and therefore are not as competitive computationally.

A new approach by Drallos and Wadehra (1988, 1989) circumvents this problem. A common precondition to the solution of (8-1-1) is to assume that the electron distribution function is spatially independent so that the derivative with respect to r can be neglected. Then for a one-component gas, (8-1-1) reduces to

$$\frac{\partial f(\mathbf{v}, t)}{\partial t} + \mathbf{a} \cdot \mathbf{V}_v f(\mathbf{v}, t) = C(\mathbf{v}, t), \qquad (8\text{-}1\text{-}12)$$

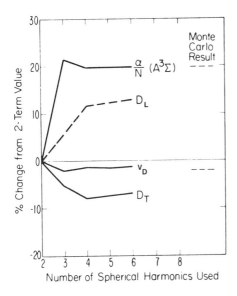

Figure 8-1-1. Percentage change from the *two-term* approximation values for the indicated transport coefficients and the A^3_2 excitation coefficient as a function of the number of spherical harmonics used. The dashed lines are Monte Carlo results, shown for comparison. The results are for N_2 at 100 Td. [From Pitchford (1981).]

where $C(\mathbf{v}, t)$ now denotes the collision integral. Multiplying both sides of (8-1-12) by a finite time interval Δt and then adding $f(\mathbf{v}, t)$ to each side yields

$$f(\mathbf{v},\ t) + \left(\Delta t \frac{\partial}{\partial t} + \Delta \mathbf{v} \cdot \mathbf{V}_v\right) f(\mathbf{v}, t) = f(\mathbf{v}, t) + C(\mathbf{v}, t)\, \Delta t, \qquad (8\text{-}1\text{-}13)$$

with $\Delta \mathbf{v} = \mathbf{a}\, \Delta t$. The terms on the left-hand side can be combined to yield the final result:

$$f(\mathbf{v} + \Delta \mathbf{v},\ t + \Delta t) = f(\mathbf{v}, t) + C(\mathbf{v},\ t)\, \Delta t. \qquad (8\text{-}1\text{-}14)$$

Equation (8-1-14) is exactly equivalent to (8-1-1) and is perhaps more fundamental since it is one of the precursor steps in the derivation of the Boltzmann equation. Solution of (8-1-14) is easily achieved numerically and precludes the need for the numerical evaluation of derivatives. By this technique, the velocity distribution function may be evaluated rapidly within modest computing times, yet with very high accuracy.

Since there is no necessity to eliminate time-dependent derivatives from the Boltzmann equation in this approach (thus assuming that equilibrium has been reached), this method can be used to study the time evolution of the electron swarm.

An alternative approach to the determination of the electron velocity distribution function is to calculate a series of electron trajectories through a gas in a Monte Carlo calculation. Such calculations have been performed by a number of workers [see Christophorou and Hunter (1984)]. Although they yield accurate results and indeed provide a good baseline for comparison between various approximation methods, the required numbers of Monte Carlo trajectories are notoriously expensive to compute. Recent improvements in the method (Braglia 1981) have, however, led to improved efficiency comparable to that for other numerical methods.

8-2. EXPERIMENTAL METHODS

The primary goal of experimental electron swarm studies is to measure electron transport coefficients as a function of the energy parameter, E/N. As we saw in Chapter 7, the measurement of ion mobilities is complicated by the fact that the ions can undergo chemical reactions during their passage through the gas, and so can change their identity. This behavior necessitated the use of mass spectrometer–assisted detection in order to produce meaningful results. In the case of electrons, the use of mass spectrometers is not necessary. It is, however, important to be able to discriminate between electrons and negative ions. Several methods of accomplishing discrimination are discussed below.

Essentially two types of experiments are performed: measurements of drift velocity and measurements of the diffusion coefficients. While these experiments might seem to be very simple in concept, in practice they require great experimental skill to perform. The reason for this is that to obtain results that will subsequently be useful for the determination of, for example, collision cross sections, high precision and accuracy must be achieved. In fact, typical modern measurements will have accuracies of ± 0.5 to 1.0%. The reviews of Elford (1972) and Hunter and Christophorou (1984) contain extensive details of the various techniques employed. In this section we provide a brief overview of some of the experimental details involved in the determination of drift velocities and diffusion coefficients.

A. Bradbury–Nielsen Method

This is one of the most successful techniques for the measurement of electron drift velocities. Figure 8-2-1 shows a schematic of the method. Electrons produced by photoemission, thermionic emission, or volume ionization by a radioactive source drift in a uniform electric field toward a collector. Two *Bradbury–Nielsen electrical shutters*, separated by a known distance d, are interposed between source and collector. These shutters consist of series of parallel fine wires arranged so that alternate wires are connected together as shown. A pair of sinusoidal potentials 180° out of phase with each other are applied to each set of wires in each shutter. These potentials are superimposed on the mean potential applied to the wires of each shutter to establish a uniform electric field between the source and collector, the value of the field being determined by the chosen value of E/N. When the potential on a given shutter is near zero, electrons can pass through. As the magnitude of the potential increases, the shutters become opaque. The measurement is performed by

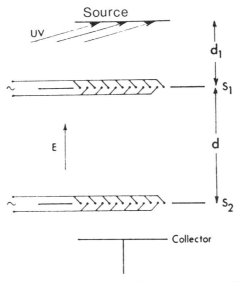

Figure 8-2-1. Schematic diagram of the Bradbury–Nielsen method for measuring electron drift velocities. [Adapted from Elford (1972).]

varying the frequency of the alternating voltage and measuring the electron current at the collector. A typical plot of current versus frequency thus produced is shown in Fig. 8-2-2. The maxima occur when electrons passing through one shutter have a drift velocity such that they arrive at the second shutter just as it opens.

The *measured drift velocity* v_d' is related to the applied frequency by the equation

$$v_d' = \frac{2F^n d}{n},\tag{8-2-1}$$

where F^n is the frequency at which the nth maximum occurs and d is the distance between the shutters.

To obtain the actual drift velocity, a small correction must be made which takes account of diffusion, so that

$$v_d' = v_d\left(1 + \frac{C}{N}\right),\tag{8-2-2}$$

where C is a constant whose exact value is determined experimentally by a procedure that involves making sure that at a given E/N, the value of v_d, so derived is independent of N over a wide range. It should be stressed that the

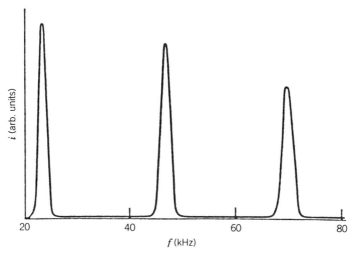

Figure 8-2-2. Typical current–frequency curve obtained using the Bradbury–Nielsen method. [From Cheung and Elford (1990).]

correction to the value of v'_d measured at the highest pressure is usually less than 0.5% and often less than 0.1%.

Figure 8-2-3 shows a typical apparatus used for a drift velocity measurement. Electrons are produced from a radioactive source (S). In this source, alpha particles are emitted from an americium foil and produce electrons by ionizing the background gas. This technique has the advantage of producing an extremely stable supply of electrons with no heat load. Since no filament is used, surface reactions that could contaminate the gas or could lead to filament deterioration are avoided. Typical electron currents used are 1×10^{-11} to 1×10^{-13} A, low enough to ensure that interelectron interactions are minimal. Indeed, during such measurements it is usual that only one electron is present in the drift region at any one time. Electrons passing through the second Bradbury–Nielsen shutter are collected and measured using a sensitive electrometer.

The electric field in the drift region is maintained to very high accuracy with the use of thick guard rings that are connected to a precision voltage divider. Critical to the success of the measurement is maintenance of constant gas temperature, achieved by immersing the entire apparatus in a Dewar filled with an appropriate liquid. For room-temperature measurements the liquid is usually water, but cryogens such as liquid nitrogen are used for low-temperature studies. Gas pressure is determined by using an absolute pressure gauge such as a quartz spiral manometer or a Baratron gauge. These instruments are routinely calibrated against standards. Of prime concern is gas purity, allowable impurity concentrations being less than a few parts per million. Ultrahigh-vacuum-compatible procedures are therefore used in both the drift apparatus and in the associated gas handling manifold.

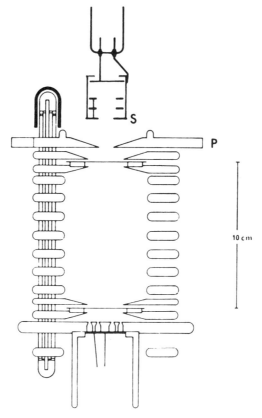

Figure 8-2-3. Electrode system used by Crompton et al. to measure electron drift velocities. [From Crompton et al. (1970).]

B.   Townsend–Huxley Method

The apparatus used in this method is rather similar to that used for the electron drift velocity measurements. Again there is an electron source and a guard ring–shielded drift region. Precautions taken to ensure accurate measurements are also similar, but there is much greater stringency on the uniformity of the electric field since the measurements can be affected significantly by the introduction of a radial component. Here it is not the timing of the electron drift that is monitored but rather the spatial extent of the arriving electron swarm. Thus a simple aperture allows the electrons to enter the drift region, but a segmented electron collector is used, as illustrated in Fig. 8-2-4. The D_T/K ratio is obtained by measuring the ratio of the current to the inner collector, to the total collected current. It can be shown that the two ratios are related by the formula

$$R = 1 - \frac{h}{d} \exp[-\lambda(d - h)], \qquad (8\text{-}2\text{-}3)$$

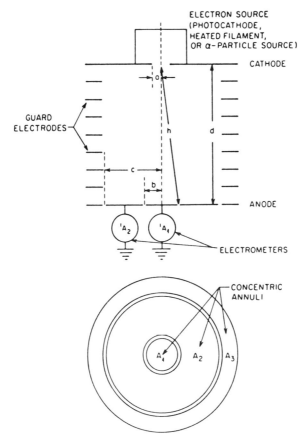

ELECTRON SOURCE
(PHOTOCATHODE,
HEATED FILAMENT,
OR α-PARTICLE SOURCE)

CATHODE

GUARD
ELECTRODES

ANODE

ELECTROMETERS

CONCENTRIC
ANNULI

Figure 8-2-4. Schematic diagram of the Townsend–Huxley transverse diffusion apparatus. [From Hunter and Christophorou (1984).]

where $\lambda = v_d/2D_T$, h is the drift length, and $d = (h_2 + b_2)^{1/2}$, b being the radius of the inner collector. It is an odd fact that the derivation of this formula by Huxley (1940) contained an algebraic error and was based on the assumption of isotropic diffusion prior to the discovery of its anisotropy, yet it gives correct results. Fortuitously, (8-2-3) is a close approximation to the formula that results from a treatment based on anisotropic diffusion when the ratio $D_L/D_T \sim 0.5$, as occurs when σ_m is approximately constant (Lowke, 1971). A detailed discussion of this point has been given by Crompton (1972).

So as not to distort the electric field near the anode, it is important that the segments of the electron collector assembly be held close to ground potential during measurement of the ratio of the currents to the segments, each of which is typically less than 10^{-12} A. This requirement has been met and the desired accuracy achieved by using a modification of the Townsend induction-balance technique (Crompton et al., 1965).

C. Time-of-Flight Methods

Another technique that is related to the Bradbury–Nielsen method but which differs in many respect from it is the *time-of-flight* approach, developed by Stevenson (1952) and later applied extensively at Oak Ridge National Laboratory by Hurst et al. (1963, 1966) and by Christophorou and coworkers (Christophorou et al., 1966, 1973, 1976; Christophorou and Christodoulides, 1969; Christophorou and Pittman, 1970). It has been used successfully to measure not only drift velocities but also longitudinal diffusion coefficients. Indeed, recognition of the anisotropy of electron diffusion came from a comparison of values of D_L from these measurements with values of D_T in gases, where these values had been established beyond reasonable doubt (Wagner et al., 1967; Parker and Lowke, 1969).

The principle of the method is illustrated in Fig. 8-2-5. A pulse of electrons is produced by illuminating a photocathode, and these electrons drift through the tube, finally being detected using a Geiger or proportional counter. Initially, this method was less accurate than the Bradbury–Nielsen technique because the pulse width emitted by the detector was broad, making accurate timing difficult. However, this problem can be overcome by using a fast electron multiplier to detect the electrons.

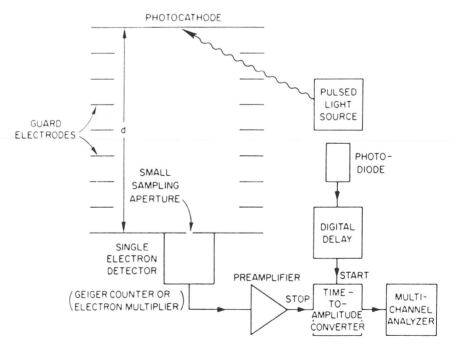

Figure 8-2-5. Schematic diagram of the apparatus used for the time-of-flight method of measuring transport coefficients. [From Hunter and Christophorou (1984).]

As in the Bradbury–Nielsen method, the measured drift velocity must be corrected for longitudinal diffusion effects. By modeling the shape of the arrival-time spectrum obtained with an assumed value of D_L, and by iterating, it is possible to obtain values for D_L as a function of E/N. This procedure can also be followed with the Bradbury–Nielsen method, but not many such measurements have been performed because of the limited accuracy attainable (Elford, private communication, 1991).

In a new time-of-flight apparatus constructed by Schmidt (1991), electrons are produced within the gas at the focus of a pulsed (0.8-ns FWHM) UV laser beam. The absence of a solid cathode means that no boundary corrections have to be applied. After drifting through a variable distance, the electrons pass through a small hole, finally being detected by a proportional counter. Boundary effects near the anode and the effect of the detector-time resolution can be canceled out by performing a series of measurements using differing drift lengths. The drift velocity and longitudinal diffusion coefficients are determined from the arrival-time spectrum of the electrons, as described above. The transverse diffusion coefficient is determined by scanning the laser spot across the axis of the drift space while observing the count rate. The entire system is controlled by a microprocessor, and values of the transport coefficients for a complete set of E/N values are determined automatically. Drift velocities can be measured to a routine accuracy of $\pm 1\%$ and diffusion coefficients to $\pm 5\%$.

It is found that as E/N increases, D_T and D_L no longer have the same value, and indeed this divergence comes on very rapidly in some gases. For example, for H_2, He, and N_2, $D_T/D_L \sim 2$ near 1 Td, while for Ar this ratio is ~ 7 at 0.1 Td. This diffusion anisotropy is a much more sensitive indicator of anisotropic elastic scattering than is the drift velocity. Knowledge of the longitudinal diffusion coefficient, as well as the transverse diffusion, is therefore very important in determinations of electron collision cross sections in situations where anisotropic elastic scattering is likely to occur (Schmidt, 1991).

8-3. ELECTRON TRANSPORT DATA

As mentioned earlier, reviews of electron transport coefficients have been given by Dutton (1975), Beaty and Pitchford (1979), and Gallagher et al. (1983). The book by Huxley and Crompton (1973) contains tabulated and graphical data for the atomic gases and some simple molecular species; further information concerning hydrocarbon species is given in the review by Hunter and Christo-phorou (1984). In this section a few representative examples of transport measurements are discussed.

A. Pure Gases

Figure 8-3-1 shows experimental results for the electron drift velocity in a variety of gases as a function of E/N. The accuracy of these particular

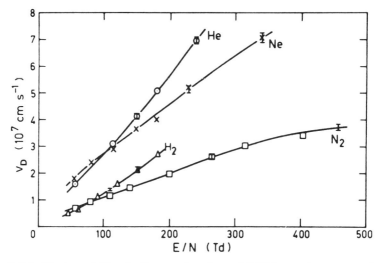

Figure 8-3-1. Electron drift velocity as a function of E/N for various gases at room temperature. [From Fletcher (1981).]

measurements is typically ± 2 to 4%. Figures 8-3-2 and 8-3-3 show corresponding results for the transverse and longitudinal diffusion coefficients. Here the accuracy is between ± 8 and $\pm 15\%$. Although these data show nearly linear increases of the coefficients with increasing E/N, such behavior is not always observed. For example, in more complex molecular gases, the availability of

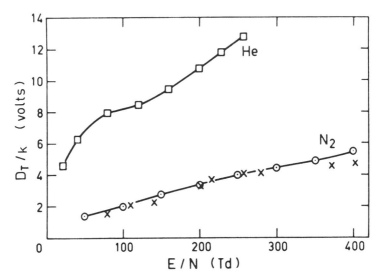

Figure 8-3-2. Ratio of transverse diffusion coefficient to electron mobility as a function of E/N in helium and nitrogen at room temperature. [From Fletcher (1981)]

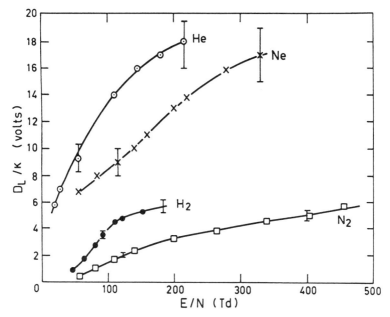

Figure 8-3-3. Ratio of longitudinal diffusion coefficient to electron mobility as a function of E/N for a variety of gases at room temperature. [From Fletcher (1981).]

excitation modes is so great that the electrons remain almost in thermal equilibrium with the gas for a significant range of E/N (Fig. 8-3-4).

As mentioned in Section 8-2, there is a considerable controversy regarding swarm-derived collision cross sections for H_2, so it is fitting to present the basic data here. Figure 8-3-5 shows drift velocity and diffusion coefficients for H_2 measured by a number of authors. It is found that there is very good agreement between the various measurements, suggesting that the source of the controversy is due to the interpretation of the data and not to experimental factors.

B. Gas Mixtures

Addition of very small quantities of a molecular gases to a rare gas with a Ramsauer–Townsend minimum can produce a substantial change in the drift velocity compared that for the pure gas (see Fig. 8-3-6). This fact has a number of consequences. First it means that one must take great care to ensure high gas purity in order to obtain meaningful results when performing swarm measurements. It can, however, be advantageous to use gas mixtures deliberately when making such measurements. For example, England and Elford (1988) have recently discussed how the addition of from 0.5 to 1.5% of hydrogen to a krypton-filled apparatus allows one to obtain greater sensitivity of the drift velocity to the Ramsauer minimum in krypton.

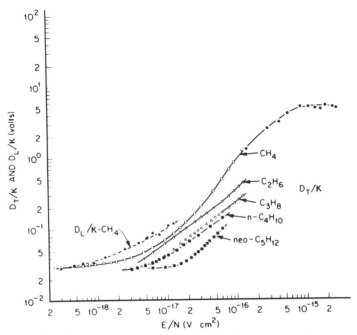

Figure 8-3-4. Ratio of transverse and longitudinal diffusion coefficients to electron mobility as a function of E/N for various hydrocarbon gases. [Adapted from Hunter and Christophorou (1984).]

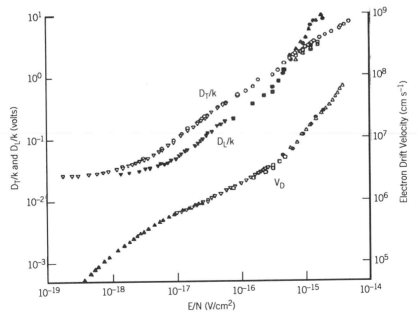

Figure 8-3-5. Electron drift velocity and ratio of transverse and longitudinal diffusion coefficients to mobility as a function of E/N for H_2 ($T \approx 293$ K). [Adapted from Hunter and Christophorou (1984).]

538

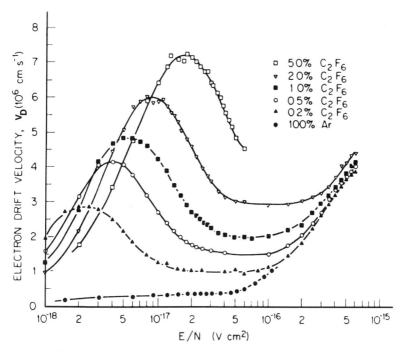

Figure 8-3-6. Electron drift velocity as a function of E/N in C_2F_6/Ar mixtures ($T \approx 298$ K). [From Christophorou et al. (1982).]

Addition of a known quantity of hydrogen also alleviates problems that arise due to small (and possibly unknown) amounts of other molecular impurities. It also lessens diffusion loss effects in the experiment. The use of gas mixtures to increase drift velocities has also been proposed for use in fast counters for nuclear particle detection and in switch technology (Christophorou, 1984).

C. High-Pressure Gases

At elevated gas pressures and in liquids, the possibility exists for scattering events involving more than one molecule at a time. Such scattering leads to a dependence of the transport coefficients on gas number density, as illustrated in Fig. 8-3-7. A number of theories have been proposed to model this behavior. Effects such as negative-ion formation, bubble or pseudo-bubble formation, quantum interference, multiple scattering, and the influence of the uncertainty principle on electron energy distributions have been proposed to explain the observations, but so far no one theory is universally successful. Detailed discussions of electron mobility in high-pressure gases and in liquids are given in Hunter and Christophorou (1984) and Christophorou and Siomos (1984).

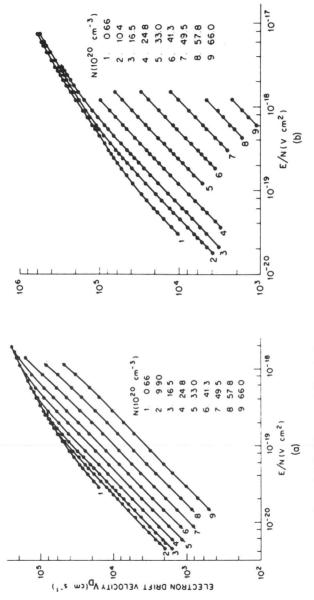

Figure 8-3-7. Electron drift velocity as a function of E/N and N for He (a) and H$_2$ (b) at 77.6 K. [From Bartels (1975).]

8-4.   APPLICATION OF TRANSPORT DATA

A number of mean physical quantities can be derived from electron transport coefficients. These quantities are useful in selecting particular gases or gas mixtures for a number of applications. Such applications include gaseous dielectric (Christophorou et al., 1982; Christophorou and Hunter, 1984) and gas laser (Garscadden, 1981) technologies.

A.   Mean Electron Energy

The ratio D_T/K has the units of volts and when multiplied by the electron charge, e, has the units of electron volts. It thus represents an energy, usually called the *characteristic energy* of the electron swarm, that is related to the *mean electron energy* by

$$\langle \varepsilon \rangle = \frac{1}{2} m \overline{v^2} = Ae \frac{D_T}{K}, \qquad (8\text{-}4\text{-}1)$$

where A is a constant. It can be shown that $A = 1.311$ when all electron collisions are elastic and σ_m is independent of v, so that the electron velocity distribution is given by a Druyvesteyn distribution. For cases where the Maxwellian distribution function more accurately describes the electron velocity distribution (i.e., when both elastic and inelastic collisions occur but E/N is small), $A = 1.5$.

The mean electron energy $\langle \varepsilon \rangle$ provides an indication of the effectiveness of a given gas as an *electron thermalizer*. The primary property of a gaseous dielectric is its ability to reduce the mobility of electrons by attaching them to an appropriate electrophillic molecule. Since electron attachment rates for such molecules are generally large near zero energy (see Chapter 6), it is often useful to include a gas in the dielectric mixture which has the property of rapidly reducing the mean electron energy. Such gases are N_2 and CO_2. Values of $\langle \varepsilon \rangle$ versus E/N for a number of gases are illustrated in Fig. 8-4-1.

B.   Electron Thermalization Times

The time required for the mean energy of electrons to cascade down to thermal values is strongly dependent on the availability of appropriate processes through which the electrons can lose energy. For low-energy electrons, such processes are elastic scattering and rotational and vibrational excitation. Higher-energy electrons can be thermalized quickly by electronic excitation and dissociation processes. One can define the *characteristic thermalization time* as the time that characterizes the exponential decay of the excess electron temperature. Figure 8-4-2 shows the characteristic thermalization time for electrons in hydrogen gas at a density of 100 torr as a function of temperature, calculated by Mozumder (1982). Individual contributions to the thermalization

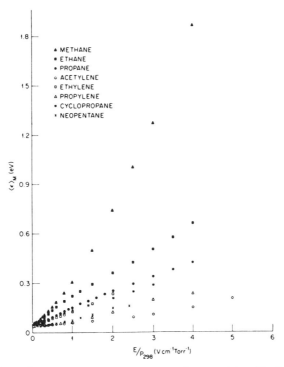

Figure 8-4-1. Mean electron energy versus pressure for a variety of hydrocarbon gases ($T \approx 298$ K). [From Christophorou (1977).]

time due to elastic scattering, and to rotational and vibrational excitation, are also shown and illustrate the relative importance of these processes. Each process contributes to the total relaxation time τ according to

$$\frac{1}{\tau} = \frac{1}{\tau_M} + \frac{1}{\tau_R} + \frac{1}{\tau_v}, \qquad (8\text{-}4\text{-}2)$$

where M, R, and v refer to momentum transfer, rotational excitation, and vibrational excitation, respectively.

Knowledge of electron thermalization times is very important in situations where high-energy electrons are injected into a gas, such as in radiation chemistry, and for the modeling of plasma phenomena.

C. Derivation of Collision Cross Sections

A point of great importance in swarm research is that the mean electron energy, $\langle \varepsilon \rangle$, is a function not just of E, but of E/N. This means that it is possible to control the energies of the electrons in a swarm by adjusting N. This is the basis

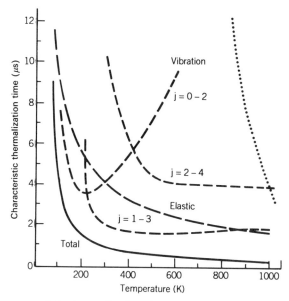

Figure 8-4-2. Characteristic thermalization time (μs) of electrons in H_2 at a density of 3.31×10^{16} molecules/cm^3 as a function of the gas temperature (K), constructed from the data of Mozumder (1982). The total relaxation time is shown by the solid curve, that due to elastic scattering by the long-dashed line, that due to rotational transitions by the short-dashed curves, and that due to vibration by the dotted curve. [From Csanak et al. (1984).]

for using swarm experiments to obtain information about collision processes occurring at thermal and subthermal energies while using relatively large values of E, which are less affected by stray field effects. Collision phenomena at these energies are difficult to study by electron beam methods. Swarm-derived data have been an important source of low-energy electron scattering cross sections, and continued improvements in theoretical methods, as outlined in Section 8-2, will ensure that this approach remains a competitive means of deriving this information. In this section, examples of swarm-derived cross sections will be given, with emphasis on both the strengths and weaknesses of the approach. A recent review of this subject has been given by Crompton (1983), and comprehensive discussions of low-energy electron scattering are given in the companion volume and in Christophorou (1984).

The basic input to the derivation of collision cross sections from swarm-derived data are the transport coefficients v_d, D_T, and D_L as defined by (8-1-9) through (8-1-11). It is seen from these formulas that the coefficients depend directly on the total momentum transfer cross section, which itself is a compound entity, being determined by the sum of competing collision processes, and on the electron velocity distribution.

The basic procedure for unfolding the collision cross sections from the transport coefficients is to begin at low energy (i.e., low E/N) and to use a set of trial cross sections to calculate the transport coefficients via (8-1-9) through (8-1-11) and then to compare the calculated values with the measured values. The values of the cross sections in the set are then adjusted to obtain a close fit between the calculated and measured transport coefficients. The value of E/N is incremented and the procedure repeated. As the energy rises, new collision processes begin to dominate and others fade in importance, as illustrated in Fig. 8-4-3.

Figure 8-4-4 shows the momentum transfer cross section for electrons in helium as a function of electron energy, derived from swarm data. Also shown are the *ab initio* theoretical results of Nesbet (1979) and O'Malley et al. (1979). It can be seen that the agreement is excellent (± 1 or 2%) over the entire energy range. When there is structure in the cross section such as in the case of argon, which has a deep Ramsauer–Townsend minimum, the analysis of the data becomes more complex but can still be tractable. O'Malley and Crompton (1980) have developed a method for analyzing transport data in argon based on modified effective range theory (MERT). The method generally starts with the

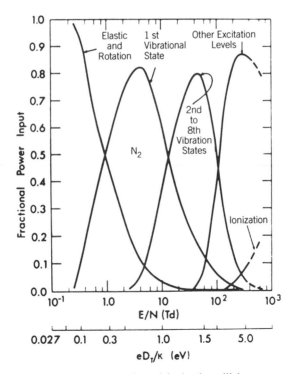

Figure 8-4-3. Distribution between elastic and inelastic collision processes of the total power input into an electron swarm in N_2. (From Engelhardt et al. (1964).]

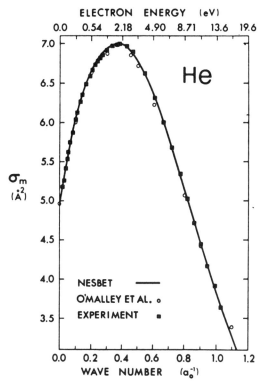

Figure 8-4-4. Comparison of the results of the swarm-derived momentum transfer cross section in He with the theoretical results of Nesbet and O'Malley et al. [From Crompton (1983).]

parametcrizcd formulas for the phase shifts given by MERT (see p. 212 of the companion volume). An initial set of parameters provides an initial set of phase shifts (as functions of electron energy) which can be used to generate an initial trial σ_m, which, in turn, can be used to generate the transport coefficients as functions of E/N. The calculated values are compared with the experimental data and a search routine used that adjusts the coefficients systematically until a final set is found that provides the least-squares fit to the data. Figure 8-4-5 shows differential cross sections for electron–argon scattering determined from swarm data–fitted phase shifts. Also shown are crossed-beam-derived differential cross sections of Andrick (private communication), and it can be seen that there is excellent agreement between the two results. Swarm-derived total cross sections and those measured using beam techniques are shown in Fig. 8-4-6 together with the swarm-derived momentum transfer cross section. (*Note:* The difference between the integral elastic scattering cross section and the momentum transfer cross section is discussed in the companion volume, p. 26.)

A case that is very puzzling is that of the cross section for the vibrational

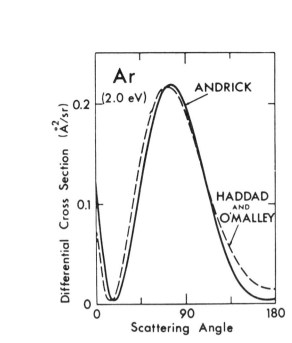

Figure 8-4-5. Comparison of the differential cross section in Ar at 2 eV derived from the MERT phase shifts with the experimental results of Andrick. [From Crompton (1983).]

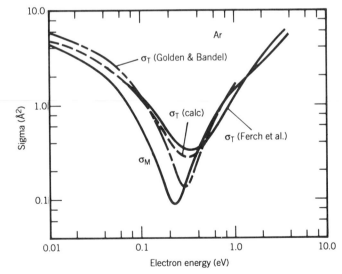

Figure 8-4-6. Momentum transfer cross section in Ar derived from swarm data with the aid of modified effective range theory (MERT). The total cross section σ_T derived from the MERT phase shifts [σ_T(calc)] is compared with the results of two experiments that measure σ_T directly. [From Crompton (1983).]

excitation of H_2 by electrons. This process has been discussed at length by Morrison et al. (1987). The problem can be seen in Fig. 8-4-7, which shows cross sections obtained from swarm-derived data (England et al., 1988) and from two separate types of conventional electron beam–gas target apparatus (Erhardt et al., 1968; Brunger et al., 1990) in which differential cross sections are measured and then integrated to give the integral cross section. In the region from threshold to 2 eV, the swarm-derived data are consistently lower, the maximum difference being 60%. Theoretical results of Buckman et al (1990) are also shown, and it can be seen that they too agree well with the beam data. There is also good agreement between the theoretical differential cross sections and those of Brunger et al. (1990) and Linder and Schmidt (1971) (see Fig. 8-4-8). The evidence thus presented would seem to imply that the swarm results are in error, but the problem is that such an error cannot be identified at this time. Previous swarm measurements of rotational excitation of H_2 are in excellent agreement with theory, as are swarm measurements of the total momentum transfer cross sections (England et al., 1988). The problem is compounded by the fact that the swarm measurements in dilute mixtures of hydrogen in helium (Petrovic and Crompton, 1987) and hydrogen in neon (England et al., 1988), which are more easily interpreted than those in pure hydrogen, lead to essentially the same results as those in the pure gas that first identified the problem.

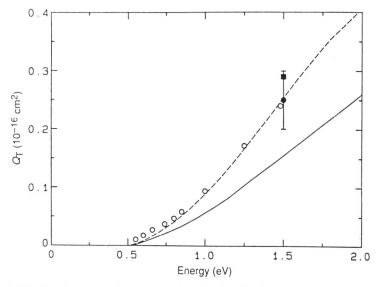

Figure 8-4-7. Total cross sections for ro-vibrational excitation ($v = 0$ to 1) of H_2 (in units of 10^{-16} cm^2) at 1.5 eV. Solid curve, swarm results of England et al. (1988); dashed curve, theoretical results Buckman et al (1990); solid square, Linder and Schmidt (1971); open circles, Erhardt et al. (1968); solid circle, Brunger et al. (1990). [From Brunger et al. (1990).]

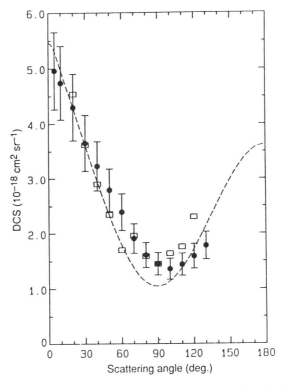

Figure 8-4-8. Absolute differential cross section for ro-vibrational excitation of H_2 ($v = 0$ to 1) (in units of 10^{-18} cm^2/sr) at 1.5 eV. Solid circles, Brunger et al. (1990); open squares, Linder and Schmidt (1971); dashed curve, Buckman et al. (1990). [From Brunger et al. (1990).]

Collision cross sections for many other species have been determined from swarm-derived data [see Christophorou (1984) and Crompton (1983)]. As discussed above, use of the two-term approximation can cause problems when anisotropic scattering occurs and when inelastic events become a significant fraction of the total scattering. Therefore, when this situation arises one must resort to more accurate solutions to the Boltzmann equation, as discussed in Section 8-2-D.

8-5. PROBLEMS

8-1. Refer to Section 7-2 and explain how you calculate K_0 (the reduced or standard mobility) from a plot of v_D versus E/N.

8-2. Using Figs. 7-3-5 and 8-3-1, compare the drift velocities for electrons and K^+ ions in helium at $E/N = 50$ Td.

8-3. Using Figs. 8-3-1 and 8-3-3, calculate D_L for electrons in N_2 at $E/N = 50$, 100, and 300 Td. Consult Fig. 7-3-6, and compare these results with D_L for K^+ ions in N_2 at the same value of E/N.

8-4. Expand the discussion in Section 8-1 to include other similarities and dissimilarities of electron and ion swarms.

8-5. **(a)** Use (7-4-8) to calculate the average energy of Ne^+ ions and of electrons in He at $E/N = 120$ Td.

(b) Use (8-4-1) to recalculate the average energy for electrons in He at $E/N = 120$ Td, and compare the result with the value calculated for electrons in part (a).

8-6. Margenau (1946) used the two-term approximation in solving the Boltzmann equation to obtain the energy distribution and drift velocity of electrons in a gas for the case where no inelastic collisions occur. His analysis is reproduced in McDaniel (1964, pp. 551–555). Fill in all of the intermediate steps in Margenau's analysis.

REFERENCES

Bartels, A. (1975), *Appl. Phys.* **8**, 59.

Beaty, E. C., and L. C. Pitchford (1979), *A Bibliography of Electron Swarm Data*, JILA Information Center Report 20, Joint Institute for Laboratory Astrophysics, Boulder, Colo.

Braglia, G. L. (1981), *Lett. Nuovo Cimento* **31**, 183.

Braglia, G. L., and G. L. Caraffini (1974), *Riv. Mat. Univ. Parma* **3**, 81.

Braglia, G. L., G. L. Caraffini, and M. Iori (1977), *Lett. Nuovo Cimento* **19**, 193.

Brunger, M. J., S. J. Buckman, and D. S. Newman (1990), *Aust. J. Phys.* **43**, 665.

Buckman, S. J., M. J. Brunger, D. S. Newman, G. Snitchler, S. Alston, D. W. Norcross, M. A. Morrison, B. C. Saha, G. Danby, and W. K. Trail (1990), *Phys. Rev. Lett.* **65**, 3253.

Cavalleri, G. (1981), *Aust. J. Phys.* **34**, 361.

Cheung, B., and M. T. Elford (1990), *Aust. J. Phys.* **43**, 755.

Christophorou, L. G., G. S. Hurst, and A. Hadjiantoniou (1966), *J. Chem. Phys.* **44**, 3506.

Christophorou, L. G. and A. A. Christodoulides (1969), *J. Phys. B.* **2**, 71.

Christophorou, L. G., R. P. Blaunstein, and D. Pittman (1973), *Chem. Phys. Lett.* **22**, 41.

Christophorou, L. G., M. W. Grant, and D. Pittman (1976), *Chem. Phys. Lett.* **38**, 100.

Christophorou, L. G. (1977), in *Proc. XIII Conf. Phenomena in Ionized Gases*, Berlin (Leipzig VEB Export-Import), Invited Lectures, p. 51.

Christophorou, L. G. (1981), *Electron and Ion Swarms*, Pergamon Press, New York.

Christophorou, L. G., Ed. (1984), *Electron–Molecule Interactions and Their Applications*, 2 vols., Academic Press, New York.

Christophorou, L. G., and S. R. Hunter (1984), in L. G. Christophorou, Ed., *Electron–Molecule Interactions and Their Applications*, Vol. 2, Academic Press, New York, p. 318.

Christophorou, L. G., and D. Pittman (1970), *J. Phys. B* **3**, 1252.

Christophorou, L. G., and K. Siomos (1984), in L. G. Christophorou, Ed., *Electron–Molecule Interactions and Their Applications*, Vol. 2, Academic Press, New York, p. 221.

Christophorou, L. G., S. R. Hunter, J. G. Carter, and R. A. Mathis (1982), *Appl. Phys. Lett.* **41**, 147.

Crompton, R. W. (1972), *Aust. J. Phys.* **25**, 409.

Crompton, R. W. (1983), in W. Botticher et al., Eds., *Proc. XVI Int. Conf. Phenomena in Ionized Gases, Dusseldorf*, University of Dusseldorf, p. 58.

Crompton, R. W., M. T. Elford, and J. Gascoigne (1965), *Aust. J. Phys.* **18**, 409.

Crompton, R. W., M. T. Elford, and A. G. Robinson (1970), *Aust. J. Phys.* **23**, 667.

Csanak, G., D. C. Cartwright, S. K. Srivastava, and S. Trajmar (1984), in L. G. Christophorou, Ed., *Electron–Molecule Interactions and Their Applications*, Vol. 1, Academic Press, New York, p. 1.

Drallos, P. J., and J. M. Wadehra (1988), *J. Appl. Phys.* **63**, 5601.

Drallos, P. J., and J. M. Wadehra (1989), *Phys. Rev.* **40**, 194.

Dutton, J. (1975), *J. Phys. Chem. Ref. Data* **4**, 577.

Elford, M. T. (1972), in E. W. McDaniel and M. R. C. McDowell, Eds., *Case Studies in Atomic Collision Physics II*, North-Holland, Amsterdam, p. 91.

Engelhardt, A. G., A. V. Phelps, and C. G. Risk (1964), *Phys. Rev.* **135**, A1566.

England, J. P., and M. T. Elford (1988), *Aust. J. Phys.* **41**, 701.

England, J. P., M. T. Elford, and R. W. Crompton (1988), *Aust. J. Phys.* **41**, 573.

Erhardt, H., L. Langhans, F. Linder, and H. S. Taylor (1968), *Phys. Rev.* **173**, 222.

Ferrari, L. (1975), *Physica* **A81**, 276.

Ferrari, L. (1977), *Physica* **A85**, 161.

Fletcher, J. (1981), in L. G. Christophorou, Ed., *Electron and Ion Swarms*, Pergamon Press, Elmsford, N.Y., p. 1.

Gallagher, J. W., E. C. Beaty, J. Dutton, and L. C. Pitchford (1983), *J. Phys. Chem. Ref. Data* **12**, 109.

Garscadden, A. (1981), in L. G. Christophorou, Ed., *Electron and Ion Swarms*, Pergamon Press, Elmsford, N.Y., p. 251.

Hunter, S. R., and L. G. Christophorou (1984), in L. G. Christophorou, Ed., *Electron–Molecule Interactions and Their Applications*, Vol. 2, Academic Press, N.Y., p. 90.

Hurst, G. S., and J. E. Parks (1966), *J. Chem. Phys.* **45**, 282.

Hurst, G. S., L. B. O'Kelly, E. B. Wagner, and J. A. Stockdale (1963), *J. Chem. Phys.* **39**, 1341.

Huxley, L. G. H. (1940), *Philos. Mag.* **30**, 396.

Huxley, L. G. H., and R. W. Crompton (1974), *The Diffusion and Drift of Electrons in Gases*, John Wiley and Sons, New York.

Kitamori, K., H. Tagashira, and Y. Sakai (1978), *J. Phys. D* **11**, 283.

Kitamori, K., H. Tagashira, and Y. Sakai (1980), *J. Phys. D* **13**, 535.

Kleban, P., and H. T. Davis (1977), *Phys. Rev. Lett.* **39**, 456.

Kleban, P., and H. T. Davis (1978), *J. Chem. Phys.* **68**, 2999.

Kleban, P., L. Foreman, and H. T. Davis (1980), *J. Chem. Phys.* **73**, 519.

Kumar, K. (1984), *Phys. Rep.* **112**, 319.

Lin, S. L., R. E. Robson, and E. A. Mason (1979), *J. Chem. Phys.* **71**, 3483.

Linder, F., and H. Schmidt (1971), *Z. Naturforsch.* **26a**, 1603.

Loeb, L. B. (1961), *Basic Processes of Gaseous Electronics*, University of California Press, Berkeley.

Lowke, J. J. (1971), *in Proc. 10th Int. Conf. Phenomena in Ionized Gases,* Oxford.

Makabe, T., and T. Mori (1978), *J. Phys. B* **11**, 3785.

Makabe, T., and T. Mori (1980), *J. Phys. D* **13**, 387.

Makabe, T., and T. Mori (1982), *J. Phys. D* **15**, 1395.

Margenau, H. (1946), *Phys. Rev.* **69**, 508.

Mason, E. A., and E. W. McDaniel (1988), *Transport Properties of Ions in Gases*, Wiley, New York.

McDaniel, E. W. (1964), *Collision Phenomena in Ionized Gases*, Wiley, New York.

Milloy, H. B., and R. O. Watts (1977), *Aust. J. Phys.* **300**, 73.

Morrison, M. A., R. W. Crompton, B. C. Saha, and Z. Lj. Petrovic (1987), *Aust. J. Phys.* **40**, 239.

Mozumder, A. (1982), *J. Chem. Phys.* **76**, 3277.

Nesbet, R. K. (1979), *Phys. Rev.* **A20**, 58.

O'Malley, T. F., and R. W. Crompton (1980), *J. Phys. B* **13**, 3451.

O'Malley, T. F., P. G. Burke, and K. A. Berrington (1979), *J. Phys. B* **12**, 953.

Parker, J. H., and J. J. Lowke (1969), *Phys. Rev.* **181**, 290.

Petrovic, Z. Lj., and R. W. Crompton (1987), *Aust. J. Phys.* **40**, 347.

Pitchford, L. C. (1981), in L. G. Christophorou, Ed., *Electron and Ion Swarms*, Pergamon Press, Elmsford, N.Y., p. 45.

Pitchford, L. C., S. V. O'Neil, and J. R. Rumble (1981), *Phys. Rev.* **A23**, 294.

Robson, R. E., and K. Kumar (1971), *Aust. J. Phys.* **24**, 835.

Schmidt, B. (1991), in *Proc. Joint Symp. Electronic and Ionic Swarms and Low Energy Scattering*, Bond University, Queensland, Australia, p. 34.

Stevenson, A. (1952), *Rev. Sci. Instrum.* **23**, 93.

Tagashira, H., T. Taniguchi, K. Kitamori, and Y. Sakai (1978), *J. Phys. D* **11**, L43.

Wagner, E. B., F. J. Davis, and G. S. J. Hurst (1967), *J. Chem. Phys.* **47**, 3138.

Wilhelm, J., and R. Winkler (1969), *Ann. Phys.* **23**, 28.

9

ELECTRON–ION AND ION–ION RECOMBINATION

9-1. INTRODUCTION

Electron–ion recombination is a process in which a positive ion captures a free electron to form a product or products, chemically distinct from the parent ion. Examples of such processes are:

(a) $e + A^{z+} \rightarrow A^{(z-1)+*} + h\nu$ *Radiative recombination*

(b) $e + A^{z+} \rightarrow A^{(z-1)+**} \rightarrow A^{(z-1)+} + h\nu$ *Dielectronic recombination*

(c) $e + e + A^{z+} \rightarrow A^{(z-1)+} + e + h\nu$ *Collisional–radiative recombination*

(d) $e + A^{z+} + B \rightarrow A^{(z-1)+} + B$ *Termolecular electron–ion recombination*

(e) $h\nu + e + A^{z+} \rightarrow A^{(z-1)+} + 2h\nu$ *Stimulated radiative recombination*

(f) $e + AB^{+} \rightarrow A + B$ *Dissociative recombination*

         $\rightarrow A^{+} + B^{-}$ *Ion-pair formation*

In each case, the product or products may be formed in a variety of excited states, and in the case of dissociative recombination a variety of different chemical products are possible.*

*Multiply charged atomic ions are formed in high-temperature plasmas and their recombination processes are extremely important. Mulitply charged molecular ions do exist but are rare. No studies of their recombination have been performed, so throughout this chapter, molecular ion recombination will refer exclusively to singly charged species.

Similar processes can occur when a positive ion interacts with a negative ion, again producing distinct products, and these are referred to as *ion–ion recombination*, thus:

(g) $A^+ + B^- \rightarrow A + B$ *Mutual neutralization*

(h) $A^+ + B^- + C \rightarrow A + B + C$ *Termolecular ion–ion recombination*

(i) $A^+ + B^- \rightarrow A^- + B^+$ *Charge-changing processes*

$ \rightarrow A^+ + B + e$

$ \rightarrow A + B^+ + e$

Electron–ion and ion–ion recombination are important processes in ionized media, for they lead to an alteration of the energy partitioning and the chemical nature of the media. Often the products of the recombination are excited, and their subsequent relaxation must be accounted for if spectroscopy is being used to diagnose the plasma. The emission of radiation during recombination may represent an energy-loss process if the plasma is optically thin. Recombination may result in the formation of new chemical species which play an active role in plasma chemistry. Recombination reactions are some of the most difficult atomic and molecular processes to study experimentally, and only in recent years is a clear understanding of some of these phenomena emerging.

Energy is required to ionize an atom, molecule, or ion. This energy can be supplied by photon, electron, or heavy particle impact and the details of such ionization processes are given in the companion volume and in this volume, Chapter 4. For recombination to occur, there must be an efficient mechanism for removal of this energy from the electron–ion or ion–ion system. Generally, this can be accomplished through the emission of radiation or, if possible, via the conversion of recombination energy into kinetic and potential energy of products or collision partners.

In this chapter we examine the details of specific recombination mechanisms and review the experimental methods used to study the phenomena. Recombination theories are mathematically complex and for further details the interested reader is directed to the references listed throughout the text. A number of review articles covering specific recombination processes are listed under the appropriate headings, but particular attention is drawn to a recent book covering the field of dielectronic recombination research by Graham et al. (1992) and to two books on dissociative recombination by Mitchell and Guberman (1989) and Rowe and Mitchell (1993). A recent general overview of recombination research has been given by Flannery (1990).

PART 𝒜.   ELECTRON–ION RECOMBINATION

9-2.   RADIATIVE RECOMBINATION

If we consider a thermal energy electron making a close approach to a bare nucleus, for a brief instant of time ($\sim 10^{-15}$ s) the system can be visualized as an atom in an excited state with its outer electron in an extended orbit. Usually, the electron will simply continue on its way, being deflected by the Coulomb field of the ion but otherwise having its speed unchanged. If, however, during its close approach, the electron captured in the quasi-atom makes a transition to a lower orbit, emitting a photon in the process, it loses velocity and can end up bound to the ion core. This can be represented by

$$e + A^{z+} \rightarrow A^{(z-1)+} + h\nu,$$

a process known as *radiative recombination*, and the inverse of *photoionization*.

The emission of radiation is, however, a slow process, so the probability of this happening is very small. *Radiative recombination* therefore typically occurs at a very low rate.

The cross section for capture of an electron with energy E_e into a state with principal quantum number n is given by

$$\sigma_n = 2.1 \times 10^{-22} \, \text{cm}^2 \, \frac{Z^4 E_0^2}{n E_e (Z^2 E_0 + n^2 E_e)}, \qquad (9\text{-}2\text{-}1)$$

where Z is the atomic number and $E_0 = 13.6 \, \text{eV}$ is the ionization energy of atomic hydrogen (Bethe and Salpeter, 1977). This simple formula is derived using the assumption that the oscillator strength for the electron transition is continuous across the continuum limit and so is expected to be accurate for high n values. It predicts that at low electron energies, the cross section should be inversely proportional to n (see Fig. 9-2-1a) and to E_e (Fig. 9-2-1b).

A more accurate calculation by Stobbe (1930) produces values 20% lower for $n = 1$, 12% lower for $n = 2$, and 9% for $n = 3$. Stobbe's analysis also takes into account the dependence of the cross section on l, and this is illustrated in Fig. 9-2-2. Experimental verification of this analysis has been given by Anderson and Bolko (1990)* using a merged electron–ion beam apparatus†, shown in Fig. 9-2-3. The energy resolution in this apparatus is on the order of 0.15 eV, and the

---

*This reference contains a table of calculated cross sections for n and l values up to 60 and 59.
†The merged beam technique is described in detail in Section 9-6-4.

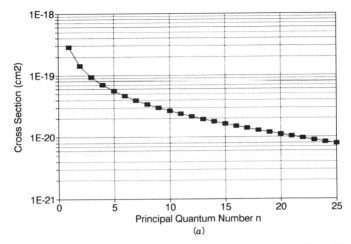

Figure 9-2-1. (*a*) Cross sections for the radiative recombination of 10-meV electrons with protons as a function of principal quantum number *n*, calculated using the Bethe–Salpeter formula.

results are shown in Fig. 9-2-4 in the form of rate coefficients, $\langle \sigma v \rangle$ versus *electron energy*, where

$$\langle \sigma v \rangle = \int \sum_n \sigma_n v f(v) \, dv. \tag{9-2-2}$$

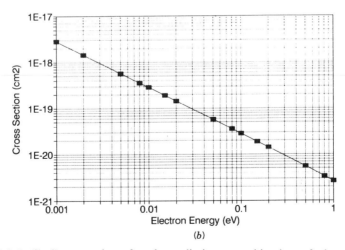

Figure 9-2-1. (*b*) Cross sections for the radiative recombination of electrons with protons to form H atoms in the *n* = 1 state, versus electron energy, calculated using the Bethe–Salpeter formula.

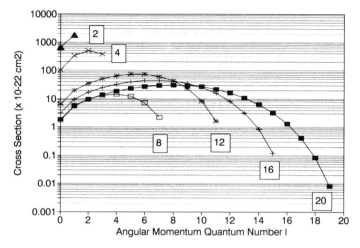

Figure 9-2-2. Variation of the cross section for the radiative recombination of 10-meV electrons and protons to form hydrogen atoms in specified n states (shown in the boxes) as a function of angular momentum quantum number l. [From the calculation of Anderson and Bolko (1990) based on Stobbe's analysis; see text.

In this experiment the electron velocity distribution $f(v)$ is taken as the product of Maxwellian distributions for the transverse and longitudinal components of electron velocity, respectively.

When an electron collides with an ion to which electrons are bound, the nuclear charge is partially screened and n should be replaced by an effective principal quantum number, n_{eff}. Anderson and Bolko (1990) found that for low

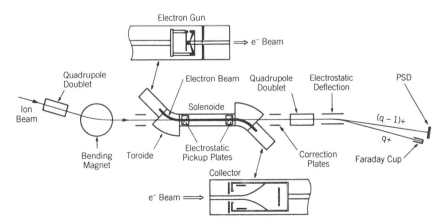

Figure 9-2-3. Merged electron–ion beam apparatus at the University of Aarhus. [From Anderson and Bolko (1990).]

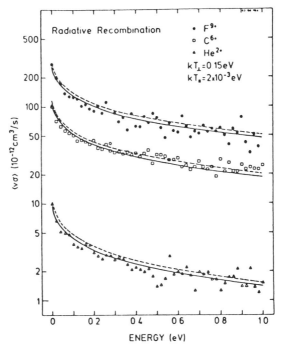

Figure 9-2-4. Experimental rate coefficients $\langle v\sigma \rangle$ as a function of electron energy for the radiative recombination of hydrogenic He, C, and F ions. The solid curves are calculations based on (9-2-1) but including the corrections of Stobbe. The dashed curves are without Stobbe's corrections. [From Anderson and Bolko (1990).]

electron energies, very good agreement was found between their results and those of Stobbe if Z_{eff} was set equal to Z_I, the ionic charge. For higher energies, McLaughlin and Hahn (1989) recommend a value of

$$Z_{\text{eff}} = \frac{Z_N + Z_I}{2}, \tag{9-2-3}$$

where Z_N is the nuclear charge.

9-3. LASER-STIMULATED RADIATIVE RECOMBINATION

If the electron–ion collision takes place within an intense photon field such as that created by a laser, stimulated emission of radiation can occur leading to enhanced stabilization. This phenomenon, known as *laser-stimulated radiative recombination*, has recently been demonstrated experimentally by Yousif et al. (1991a) and Schramm et al. (1991).

One can define a gain factor as being the ratio of the stimulated cross section to the spontaneous cross section. Neuman et al. (1983) have shown that this factor is given by

$$G = \frac{\sigma^{\text{stim}}}{\sigma^{\text{spon}}} = \frac{Pc^2}{F \Delta v \, 8\pi h v^3}, \qquad (9\text{-}3\text{-}1)$$

where P is the laser power in watts, c the velocity of light, F the area of the laser beam, v the laser frequency, and Δv is a term that takes account of the match between the energy distributions of the electron beam, the laser beam, and the width of the level into which capture is stimulated. In practice this is inevitably dominated by the energy spread of the electrons. One can express the velocity spread as a frequency spread: thus

$$\Delta v = \frac{mv_e}{h} \Delta v_e. \qquad (9\text{-}3\text{-}2)$$

Cross sections for stimulated capture to the $n = 11$ and $n = 12$ states of atomic hydrogen are shown in Fig. 9-3-1a and b. This is a resonant process and the enhancement occurs when the electron energy plus the energy of the particular level below the ionization limit exactly matches the energy of the stimulating photons. The results of Yousif et al. are in general agreement with (9-3-1) when account is taken of the overlap of the laser beam with the electron and ion beams. These results also beautifully illustrate the very high energy resolutions achievable with the merged beams technique.

Laser-stimulated radiative recombination has been suggested as a possible means of producing exotic species such as antihydrogen (Neumann et al., 1983) and may also have application for the production of specific energy states in plasmas.

9-4. DIELECTRONIC RECOMBINATION

An additional stabilization mechanism is possible for ions that contain electrons prior to recombination. This may be represented as follows:

$$e + A^{z+} \rightarrow A^{(z-1)+**} \rightarrow A^{(z-1)+*} + hv.$$

The incoming electron excites a bound electron that moves to an orbit farther from the nucleus. This reduces the nuclear screening, temporarily allowing the projectile electron to be captured into an excited (Rydberg) orbit. The most likely outcome is for the electron–electron repulsion to eject this electron (i.e., autionization occurs), and this generally happens in a time on the order of 10^{-13} s. If, however, the inner electron drops down to its initial state before

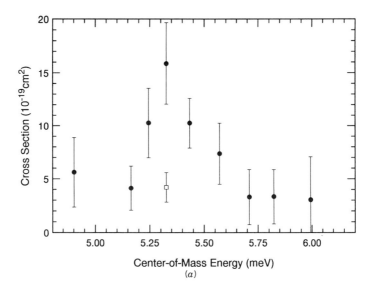

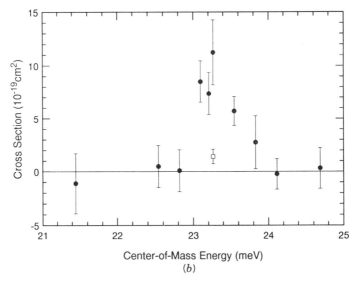

Figure 9-3-1. Measured effective electron–proton radiative recombination cross section with ●, laser on, and □, laser off. (*a*) Peak at 5.33 meV corresponds to stimulated recombination to the $n = 11$ level. (*b*) Peak at 23.3 meV indicates stimulated recombination to the $n = 12$ level. [From Yousif et al. (1991a).]

autoionization occurs (emitting a photon in the process), there is no longer sufficient energy available to eject the second electron, and it remains bound in its excited orbit. The recombination has therefore been stabilized. This process is known as *dielectronic recombination* (DER). Unlike radiative recombination, DER is a resonant process and its cross section is characterized by a series of (often unresolved) peaks lying below an allowed ionic excitation threshold. It usually exhibits a much larger rate than radiative recombination because of the much greater density of states in the continuum into which the electron can be captured [see Bates (1988) and Chapter 7 of the companion volume.]

Some recent experimental results obtained using the merged beam apparatus shown in Fig. 9-2-2 (Anderson et al., 1989) deserve special mention here and are shown in Fig. 9-4-1. The high-energy resolution of this experiment allows many of the resonance features associated with this process to be clearly visible. Experimental advances are occurring very rapidly in this area, and a thorough description of the state of the art is given by Graham et al. (1992).

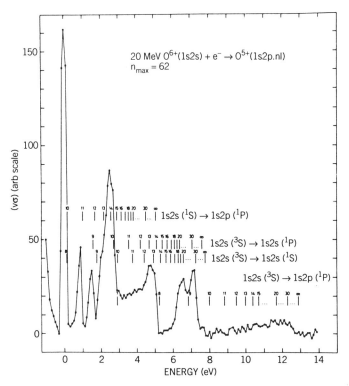

Figure 9-4-1. Rate coefficient $\langle v\sigma \rangle$ as a function of electron energy for dielectronic recombination of O^{6+}. To obtain a lower limit for the rate coefficient, multiply the ordinate by $10^{-11}\,\text{cm}^3/\text{s}$. [From Anderson et al. (1989).]

9-5. COLLISIONAL RADIATIVE AND TERMOLECULAR RECOMBINATION

If the electron density in the ionized medium is sufficient, the recombination energy can also be dissipated via a series of superelastic and inelastic collisions with other electrons. One can define a compound process known as collisional radiative recombination (CRR) thus:

$$e + e + A^{z+} \rightarrow A^{(z-1)+} + e + hv.$$

This process was first discussed by Bates et al. (1962), who developed a statistical theory to describe the radiative and collisional excitation/deexcitation and ionization transitions involved in the passage of the free electron down to a stabilized bound atomic state. If the temperature of the plasma is sufficiently low, radiative stabilization dominates. If the electron density is high, collisional stabilization dominates.

One can visualize this process as the electrons moving up and down an energy ladder, their passage being driven by collisional excitation/deexcitation and by radiative processes. As the principal quantum number of the "temporary" recombined atom decreases, the rate of collisional deexcitation decreases while the radiative deexcitation increases. There exists a minimum, therefore, in the rate of deexcitation of recombined but still excited atoms and this occurs at an n value that we can designate n_c. Excited atoms with n values less than n_c have a far greater probability of making a transition to the ground state than being reexcited back toward the ionization continuum. Thus the recombination rate can be defined as the net deexcitation rate down to n_c. This is referred to as the *recombination bottleneck*.

Bates et al. used the classical theory of Gryzinski (1959) to estimate the excitation/deexcitation rates and formulas based on the work of Seaton (1959, 1962) for the radiative recombination and collisional ionization rates. Spontaneous radiative transition rates were taken from the accurate tables of Baker and Menzel (1938) and Green et al. (1957). Later, Mansbeck and Keck (1969) used a Monte Carlo approach to evaluate the validity of the Gryzinski theory and while they found that the classical approximation was not strictly valid for transitions involving a small energy change, the results obtained by Bates et al. were similar to those obtained using the more exact treatment. This work was followed by Stevefelt et al. (1975), who incorporated radiative transitions into the analysis of Mansbeck and Keck and set down an analytical expression for the CRR recombination rate: thus

$$\alpha = 1.55 \times 10^{-10} T^{-0.63} + 6.0 \times 10^{-9} T^{-2.18} n(e)^{0.37}$$
$$+ 3.8 \times 10^{-9} T^{-4.5} n(e) \, \text{cm}^3/\text{s}, \tag{9-5-1}$$

where T is the electron temperature and n the electron density. This is illustrated

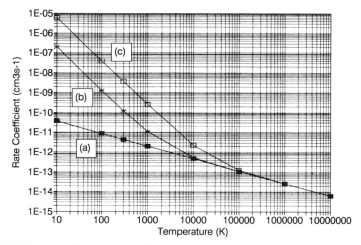

Figure 9-5-1. Rate coefficients for collisional radiative recombination of electrons and protons for electron densities (a) $n_e = 0$, (b) $n_e = 10^{10} \, cm^3/s$, and (c) $n_e = 10^{14} \, cm^3/s$. [Calculated from analytical formula of Stevefelt et al. (1975).]

graphically in Fig. 9-5-1. It should be noted that this relation exhibits an extremely steep $T^{-4.5}$ temperature dependence. Experimental studies of this process have been rare, but support for (9-5-1) is generally fairly good within the temperature range (300 to 3000 K) studied (Stevefelt et al., 1975; Boulmer et al., 1977). Pert (1990) has recently reviewed CRR theory, but his findings differ little from those of the previous studies.

Termolecular electron–ion recombination,

$$e + A^+ + M \rightarrow A(n) + M,$$

is generally a much less efficient process for only a small fraction $(2m/M)$ of the electron's energy can be transferred to the atom M. Cao and Johnsen (1991) have studied the termolecular recombination of a mixture of N_2^+, O_2^+, and NO^+ ions in helium gas at densities from 2.5×10^{19} to $2.9 \times 10^{20} \, cm^{-3}$ and temperatures from 77 to 150 K using the apparatus shown in Fig. 9-5-2.

The vacuum chamber, which is suspended in a cryogenic Dewar, is filled with helium gas to the desired pressure and this is ionized by discharging a large capacitor across the spark gap. Ultraviolet light from this spark produces a weakly ionized gas $(n_e = 10^{10} \, cm^{-3})$ by photoionizing the helium gas and gaseous impurities within this gas. At low temperatures, the primary ions so produced are He_2^+, N_2^+, O_2^+, and some NO^+. The He_2^+ ions react rapidly with gaseous impurities so that after about $100 \, \mu s$, the plasma is dominated by the latter ions.

The decay of the electron density is monitored by measuring the electrical conductivity of the plasma between the inner and outer electrodes, as a function

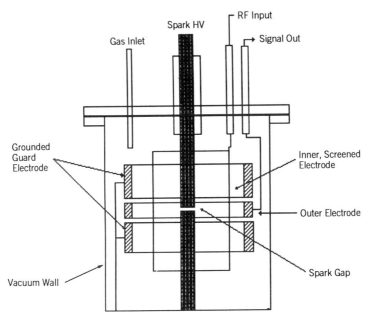

Figure 9-5-2. Schematic diagram of the plasma chamber in the apparatus of Cao and Johnsen (1991) containing an ionizing spark gap and a radio-frequency probe. The entire chamber is housed in a liquid helium dewar.

of time. This is accomplished by applying a radio-frequency voltage to the inner electrode and measuring the current to the outer electrode using tuned amplifiers. Details of this technique are given in the paper. Recombination coefficients are obtained from the slopes of graphs of $1/n_e$ versus time. The measured rates were found to be in close agreement with theoretical estimates of termolecular recombination by Pitaevskii (1962a) and Bates and Khare (1965), although these theories applied specifically to atomic ions. These theoretical models predicted that the rate coefficient would be largely independent of the chemical identity of the ion, would be inversely proportional to the reduced mass of the ion and the neutral, and would scale as $T^{-2.5}$ in the limit of low gas temperature.

In principle, when M is a molecular gas, the energy transfer can be more efficient by involving rotational or vibrational excitation. Bates et al. (1971) have examined this process using a semiquantal approach and found that indeed the termolecular recombination coefficients were orders of magnitude higher for molecular third bodies than for atomic. Furthermore, the calculated rates agreed well with measured recombination coefficients for atomic ions in flame gases (Kelly and Padley, 1969). Bates et al. predicted that the rate would fall off rapidly with temperature, and this was also observed in the flame experiments. This approach was found to fail, however, when applied to molecular ions recombining in a high-pressure molecular gas.

Using a pulse radiolysis technique,* Warman et al. (1979) and Sennhauser et al. (1980a) found that the recombination rates in CO_2 and H_2O were considerably larger than predicted by the theory of Bates et al. (1971), and moreover, they fell off much more slowly with energy. It was found that the magnitude of the results could be reproduced fairly well using classical theory based on the *trapping radius* concept of Thomson (1924) [extended to electron–ion recombination by Massey and Burhop (1952) and Pitaevskii (1962b)] but including the effects of rotational and vibrational energy transfer (Dalidchik and Sayasov, 1966, 1967; Denisov and Kuznetsov, 1972).

Bates (1980) extended this theory to examine the temperature dependence of the rate coefficient, predicted by this approach. The total recombination rate coefficient can be expressed as

$$\alpha = \alpha_2 + \alpha_3 N, \tag{9-5-2}$$

where α_2 is the two-body recombination rate and α_3 takes account of three-body interactions. The experimental values were found to be

$$\alpha_3(CO_2) = 5.5 \times 10^{-25} \left(\frac{300}{T_g}\right)^{0 \pm 0.5} \text{cm}^6/\text{s}$$

$$\alpha_3(H_2O) = 2.6 \times 10^{-23} \left(\frac{300}{T_g}\right)^{2 \pm 0.5} \text{cm}^6/\text{s}$$

where T_g is the gas temperature. Bates (1980) theory was found to yield values of

$$\alpha_3(CO_2) = 5.7 \times 10^{-25} \left(\frac{300}{T_g}\right)^{1.3} \text{cm}^6/\text{s}$$

$$\alpha_3(H_2O) = 8 \times 10^{-23} \left(\frac{300}{T_g}\right)^{4.2} \text{cm}^6/\text{s}$$

Thus although the predicted magnitudes are close to that found experimentally, the predicted temperature dependence is clearly at odds with the observations. Bates (1981a) was able to rectify this situation by introducing the concept of *collisional dissociative recombination*, which he predicted would enhance the termolecular recombination rate for molecular ions in molecular gases at high gas densities. In this process the electron is captured into a high-lying, neutral Rydberg state through energy-loss processes as in other three-body recombination mechanisms. The ability of the Rydberg state to decay via dissociation (in addition to radiation) leads to greatly enhanced recombination efficiency. Inclusion of this mechanism is found to yield rate coefficients and

---

*In this technique, a beam of high-energy electrons is injected into a gas, producing ionization. The electron density is measured by microwave scattering methods and the recombination rate is determined by monitoring the decay of the electron density with time.

temperature dependencies which are in good agreement with the results of the pulse radiolysis experiments.

9-6.   DISSOCIATIVE RECOMBINATION

As mentioned in Section 9-5, when the ion is molecular, it is possible, in principle, for the resulting neutral to dissociate and for the recombination energy to be converted into kinetic and potential energy of dissociation products. This process, known as two-body *dissociative recombination* (DR), was proposed by Massey and Bates [see Bates (1988)] to explain observed high electron–ion recombination rates in the ionosphere. Bates (1950) showed that high recombination rates can occur when the electron is captured by the ion into an autoionizing intermediate neutral complex which subsequently dissociates, thus:

$$e + AB^+ \rightarrow AB^{**} \rightarrow A^* + B.$$

Experimental studies have since shown that DR does indeed usually display a large recombination rate.

Figure 9-6-1 shows the potential energy diagram for H_2^+ and H_2. It can be seen that there is a "resonance" state that intersects the ion curve in the vicinity of the $v = 1$ level. In fact, this is a diabatic* rather than an adiabatic state, so the intersection of two molecular curves of the same symmetry, otherwise forbidden by the noncrossing rule (Wigner, 1948), is permitted.

It is evident from the diagram that if an electron approaches the ion with essentially any kinetic energy, the total energy of the ion + free electron system is degenerate with the energy of this resonance. The electron can therefore form this state by inducing an excitation of the bound electron from a bonding to a nonbonding orbital, being itself captured. As in the case of DER, this complex is thus doubly excited and autoionizing. The lifetime against autoionization is on the order of 10^{-13} s. At small R values, however, the two protons in the ion core repel each other, increasing their kinetic energy at the expense of potential energy. Within a time of about 10^{-14} s, the potential energy has fallen to a value less than the ionization potential of the neutral molecule, and the recombination is stabilized. The process can be written

$$e + H_2^+ \rightarrow H_2^{**} \rightarrow H(1s) + H(nl). \tag{I}$$

---

*A normal adiabatic or stationary state is derived from the diagonalization of the electronic Hamiltonian for the molecule. A diabatic state is derived from the diagonalization of the entire molecular Hamiltonian (i.e., the combination of the electronic and nuclear parts). However, the electronic Hamiltonian is not itself diagonal in this case. A diabatic state can be thought of as a scattering state (i.e., for the case of H_2 it describes the potential energy of a pair of hydrogen atoms as they approach).

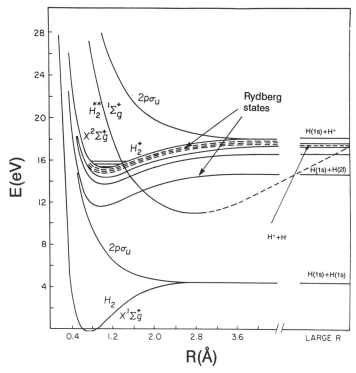

Figure 9-6-1. Potential energy curves for H_2 and H_2^+. The $^1\Sigma_g^+$ state is diabatic in form.

The intermediate state in this case actually tends toward the limit of $H^- + H^+$ at large R, but experiments have shown that this is a minor exit channel for the recombination of H_2^+ (Peart and Dolder, 1975). The dissociating system must therefore undergo a secondary transition to a lower Rydberg level, terminating in an atom pair, $H(1s) + H(nl)$.

Because of the rapidity of the dissociation process compared to auto-ionization, DR exhibits a fast rate in this instance. If the curve crossing is less optimally placed (i.e., if the overlap between the vibrational wavefunction of the ion and that of the resonance is poor), the rate for direct recombination will be lower, though not necessarily negligible.

A.   Dissociative Recombination Mechanisms

Having identified the states involved in the recombination process, one can, in principle, determine the cross section by calculating the probability of a transition from the electron–ion continuum to an atom–atom dissociation continuum. In practice, this involves calculating the probability of the capture of a free electron into a doubly excited resonance state followed by the determination of the probability of dissociation of this state. The recombination cross section is often represented as

$$\sigma_{DR} = \sigma_c \times S_F, \tag{9-6-1}$$

where S_F, the *survival factor*, represents the dissociation probability having a value less than or equal to unity and σ_c is the capture cross section. The coupling between the initial scattering state, the intermediate resonance, and the final dissociation continuum is complicated by the presence of an infinite number of Rydberg levels of the neutral molecule whose potential energy curves lie beneath that of the molecular ion. Initially we shall ignore these and consider only direct dissociative recombination, which involves only the three states already alluded to.

σ_{DR} is given by

$$\sigma_{DR} = \frac{4\pi^2}{k^2} \, g|T|^2, \tag{9-6-2}$$

where $k = p/\hbar$ is the electron wave number (p being the momentum, $\hbar = h/2\pi$), g is the ratio of the statistical weights between the two states, and T is given by

$$T = \exp(i\rho)\langle\chi_i|V_a|\chi_r\rangle. \tag{9-6-3}$$

Here χ_i is the vibrational wave function of the initial ion, χ_r is that of the resonance state, and V_a is the electronic transition matrix element

$$V_a = \langle\phi_i|H_e|\phi_r\rangle, \tag{9-6-4}$$

H_e being the electronic Hamiltonian, ϕ_i and ϕ_r the electronic wave functions of the initial and resonance states. ρ is the complex phase shift for the resonant state and the square of the $\exp(i\rho)$ term is the survival factor. The resonance width Γ_a is given by

$$\Gamma_a = 2\pi|V_a|^2, \tag{9-6-5}$$

where the autoionization lifetime of the resonance is $\hbar/\Gamma_a$. One can thus express the cross section for DR via the direct mechanism:

$$\sigma = \frac{\Gamma_A}{E} \times FC \times S_F, \tag{9-6-6}$$

where

$$FC = \langle\chi_i|\chi_r\rangle, \tag{9-6-7}$$

the *Franck–Condon* factor, is the overlap integral between the initial and resonance vibrational wave functions and E is the energy of the incoming electron.

It is interesting to note that if the overlap is good, the cross section should vary as E^{-1}. Details of calculations of the parameters in (9-6-6) are given in papers by Bardsley (1968).

There are alternative paths that can play an important role in the recombination process. Resonant capture of an electron can occur into a vibrationally excited Rydberg state which is subsequently predissociated, thus:

$$e + H_2^+ \to H_2^R(v') \to H_2^{**} \to H(1s) + H(nl). \tag{II}$$

This is referred to as *indirect dissociative recombination* and occurs when the electron has exactly the correct energy to induce a transition involving a change in vibrational quantum number, v. Figure 9-6-2 shows a typical ion and neutral Rydberg state of a diatomic molecule. It can be seen that they are very similar in shape, the Rydberg state being shifted to lower potential energy. Also shown is the manifold of vibrational energy levels for a progression of principal quantum number states of H_2. When first proposed by Bardsley (1967) and by Chen and

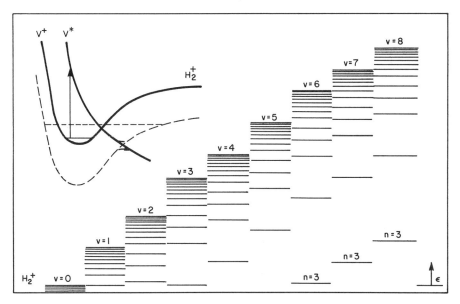

Figure 9-6-2. The inset shows the ground state of H_2^+ (solid curve) and a Rydberg state of H_2 (dashed curve). (The vertical separation is exaggerated for clarity.) The energy-level diagram shows the relative separations of the Rydberg series going to the limits of $v = 0$, $1, 2, \ldots, 8$ of H_2^+. [From Hickman (1989).]

Mittelman (1967), it was believed that the most likely transition would involve a change of v of one quantum (i.e., $\Delta v = +1$). This is because the vibronic coupling between the electron and the nucleus is very weak (i.e., it is difficult for the light electron to induce the massive nuclei to move). The differences in nuclear and electron motion are embodied in the *Born–Oppenheimer approximation*. Usually, electron transitions are so rapid that the nuclei are essentially frozen in place during the transition. Hence electronic transitions are essentially independent of transitions involving a change in nuclear motion, in this approximation.

A number of theoretical calculations, based mainly on *multichannel quantum defect theory* (MQDT), have been performed, in which the direct and indirect processes have been included simultaneously (Giusti, 1977; Giusti et al., 1983; Nakashima et al., 1986; Hickman, 1987). For the case of H_2^+, reaction II is manifested by a series of deep window resonances (Fig. 9-6-3). This is because the electronic coupling responsible for the direct process, reaction I, is much stronger than the vibronic coupling responsible for the indirect process. The Rydberg state is, however, a bound state and therefore the electron's wave function is strongly localized in this configuration at the resonance energy. The result is that the recombination cross section essentially disappears at the resonance.

An alternative reaction path first proposed by O'Malley (1981) has been shown in calculations by Hickman (1989) to strongly influence the recombination process. This process involves large changes in vibrational quantum number. Figure 9-6-4 shows the transition probability, σE versus E for the $v = 0$ and $v = 1$ states of H_2^+. The overlap between the $H_2^+(v = 0)$ and the H_2^{**} state is

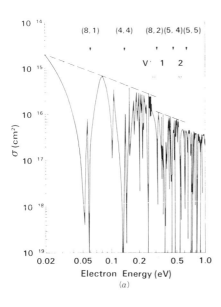

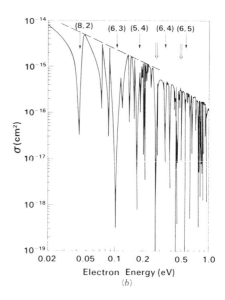

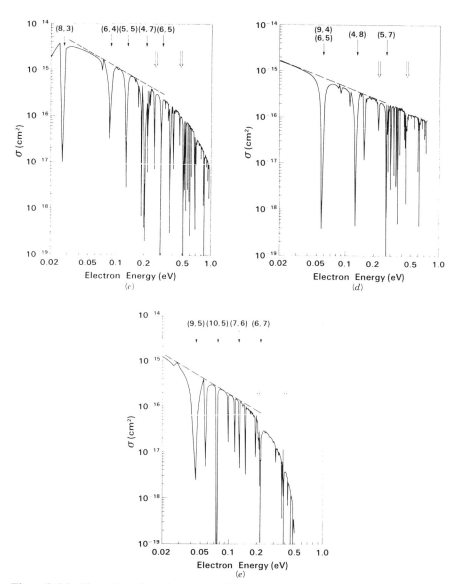

Figure 9-6-3. Theoretical dissociative recombination cross sections for H_2^+ with (a) $v = 0$, (b) $v = 1$, (c) $v = 2$, (d) $v = 3$, and (e) $v = 4$. [From Nakashima et al. (1986).]

low, so the direct process is not as strong in this case. It can be seen that incorporation of low n interactions leads to considerable enhancement of the transition probability. For $H_2^+(v = 1)$, the direct process is much stronger and the influence of the low n states leads to a reduction in the transition probability as it interrupts the dissociation process, providing an autoionization path back to the ion state.

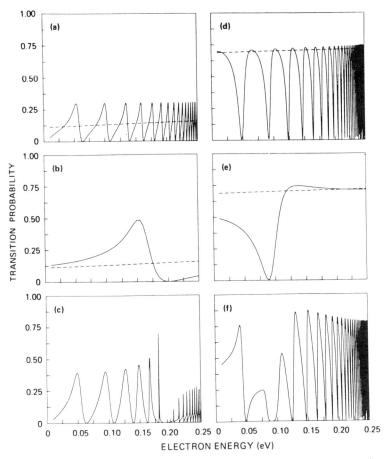

Figure 9-6-4. Dissociative recombination transition probabilities (σE_e) for H_2^+ $(v = 0)$ due to (a) the direct process (dashed line) and the indirect mechanism (solid line) (b) with the incorporation of low-n state interactions and (c) combined results. (d), (e), and (f) show the same for $H_2^+(v = 1)$. [From Hickman (1989).]

Bates (1991a) has addressed the problem of the dissociative recombination of polyatomic ions. It has often been supposed that the larger recombination rates exhibited by such ions, compared to light diatomic ions, is explicable in terms of the indirect process. Given the greater number of vibrational modes available with polyatomic molecules, it might be assumed that many more channels for indirect recombination would be open. In an argument based on the $\Delta v = 1$ propensity rule for autoionization of Berry and Nielsen (1970), Bates has proposed that in fact this is not so and that only a change of one normal mode is allowed. Instead, it is argued that the higher recombination rates are explicable in terms of the number of available bonds that can be broken. Unusually large recombination rates are seen for heavy, diatomic rare gas ions and for cluster ions. In a subsequent paper, Bates (1991b) introduced the concept of *super dissociative recombination* to explain this phenomenon. Analysis of the gradient of the resonance states reveals that for the diatomic ions, the resonance states have very shallow slopes in the vicinity of the crossing region. This means that there is a greatly enhanced overlap between the wave functions of the ion and the neutral resonance state leading to enhanced electron capture probability. The shallowness also means that the dissociation limit of the state coincides with the region where there is a high density of neutral Rydberg states, so that there are a wide variety of final states available, following dissociation. In a cluster ion, the intermolecular bonds are weak, and again the resonance states are found to have shallow slopes.* According to Bates, these factors together can account for the large observed recombination rates, again without resorting to indirect mechanisms and high vibrational mode densities. In the case of H_3^+, however, discussed in Section 9-6-C, the direct mechanism is essentially ruled out and yet the recombination rate is found to be large. The electron capture must proceed via the Rydberg channels and may also involve large changes in Δv.

B. Measurement Techniques for Dissociative Recombination

An examination of Fig. 9-6-1 shows some of the problems associated with measurement of this complex process. First among these is the fact that the process is most important at low electron energies. One way to examine DR is to study the decay characteristics of plasmas and most of the work performed has utilized this approach. Such techniques, however, measure rate coefficients and fine details of the recombination mechanism are washed out. To make comparisons between theory and experiment it is preferable to use the merged beam approach, which allows electron–ion collision cross sections to be measured at very low energies with very high energy resolution.

The second problem is that molecular ions are usually formed with substantial amounts of internal energies associated with vibrational and rotational motions. Occasionally, electronic excitation can also be a factor. In astrophysical applications, the molecular ions are almost always in their ground states, since collision times are long and experiments should therefore be performed on ground-state ions. This means that sufficient time should elapse for deexcitation processes to cool the ions before recombination measurements are performed. In practice, this is not always accomplished, so one should be careful in the interpretation of data obtained from experiments.

A third complicating factor with DR is that many decay channels, producing a variety of neutral species in various states of excitation, may be energetically possible. Very little experimental and theoretical information is currently available concerning branching ratios for different decay channels, although a

---

*In this case, however, the final products must be in their ground states, for the recombination energy of the cluster ion is too low to support excited states.

number of new techniques for approaching this problem have recently been developed.

A problem that plagued early measurements of DR concerned identification of the ion actually undergoing recombination. This is especially serious in afterglows, and proper use of mass spectrometric analysis is vital to the understanding of such experiments. Merged beam experiments use mass-analyzed ion beams, and this problem is not as serious unless there exists the possibility of producing two ions with the same mass. In such experiments, however, the products are identified directly and this greatly aids primary ion identification. For example, if an energy-sensitive detector is used to detect the products, one can distinguish between dissociation processes, leading to different numbers of products or between products with different masses. Hence one can distinguish between a diatomic molecule such as O_2^+ and a polyatomic molecule such as CH_4^+ even though they have the same mass. This feature has proved to be invaluable in ensuring the efficacy of merged beam measurements.

1. *Flowing Afterglow Langmuir Probe Methods*

The flowing afterglow method for the study of ion–molecule reactions was discussed in Chapter 7. This technique has been adapted to the measurement of electron–ion recombination processes by adding a movable Langmuir probe to the apparatus (Smith and Adams, 1983). This probe samples the electron density along the axis of the drift tube. This flowing afterglow Langmuir probe (FALP) apparatus is shown in Fig. 9-6-5.

If the loss rate for ions via recombination is much greater than that due to diffusion, the rate of change of the density of a given ion A^+ with time is given by

$$\frac{dn(A^+)}{dt} = -\alpha n(A^+)n(e). \tag{9-6-8}$$

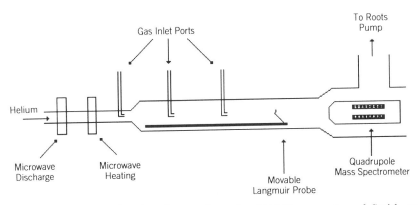

Figure 9-6-5. Flowing afterglow Langmuir probe (FALP) apparatus of Smith and Adams (1983).

If the plasma is dominated by only one ion, it is usual that the ion and electron densities are approximately equal; that is,

$$n(A^+) = n(e) \qquad (9\text{-}6\text{-}9)$$

One can rewrite (9-6-2) in terms of $n(e)$ and the distance x along the flow tube thus:

$$v\frac{dn(e)}{dx} = -\alpha n^2(e). \qquad (9\text{-}6\text{-}10)$$

By integrating this equation with respect to x, one obtains

$$\frac{1}{n_x(e)} = \frac{1}{n_0(e)} + \frac{\alpha}{v}x. \qquad (9\text{-}6\text{-}11)$$

Thus a plot of $1/n(e)$ versus x should yield a straight line with slope α/v, where v is the flow velocity. This can easily be measured by modulating the microwave discharge and measuring the time delay of the discharge pulse arriving at several points along the flow tube using the Langmuir probe.

Great care must be exercised when using an apparatus such as this to ensure that all the gases introduced into the flow tube are extremely pure. Even small quantities of an easily ionized species can lead to large impurity ion concentrations. Since the electron density is typically six or seven orders of magnitude less than the neutral gas concentration in the tube, impurity densities in concentrations of around 0.1% can lead to strong ion loss competition due to ion–molecule reactions. Mass spectroscopic analysis of the plasma is essential to ensure that such impurity levels are negligible.

2. The Rennes Afterglow Apparatus

A new type of flowing afterglow apparatus has recently been constructed at the University of Rennes in France (Rowe et al., 1992). This is shown in Fig. 9-6-6 and features both a movable Langmuir probe and a movable mass spectrometer. One can write a more complete description for ion loss taking account of ion-impurity molecule reactions thus:

$$v\frac{dn(A^+)}{dx} = -\alpha n(A^+)n(e) - kn(A^+)n(I), \qquad (9\text{-}6\text{-}12)$$

Integrating, one obtains

$$\log\frac{n_x(A^+)}{n_0(A^+)} = -\frac{\alpha}{v}\int n(e)\,dx + kn(I)\,\frac{x}{v}. \qquad (9\text{-}6\text{-}13)$$

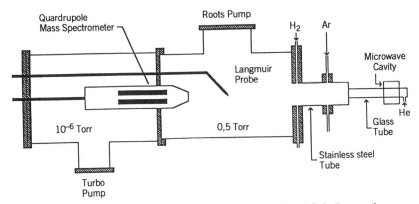

Figure 9-6-6. New flowing afterglow apparatus at the Université de Rennes incorporating a movable mass spectrometer in addition to a movable Langmuir probe. [From Rowe et al. (1992).]

Here $n_x(A^+)$ and $n_0(A^+)$ are the ion densities measured at a distance x along the flow tube and at some chosen origin downstream from the gas entry port. There are two ways in which to measure the recombination rate. One can measure $n_x(A^+)$ and $n(e)$ at various x values and produce a plot of

$$\log \frac{n_x(A^+)}{n_0(A^+)} \text{ versus } \int n(e)\, dx.$$

If $n(I)$ and the diffusion rate are negligible, the plot is linear. The slope of this line then yields α if the plasma flow velocity is known. An alternative approach is to vary the electron density at a given value of x by changing the position and power of the microwave discharge and to plot the same parameters. In this case, however, the plot will necessarily be unaffected by the effects of impurities or diffusion since these are not affected by the microwave power. This is therefore a more reliable method of determining α. Figure 9-6-7a and b show plots for H_3^+ obtained using these two different approaches, and it can be seen that, in this instance, the two methods agree well.

3. Amano's Method

This technique (Amano, 1988, 1990) employs an apparatus such as that shown in Fig. 9-6-8. A plasma is generated by applying a high-voltage pulse to a cooled hollow cathode assembly. The plasma is illuminated by the beam from a frequency-difference infrared laser system, and the rate coefficient for the recombination of a specific ion in a specific ro-vibration state is measured by observing the decay of the infrared absorption signal associated with that state, as a function of time (Fig. 9-6-9). A multipass optical system is used to extend the absorption path length to nearly 30 m. The transmitted laser radiation is

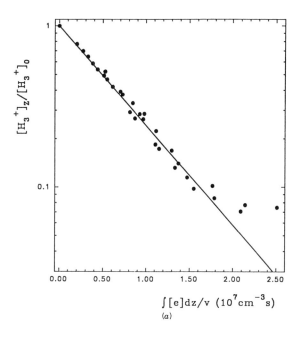

(a)

$\int [e] dz/v \ (10^7 cm^{-3} s)$

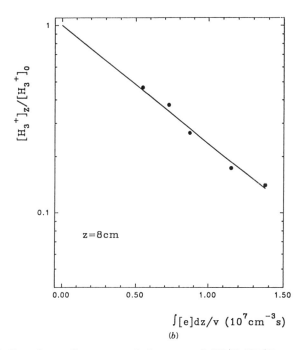

(b)

$\int [e] dz/v \ (10^7 cm^{-3} s)$

Figure 9-6-7. Experimentally measured decrease of $[H_3^+]_z/[H_3^+]_0$ versus $\int [e] \, dz/v$, measured with the Rennes flowing afterglow apparatus. (a) The integral is varied by changing z as well as $[e]$. (b) The decrease measured at a fixed z but with the electron density being varied by changing the microwave cavity conditions. [From Canosa et al. (1992).]

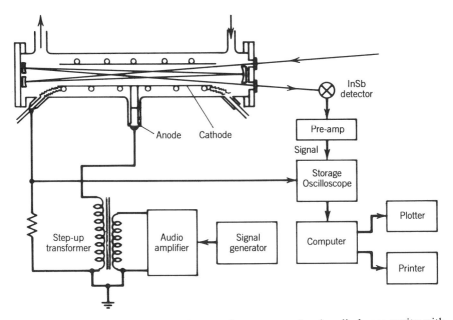

Figure 9-6-8. Schematic diagram of Amano's apparatus showing discharge cavity with the multipass mirror arrangement. [From Amano (1990).]

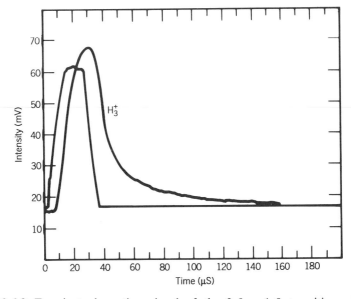

Figure 9-6-9. Transient absorption signal of the $2\,0_1 \leftarrow 1\,0$ transition of the v_2 fundamental band of H_3^+ (channel a) and the discharge current (channel b) recorded at liquid nitrogen temperature. [From Amano (1990).]

detected using a liquid-nitrogen-cooled InSb photovoltaic detector. The temperature of the apparatus can be varied by passing a variety of cryogenic liquids through the cooling coils surrounding the discharge, and Amano has reported measuring rate coefficients over a temperature range from 110 to 300 K. Care is taken to ensure that the dominant decay process is two-body electron–ion recombination and not diffusion, impurity reaction, or three-body recombination.

By filling the apparatus with hydrogen at a pressure of 200 to 600 mtorr, the major ion present is H_3^+. If small amounts (20 to 30 mtorr) of nitrogen or carbon monoxide are added to the discharge, the H_3^+ is rapidly converted to HN_2^+ or HCO^+, respectively, so those ions can be studied. In principle the same types of ions could be studied as in the FALP apparatus, provided that the available laser system can be tuned to the absorption transitions appropriate to those ions and provided that these transitions can be observed unambiguously. The method has the obvious advantage that the ro-vibrational state of the ions is a well-known quantity, so problems associated with vibrationally excited ions and impurities can be avoided. There have, however, been a number of concerns raised about the possible influence of *collisional radiative recombination* and about the electron temperatures in the discharge. Despite these concerns, the method shows great promise and will undoubtedly see extended usage.

4. *Merged Beams Technique*

The principle behind this technique is to form a beam of electrons and a beam of ions and to merge them so that the relative velocities of the particles in the two beams can be made very small. The merged beams apparatus at the University of Western Ontario in Canada (Auerbach et al., 1977) is shown in Fig. 9-6-10. In this experiment the ions are produced in the terminal of a 400-keV van de Graaff accelerator, mass analyzed, and injected into the ultrahigh-vacuum interaction chamber. Inside this chamber the beam is deflected twice so that neutrals produced by collisions in the injection line are removed, whereupon the beam enters the collision region. A magnetically confined electron beam is produced inside the interaction chamber from an indirectly heated barium oxide cathode, and this beam is made to merge with the ion beam by means of a trochoidal analyzer. In this device a transverse electric field causes the electrons to spiral about the axial magnetic field lines. The magnetic field is tuned so that the electrons will have completed a spiral just as they leave the electric field. At this point they stop spiraling and proceed along a straight-line path that is shifted laterally from their original path. It is arranged that this new path coincides with that of the ion beam so that the two beams are now merged.

The profiles of the ion and electron beams are measured at several places along the interaction length, using movable scanners, in order to determine the effective collision area and after a distance of 8.6 cm, the electron beam is demerged from the ion beam using a second trochoidal analyzer. The electrons are collected in a Faraday cup and their current is measured. The ion beam is

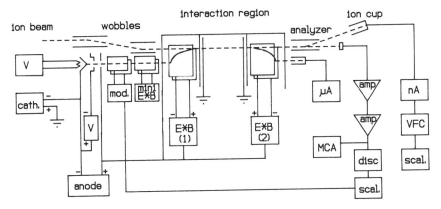

Figure 9-6-10. Schematic diagram of the merged electron–ion beam experiment (MEIBE) at the University of Western Ontario. The electron beam is produced by the cathode and is merged with and demerged from the ion beam using the $E \times B$ trochoidal analyzers. The ion beam is produced by a 400-keV van de Graaf accelerator. After the interaction region, the ion beam is deflected into a Faraday cup and the neutrals pass on to a surface barrier detector.

deflected using a transverse electric field into a second Faraday cup and its current is measured. The neutrals formed in the interaction region are detected using a surface barrier detector.

The collision cross section is given by

$$\sigma = \frac{c_n e^2}{I_e I_i L} \left| \frac{v_e v_i}{v_i - v_e} \right| F, \qquad (9\text{-}6\text{-}14)$$

where I_i, I_e. v_i, and v_e are the ion and electron currents and velocities, respectively, e the electronic charge, L the length of the interaction region, and F the form factor or effective collision area:

$$F = \left[\iint i_e(x, y)\, dx\, dy \iint i_i(x, y)\, dx\, dy \right]$$

$$\times \left[\iint i_e(x, y) i_i(x, y)\, dx\, dy \right]^{-1}. \qquad (9\text{-}6\text{-}15)$$

The collision energy in the center-of-mass frame of reference is given by

$$E_{\text{cm}} = \tfrac{1}{2}\, \mu v_r^2 = E_+ + E_e - 2(E_+ E_e)^{1/2} \cos \theta, \qquad (9\text{-}6\text{-}16)$$

where μ is the reduced mass:

$$\mu = \frac{m_e m_i}{m_e + m_i} \approx m_e, \tag{9-6-17}$$

m_i and m_e being the ion and electron masses, respectively, and E_+ is the reduced ion energy:

$$E_+ = \frac{m_e}{m_i} E_i. \tag{9-6-18}$$

In a merged beam experiment θ is small, so one can make the approximation

$$\cos \theta \approx 1 - \frac{\theta^2}{2}, \tag{9-6-19}$$

and (9-6-16) can be written

$$E_{cm} \approx (E_+^{1/2} - E_e^{1/2})^2 + (E_e E_+)^{1/2} \theta^2. \tag{9-6-20}$$

Since all experimental parameters are measured and the collision energy is well determined, this apparatus can be used to determine absolute cross sections for electron–ion collision processes in the energy range from a few meV to tens of electron volts. If one differentiates (9-6-20) with respect to E_+, E_e, and θ, one obtains an expression for the uncertainty in the center-of-mass collision energy as a function of the uncertainties in the reduced ion energy, electron energy, and collision angle:

$$\Delta E_{cm} = \left(\left\{ \left[1 - \left(\frac{E_e}{E_+} \right)^{1/2} \right] \Delta E_+ \right\}^2 \right.$$
$$\left. + \left\{ \left[1 - \left(\frac{E_+}{E_e} \right)^{1/2} \right] \Delta E_e \right\}^2 + [2(E_e E_+)^{1/2} \theta \Delta \theta]^2 \right)^{1/2}. \tag{9-6-21}$$

It can be seen from (9-6-21) that when $E_+ \approx E_e$, the contribution of the first two terms becomes negligible and the energy resolution is determined primarily by the uncertainty in the intersection angle. The results shown in Fig. 9-3-1a and b illustrate the extremely high-energy resolution achievable using the merged beam technique.

Neutral product detection in a merged beam experiment is performed using particle detectors such as surface barrier detectors or channel plate multipliers. The advantage of the former is that these detectors are energy sensitive and so in some circumstances can distinguish between different collision channels which give rise to neutral products with differing energies. For example, when using a triatomic ion such as H_3^+, the following reactions can occur in the interaction region:

$$H_3^+ + X \rightarrow H + H_2^+ + X \tag{III}$$

$$\rightarrow H_2 + H^+ + X \tag{IV}$$

$$\rightarrow (H + H) + H + X^+ \tag{V}$$

$$H_3^+ + e \rightarrow H + H + H \tag{VIa}$$

$$\rightarrow H_2 + H. \tag{VIb}$$

Since the products travel with essentially the same velocity as the primary ion beam, the energy carried by individual products is proportional to their mass. Thus the H atom produced by reaction (III) carries one-third of the total beam energy, the H_2 produced by (IV) carries two-thirds of the total beam energy, while the products of reactions (V), (VIa), and (VIb) all arrive simultaneously at the detector and appear as a single full-beam energy particle. The surface barrier detector, being energy sensitive, produces an output pulse with a height proportional to the input energy, so it is possible, using single-channel analyzers, to discriminate between the three cases. This is extremely useful as it improves the signal-to-noise ratio in the signal channel by eliminating neutral particle contributions from unwanted reactions. In the merged beam experiments, signals from reactions such as (III), (IV), and (V) are distinguished from those from reactions such as (VIa) and (VIb), by modulating the electron beam and counting the products, in and out of phase with the modulation.

5. Other Techniques

A number of other methods have been used for the determination of recombination rates or cross sections and these include stationary afterglow techniques (Mehr and Biondi, 1969; Johnsen, 1987), shock tubes (Ogram et al., 1980), flames (Hayhurst and Telford, 1974), electron beam traps (Hasted, 1983), ion traps (Walls and Dunn, 1974), and crossed and inclined beam experiments (Phaneuf et al., 1975; Peart and Dolder, 1974). A review of these techniques contrasting their strengths and weaknesses has been given by Mitchell and McGowan (1983).

6. Determination of Internal Energies

As mentioned earlier, a prime concern in measurements of DR is to determine the internal energy of the ions. In a merged beam system, this can be achieved by examining the energy of the thresholds for dissociative excitation processes, such as

$$e + H_3^+ \rightarrow H + H_2^+ + e. \tag{VII}$$

Typical excitation energies are a few electron volts, and it is simply a matter of increasing the electron energy and looking for the appearance of neutral products arising from the dissociation of the excited molecular ion following electron impact excitation. At these energies the cross section for DR is usually very small and can be neglected. For the particular case illustrated in (VII), the H atom can be identified by using a single-channel analyzer to select only pulses with energies corresponding to one-third beam energy particles. Figure 9-6-11 shows the results of such a measurement. The thresholds are often obscured by Feshbach resonances, but it is usually possible to ensure that one is dealing only with ground-state ions.

The state of excitation of ions in afterglow experiments has usually been assumed rather than measured. Deexcitation rates have been measured for many ions, and these have been incorporated into models to predict the excitation states of the ions used in such experiments. The dangers associated with this have been discussed by Mitchell and Rowe (1992), and it is recommended that in such experiments, attempts be made to use infrared absorption techniques to hunt for hot-band absorptions from excited vibrational levels in an effort to define more properly the internal energies of the ions.

C. Dissociative Recombination Results

1. *Diatomic Ions*

Figure 9-6-12 shows high-resolution experimental results for the dissociative recombination of $H_2^+(v = 0)$ obtained using the merged beam technique. Also shown are MQDT calculated results for the same system. It can be seen that

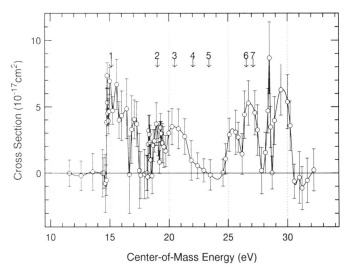

Figure 9-6-11. Dissociative excitation cross section for $H_3^+(v = 0)$. The threshold for this reaction is at 15 eV. [From Yousif et al. (1991b).]

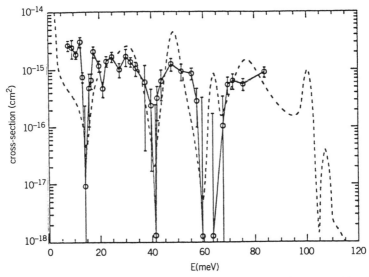

Figure 9-6-12. Experimental cross sections for the dissociative recombination of $H_2^+ (v = 0)$ (Van der Donk et al. 1991). The dashed line shows the results of a multichannel quantum defect theory calculation by Schneider et al. 1991, convoluted with a 2.5-meV triangular apparatus resolution.

there is excellent agreement between the two, indicating the high degree of sophistication of both theory and measurements for this system. High-resolution measurements have been performed for HeH^+ (Fig. 9-6-13), but corresponding theoretical calculations are not available. Detailed calculations have been performed for CH^+ (Takagi et al., 1991), O_2^+ (Guberman and Giusti-Suzor, 1991), and NO^+ (Sun and Nakamura, 1990), but these await high-resolution experimental verification. High-resolution experimental measurements of N_2^+ DR (Noren et al., 1989) displayed a resonance structure similar to that found in recent calculations for N_2^+ (Guberman, 1991).

Low-resolution measurements of cross sections or reaction-rate coefficients do exist for about 75 systems, and these have been tabulated by Mitchell (1990). Typical room-temperature rate coefficients are on the order of a few $\times 10^{-7} \text{cm}^3/\text{s}$ for light diatomics and polyatomics, while more complex species can exhibit rates as high as $10^{-5} \text{cm}^3/\text{s}$. Discussions of many of these measurements have been given in reviews by Mitchell and McGowan (1983), Bardsley and Biondi (1970), McGowan and Mitchell (1984), Mitchell (1990), and Eletskii and Smirnov (1982). Only one system, H_3^+, is discussed in detail here. This has been a source of much controversy, and recent findings have important implications for our understanding of the recombination process.

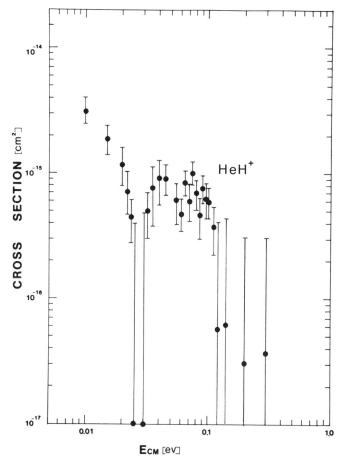

Figure 9-6-13. Measured dissociative recombination cross sections for HeH^+ ions prepared under low-extraction source conditions. [From Yousif and Mitchell (1989).]

2. H_3^+ *Recombination*

Prior to 1983, a number of measurements had yielded rate coefficients for this process which agreed with each other to within a factor of 2 with the accepted value being 2.3×10^{-7} cm^3/s. The techniques used in these experiments included microwave afterglow (Leu et al., 1973; MacDonald et al., 1984), inclined beam (Peart and Dolder, 1974), merged beam (Auerbach et al., 1977; Mitchell et al., 1985) and electron beam trap (Mathur et al., 1978). In 1984, a new measurement by Adams et al. (1984) (Fig. 9-6-14) employing the flowing afterglow Langmuir probe (FALP) method set an upper limit for the recombination rate coefficient of 2×10^{-8} cm^3/s for what was believed to be the ground vibrational state of H_3^+.

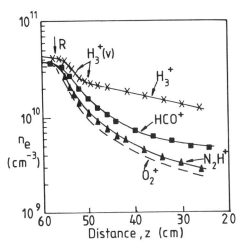

Figure 9-6-14. Decay curves for the ions H_3^+, HCO^+, N_2H^+, and O_2^+ in the FALP apparatus of Smith and Adams. (From Adams et al. (1984).]

Theoretical support for this value came from calculations of the $^2A_1$ resonance state (Michels and Hobbs, 1984; Kulander and Guest, 1975) through which the recombination proceeds. This state is illustrated in Fig. 9-6-15. It can be seen that it intersects the ion state in the vicinity of the $v = 3$ level so that a reduction of the capture cross section is to be expected for the lower v states. Earlier merged beam measurements were known to have employed ions with a considerable population of excited vibrational levels, and this was believed to explain the higher measured cross sections. The FALP measurements were believed to apply to ground-state ions due to the rapid deexcitation of H_3^+ via proton transfer collisions with H_2, thus:

$$H_3^+(v) + H_2 \rightarrow H_3^+(v = 0) + H_2 \tag{VIII}$$

It was decided to repeat the experiment using ions prepared in a radio-frequency trap ion source. Sen and Mitchell (1986) demonstrated that this source was capable of producing H_3^+ ions with low internal energies and measurements of dissociative excitation allowed the internal energy of the ions to be determined (see earlier). The results of this measurement (Hus et al., 1988) for ions produced under a range of conditions are shown in Fig. 9-6-16. It can be seen that it is possible to control the number of vibrational states populated by varying the source pressure and extraction potential. Figure 9-6-17 shows the corresponding dissociative recombination measurements. The lowest cross sections measured in this experiment are about an order of magnitude smaller than previous merged beam results and yield a rate coefficient of about 2×10^{-8} cm^3/s at 300 K (i.e., in agreement with the FALP results).

Following recalibration of their apparatus, however, Smith and Adams

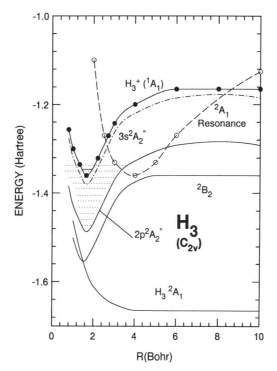

Figure 9-6-15. Potential energy curves for H_3^+ in C_{2v} symmetry. The dashed curve is the resonance through which the direct recombination would proceed if the ions were vibrationally excited. [Adapted from Michels and Hobbs (1984).]

(1987) revised their estimate for the rate coefficient down to $< 10^{-11}$ cm^3/s. They based their reevaluation on results obtained at very low hydrogen gas flows where the dominant ions were He$^+$ and HeH$^+$. They found that they could observe no change in the rate of electron decay with these ions and concluded that

$$\alpha(H_3^+) \approx \alpha(He^+) \approx \alpha(HeH^+)$$

Since He$^+$ is atomic, it can only undergo radiative recombination in a two-body collision for low electron energies and the calculated rate for this process is $< 10^{-11}$ cm^3/s. The absence of a suitable curve crossing for HeH$^+$ was believed to preclude dissociative recombination for this ion, so the estimated rate was also $< 10^{-11}$ cm^3/s. Smith et al. therefore concluded that $\alpha(H_3^+)$ would also have this value. It is evident, however, from Fig. 9-6-13 that the sensitivity of the FALP method is insufficient to really justify such a statement. Yousif and Mitchell (1989) used the merged beams technique to measure the recombination of HeH$^+$ and found that it actually proceeds with a rate of about 1×10^{-8} cm^3/s

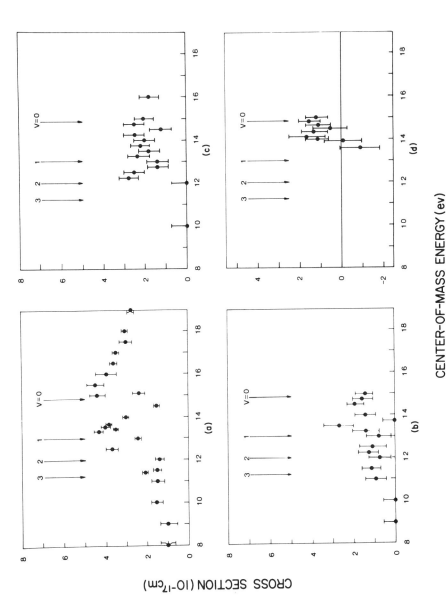

Figure 9-6-16. Dissociative excitation cross sections for H_3^+ in collision with electrons: (a) ions produced in a conventional radio-frequency source; (b) ions produced in a trap source at 10 mtorr, pure H_2; (c) trap ion source, 30 mtorr, pure H_2; (d) trap ion source, low extraction 10:2:1 $H_2 \cdot Ar \cdot He$ mixture. [From Hus et al. (1988).]

CENTER-OF-MASS ENERGY (ev)

CROSS SECTION (10^{-17}cm)

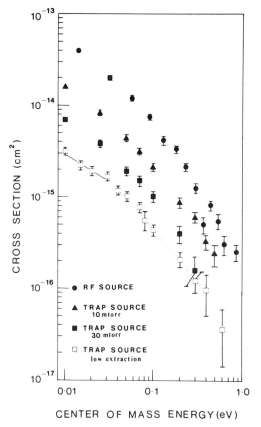

Figure 9-6-17. Cross sections for $e-H_3^+$ dissociative recombination. The four sets of results shown were measured using ions formed under conditions that yielded the corresponding four excitation functions shown in Fig. 9-6-16. [From Hus et al. (1988).]

at 300 K, so this throws the calibration argument used by Smith and Adams into doubt.

There have been a number of recent developments in this area which shed new light on H_3^+ recombination. First, Amano (1988, 1990), using the laser absorption technique, found a rate coefficient of 2×10^{-7} cm^3/s! A similar result has also been found by Canosa et al. (1992) using the new flowing afterglow apparatus (described earlier) at the University of Rennes in France. Both these measurements refer to ground-state ions, but it was pointed out by Mitchell that Adams et al. had assumed too high a rate for reaction (VIII), so in fact their data referred to ions with at least up to $v = 3$ populated. An overlooked measurement by Blakely et al. (1977) had shown that in fact the rate coefficient for this reaction, for low vibrational states, was between one and two orders of magnitude lower than previously believed. At the pressures and flow rates used in the Adams et al. experiment, there would not have been sufficient time for reaction (VIII) to depopulate excited ions in the discharge.

Given the veracity of the results of Amano and Rowe et al., one is left with the need to provide an explanation for the large reaction rate in light of the unfavorable position of the curve crossing. Mitchell and Rowe (1991) have proposed that the recombination could proceed via transitions from the ion state to low-n, high-v Rydberg states of H_3 which are subsequently predissociated by the ground state (see Fig. 9-6-15). Experimental evidence for such transitions has recently been presented by Bordas et al. (1991), who have found that certain high Rydberg states, H_3^*, laser excited from a lower, metastable H_3^* state, are predissociated. This can occur only through transitions involving large changes in vibrational quantum number to a lower state. Two merged beam studies (Mitchell and Yousif, 1989; Yousif et al., 1993), using quite different detection methods, have shown that long-lived (10^{-7} s) H_3^* molecules are formed during the recombination process. It may be that some of these molecules are field ionized by the electric fields employed in the merged beam technique and that this might explain the low cross sections for $H_3^+(v = 0)$ measured by Hus et al. (1988). Alternatively, the electric fields in the apparatus might influence the transitions between the various Rydberg states during the passage toward stabilization leading to a diminishing of the measured recombination cross section (Helm, 1992).

This strange saga of H_3^+ has shown us that the mechanism for recombination is not as strightforward as previously believed. It is a subject ripe for theoretical analysis and it is hoped that the next few years will see a complete calculation of H_3^+ recombination. Not until we can successfully predict the recombination of a relatively simple molecule such as H_3^+ can we hope to fully understand the behavior of the more complex polyatomic species that play so important a role in both terrestrial and interstellar plasma chemistry.

3.   *Branching Ratios for Dissociative Recombination*

As mentioned earlier, when a polyatomic molecule undergoes DR there may be several different final product channels, and information concerning which channels are favored is of great importance to interstellar chemistry (Millar et al., 1988). A number of recent experimental studies (Mitchell et al., 1983; Mitchell and Yousif 1989; Adams et al. 1989, 1991; Herd et al., 1990; Turner, 1989) have shed some light on this subject, and this has been complemented by theoretical studies based on statistical, phase space methods (Herbst, 1985; Galloway and Herbst, 1991) and upon considerations of curve crossings and molecular bonding (Bates 1991c).

The experiments of Mitchell et al. were based on a technique that can separate different final channels based on the numbers of particles produced. This is a difficult method and has only been applied to H_3^+, although in principle it is applicable to a wide range of other species (Forand et al., 1985). The afterglow measurements are more direct and use techniques such as *laser-*

induced fluorescence (LIF) and VUV absorption spectroscopy to determine the concentrations of OH radicals and H atoms, respectively. The details of these techniques and the calibration procedures that must be applied are reviewed in the paper by Adams et al. (1991).

For the case of H_3O^+ recombination, the following exit channels must be considered:

$$e + H_3O^+ \rightarrow H_2O + H + 6.4\,eV \tag{a}$$
$$\rightarrow OH + H_2 + 5.8\,eV \tag{b}$$
$$\rightarrow OH + H + H + 1.3\,eV \tag{c}$$
$$\rightarrow O + H_2^* + H - 1.6\,eV. \tag{d}$$

Adams et al. (1991) found that, on average, 1 H atom and 0.65 OH radical were produced per recombination. This is usually expressed

$$f\left(\frac{H_3O^+}{OH}\right) = 0.65 \qquad f\left(\frac{H_3O^+}{H}\right) = 1.$$

Channel (d) can be eliminated since it is exothermic. This means that

$$R((b) + R(c) = 0.65$$
$$R(a) + 2R(c) = 1.0$$

so

$$R(a) = 0.35$$
$$R(b) = 0.32$$
$$R(c) = 0.33$$

where $R(a)$ is the branching ratio to channel (a), and so on. Bates (1991c) has analyzed these results and has argued that they can be explained in the following terms. Channel (a) proceeds via a radiationless transition in which the electron is captured into an orbital of the O atom forming a lone pair which then repels the H atom. Channel (a) then proceeds via a unique repulsive potential energy curve. The chance of such a curve making a favorable crossing with the ground state of the ion is not large, so the probability of the system proceeding via this channel is small (0.35). An alternative electron capture process involves the conversion of a bonding orbital in the ion being converted to an antibonding orbital. Channels (b) and (c) represent such transitions with the H_2 molecule being formed in a bound singlet state in channel (b) and in an unbound triplet state in channel (c). There is little difference between these two transitions, so they have similar probabilities of occurring. The force separating the products of these two channels is not large and this means that the gradients of the potential energy curves for these channels are low. As mentioned earlier, this implies a favorable overlap between the wave functions of the ion state and those of the dissociating states, so the recombination rates are likely to be rapid.

Adams et al. have measured branching ratios for a number of other polyatomic molecules, including N_2H^+, HCO^+, HCO_2^+, N_2OH^+, $OCSH^+$, H_2CN^+, H_3S^+, NH_4^+, and CH_5^+, and again Bates (1991c) has analyzed these measurements and provided a qualitative description of the processes. Although Bates's analysis is instructive, it cannot be considered to be predictive at this stage. The phase-space approach of Herbst is also unable to confidently predict experimentally determined branching ratios (Galloway and Herbst, 1991). On the other hand, the accuracy of the measurements is only in the range 20 to 30%, and this can allow for large swings in the branching ratios. There is, therefore, a strong need for more accurate experimental work covering many more polyatomic species before a clear understanding of branching ratios is possible. Coupled with this is a need to more fully understand the mechanisms through which dissociative recombination proceeds.

PART $\mathscr{B}$.   ION–ION RECOMBINATION

The formation of negative ions was discussed in Chapter 6. It was seen that the rate for electron attachment to electronegative species could be very high, especially in low-temperature plasmas. In such plasmas, negative ion–positive ion recombination can occur and can compete effectively with electron–ion recombination processes. Reviews of ion–ion recombination processes have been given by Mahan (1973), Moseley et al. (1975), Flannery (1976, 1982a, 1990), Smith and Adams (1983), Brouillard (1984), and Dolder and Peart (1985).

9-7.   MUTUAL NEUTRALIZATION

At low gas densities, ion–ion recombination is dominated by the process of mutual neutralization, thus:

$$A^+ + B^- \rightarrow A^* + B^*,$$

where * denotes excited states of the products. The internal energy that is released by the transfer of the electron from the negative ion to the positive ion is converted into kinetic and potential energy of the products. Bates and Massey (1943) first examined this process and showed that this conversion of energy could be accomplished via a *pseudocrossing* of the initial and final potential energy surfaces, as illustrated in Fig. 9-7-1a.

The reactants A^+ and B^- approach each other at a finite speed v along the

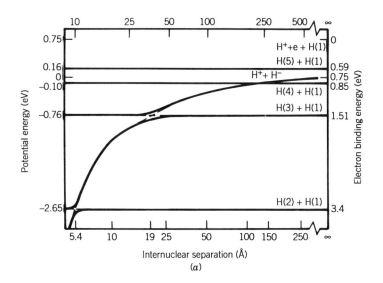

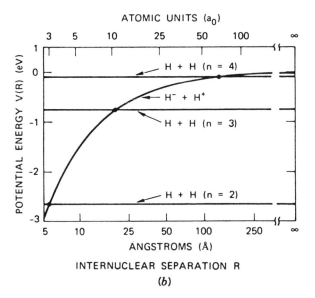

Figure 9-7-1. (a) Adiabatic and (b) diabatic potential energy curves for ionic $H^+ + H^-$ and covalent $H(1s) + H(nl)$ systems. [From Flannery (1976).]

593

potential energy curve as shown. In the vicinity of the pseudocrossing at R_x there is a probability P_{if} that the system will make a transition to the state AB, whereupon the electron transfer occurs and the final products recede from each other toward the A* + B* asymptote. The probability for no transition to occur is $(1 - P_{tf})$. As the ions recede there is the probability of a transition back to the original $A^+ + B^-$ state, so the overall probability for mutual neutralization is given by

$$P = 2P_{if}(1 - P_{if}). \qquad (9\text{-}7\text{-}1)$$

Crossings between ionic and covalent states take place at fairly large separations $R_x \sim 10$ to $50\,\text{Å}$, and if the strong Coulomb attraction between the reactants is taken into account, large cross sections and thus large recombination rates

$$\alpha = P(C\pi R_x^2)v \qquad (9\text{-}7\text{-}2)$$

occur. C is called the electrostatic focusing factor and has a value of ~ 25. At room temperature, v at $R_x \sim 20\,\text{Å}$ is 10^5 cm/s so that

$$\alpha = 3 \times 10^{-7}P. \qquad (9\text{-}7\text{-}3)$$

A simple approximation for P is given by the Landau–Zener formula (Landau, 1932; Zener, 1932; Stuckleberg, 1932). This approximation makes the assumption that the two potential energy curves shown in the figure are in fact linear combinations of diabatic states that cross at R_x (Fig. 9-7-1b). These states are taken to be linear in form in the vicinity of R_x, and the potential that couples the two states is taken to be a constant.

At large ion–ion separations, the diabatic potential for the ion-pair state can be approximated by

$$V_i(R) = -\frac{e^2}{R}, \qquad (9\text{-}7\text{-}4)$$

with zero potential taken as the ion–ion asymptote at infinite separation. The potential intersects the A* + B* potential at R_x. Since the neutral–neutral state is essentially flat at large R, its energy at R_x is

$$V_f(R) = -[I(A^*) - EA(B^*)], \qquad (9\text{-}7\text{-}5)$$

where $I(A^*)$ is the ionization potential of A* and $EA(B^*)$ is the electron affinity of B*. The crossing point is then given by

$$R_x = \frac{e^2}{\Delta E}, \qquad (9\text{-}7\text{-}6)$$

where ΔE is the energy converted to translational motion:

$$\Delta E = [I(A^*) - EA(B^*)].\qquad(9\text{-}7\text{-}7)$$

The Landau–Zener probability for transition at a single crossing is

$$P_{ab}(R_x) = \exp\left(-\frac{\eta}{v_x}\right),\qquad(9\text{-}7\text{-}8)$$

where

$$\eta = \frac{\pi}{2\hbar}\frac{|\Delta U(R_x)|^2}{\left[\dfrac{d}{dR}(V_i - V_f)\right]_{R=R_x}} \approx \frac{\pi}{2\hbar e^2} R_x^2 |\Delta U(R_x)|^2.\qquad(9\text{-}7\text{-}9)$$

The $\Delta U(R_x)$ term is the energy difference between the adiabatic potential curves for $A^+ + B^-$ and $A^* + B^*$ at R_x.

The speed of approach of the colliding ions v_x at R_x is given by

$$v_x = \left[1 - \frac{V_i(R_x)}{E} - \frac{\rho^2}{R_x^2}\right]^{1/2} v_i,\qquad(9\text{-}7\text{-}10)$$

where E and v_i are the incident kinetic energy and velocity and ρ is the impact parameter. $\rho_{\max}$ is given by

$$\rho_{\max} = \left[1 - \frac{V_i(R_x)}{E}\right]^{1/2} R_x \approx \left(1 + \frac{\Delta E}{E}\right)^{1/2} R_x.\qquad(9\text{-}7\text{-}11)$$

(The term in brackets is the Coulomb focusing factor.) The cross section for mutual neutralization is obtained from

$$\sigma = 4\pi \int_0^{\rho_{\max}} P_{if}(1 - P_{if})\rho \, d\rho,\qquad(9\text{-}7\text{-}12)$$

which decreases as E^{-1} at low E and as $E^{-1/2}$ at high E. In fact, this is incorrect, for the cross section should decrease as E^{-1} at large E. The corresponding rate coefficient, assuming a Maxwellian distribution of ion speeds, is

$$\alpha = 4\pi \left(\frac{\mu}{2mkT}\right)^{3/2} \int_0^\infty \sigma \exp\left(-\frac{\frac{1}{2}\mu v_i^2}{kT}\right) v_i^3 \, dv_i.\qquad(9\text{-}7\text{-}13)$$

To calculate σ and α one must determine V_i, V_f, R_x, and ΔU. The first three terms can be obtained from (9-7-4) through (9-7-6).

Bates and Lewis (1955) and Bates and Boyd (1956) used a variational approach for determining ΔU coupled with the L–Z approximation to calculate cross sections for O^+-O^- and $H^--(H^+, Li^+, Na^+,$ and $K^+)$ collisions. They found values of the order of 10^{-12} cm^2, corresponding to rate coefficients of the order of 10^{-7} cm^3/s. Discussions of the limitations of Landau–Zener theory and of more recent calculations have been given by Moseley et al. (1975), Janev (1976), and Sidis et al. (1983). The results of these calculations for $H^+ + H^-$ are discussed later.

When either or both of the colliding ions is molecular, the situation is more complex for now there is a myriad of curve crossings. It is not reasonable to consider treating the contribution from each one individually, and Olson (1972) developed an absorbing sphere model to deal with this situation. In this model it is considered that when the two ions approach within a specified radius, the recombination will proceed with unit probability. The details of how this radius is determined are given in the reference mentioned above and summarized in Moseley et al. (1975). The cross section is given by

$$\sigma = \left(1 + \frac{e^2}{R_0 E}\right) \pi R_0^2. \tag{9-7-14}$$

It is found that the cross section has a very weak dependence on the masses of the colliding ions, so that one obtains almost a universal neutralization curve. This is illustrated in Fig. 9-7-2, which shows experimental data for a number of ion combinations together with the results of an absorbing sphere calculation. Given the simplicity of the argument, the latter approach is seen to give reasonable agreement with experiment. The absorbing sphere model also

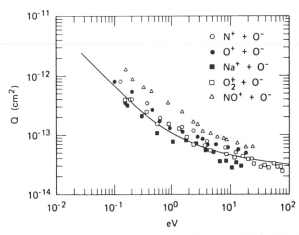

Figure 9-7-2. Cross sections for the mutual neutralization of N^+, O^+, Na^+, O_2^+, and NO^+ with O^-. The solid line is a universal curve derived from an absorbing sphere model. [From Olson (1972).]

predicts that the cross section should vary as $T_e^{-1/2}$ and that it should be larger for negative ions with small vertical detachment energies. This model leads to the prediction that an enhanced cross section will result when the negative ion is in an excited state.

A. Experimental Methods

Mutual neutralization is a relatively low energy collision process, so experiments must either be performed in a plasma or using low collision energy intersecting beam techniques. At the present time, only one group is currently studying ion–ion mutual neutralization. This is the group of Brouillard at the Université de Louvain in Belgium. In the late 1960s and early 1970s, Peterson's group at SRI International in California used the merged beam approach to study several mutual neutralization reactions, and this work was complemented by the inclined beam experiments of Harrison et al. at the Culham Laboratory in the United Kingdom and later by Dolder and coworkers. In the late 1970s and early 1980s, Smith and Adams, at the University of Birmingham in the United Kingdom employed the flowing afterglow Langmuir probe (FALP) technique to study both two- and three-body recombination between molecular species. In this section we discuss Brouillard's apparatus and give a brief description of how the FALP technique, which has been discussed in more detail earlier in this chapter, can be used for ion–ion studies. Descriptions of the earlier experimental techniques may be found in Moseley et al. (1975), Rundel et al. (1969), and Dolder and Peart (1985).

1. *Brouillard's Apparatus*

This is a merged beam apparatus and is illustrated schematically in Fig. 9-7-3. The positive- and negative-ion beams are formed by duoplasmotron ion

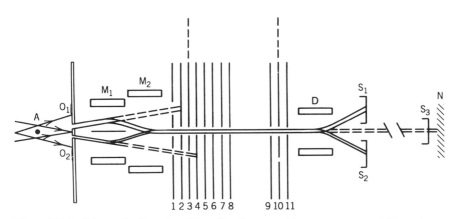

Figure 9-7-3. Schematic diagram of Brouillard's merged beams apparatus. [From Szucs et al. (1984).]

sources, and after acceleration to a few keV and mass analysis, are directed toward the entrance aperture of the interaction region. Each beam initially makes an angle of 4.5° with the axis of the interaction region as shown. They are merged by means of a pair of electrostatic deflector plates, M_1 and M_2, and interact over a distance of 10 cm before being separated by a third deflector, D. The ion beams are collected in Faraday cups while the neutrals pass undeflected to a multiplier. There are apertures in plates 1, 2, and 3 that are used for alignment purposes. The overlap of the two beams is checked by displacing plates 3 and 10 using micrometer devices and tracing out the profiles of the beams. Mutual neutralization gives rise to a pair of time correlated neutrals, and this fact allows true signals to be distinguished from neutrals produced via collisions with the background gas which produce uncorrelated neutrals. The apparatus can measure collisions over the energy range from about 5 to 1000 eV. The energy is varied by changing the potential on plates 2 to 10, plates 1 and 11 being maintained at earth potential.

The Louvain group also operates other similar merged beam apparatus. Experiments have been performed on *associative ionization* processes such as:

$$H(1s) + H(nl) \;\rightarrow\; H_2^+ + e \qquad \text{(Poulaert et al., 1978; Urbain et al., 1990)}$$
$$\rightarrow\; H^+ + H^- \qquad \text{(Claeys et al., 1977; Fussen et al., 1982)}$$

as well as charge-changing processes:

$$H^+ + H^- \rightarrow H^+ + H + e \qquad \text{(Fussen and Claeys, 1984)}$$
$$\rightarrow H^- + H^+ \qquad \text{(Brouillard et al., 1979)}$$
$$\rightarrow H + H^+ + e. \qquad \text{(Schon et al., 1987)}$$

(The latter reaction was also studied at higher energies by Salzborn's crossed beam group at the University of Giessen.)

2. *FALP Studies*

Application of the FALP technique for the measurement of electron–ion recombination rate coefficients was discussed in Section 9-5. This technique can also be adapted for studying ion–ion recombination processes. To understand how this is done, one must remember that in the discharge, electrons and positive and negative ions are very much minority species. Fast reactions such as electron attachment to electronegative gases will occur much more rapidly than electron–ion recombination. Thus it is possible, by adding an electronegative gas such as NO_2, Cl_2, or SF_6, to convert the discharge from being electron dominated to being negative-ion dominated, through attachment processes. The decay of the discharge then proceeds via positive ion–negative ion recombination. As in the case of electron–ion recombination, a Langmuir probe measures the charge density, the electron mass being replaced by the ion mass in the relation between probe current and charge number density. Because of the chemical complexity of the discharge, a mass spectrometer is essential to determine the exact identity of the ions under investigation.

Smith and Adams (1983) have presented a comprehensive review of the application of the FALP method to studies of ion–ion recombination. They have found that the rate coefficients for ion–ion mutual neutralization processes, which can lead to many product states, vary little, falling between 5 and 10×10^{-8} cm^3/s. For the case of $NO^+ + NO_2^-$ and $NO^+ + NO_3^-$, the only reactions to be measured both by the FALP and the merged beam methods, Smith and Adams' values of 6.4×10^{-8} and 5.7×10^{-8} cm^3/s are, however, about an order of magnitude smaller than those of Peterson et al. (1971) and Aberth et al. (1971) measured with the latter technique (51×10^{-8} and 81×10^{-8} cm^3/s, respectively). This discrepancy has been attributed to the presence of excited ions in the merged beam experiments. Stationary afterglow studies (Eisner and Hirsh, 1971) of the same reactions found corresponding values of 17.5×10^{-8} and 3.4×10^{-8} cm^3/s. These reactions are not the only ones to generate controversy. Indeed, the most striking debate has concerned the case of H^+/H^- recombination, discussed below.

3. H^+/H^- *Mutual Neutralization*

In the mid-1980s, a remarkable controversy arose over the energy dependence for the cross section for $H^+ - H^-$ mutual neutralization. This had been the subject of four separate investigations by three separate groups: Moseley et al. (1970), Rundel et al. (1969), Gaily and Harrison (1970), and Peart et al. (1976). The data from these studies are shown in Fig. 9-7-4. The most striking feature of this cross section is the cusplike structure at ~ 140 eV. Also shown are theoretical calculations by Bates and Lewis (1955), Dalgarno et al. (1971), and Olson et al. (1970). These theories indicated that the main contribution to the cross section came from the interaction between the ion-pair state and the $H(1s) + H(n = 2)$ state except for energies < 30 eV, where the $H(1s) + H(n = 3)$ state begins to contribute.

Clearly, agreement between the experimental data and theory is poor above 20 eV. Because of this situation, Szucs et al. (1984) set out to repeat the experiment using a merged beam approach. The results that they obtained are shown in Fig. 9-7-5 together with a selection of the earlier results and an ab initio semiclassical theoretical calculation by Sidis et al. (1983). These authors showed that crossings between the ion-pair state and the $H(n = 3) + H(1s)$ state substantially augmented the neutralization cross section, as did interactions far from the intersection points.

There is remarkably good agreement between the new experimental results and this theory, although the results show no evidence of the cusp seen by other workers. This puzzling situation prompted Peart et al. (1985) to repeat their inclined beams measurement, and remarkably, they were unable to see the cusp structure despite repeated attempts. In fact, their data now agree well with those

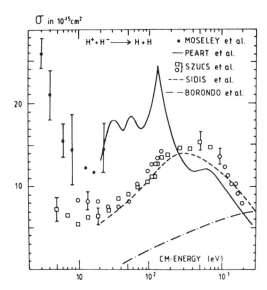

Figure 9-7-4. Cross sections for the mutual neutralization of $H^+ + H^-$ showing the early results of Moseley et al. (1970) and Peart et al. (1976) together with the later results of Szucs et al. (1984). Also shown is a theoretical calculation by Sidis et al. (1983).

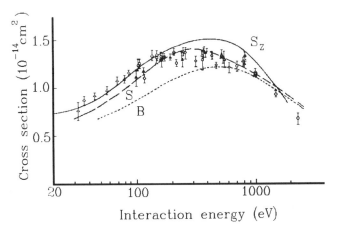

Figure 9-7-5. The triangles, squares, and circles show measurements of Peart et al. (1985) for the mutual neutralization of $H^+ + H^-$ ions, obtained by counting H atoms formed from the H^+ beam. The solid curve Sz denotes the results of the coincidence measurement by Szucs et al. 1984 whilst the dashed curve S is a theoretical calculation by Sidis et al. (1983).

of Szucs et al. What is especially remarkable about this whole situation is that measurements of He^+/H^- mutual neutralization performed by all the parties mentioned here have consistently agreed with each other. It is tempting to speculate about the reason for this dilemma. Effects concerning the excitation states of the reactants can be neglected immediately, for H^- exists in its ground state only. The structure does occur at the high- and low-energy limits of the merged beam and inclined beam techniques, respectively, and perhaps the signal quality was suspect in these cases. If this was the case, however, it is surprising that similar problems did not arise in the He^+/H^- experiments. An alternative suggestion is that the mutual neutralization process is more complex than expected and that rather than being a direct transition from initial to final state, it might pass through high Rydberg intermediate states. Such states would be susceptible to destruction in electric fields and the strength of such fields could vary widely from one experiment to another.

9-8.   TERMOLECULAR ION–ION RECOMBINATION

At pressures in excess of a few torr, mutual neutralization becomes overshadowed by *termolecular recombination* processes of the form

$$A^+ + B^- + X \rightarrow [AB]^* + X,$$

where X is a gas molecule and the resulting product AB may or may not be dissociated. This reaction proceeds via nonresonant scattering, the third body removing energy gained by A^+ and B^- by acceleration in the Coulomb field so that the highly excited bound levels of AB^*, so formed, are cascaded down to low-lying stable levels. If one compares the kinetic energy gained by the ion pair due to Coulombic attraction, to that due to thermal energy alone, one finds that for room temperature they become equal at $R \sim 370$ Å. The curve crossings necessary for mutual neutralization to proceed occur at internuclear separations of 10 to 50 Å. When sufficient numbers of third bodies are available, however, to remove energy continuously as the ions approach, one can have the situation where the two ions come very close together while having a low relative kinetic energy. This situation is conducive to molecule formation and electron transfer (in addition to mutual neutralization), so the recombination is stabilized. In essence, then, at room temperature, termolecular recombination can sample ion pairs that approach within ~ 370 Å, where the ions begin to attract and associate with each other, while the mutual neutralization process can only sample those ions that, through their field-induced motion, come within 10 to 50 Å of each other. The termolecular recombination process is able therefore to exhibit a much larger rate coefficient than mutual neutralization, and for low densities this is expected to increase linearly with number density, N.

Like collisional radiative recombination discussed in Section 9-5, termolecular ion–ion recombination proceeds through a mechanism in which the energy of the recombining system moves up and down an energy ladder. Again one can define an energy bottleneck below which the recombination is stabilized. In this case the bottleneck occurs within a few kT of the dissociation limit, where T is the gas temperature (see Fig. 9-8-1). Figure 9-8-2 illustrates how the dissociation probability decreases and the association probability increases as the potential energy of the system moves down below the dissociation limit.

The recombination is given by the expression

$$\alpha = \int_{-E}^{\infty} dE_i \int_{-D}^{-E} n_i v_{if} P_f^s \, dE_f, \tag{9-8-1}$$

where v_{if} is the collision frequency to move from state i to state f, n_i the number of associated ion pairs in state i, and P_f^s is the probability that level f is stabilized. $-D$ is the energy of the lowest state to which the system proceeds.

Two approaches have been taken to evaluate this expression. Bates and Mendas (1982) have used a Monte Carlo method while Flannery (1982a) has set out to develop a microscopic theory. Recently, Flannery (1987) has applied the variational principle to the problem. This approach is based on the assumption that the association rate is a minimum at any given time and so involves varying P_i so as to obtain a minimum α.

If the gas density N is increased, the recombination rate will rise initially with N but will eventually saturate as shown in Fig. 9-8-3. On the other hand, as the

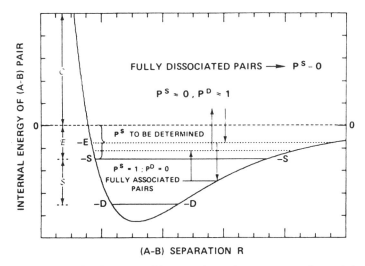

Figure 9-8-1. Schematic diagram to illustrate the various energy regions of significance to termolecular ion–ion recombination. $-D$, lowest energy level of the recombined system; $-S$, energy below which recombination is stabilized; $-E$, variable energy of the ion pair during its passage along the energy ladder toward stabilization; P^s, probability of the system being stabilized; P^D, probability of the system dissociating. [From Flannery (1990).]

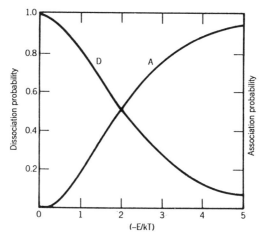

Figure 9-8-2. Plot of the stabilization and dissociation probabilities as a function of $-E/kT$, where $-E$ is the potential energy of the ion pair.

density increases it becomes more and more difficult for the ions to approach each other. The transport rate therefore decreases as $1/N$. Eventually, the recombination rate will be limited by this transport rate and the recombination will begin to decrease, ultimately as $1/N$. The overall form of the recombination coefficient is shown in Fig. 9-8-4, which presents theoretical results from both the microscopic theory of Flannery and the Monte Carlo approach of Bates and Mendas.

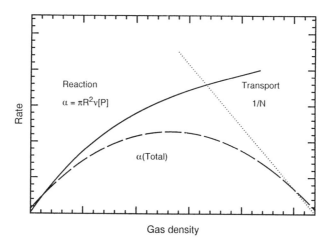

Figure 9-8-3. Plot of the recombination rate, neglecting transport effects, and the transport versus gas density. The dotted line is the overall recombination rate.

Extensive theoretical reviews of this process have been given by Flannery (1976, 1982a, b, 1990). Experimental studies are notably scarce, consisting primarily of a few early investigations (Sayers, 1962; Mahan and Person, 1964; McGowan, 1967) and some more recent measurements by Sennhauser et al (1980a, b), Smith and Adams (1983), Lee and Johnsen (1989), and Mezyk et al. (1989). A problem that arises in experimental investigations of three-body recombination is that at high pressure, extensive ion clustering arises so that more than one ion species is typically recombining at any given time and that species is often complex in form.

Examples of theoretical and experimental data for termolecular recombination are given in Figs. 9-8-5 and 9-8-6, which show the rate coefficients for $(O_4^+ + O_2^- + O_2)$ and $(O_4^+ + O_4^- + O_2)$ as a function of O_2 number density (McGowan, 1967) and of $H_3O^+(H_2O)_n + NO_3^- + Ar$ (Lee and Johnsen, 1989) as a function of argon density. It can be seen that the agreement between the experimental results and the Monte Carlo theory of Bates and Mendas is reasonable. Lee and Johnsen have found that inclusion of stabilization by mutual neutralization (Bates, 1985) improves the agreement between theory and experiment as shown by the full curve in the figure.

For the case of Xe_2^+ ions recombining with F^- in SF_6 gas, the Monte Carlo results were much smaller than experimentally measured. Bates (1981b) has shown that for diatomic ions and neutrals, the effects of the interchange of translational and rotational motion had only a small ($\sim 10\%$) effect on the recombination coefficient, so this cannot explain the discrepancy. Bates and

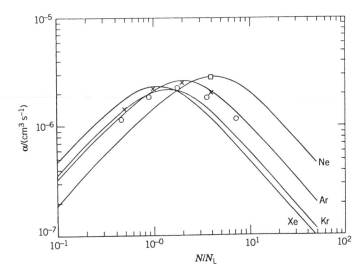

Figure 9-8-4. Recombination rate coefficient α at 300 K for $K^+ - F^-$ in rare gases (Ne, Ar, Kr, Xe) as a function of gas density (in multiples of Loschmidt number $N_L = 2.69 \times 10^{19}$ at STP). Solid curves, results of Flannery (1982); $\times$, $\bigcirc$, universal Monte Carlo calculation (Bates 1980) for Ar and Ne, respectively. [From Flannery (1982a).]

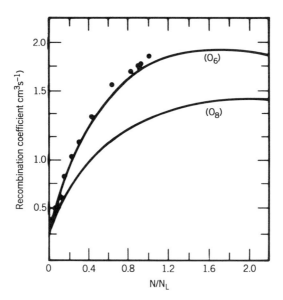

Figure 9-8-5. Rate coefficient α (cm³/s) at 298 K for recombination in both $(O_4^+ + O_2^- + O_2)$ and $(O_4^+ + O_4^- + O_2)$ systems as a function of O_2 number density. Solid curves, theoretical calculation by Bates and Flannery (1969); ●, McGowan (1967). [From Flannery (1976).]

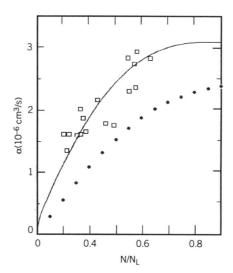

Figure 9-8-6. Measured (square symbols) rate coefficients for the recombination of $H_3O^+(H_2O)_n$ in argon as a function of number density (in units of Loschmidt's number). The dots are results from the Monte Carlo calculation of Bates and Mendas (1982). The solid curve is from a theoretical treatment that includes the effects of both collisional stabilization and enhanced mutual neutralization. [From Lee and Johnsen (1989).]

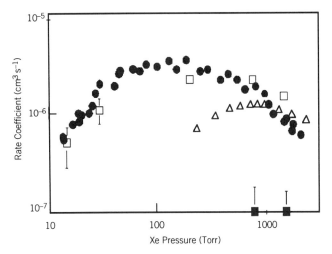

Figure 9-8-7. Comparison of calculated rate coefficients with experimental values (●) of Mezyk et al. (1989). Results calculated from a simple theory (treating the ions as atomic) are represented by △; those from the tidal theory by □ for the formation of XeCl* and by ■ for the formation of Xe_2Cl^*. [From Bates and Morgan (1990).]

Morgan (1990) have now explained this discrepancy between theory and experiment in terms of a *tidal recombination* mechanism. The internuclear separation of the Xe_2^+ dimer is very large ($6.18a_0$), and the neighboring rotational and vibrational energy levels are so close as to be essentially a continuum. As the Cl^- orbits about the Xe_2^-, the difference, at perihelion, in the force between the Cl^- and Xe^+, and the Cl^- and the Xe is sufficient to cause internal energy excitation of the dimer, and this represents an efficient mechanism for reducing the relative translational energy, and hence the orbital radius of the ion pair, leading to enhanced recombination. When this effect is included in the calculation, much better agreement with experiment is found (Fig. 9-8-7).

9-9. PROBLEMS

9-1. Using equations from Chapter 7 of the companion volume, derive (9-2-1), a simple expression for the radiative recombination cross section that applies to a bare nucleus capturing an electron.

9-2. Examine (9-2-1) and choose $Z = 1$. Draw three-dimensional graphs (similar to those on pp. 408 and 410 of our companion volume) showing the dependence of the cross section on n and E_e.

9-3. Choose $n = 1$ and draw graphs as in Problem 9-2 that show the dependence of the cross section on Z and E_e. Let Z assume the values 1, 10, 30, and 90 and E_e the values 0.01, 0.1, 1.0, 5.0, and 10.0 eV.

9-4. Study the derivation of (9-2-2) as given in the literature.

9-5. Equation (9-5-1) gives the rate coefficient for collisional radiative recombination. Use it to plot α versus T_e for various values of n_e.

9-6. Draw suitable potential energy curves and justify the form of (9-6-1) for the dissociative recombination cross section.

9-7. Justify (9-6-14) through (9-6-21), which apply to an electron–ion merged beam experiment.

9-8. Assume reasonable values for the relevant physical quantities and calculate the path of electrons through the trochoidal analyzer of Fig. 9-6-10.

9-9. Write a short description of surface barrier detectors and channel plate detectors.

9-10. Many applications involving electron–ion recombination require as input data, not cross sections, but rate coefficients, frequently as functions of temperature. M. A. Biondi and his associates have made a valuable contribution by developing and utilizing microwave cavity techniques to supply such data. Their work is described in a long series of papers stretching over several decades and is briefly discussed in McDaniel (1964, pp. 604–610) and Mason and McDaniel (1988, pp. 131–133). The basis of the microwave cavity method is the measurement, as a function of time, of the electron number density in a decaying afterglow. Various microwave, spectroscopic, and mass spectrometric techniques serve the vital functions of extending the temperature range of the measurements, and of revealing the atomic and molecular processes at work.

(a) Review Problems 7-4, 7-5, and 7-6.

(b) Study the methods used to obtain information on electron–ion recombination in the presence of ambipolar diffusion, which is another loss mechanism. [See E. P. Gray and D. E. Kerr's analysis of the effects of diffusion on recombination experiments. A summary of this important paper appears in McDaniel (1964, pp. 611–623).]

(c) In some, but not all, of the analyses of the Biondi experimental data, certain effects were not properly taken into account. These data did, however, yield accurate results when reanalyzed by improved techniques. See Dulaney et al. (1987) for details.

REFERENCES

Aberth, W., J. T. Moseley, and J. R. Peterson (1971), *Two-Body Ion–Ion Neutralization Cross Sections*, AFCRL Report 71-0481, Air Force Cambridge Research Laboratories, Bedford, Mass.

Adams, N. G., D. Smith, and E. Alge (1984), *J. Chem. Phys.* **81**, 1778.

Adams, N. G., C. R. Herd, and D. Smith (1989), *J. Chem. Phys.* **91**, 963.

Adams, N. G., C. R. Herd, M. Geoghegan, D. Smith, A. Canosa, J. C. Gomet, B. R. Rowe, J. L. Queffelec, and M. Moralis (1991), *J. Chem. Phys.* **94**, 4852.

Amano, T. (1988), *Astrophys. J.* **329**, L121.

Amano, T. (1990), *J. Chem. Phys.* **92**, 6492.

Anderson, L. H., and J. Bolko (1990), *J. Phys. B* **23**, 3167.

Anderson, L. H., P. Hvelplund, H. Knudsen, and P. Kvistgaard (1989), *Phys. Rev. Lett.* **62**, 2656.

Auerbach, D., R. Cacak, R. Caudano, T. D. Gaily, C. J. Keyser, J. W. McGowan, J. B. A. Mitchell, and S. F. J. Wilk (1977), *J. Phys. B* **10**, 3797.

Baker, J. G., and D. H. Menzel (1938), *Astrophys. J.* **88**, 52.

Bardsley, J. N. (1967), in I. P. Flaks and E. S. Solovyov, Eds., *Proc. 5th Int. Conf. Physics of Electronic and Atomic Collisions, Leningrad,* Nauka, Leningrad, abstracts, p. 338.

Bardsley, J. N. (1968), *J. Phys. B* **1**, 349, 365.

Bardsley, J. N., and M. A. Biondi (1970), *AAMP* **6**, 1.

Bates, D. R. (1950), *Phys. Rev.* **77**, 718.

Bates, D. R. (1981a), *J. Phys. B* **14**, 3525.

Bates, D. R. (1981b), *J. Phys. B* **14**, 2853.

Bates, D. R. (1985), *AAMP* **20**, 1.

Bates, D. R. (1987), *ICPEAC XV*, Brighton, p. 3.

Bates, D. R. (1991a), *J. Phys. B* **24**, 695.

Bates, D. R. (1991b), *J. Phys. B* **24**, 703.

Bates, D. R. (1991c), *J. Phys. B* **24**, 3267.

Bates, D. R., and T. J. M. Boyd (1956), *Proc. Phys. Soc. London A* **69**, 910.

Bates, D. R., and M. R. Flannery (1969), *Proc. R. Soc. A* **302**, 367.

Bates, D. R., and S. P. Khare (1965), *Proc. R. Soc.* **85**, 231.

Bates, D. R., and J. T. Lewis (1955), *Proc. Phys. Soc. London A* **68**, 173.

Bates, D. R., and H. S. W. Massey (1943), *Philos. Trans. R. Soc. A* **239**, 269.

Bates, D. R., and I. Mendas (1982), *J. Phys. B* **15**, 1949.

Bates, D. R., and W. L. Morgan (1990), *Phys. Rev. Lett.* **64**, 2258.

Bates, D. R., A. E. Kingston, and R. W. P. McWhirter (1962), *Proc. R. Soc.* **267**, 297; **270**, 155.

Bates, D. R., V. Malaviya, and N. A. Young (1971), *Proc. R. Soc. A* **320**, 437.

Berry, R. S., and S. E. Nielsen (1970), *Phys. Rev. A* **1**, 395.

Bethe, H., and E. Salpeter (1977), *Quantum Mechanics of One- and Two-Electron Atoms,* Plenum Press, New York.

Blakley, C. R., M. L. Vestal, and J. H. Futrell (1977), *J. Chem. Phys.* **66**, 2392.

Bordas, M. C., L. J. Lembo, and H. Helm (1991), *Phys. Rev. A* **44**, 1817.

Boulmer, J., F. Devos, J. Stevefelt, and J.-F. Delpech (1977), *Phys. Rev. A* **15**, 1502.

Brouillard, F. (1983), *ICPEAC XIII*, Berlin, p. 343.

Brouillard, F., W. Claeys, G. Poulaert, G. Rahmat, and G. Van Wassenhove (1979), *J. Phys. B* **12**, 1253.

Canosa, A., J. C. Gomet, B. R. Rowe, J. B. A. Mitchell, and J. L. Queffelec (1992), *J. Chem. Phys.*, **97**, 1028.

Cao, Y. S., and R. Johnsen (1991), *J. Chem. Phys.* **94**, 5443.

Chen, J. C. Y., and M. A. Mittelman (1967), in I. P. Flaks and E. S. Solovyov, Eds., *Proc. 5th Int. Conf. Physics of Electronic and Atomic Collisions,* Leningrad, Nauka, Leningrad, abstracts, p. 329.

Claeys, W., F. Brouillard, and G. Van Wassenhove (1977), *Proc. 10th Int. Conf. Physics of Electronic and Atomic Collisions,* Commissariat à l'Energie Atomique, Paris, abstracts, p. 460.

Dalgrano, A., G. A. Victor, and P. Blanchard (1971), *Theoretical Studies of Atomic Transitions and Interatomic Interactions,* AFCRL Report 71-0342, Air Force Cambridge Research Laboratories, Bedford, Mass.

Dalidchik, I. G., and Yu. S. Sayasov (1966), *Sov. Phys. JETP* **22**, 212.

Dalidchik, I. G., and Yu. S. Sayasov (1967), *Sov. Phys. JETP* **25**, 1059.

Denisov, Y. P., and N. M. Kuznetsov (1972), *Sov. Phys. JETP* **34**, 1231.

Dolder, K. T., and B. Peart (1985), *Rep. Prog. Phys.* **48**, 1283.

Dulaney, J. L., M. A. Biondi, and R. Johnsen (1987), *Phys. Rev. A* **36**, 1342.

Eisner, P. M., and M. N. Hirsh (1971), *Phys. Rev. Lett.* **26**, 874.

Eletskii, A. V., and B. M. Smirnov (1982), *Sov. Phys. Usp.* **25**, 13.

Flannery, M. R. (1976), in P. G. Burke and B. L. Moiseiwitsch, Eds., *Atomic Processes and Applications,* North-Holland, Amsterdam, p. 407.

Flannery, M. R. (1982a), in E. W. McDaniel and W. L. Nighan, Eds., *Applied Atomic Collisions Physics,* Vol. 3, Academic Press, New York, p. 141.

Flannery, M. R. (1982b), *Philos. Trans. R. Soc. London A* **304**, 447.

Flannery, M. R. (1987), *J. Phys. B* **20**, 4929.

Flannery, M. R. (1990), in T. Watanabe, I. Shimamura, M. Shimizu, and Y. Itikawa, Eds., *Molecular Processes in Space,* Plenum Press, New York, p. 145.

Forand, J. L., J. B. A. Mitchell, and J. W. McGowan (1985), *J. Phys. E* **18**, 623.

Fussen, D., and W. Claeys (1984), *J. Phys. B* **17**, L89.

Fussen, D., W. Claeys, A. Cornet, J. Jereta, and P. Defrance (1982), *J. Phys. B* **15**, L715.

Gaily, T. D., and M. F. A. Harrison (1970), *J. Phys. B* **3**, L27.

Galloway, E. T., and E. Herbst (1991), *Astrophys. J.* **376**, 531.

Giusti, A. (1977), *J. Phys. B* **13**, 3867.

Giusti-Suzor, A., J. N. Bardsley, and C. Derkits (1983), *Phys. Rev. A* **28**, 682.

Graham, W. G., Y. Hahn, and J. Tanis (1992), *The Recombination of Atomic Ions,* Plenum Press, New York.

Green, L. C., P. P. Rush, and C. D. Chandler (1957), *Astrophys. J. Suppl.* **3**, 37.

Gryzinski, M. (1959), *Phys. Rev.* **115**, 374.

Guberman, S. L. (1991), *Geophys. Res. Lett.* **18**, 1051.

Guberman, S. L., and A. Giusti-Suzor (1991), *J. Chem. Phys.* **95**, 2602.

Hasted, J. B. (1983), in F. Brouillard and J. W. McGowan, Eds., *Physics of Ion–Ion and Electron–Ion Collisions,* Plenum Press, New York, p. 461.

Hayhurst, A. N., and N. R. Telford (1974), *J. Chem. Soc. Faraday Trans. I* **322**, 1999.

Helm, H. (1992), in B. R. Rowe and J. B. A. Mitchell, Eds., *Dissociative Recombination*, Plenum Press, New York.

Herbst, E. (1985), *Astron. Astrophys.* **153**, 151.

Herd, C. R., N. G. Adams, and D. Smith (1990), *Astrophys. J.* **349**, 388.

Hickman, A. P. (1987), *J. Phys. B* **20**, 2091.

Hickman, A. P. (1989), in J. B. A. Mitchell and S. L. Guberman, Eds., *Dissociative Recombination: Theory, Experiment and Applications*, World Scientific, Singapore, p. 35.

Hus, H., F. B. Yousif, A. Sen, and J. B. A. Mitchell (1988), *Phys. Rev. A* **38**, 658.

Janev, R. K. (1976). *AAMP* **12**, 1.

Johnsen, R. (1987), *Int. J. Mass Spectrom. Ion Process.* **81**, 67.

Kelly, R., and P. J. Padley (1969), *Trans. Faraday Soc.* **65**, 355.

Kramers, H. A. (1923), *Philos. Mag.* **46**, 836.

Kulander, K. C., and M. F. Guest (1975), *J. Phys. B* **12**, L501.

Landau, L. (1932), *J. Phys. (USSR)* **2**, 46.

Lee, H. S., and R. Johnsen (1989), *J. Chem. Phys.* **90**, 6328.

Leu, M. T., M. A. Biondi, and R. Johnsen (1973), *Phys. Rev. A* **8**, 413.

MacDonald, J. A., M. A. Biondi, and R. Johnsen (1984), *Planet. Space Sci.* **32**, 651.

Mahan, B. H. (1973), in I. Prigogine and S. A. Rice, Eds., *Advances in Chemical Physics*, Wiley, New York, p. 1.

Mahan, B. H., and J. C. Person (1964), *J. Chem. Phys.* **40**, 392.

Mansbeck, P., and J. Keck (1969), *Phys. Rev.* **181**, 275.

Mason, E. A., and E. W. McDaniel (1988), *Transport Properties of Ions in Gases*, Wiley, New York.

Massey, H. S. W., and E. H. S. Burhop (1952), *Electronic and Ionic Impact Phenomena*, Oxford University Press, Oxford.

Mathur, D., S. U. Khan, and J. B. Hasted (1978), *J. Phys. B* **11**, 3615.

McDaniel, E. W. (1964), *Collision Phenomena in Ionized Gases*, Wiley, New York.

McGowan, S. (1967), *Can. J. Phys.* **45**, 439.

McGowan, J. W., and J. B. A. Mitchell (1984), in L. G. Christophorou, Ed., *Electron–Molecule Interactions and Their Applications*, Vol. 2, Academic Press, New York, p. 65.

McLaughlin D. J., and Y. Hahn (1989), *Phys. Rev.* **43**, 1313.

Mehr, F. J., and M. A. Biondi (1969), *Phys. Rev.* **181**, 264.

Mezyk, S. P., R. Cooper, and J. Sherwell (1989), *J. Phys. Chem.* **93**, 8187.

Michels, H. H., and R. H. Hobbs (1984), *Astrophys. J.* **286**, L27.

Millar, J. G., D. J. De Frees, A. D. McLean, and E. Herbst (1988), *Astron. Astrophys.* **194**, 250.

Mitchell, J. B. A. (1990), *Phys. Rep.* **186**, 215.

Mitchell, J. B. A., and S. L. Guberman, Eds. (1989), *Dissociative Recombination: Theory, Experiment and Applications*, World Scientific, Singapore.

Mitchell, J. B. A., and J. W. McGowan (1983), in F. Brouillard and J. W. McGowan, Eds., *Physics of Ion–Ion and Electron–Ion Collisions*, Plenum Press, New York, p. 279.

Mitchell, J. B. A., and B. R. Rowe (1991), in C. Cisneros, T. J. Morgan, and I. Alvarez, Eds., *Atomic and Molecular Physics, 3rd US/Mexico Symposium,* World Scientific, Singapore, p. 16.

Mitchell, J. B. A., and F. B. Yousif (1989), in H. E. Brandt, Ed. *Microwave and Particle Beam Sources and Directed Energy Concepts,* Vol. 1061, (SPIE-International Society for Optical Engineering, Bellingham, WA.

Mitchell, J. B. A., J. L. Forand, C. T. Ng, D. P. Levac, R. E. Mitchell, P. M. Mul, W. Claeys, A. Sen, and J. W. McGowan (1983), *Phys. Rev. Lett.* **51**, 885.

Mitchell, J. B. A., C. T. Ng, J. L. Forand, R. Janssen, and J. W. McGowan (1985), *J. Phys. B* **17**, L909.

Moseley, J. T., W. Aberth, and J. R. Peterson (1970), *Phys. Rev. Lett.* **24**, 435.

Moseley, J. T., R. E. Olson, and J. R. Peterson (1975), in M. R. C. McDowell and E. W. McDaniel, Eds., *Case Studies in Atomic Physics,* Vol. 5, North-Holland, Amsterdam, p. 1.

Nakashima, K., H. Takagi, and H. Nakamura (1986), *J. Chem. Phys.* **86**, 726.

Neumann, R., H. Poth, A. Winnacker, and A. Wolf (1983), *Z. Phys. A* **313**, 253.

Noren, C., F. B. Yousif, and J. B. A. Mitchell (1989), *J. Chem. Phys. Faraday Trans. 2* **85**, 1697.

Ogram, G. L., J. S. Chang, and R. M. Hobson (1980), *Phys. Rev. A* **21**, 982.

Olson, R. E. (1972), *Phys. Rev. A* **6**, 1822.

Olson, R. E., J. R. Peterson, and J. T. Moseley (1970), *J. Chem. Phys.* **53**, 3391.

O'Malley, T. F. (1981), *J. Phys. B* **14**, 1229.

Peart, B., and K. T. Dolder (1974), *J. Phys. B* **7**, 1948.

Peart, B., and K. T. Dolder (1975), *J. Phys. B* **8**, 1570.

Peart, B., R. Grey, and K. T. Dolder (1976), *J. Phys. B* **9**, 3047.

Peart, B., M. A. Bennett, and K. T. Dolder (1985), *J. Phys. B* **18**, L439.

Pert, G. J. (1990), *J. Phys. B* **23**, 619.

Peterson, J. R., W. Aberth, J. T. Moseley, and J. R. Sheridan (1971), *Phys. Rev. A* **3**, 1651.

Phaneuf, R. A., D. H. Crandall, and G. H. Dunn (1975), *Phys. Rev. A* **11**, 528.

Pitaevskii, L. P. (1962a), *Sov. Phys. JETP* **15**, 5.

Pitaevskii, L. P. (1962b), *Sov. Phys. JETP* **15**, 919.

Poulaert, G., F. Brouillard, W. Claeys, J. W. McGowan, and G. Van Wassenhove (1978), *J. Phys. B* **11**, L671.

Rowe, B. R., and J. B. A. Mitchell (1993), *Dissociative Recombination: Theory, Experiment and Applications, II,* Plenum Press, New York.

Rowe, B. R., J. C. Gomet, A. Canosa, C. Rebrion, and J. B. A. Mitchell (1992), *J. Chem. Phys.* **96**, 1105.

Rundel, R. D., R. L. Aitken, and M. F. A. Harrison (1969), *J. Phys. B* **2**, 954.

Sayers, J. (1962), in D. R. Bates, Ed., *Atomic and Molecular Processes,* Academic Press, New York, p. 272.

Schon, W., S. Krudner, F. Melchert, K. Rinn, M. Wagner, E. Salzborn, M. Karemera, S. Szucs, M. Terao, D. Fussen, R. Janev, X. Urbain, and F. Brouillard (1987), *Phys. Rev. Lett.* **59**, 1565.

Schramm, U., J. Berger, M. Greiser, D. Habs, E. Jaeschke, G. Kilgus, D. Schwalm, A. Wolf, R. Neumann, and R. Schuch (1991), *Phys. Rev. Lett.* **67**, 22.

Schneider, J., O. Dulieu and A. Giusti-Suzor (1991) *J. Phys. B.* **24**, L289.

Seaton, M. J. (1959), *Mon. Not. Astron. Soc.* **119**, 81.

Seaton, M. J. (1962), in D. R. Bates, Ed., *Atomic and Molecular Processes*, Academic Press, New York, p. 375.

Sen, A., and J. B. A. Mitchell (1986), *J. Phys. B* **19**, L545.

Sennhauser, E. S., D. A. Armstrong, and J. M. Warman (1980a) *Radiat. Phys. Chem.* **15**, 479.

Sennhauser E. S., D. A. Armstrong, and J. M. Warman, (1980b), *J. Phys. Chem.* **84**, 123.

Sidis, V., C. Kubach, and D. Fussen (1983), *Phys. Rev. A* **27**, 2431.

Smith, D., and N. G. Adams (1983), in F. Brouillard and J. W. McGowan, Eds., *Physics of Ion–Ion and Electron–Ion Collisions*, Plenum Press, New York, p. 501.

Smith, D., and N. G. Adams (1987), *J. Chem. Soc. Faraday Trans. 2* **83**, 149.

Stevefelt, J., J. Boulmer, and J. F. Delpech (1975), *Phys. Rev. A* **12**, 1246.

Stobbe, M. (1930), *Ann. Phys. Lpz.* **7**, 661.

Stuckelberg, E. C. G. (1932), *Phys. Acta* **5**, 369.

Sun, H., and H. Nakamura (1990), *J. Chem. Phys.* **93**, 6491.

Szucs, S., M. Karemera, M. Terao, and F. Brouillard (1984), *J. Phys. B* **17**, 1613.

Takagi, H., N. Kosugi, and M. Le Dourneuf (1991), *J. Phys. B* **24**, 711.

Thomson, J. J. (1924), *Philos. Mag.* **47**, 337.

Turner, B. (1989), in J. B. A. Mitchell and S. L. Guberman, Eds., *Dissociative Recombination: Theory, Experiment and Applications*, World Scientific, Singapore, p. 329.

Urbain, X., A. Cornet, F. Brouillard, and A. Giusti (1990), *Phys. Rev. Lett.* **66**, 1685.

Van der Donk, P., F. B. Yousif, J. B. A. Mitchell, and A. P. Hickman (1991), *Phys. Rev. Lett.* **67**, 42.

Walls, F. L., and G. H. Dunn (1974), *J. Geophys. Res.* **79**, 1911.

Warman, J. M., E. S. Sennhauser, and D. A. Armstrong (1979), *J. Chem. Phys.* **70**, 995.

Wigner, E. P. (1948), *Phys. Rev.* **73**, 1002.

Yousif, F. B., and J. B. A. Mitchell (1989), *Phys. Rev. A* **40**, 4318.

Yousif, F. B., P. Van der Donk, Z. Kucherovsky, J. Reis, E. Brannen, J. B. A. Mitchell, and T. J. Morgan (1991a), *Phys. Rev. Lett.* **67**, 26.

Yousif, F. B., P. J. T. Van der Donk, M. Orakzai, and J. B. A. Mitchell (1991b), *Phys. Rev. A* **44**, 5653.

Yousif, F. B., P. J. T. Van der Donk, and J. B. A. Mitchell (1993), in preparation.

Zener, C. (1932), *Proc. R. Soc. London A* **137**, 696.

APPENDIX I

BACKGROUND INFORMATION ON ATOMIC STRUCTURE AND SPECTRA

The main goals in the study of atomic structure have been to determine the atomic energy levels, wave functions, selection rules, and transition probabilities; to understand the effects of externally applied electric and magnetic fields; and to ascertain the role of polarization in the emission and absorption of radiation. Clearly, there is a close connection between these topics and atomic collisions, especially processes involving photons, so we now undertake a brief review of some of the essential features of atomic structure. Most of the experimental information available on this subject comes from spectroscopic investigations. In the X-ray, ultraviolet (UV), visible, and infrared (IR) regions (see Fig. 8-1-1 in the companion volume), the *wavelengths* of the transitions are usually the measured quantities, whereas in the microwave and RF regions, it is the transition *frequencies* that are measured. The X-ray, UV, and visible regimes are usually studied by obtaining *emission* spectra, but *absorption* methods are required in microwave and RF spectroscopy, and sometimes for IR studies. Absorption techniques are also frequently employed at the higher frequencies to avoid the clutter of lines that often characterizes the emission spectrum.

1. MAINLY THE HYDROGEN ATOM

Of great importance in the accretion of knowledge concerning atomic structure have been studies of *optical* line spectra, covering the visible region and extending toward both higher and lower frequencies. Such spectra are "simple" only for atoms with one or two valence electrons outside closed shells (hydrogen, alkalis, and alkaline earths). Another requirement for obtaining "simple" spectra is that the studies be made at low resolution, so that complicating aspects such as fine structure (FS) and hyperfine structure (HFS) not be in evidence. Studies of the "gross spectrum" of the H atom that revealed only its "gross structure"

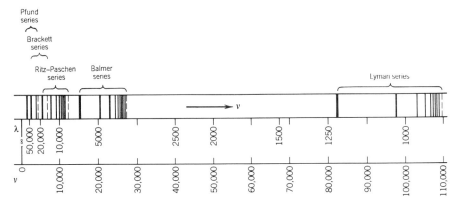

Figure I-1. Schematic representation of the gross spectrum of the hydrogen atom. The intensities of the lines are roughly indicated by their thicknesses. The dashed lines correspond to the *line series limits* (i.e., the limits of series of lines). [From Herzberg (1944).]

were of enormous importance historically because they provided just the right mixture of information and tantalizing challenge to advance our understanding early in this century.

As is well known, both simple and complicated optical spectra consist in part of *series of lines* (or series of line groups) whose separation and intensity decrease regularly toward higher frequencies in the manner shown in Fig. I-1.* (Band spectra, consisting of extremely closely spaced lines or even true continua, are associated with molecules.) The first regularity in a spectrum to be formulated mathematically was the Balmer series of lines in atomic hydrogen (1885). Actually, Balmer discovered a special case of the general, empirical expression later found by Rydberg in 1896, which describes all of the line series of the H atom (see Fig. I-2) as far as its gross spectrum is concerned. In 1913, Bohr gave the Rydberg equation a theoretical basis, which was improved in 1916 by Sommerfeld, who extended Bohr's semiclassical analysis by including the relativistic change in electron mass and by allowing elliptical, not merely circular, orbits. The gross structure for the hydrogen atom was solved fully and correctly in 1926 by Schrödinger through application of his nonrelativistic quantum mechanics. The explanation of the fine details of the structure of the H-atom had to await the introduction of the electron spin into the theory by Pauli in 1927, Dirac's relativistic quantum theory for the electron (1928), and quantum electrodynamics (Heisenberg and Pauli in 1929, and later workers). At all levels of understanding, the hydrogen atom has guided theorists in their attack on more complex atoms and has served as a prototype, much as the harmonic oscillator has in the analysis of complicated vibration problems.

---

*A continuous spectrum of lines (the *ionization continuum*) joins each series of lines at the high-frequency end.

Figure I-2 shows the gross energy levels and spectrum of the H-atom. Sommerfeld used the term "fine structure" to describe the very slight splitting of a level associated with a given value of the principal quantum number n into levels associated with different values of l, the orbital angular momentum quantum number. As we point out in Section 3, we now mean something quite different by "fine structure." In Fig. I-2, the energy is measured from a zero value corresponding to the ground state. This practice is now common, especially for complex atoms; however, for simple atoms the energy zero is frequently taken at

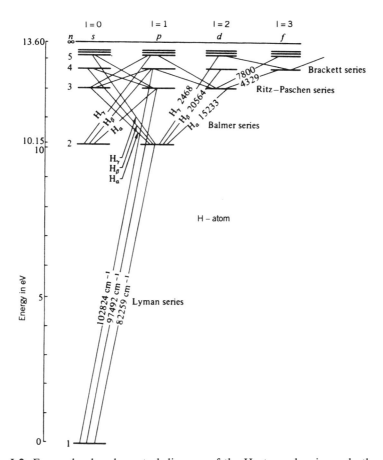

Figure I-2. Energy-level and spectral diagram of the H atom, showing only the gross features. This type of display is known as a *Grotrian diagram*. Here the positions of the energy levels along the vertical axis are determined by the principal quantum number n. The energy levels for various values of the orbital angular momentum quantum number l are spread out horizontally for clarity; the allowed energies do not depend on the value of l. The allowed energies can be expressed as terms, $T_n = -E_n/hc$; the energies in a given column then constitute a *series of terms*. [Adapted from Mizushima (1970).]

the level corresponding to $n = \infty$. Bohr took $E = 0$ according to the latter convention, making the energies of the quantized, bound states all negative. The ionization limit is at $E = 0$, and positive energies correspond to unquantized, unbound continuum states.

In Fig. I-2, certain radiative transitions between energy levels are allowed; others are forbidden by the electric dipole selection rule $\Delta l = \pm 1$ (see Section 8-7 of the companion volume). The vertical separation of the various l levels associated with a given value of n in the hydrogen atom is too small to be visible in this drawing, because the electron-nucleus potential is exactly proportional to $1/r$. In the alkalis, although the valence electron is also in a central force field, the potential does not vary as $1/r$. This circumstance leads to a much larger separation of levels with different l but the same n. As indicated in Fig. I-3 for the case of the sodium atom, Sommerfeld's original usage of the term "fine structure" would surely not be appropriate in the alkalis. Solution of the nonrelativistic Schrödinger wave equation without consideration of electron spin gives no dependence of the stationary state energies on l for the H atom. A slight dependence is observed in measurements, however, and the full quantum theory gives essentially exact agreement with experiment.

In the days before quantum mechanics, two different series of lines in simple gross spectra were classified according to their appearance, one as sharp (s or S), the other as diffuse (d or D). Two other were called the principal (p or P) and fundamental (f or F) series. This prequantum terminology was later correlated with the orbital angular momentum quantum number (l or L) when quantum mechanics was developed. This usage of s, p, d, f to designate the value of l for a single electron and the usage of S, P, D, F to denote the value of L for an atom

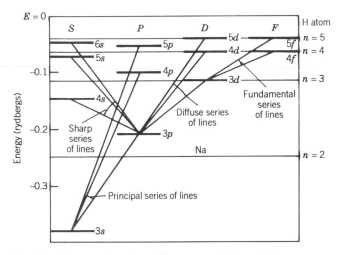

Figure I-3. Grotrian diagram for the sodium atom. Note the large energy spacings between levels with the same value of n but different values of l, as contrasted with the situation in Fig. I-2 for the H atom. [Adapted from Slater (1960), Vol. 1).]

persists to the present and presumably will be with us forever. Extending this chain of letters down the conventional alphabet gives us the spectroscopists' alphabet:

$$l = \begin{matrix} s & p & d & f & g & h & i & k & l & m & n & o & q & r & t & u & \cdots \\ 0 & 1 & 2 & 3 & 4 & 5 & 6 & 7 & 8 & 9 & 10 & 11 & 12 & 13 & 14 & 15 & \cdots \end{matrix}$$

where we omit a few letters that have been reserved to designate other quantities. When spectral lines are described by differences of *term symbols*, as in the case $2S-3P$, where the integers refer to the n values and the letters to the L values, the first term of the difference *always* indicates the level of the lower energy.

We have thus far used the words *level* and *state* in a rather cavalier fashion. Actually, they have precise meanings in the modern parlance of atomic structure, and we should be careful in their use once we get to the point (Section 3) of being able to define their exact modern meanings.

In describing radiation, spectroscopists use the frequency (v), wavelength (λ), and wave number ($\tilde{v}$). These quantities are related by the equations

$$v = \frac{c_{\text{med}}}{\lambda_{\text{med}}} = \frac{c_{\text{vac}}}{\lambda_{\text{vac}}} \tag{I-1}$$

and

$$\tilde{v} = \frac{v}{c_{\text{vac}}} = \frac{1}{\lambda_{\text{vac}}} = \frac{1}{n_{\text{med}} \lambda_{\text{med}}}, \tag{I-2}$$

where the subscript "vac" means "vacuum" and "med" indicates the medium through which the radiation is traveling. Here n is the index of refraction of the medium. We see that the wave number is the reciprocal of the wavelength in a vacuum.

The wave numbers of the lines in a given series of the hydrogen atom can be described by the *Rydberg equation*:

$$\tilde{v} = \frac{R}{n_2^2} - \frac{R}{n_1^2}, \tag{I-3}$$

where R is the Rydberg constant and n_1 and n_2 are integers ($n_1 > n_2$). The index n_2 is constant for a given series of lines. With increasing values of the order number n_1, $\tilde{v}$ approaches a limit $\tilde{v}_\infty = R/n_2^2$. That is, the separation of consecutive members of a given series of lines decreases so that $\tilde{v}$ cannot exceed a fixed limit, the *(line) series limit*. The wave number of any line in the gross spectrum of the H atom is the difference between two members of the series, $T(n) = R/n^2$, having different values of n. These members are called *terms*. The emission and absorption lines of other elements also can be represented as the differences between two terms (in the gross spectrum), although these terms for atoms other than the alkalis do not have a simple form mathematically. Thus, for any atom, we can express the wave number of a line in the gross spectrum as $\tilde{v} = T_2 - T_1$. The *Rydberg–Ritz combination principle* states that subject to certain limitations, the difference between any two terms of an atom gives the wavenumber of a line in the gross spectrum of the atom. These limitations are the *selection rules* (pp. 546–551 of the companion volume). Actually, the terms in a spectrum are related precisely to energy levels, although the exact connection was not known until Bohr developed his atomic model.

2. THEORIES OF THE HYDROGEN ATOM AND HYDROGENLIKE IONS

A. Bohr Theory (1913)

In his semiclassical model, Bohr considered a single electron moving at a nonrelativistic speed v in a circular orbit of radius r about a bare nucleus of charge $+Ze$. He quantized the angular momentum by the condition $\mu v r = n\hbar$, where $n = 1, 2, 3, \cdots$ is the principal quantum number. Here μ is the reduced mass of the electron and nucleus, having a different value for each kind of nucleus. The permissible orbits are thus restricted to having the radii

$$r_n = \frac{(4\pi\varepsilon_0)\hbar^2 n^2}{Ze^2\mu} \qquad (n = 1, 2, 3, \ldots) \qquad \text{(SI units)}. \qquad \text{(I-4)}$$

The corresponding energies are

$$E_n = -\frac{\mu}{2\hbar^2}\left(\frac{Ze^2}{4\pi\varepsilon_0}\right)^2 \frac{1}{n^2} \qquad (n = 1, 2, 3, \ldots) \qquad \text{(SI units)}. \qquad \text{(I-5)}$$

Of course, these energies relate only to the gross structure of the atom. The wave number of radiation emitted or absorbed in a transition between bound states n_1 and n_2 is

$$\tilde{v} = \frac{v}{c} = \frac{1}{2\pi\hbar c}(E_{n_1} - E_{n_2}). \qquad \text{(I-6)}$$

Now we see what the terms meant in the Rydberg equation, and we have a theoretical value for R:

$$T_n = -\frac{E_n}{2\pi\hbar c} \qquad \text{(term)} \qquad \text{(I-7)}$$

and

$$R = \frac{\mu}{4\pi c \hbar^3} \left(\frac{e^2}{4\pi\varepsilon_0} \right)^2 \qquad \text{(Rydberg constant).} \qquad \text{(I-8)}$$

The precise numerical value of R is given in Table I of Appendix III in the companion volume, along with other atomic constants. The energy levels are given approximately by the equation

$$E_n = -13.6 Z^2 (\text{eV}). \qquad \text{(I-9)}$$

Bohr's theory gave agreement with the observed gross spectra for one-electron atoms.

B. Sommerfeld Theory (1916)

In his extension of the Bohr theory, Sommerfeld allowed for elliptical orbits and quantized both the radial and angular motion, obtaining two quantum numbers. The *radial quantum number* is denoted n_r and the *azimuthal quantum number* $n_\phi \equiv k$. The *principal quantum number* is now defined as $n = n_r + k$, where $n_r = 0, 1, 2, \ldots$ and $k = 1, 2, \ldots, n$. (In this model $k = 0$ is considered to be impossible because it would require that the electron go through the nucleus.) Here n is a measure of the major axis of the ellipse, and k is related to the length of the minor axis. Also, k gives the angular momentum in units of $\hbar$. Sommerfeld obtained the Bohr expression for the bound-state energies (I-5), with only the quantum number n appearing in the result.

When Sommerfeld also allowed for the relativistic change of mass with velocity (Sard, 1970, pp. 190–196), he obtained bound-state energies that depend mainly on n but that also have a slight dependence on k:

$$E_{n,k} = -\frac{\mu}{2\hbar^2} \left(\frac{Ze^2}{4\pi\varepsilon_0} \right)^2 \frac{1}{n^2} \left[1 + \frac{\alpha^2 Z^2}{n^2} \left(\frac{n}{k} - \frac{3}{4} \right) \right] \qquad (n = 1, 2, 3, \ldots). \qquad \text{(I-10)}$$

Here α is the *fine structure constant*, which is now known to be a measure of the strength of the coupling of an electric charge to the electromagnetic field. As indicated in Table I of Appendix III in the companion volume, $\alpha = e^2/(4\pi\varepsilon_0)\hbar c \approx 1/137$. Equation (I-10) gives energies very similar to those derived from relativistic quantum mechanics, which includes not only the *change of mass with velocity*, but also *spin-orbit coupling* and the *Darwin term* [see Woodgate (1980, pp. 59–65)]. Again we point out that what Sommerfeld called fine structure is no longer called that. Good references for the Sommerfeld model are Eisberg (1961), Born (1967), and Slater (1960).

C. Schrödinger Quantum Mechanical Solution of the Gross Structure

We start with the *time-dependent wave equation for the one-electron atom,*

$$ih\frac{\partial}{\partial t}\Psi(\mathbf{r}, t) = \left[-\frac{\hbar^2}{2m}\nabla^2 + V(\mathbf{r}, t)\right]\Psi(\mathbf{r}, t), \tag{I-11}$$

where for the present case, the potential is

$$V = V(r) = -\frac{Ze^2}{(4\pi\varepsilon_0)r}. \tag{I-12}$$

We seek solutions of the form

$$\Psi(\mathbf{r}, t) = \psi_E(\mathbf{r})e^{-(i/\hbar)Et}, \tag{I-13}$$

the $\psi_E(\mathbf{r})$ being solutions of the *time-independent wave equation,* which is also the energy eigenvalue equation:

$$\left[-\frac{\hbar^2}{2m}\nabla^2 + V(r)\right]\psi_E(\mathbf{r}) = E\psi_E(\mathbf{r}). \tag{I-14}$$

The energy eigenfunctions, $\psi_E(\mathbf{r})$, correspond to the *stationary states* of the structure being considered. They have this name because the probability density associated with these states is independent of time. We can obtain particular solutions of (I-14) in the form

$$\psi_{E,l,m}(r, \theta, \phi) = R_{E,l}(r)Y_{lm}(\theta, \phi), \tag{I-15}$$

where $Y_{lm}(\theta, \phi)$ is the spherical harmonic corresponding to the orbital angular momentum quantum number l and to the magnetic quantum number m (with $m = -l, -l + 1, \dots +l$). The radial functions in (I-15) are the associated Laguerre polynomials. Each spherical harmonic can be written as the product of an associated Legendre function $P_l^m(\cos\theta)$ and an exponential function $e^{im\phi}$. The Schrödinger solution discussed here gives the same energy levels as did the Bohr model. The energy of any particular stationary state is determined by the value of the principal quantum number, so the subscript n could be used in place of the subscript E in (I-15). A full discussion of the Schrödinger solution appears in Park (1974), Bransden and Joachain (1983, 1989), and many other books on atomic structure and quantum mechanics.

The introduction of the electron spin into the Schrödinger equation via the *Pauli spin matrices* is described in the books cited above, as well as in Cohen-Tannoudji et al. (1977), Bethe and Salpeter (1957), Morrison et al. (1976), and Bates (1961, 1962a). The discovery of the electron spin is chronicled by Pais (1989).

D. Dirac's Relativistic Quantum Mechanics; Quantum Electrodynamics

The Schrödinger theory with the ad hoc insertion of electron spin suffices for most applications in atomic and molecular physics, so we make only a few remarks here concerning the more advanced theories. Dirac's relativistic quantum mechanics, and its application to the structure of the hydrogen atom, are well covered in the books by Eisele (1969) and Bates (1962b). We should emphasize that the spin arises naturally in the Dirac theory and hence does not have to be artificially introduced. Quantum electrodynamics (the theory of the interaction of light with matter) is treated by Scadron (1979), Feynman (1985), Gleick (1992), Holstein (1992), Bates (1962b), and Cohen-Tannoudji et al. (1989, 1992). The key new element here is the quantization of the electromagnetic field (Note 8-2 in the companion volume).

3. COMPLEX ATOMS; CENTRAL FIELD APPROXIMATION AND BEYOND

To describe a nonhydrogenic atom in the first approximation, we imagine that each electron is in a central force field and has a well-defined value of n and l.* We provide the value of the atomic mass A and atomic number Z of the nucleus. If then we specify the value of n and l for each electron, we have specified the *configuration* of the atom. The value of n for a given electron tells which *shell* it is in; the *subshell* occupied by the electron is indicated by its value of n and l, where we hark back to prequantum mechanics concepts. In Table VI of Appendix III in the companion volume, we give the ground-state configuration for all the atoms in the periodic table. As we build up the periodic table from hydrogen, there is perfect regularity in the assignment of n and l for each added electron up through argon ($Z = 18$). The maximum number of electrons that can go into a given subshell and shell is determined by the Pauli exclusion principle. The ground-state configuration of argon, for example, is usually written $1s^2\ 2s^2\ 2p^6\ 3s^2\ 3p^6$, where the number in front of each letter gives the value of n associated with the value of l implied by the letter, and the right superscript gives the number of electrons having that value of n and l. Note the irregularity we encounter when we go to the next atom, K, with $Z = 19$. The nineteenth electron goes into the $4s$ subshell, not the $3d$ subshell. It turns out that this assignment will give the atom the lowest possible energy.

Consider the simple case of a ground-state, nonhydrogenic atom having only one valence (or "emission") electron outside closed shells. Examples are the

*For discussions of the central field approximation, see Chapter 8, Volume I of Slater (1960) and Chapter 6 of Woodgate (1980). These authors also treat the *Thomas–Fermi method* of calculating the central part of the potential experienced by each of the atomic electrons, the *Hartree method* of the self-consistent field, and the *Hartree–Fock method* of obtaining properly antisymmetrized wave functions that satisfy the *Pauli exclusion principle*.

alkalis (Li, Na, K, Rb, and Cs) in their ground states. All of the inner electrons of such atoms are in their normal, known, closed subshells. The valence electron is in the higher subshell that corresponds to the atom as a whole having the lowest possible total energy. Now consider the atom to be excited by promotion of the valence electron to a subshell giving the atom a higher energy. To specify the configuration of the excited atom, we need give the values of n and l for only the valence electron, since n and l for all of the inner electrons are already known.

In simple atoms such as described here, the inner shells (which are closed, i.e., fully occupied) contribute no orbital angular momentum to the atom, as the resultant for each closed shell is zero. Hence the entire atom has an orbital angular momentum (L, in units of $\hbar$) equal to that (l, in units of $\hbar$) of the single valence electron. That is, $L = l$, so we can use the capital letters $S, P, D, F, \ldots$ to specify the orbital angular momentum of the valence electron if we wish. This is sometimes done. Figure I-3 shows an energy-level diagram for the gross structure of the sodium atom and indicates some of the transitions producing the sharp, principal, diffuse, and fundamental series of lines in the gross spectrum. The selection rule $\Delta l = \pm 1$, or $\Delta L = \pm 1$, applies. In the other alkalis, the lower level on which each of these series terminates has a different value of n, but the same value of l as for sodium. Each vertical column in Fig. I-3 shows a *series of energy levels*. We could alternatively provide the same information by changing the units along the vertical axis and calling horizontal lines "terms." Then each column would show a *series of terms*. In *simple* atoms, all term series converge to a common accurately known limit, which can be used to define the zero on the energy, or term value, scale. In complex atoms it is difficult to establish the series limits (here we mean the limits of the various series of terms) accurately, so the term values or energies are generally tabulated on a scale in which the ground state defines the zero point [Slater (1960, p. 243, Vol. I)]. The vertical distance between the 3s level and the $\infty \, p$ level (the P term series limit) gives the line series limit for the principal series of lines. Similar statements can be made for the other series of lines.

An energy level described by specifying only the n and l values for each electron (i.e., the configuration of the atom) is actually split into various *terms*, *levels*, and *states* by several effects which we must consider when we investigate the atom at higher levels of approximation. We shall need to specify values for each of *four* quantum numbers for each electron in an atom in order to obtain a complete specification of the energy of the atom—one for each dimension of ordinary space plus one for the spin space. The particular set of quantum numbers we use depends on the particular atom we are talking about and the state of excitation or ionization it is in. What is really involved is the choice of a particular realistic scheme for coupling the various angular momenta of all the atomic electrons. By far the most commonly used scheme is *L-S coupling* (*Russell–Saunders*); another one frequently used is *j-j coupling* (Zare, 1988).

Figure I-4 illustrates *schematically* the splitting of a configuration into terms, these terms into levels, and these level into states. The cause of each successive splitting is indicated on the drawing and is discussed in Woodgate (1980, pp. 4-

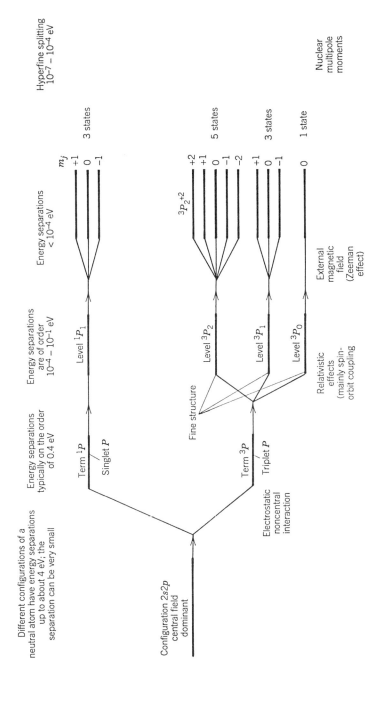

Figure I-4. Splitting of a 2s2p configuration into terms, levels, and states. The causes of the successive splittings are indicated at the bottom of the diagram. Typical magnitudes of the splittings are given at the top. The set of levels into which a term splits is often called a *multiplet*. [Adapted from Slater (1960, Vol. 1).]

7). The drawing is only schematic because it fails to show the relative magnitudes of the various stages of splitting. Here we have tacitly assumed that L-S coupling is appropriate. We are considering a *2s2p configuration* initially, on the left side of the drawing, assuming that the two electrons each experience a central electrostatic force. Then we improve the approximation by turning on an additional electrostatic force field that describes the noncentral part of the force that we previously neglected. (Of course, this noncentral force is there all of the time in the real atom, whether you consider it or not.) This act splits the *configuration* into *terms*. Then we turn on the relativistic effects, of which the spin-orbit coupling is almost always the strongest. These effects split each *term* into *levels*. Then if we turn on a not-too-strong external magnetic field, we get Zeeman splitting of each *level* into *states*. (For strong external magnetic fields, we get Paschen−Back splitting instead.) Hyperfine splitting is caused by the interaction of the nuclear moments with the electric and magnetic fields produced by the orbital electrons.

We use the following notation: L, the orbital angular momentum of the entire atom in units of $\hbar$, is indicated by a capital letter (P means that $L = 1$). S (we do not mean that $L = 0$ here) gives the spin angular momentum of the entire atom in units of $\hbar$. J gives the total angular momentum of the entire atom in units of $\hbar$. The component of the total angular momentum of the entire atom along an externally imposed magnetic field is indicated by m_J. The *multiplicity* of a term is given by $(2S + 1)$ and appears as a left superscript in the term symbol. This integral number tells you how many levels the particular term splits into. This is where the *fine structure* appears. It involves the splitting of a term into various J levels. The value of J for each level appears as a right subscript. The number of possible values of m_J for the given J value of a level determines the number of states into which that level splits. We festoon the state symbol with the value of m_J as a right superscript. A summary in the form of specific examples is given below, culminating with a state symbol.

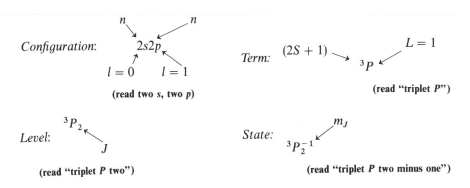

Several additional items of terminology still remain to be explained:

1. An *isolectronic sequence* is a sequence of structures having the same

number of electrons, such as Na, Mg^+, Al^{2+}, Si^{3+}, P^{4+}, S^{5+}, Cl^{6+}. This sequence is frequently written Na I, Mg II, Al III, Si IV, P V, S VI, Cl VII. The spectrum of a neutral species (I) is often referred to as the *arc spectrum*, and that of a singly ionized species (II), the *first spark spectrum*. The *second spark spectrum* is the spectrum for a doubly ionized species (III). This nomenclature has its origin in the characteristics of the electrical discharges used in light sources. An arc discharge is characterized by a low operating voltage, and the resulting radiation has the low frequencies associated with excitation of neutral atoms. On the other hand, a spark discharge operates at high voltage, and the emitted radiation can produce ions and then excite these ions.

2. Now consider *configuration interaction* (or *mixing*). Start with a single given configuration in a given approximation of the Hamiltonian operator. Now introduce an additional small interaction to make the analysis more realistic. It will act as a perturbation on the original problem. If the perturbation is not such that the given configuration wave function is an eigenfunction of the perturbation operator, the perturbation will mix in other configurations that may be nearby in energy [see Woodgate (1980, App. B)]. The wave function now is mainly the original one, but it contains a small admixture of other configuration wave functions. The amount contained depends on the nature and strength of the perturbation and the energy by which the added configuration is separated from the original configuration. Equation (B-7) on p. 207 of Woodgate (1980) gives the resulting first-order wave function.

4. ADDITIONAL TOPICS

Here we enumerate some other topics that are related to atomic structure and spectra and provide references to sources of information about them. Some of the discussions that are cited here are highly abbreviated, but references to more complete treatments are included in these discussions.

Anomalous dispersion (Section 8-14-C of the companion volume)

Atomic beams (Ramsey, 1988a, 1990)

Atomic clocks: Ramsey (1988b, 1990)

Atomic data: Appendix III of the companion volume

Beam foil spectroscopy: (Section 8-12-B of the companion volume)

Cascading of radiation: (Section 5-7-C of the companion volume)

Detailed balancing in emission and absorption of radiation: (Section 8-2-E of the companion volume)

Doppler effect: (Note 8-5 of the companion volume)

Einstein A and B coefficients: (Section 8-3 of the companion volume)

Field ionization: (Note 6-2 of the companion volume)

Forbidden transitions and metastable atoms: (Section 8-12-D of the companion volume)

Hanle effect: Corney (1977)

Laser cooling and trapping of atoms: (Note 8-4 of the companion volume)

Lifetime measurements: (Section 5-8 of the companion volume)

Line broadening and line shape: Section 8-11 of the companion volume)

Multiphoton processes: (Section 8-17 of the companion volume)

Optical double resonance: Corney (1977)

Optical pumping: Corney (1977)

Oscillator strengths and sum rules; line strengths: (Sections 5-7-F and 8-10 of the companion volume)

Photoionization: (Sections 8-9 and 8-16 of the companion volume)

Photon emission processes: (Section 8-1-A of the companion volume)

Photon absorption processes: (Sections 8-1-C and 8-13 through 8-16 of the companion volume)

Polarization of emitted radiation: (Section 5-7-D of the companion volume)

Radiative transfer, trapping of resonance radiation: (Sections 5-7-E and 8-12-E of the companion volume)

Rydberg states: (Note 6-1 of the companion volume)

Selection rules: Sections 5-7-G, 8-7, 8-8 of the companion volume)

Superradiance: (Note 8-3 of the companion volume)

Theory of radiative transitions in atoms: (Sections 8-3 through 9-11 of the companion volume)

5. PROBLEMS

Solutions to most of these problems appear in the references indicated.

I-1.   On p. 546 of the companion volume, (8-6-8) gives the Einstein A coefficient for spontaneous emission of unpolarized radiation by the hydrogen atom. Calculate approximate numerical values for this coefficient at the mid-frequencies of each of the spectral ranges shown in Fig. 8-1-1 on p. 514 (excluding the gamma-ray region). Use these numbers to justify our statement that emission spectroscopy is feasible for the higher frequencies but not the lower.

I-2.   Consider the operators $H_0, \mathbf{S} \cdot \mathbf{L}, L^2, L_x, L_y, L_z, S^2, S_x, S_y, S_z, J^2, J_x, J_y, J_z,$ H', where $H_0 = -(\hbar^2/2m)\nabla^2 + V(r)$, and $H' = f(r)\mathbf{S} \cdot \mathbf{L}$. $\mathbf{L}, \mathbf{S}$, and $\mathbf{J}$ are the orbital, spin, and total angular momentum vectors. Determine which pairs of the operators mutually commute and which do not. (See p. 168 of the companion volume for the motivation behind this problem.)

I-3. In 1937–1938, I. I. Rabi developed his *molecular beam magnetic resonance method*, which proved to be an important extension of the Stern–Gerlach method discussed in Problem 1-11. The version of Rabi's apparatus shown in Fig. I-5 consists of an oven effusion source, two identical inhomogeneous magnets (A and B) with oppositely directed gradients, a homogeneous magnet (C) with an RF loop between its pole pieces, and a tungsten filament detector. Beam particles with magnetic moments that emerge through the collimating slits are pushed in one direction by the field gradient in magnet A and, later, in the opposite direction, by the field gradient in B. Particles entering A with the right velocity and angle of inclination follow a sigmoidal path through C and B to the detector if the RF loop is not activated. (No deflection is produced by the C magnet.) However, the RF loop can provide radiation of variable frequency and can induce magnetic dipole transitions in the beam particles if tuned to resonance. Supply the missing details of this description of the Rabi method and show how it can be used to measure nuclear magnetic moments. [See Rabi et al. (1939) and Corney (1977, pp. 692–705).]

I-4. In 1949, N. F. Ramsey made another advance in molecular beam technology when he introduced his *method of successive oscillatory fields*. Here the single oscillating field in the center of the Rabi apparatus is replaced by two oscillating fields at the entrance and exit, respectively, of the space in which the magnetic moments are investigated, and the

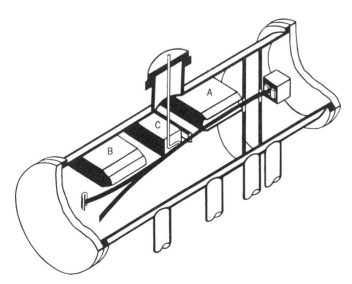

Figure I-5. Section view of the Rabi magnetic resonance beam apparatus. [Adapted from Kusch (1951).]

resonance can thereby be narrowed. By taking as the signal the difference in the counting rate when the relative phase of the two RF fields is $0°$ and $180°$, the characteristic Ramsey interference pattern is obtained, with a width determined by the separation distance of the two RF fields rather than the natural lifetime of the states. See Ramsey (1980, 1990) and write out a more complete description of his method. Ramsey was awarded a Nobel Prize in 1989 for this invention and related work.

I-5. In 1961, N. F. Ramsey and D. Kleppner developed the *hydrogen maser*, the first instrument of this type that operates on atoms rather than molecules. Here the atoms collected at the end of the apparatus are stored in a Teflon-coated quartz storage bulb for about 1 s, and consequently, the linewidth calculated from the uncertainty principle is narrower by a factor of 1000 than that obtained in a conventional atomic beam apparatus. Write out a description of the hydrogen maser and compare it with the earlier ammonia maser developed by Townes. [See Ramsey (1988b, 1990) and Corney (1977, pp. 708–714).]

I-6. Two electrons in an atom are said to be *equivalent electrons* if they have the same values for the quantum numbers n and l.

(a) Show that the terms that can arise from two equivalent p electrons are ^1S, ^3P, and ^1D.

(b) What are the six terms that can arise from two nonequivalent p electrons?

[See Herzberg (1944, Chap. 3).]

I-7. Draw energy-level diagrams for the H atom and for the He atom with only one electron excited out of the ground state. Neglect relativistic effects and consider the nuclei to be point charges. Label the energy levels with spectroscopic notation. Explain the similarities and differences between the two diagrams. [See Cronin et al. (1979, p. 45).]

I-8. Give the electron configuration and the L, S, and J quantum numbers for the ground state of the potassium atom ($Z = 19$). Illustrate the Zeeman effect for this state graphically. [See Chen (1974, p. 263).]

I-9. Neglect spin-orbit coupling and work out the three lowest terms of the carbon and nitrogen atoms. [See Cronin et al. (1979, p. 45).]

I-10. Work out the quantum numbers L, S, J for (a) ground-state boron ($Z = 5$); (b) ground-state Na^+($Z = 11$); (c) ground-state Na^{2+}; (d) first excited state of Na^+; (e) ground-state H_2. [See Chen (1974, p. 276).]

I-11. (a) Write down the electronic configuration for the ground state of the Na atom ($Z = 11$), and give its spectroscopic notation.

(b) The lowest-frequency line in the absorption spectrum is a doublet. What are the notations for the two levels to which Na is excited in this process?

(c) What causes the splitting of the pair of energy levels? [See Chen (1974, p. 275).]

I-12. Use first-order perturbation theory to calculate the energy shift of the $1S$ state of the H atom that results if the nucleus is considered to be a uniformly charged spherical shell of radius 10^{-13} cm instead of a point charge. [See Chen (1974, p. 266).]

I-13. Explain why atomic hyperfine structure is smaller than atomic fine structure by approximately the ratio m/MZ, where m is the electron mass, M the proton mass, and Z the atomic number. [See Chen (1974, p. 277).]

I-14. What is the magnetic field at the nucleus that is produced by an electron in the $2p$ state of the H atom? [See Chen (1974, p. 263).]

I-15. The 1849-Å line of mercury splits into three components separated by 0.0016 Å when a magnetic field of 1000 G is applied. Is the Zeeman effect normal or anomalous? [See Cronin et al. (1979, p. 46).]

I-16. What is the Zeeman splitting of the spectral lines of the H atom in a magnetic field of 10,000 G? [See Chen (1974, p. 284).]

I-17. How strong would a magnetic field have to be to produce the Paschen–Back effect? [See Chen (1974, p. 266).]

I-18. An argon light source at 300 K emits radiation at 0.5 μm. At what pressure do the collision broadening and Doppler broadening have the same magnitude? Take the argon atoms to be solid spheres of radius 1 Å. [See Section 8-11 of the companion volume and Cronin et al. (1979, p. 47).]

I-19. What is the dependence of the radius of the K shell of an atom on the atomic number Z? [See Chen (1974, p. 286).]

I-20. Write out the simplest formula for the binding energy of a K-shell electron. Do **(a)** relativistic corrections, **(b)** screening by other electrons, and **(c)** the finite size of the nucleus increase or decrease the binding energy? [See Chen (1974, p. 268).]

I-21. Moseley's law gives the frequency v of the K_α X-ray line as $v^{1/2} = aZ - b$.

(a) Work out an approximate formula for a in terms of fundamental constants.

(b) Why do the frequencies of X-ray lines vary in such a simple way as Z is changed?

[See Chen (1974, p. 272).]

REFERENCES

Bates, D. R., Ed. (1961), *Quantum Theory*, Vol. 1, *Elements*, Academic Press, New York.

Bates, D. R., Ed. (1962a), *Quantum Theory*, Vol. 2, *Aggregates of Particles*, Academic Press, New York.

Bates, D. R., Ed. (1962b), *Quantum Theory*, Vol. 3, *Radiation and High Energy Physics*, Academic Press, New York.

Bethe, H. A., and E. E. Salpeter (1957), *Quantum Mechanics of One- and Two-Electron Atoms*, Academic Press, New York.

Born, M. (1967), *The Mechanics of the Atom*, Ungar, New York. (Reprint of the 1927 edition.)

Bransden, B. H., and C. J. Joachain (1983), *Physics of Atoms and Molecules*, Longman, Harlow, Essex, England.

Bransden, B. H., and C. J. Joachain (1989), *Introduction to Quantum Mechanics*, Wiley, New York.

Chen, M. (1974), *Berkeley Physics Problems with Solutions*, Prentice Hall, Englewood Cliffs, N.J.

Cohen-Tannoudji, C., B. Diu, and F. Laloë (1977), *Quantum Mechanics*, 2 vols., Wiley, New York.

Cohen-Tannoudji, C., J. Dupont-Roc, and G. Grynberg (1989), *Photons and Atoms*, Wiley, New York.

Cohen-Tannoudji, C., J. Dupont-Roc, and G. Grynberg, (1992), *Atom−Photon Interactions*, Wiley, New York.

Corney, A. (1977), *Atomic and Laser Spectroscopy*, Clarendon Press, Oxford.

Cronin, J. A., D. F. Greenberg, and V. L. Telegdi (1979), *University of Chicago Graduate Problems in Physics with Solutions*, Addison-Wesley, Reading, Mass.

Eisberg, R. M. (1961), *Fundamentals of Modern Physics*, Wiley, New York.

Eisele, J. A. (1969), *Modern Quantum Mechanics*, Wiley, New York.

Feynman, R. P. (1985), *QED*, Princeton University Press, Princeton, N.J.

Gleick, J. (1992), *Genius, The Life and Science of Richard Feynman*, Pantheon Books, New York. This excellent biography deals with the development of quantum electrodynamics, among other things, and provides an extensive bibliography.

Herzberg, G. (1944), *Atomic Spectra and Atomic Structure*, 2nd ed., Dover, New York.

Holstein, B. R. (1992), *Topics in Advanced Quantum Mechanics*, Addison-Wesley Publishing Company, Redwood City, CA.

Kusch, P. (1951), *Physica* **17**, 339.

Mizushima, M. (1970), *Quantum Mechanics of Atomic Spectra and Atomic Structure*, W. A. Benjamin, New York.

Morrison, M. A., T. L. Estle, and N. F. Lane (1976), *Quantum States of Atoms, Molecules, and Solids*, Prentice Hall, Englewood Cliffs, N.J.

Pais, A. (1989), *Phys. Today* **42** (12), 34.

Park, D. (1974), *Introduction to the Quantum Theory*, 2nd ed., McGraw-Hill, New York.

Rabi, I. I., S. Millman, P. Kusch, and J. R. Zacharias (1939), *Phys. Rev.* **55**, 526.

Ramsey, N. F. (1980), *Phys. Today* **33** (7), 25.

Ramsey, N. F. (1988a), *Z. Phys. D* **10**, 121.

Ramsey, N. F. (1988b), *Am. Sci.* **76** (1), 42.

Ramsey, N. F. (1990), *Rev. Mod. Phys.* **62**, 541.

Sard, R. D. (1970), *Relativistic Mechanics*, W. A. Benjamin, New York.

Scadron, M. D. (1979), *Advanced Quantum Theory*, Springer-Verlag, New York.

Slater, J. C. (1960), *Quantum Theory of Atomic Structure*, 2 vols., McGraw-Hill, New York.

Woodgate, G. K. (1980), *Elementary Atomic Structure*, 2nd ed., Clarendon Press, Oxford.

Zare, R. N. (1988), *Angular Momentum*, Wiley, New York.

APPENDIX II

DESIGNATION OF DIATOMIC MOLECULAR STATES

According to the Born–Oppenheimer approximation (Section 5-5-A), it is possible to treat the electrons in molecules at any instant as moving about fixed nuclei located at given distances from one another. By determining the energy levels U_n for the system, we find the *electron terms* for the molecule. In atoms the energy levels and electron terms are numbers. In diatomic molecules, however, the electron terms are not numbers, but functions of parameters, the internuclear distances. It should be pointed out that the energy U_n includes the mutual electrostatic interaction energy of the nuclei, so U_n is the total energy of the molecule for a given arrangement of the stationary nuclei.

In classifying terms of atoms, we utilize the quantized values of the total orbital angular momentum L of the atom (see Section 3 of Appendix I). In molecules, on the other hand, the total orbital angular momentum of the electrons is not conserved because the electric field of the array of nuclei is not spherically symmetric. However, in a diatomic molecule, the electric field has axial symmetry about an axis passing through the two nuclei. Accordingly, the projection of the orbital angular momentum on this axis is conserved, and we can classify electron terms according to the values of this projection. The symbol $\Lambda = 0, 1, 2, \ldots$ is used to denote the absolute values of the projected orbital angular momentum along the molecular axis. The terms with different values of Λ are labeled by the capital Greek letters corresponding to the Latin letters for the atomic terms with various L. Thus we speak of $\Sigma, \Pi, \Delta, \ldots$ terms for $\Lambda = 0, 1, 2, \ldots$, respectively.

Next we characterize each electron state of the molecule by the total spin S of all the electrons in the molecules. For $S \neq 0$, there is degeneracy $(2S + 1)$ with respect to the directions of the total spin, and this number is called the *multiplicity* of the term, as in the atomic case. A term with $\Lambda = 1, S = 1$ is written $^3\Pi$.

In addition to rotations through any angle about the axis, the diatomic molecular symmetry also allows a reflection in any plane passing through the axis. In such a reflection, the energy of the molecule is unchanged, but the final state differs from the initial state in that the sign of the angular momentum about this axis is changed. All electron terms with $\Lambda \neq 0$ are doubly degenerate; for each value of the energy there are two states that differ in the direction of the projection of the orbital angular momentum on the molecular axis. There is no degeneracy for $\Sigma (\Lambda = 0)$ states, for which the state of the molecule is not changed upon reflection. The wave function of a Σ term can only be multiplied by a constant as the result of the reflection, but since a double reflection in the same plane is an identity transformation, this constant is ± 1. Σ terms whose wave functions are unchanged by reflection are denoted Σ^+; those terms whose wave functions change sign are labeled Σ^-.

In the case of a homonuclear molecule, the identity of the two nuclei introduces a new symmetry and a new characteristic of the electron terms. Now there is a center of symmetry at the midpoint of the internuclear axis, which we take to be the origin. Here the Hamiltonian is invariant with respect to a simultaneous change of sign of the coordinates of all of the electrons in the molecule (the coordinates of the nuclei remaining unchanged.) Since the operator of this transformation commutes with the orbital angular momentum operator, we can classify terms with a given value of Λ according to their parity. Thus the wave functions of *even* (gerade) states are unchanged when the coordinates of the electrons change sign, whereas for *odd* (ungerade) states the wave functions change sign. The *parity*, g or u, is usually written as a suffix on the letter designating the term, as in Π_g.

The arguments used above (based on Landau and Lifshitz, 1965) do not suffice to designate states of polyatomic molecules. More involved symmetry principles and arguments based on group theory are brought into play here. See Landau and Lifshitz (1965), Morrison et al. (1976), and Tinkham (1964).

REFERENCES

Landau, L. D., and E. M. Lifshitz (1965), *Quantum Mechanics: Nonrelativistic Theory*, Pergamon Press, Oxford.

Morrison, M. A., T. L. Estle, and N. F. Lane (1976), *Quantum States of Atoms, Molecules, and Solids*, Prentice Hall, Englewood Cliffs, N.J.

Tinkham, M. (1964), *Group Theory and Quantum Mechanics*, McGraw-Hill, New York.

APPENDIX III

GENERAL REFERENCES AND SOURCES OF DATA

1. ABBREVIATIONS AND PUBLICATION DATA FOR SOME OF THE JOURNALS, REPORTS, AND SERIAL PUBLICATIONS CITED HEREIN

AAMP: Advances in Atomic and Molecular Physics, bound volumes published yearly by Academic Press, New York. Volumes 1–9 edited by D. R. Bates and I. Estermann; Vols. 10–25 edited by D. R. Bates and B. Bederson. Starting with Vol. 26, title changed to *AAMOP: Advances in Atomic, Molecular, and Optical Physics*, edited by D. R. Bates and B. Bederson.

AD:Atomic Data, journal published by Academic, New York.

ADNDT: Atomic Data and Nuclear Data Tables, journal published by Academic, New York. ADNDT came into existence in 1973 as a combination of AD and NDT, which ceased publication at that time.

Adv. Chem. Phys.: Advances in Chemical Physics, bound volumes published at irregular intervals by Wiley, New York.

AEEP: Advances in Electronics and Electron Physics, bound volumes published at irregular intervals by Academic, New York.

Annu. Rev. Phys. Chem.: Annual Reviews of Physical Chemistry, bound volumes published annually by Annual Reviews, Inc., Palo Alto, Calif.

Atomic Physics, bound volumes—see "ICAP" in Section 2.

HCI: International Conference on the Physics of Highly-Charged Ions. Papers of the Giessen Conference in 1990 were published in *Z. Phys.* **D-21** (1991). The 1992 meeting was held in Manhattan, Kansas, and the proceedings were published by AIP in 1993.

ICPEAC: see Section 2.

IPPJ-AM, reports on atomic and molecular processes published by Institute of Plasma Physics, Nagoya University, Nagoya 464, Japan.

JAERI-M, reports on atomic and molecular processes published by Japan Atomic Energy Research Institute, Tokai-muri, Naka-gun, Ibaraki-ken 319-11, Japan.

MEP: Methods of Experimental Physics, bound volumes published by Academic Press, New York, edited by various individuals.

NDT: Nuclear Data Tables, journal published by Academic Press, New York.

X-90: Fifteenth International Conference on X-Ray and Inner-Shell Processes (T. A. Carlson, M. O. Krause, and S. T. Manson, Eds.), Am. Inst. Phys. Conf. Proc. 215. Previous conferences had different titles, and their proceedings were published by different publishers.

2. MAJOR CONFERENCE SERIES

ICAP: *International Conference on Atomic Physics*. The invited papers presented at each conference are published in volumes entitled *Atomic Physics*. The conferences held to date are listed below together with information on the published volumes.

ICAP	1: New York (1968)	*Atomic Physics*	1(1969)
	2: Oxford (1970)		2(1971)
	3: Boulder (1972)		3(1973)
	4: Heidelberg (1974)		4(1975)
	5: Berkeley (1976)		5(1977)
	6: Riga (1978)		6(1979)
	7: Cambridge (1980)		7(1981)
	8: Göteborg (1982)		8(1983)
	9: Seattle (1984)		9(1985)
	10: Kyoto (1986)		10(1987)
	11: Paris (1988)		11(1989)
	12: Ann Arbor (1990)		12(1991)
	13: Munich (1992)		13(1993)

Volumes 1–8 published by Plenum Press, New York; Vols. 9–11 by World Scientific, Singapore; Vols. 12–13 by the American Institute of Physics.

ICPEAC: *International Conference on the Physics of Electronic and Atomic Collisions*. The conferences held to date are listed below.

ICPEAC	I:	New York (1958)	V:	Leningrad (1967)
	II:	Boulder (1961)		
	III:	London (1963)	VI:	Cambridge, Mass. (1969)
	IV:	Quebec (1965)		

ICPEAC   VII: Amsterdam (1971)

VIII: Belgrade (1973)

IX: Seattle, Wash. (1975)

X: Paris (1977)

XI: Kyoto (1979)

XII: Gatlinburg, Tenn. (1981)

XIII: Berlin (1983)

XIV: Stanford, Calif. (1985)

XV: Brighton, England (1987)

XVI: New York (1989)

XVII: Brisbane (1991)

Invited papers and abstracts of contributed papers for the more recent conferences are published by North-Holland, Amsterdam, or by the American Institute of Physics in separate volumes during the year immediately following each conference. The volumes are entitled *Electronic and Atomic Collisions*. Proceedings of the earlier conferences were published by the host institution and are not generally available. In this book an entry such as *ICPEAC XV*, Brighton (1987) refers to the bound volume of *invited* papers.

3. GENERAL REFERENCES

ADNDT (*Atomic Data and Nuclear Data Tables*). Contains extensive data compilations, for the most part, with periodic author and subject indices. Essentially all branches of atomic collisions and atomic and molecular structure are covered.

AAMP (*Advances in Atomic and Molecular Physics*). Published annually through 1988. Title changed in 1989 to *AAMOP* (see Section 1 of this Appendix). Contains reviews on practically all aspects of atomic and molecular physics.

CFADC (Controlled Fusion Atomic Data Center), Oak Ridge National Laboratory, Oak Ridge, Tennessee. The CFADC data compilations and bibliographies are listed in the bibliographies by McDaniel et al. cited below.

JILA Data Center (Joint Institute for Laboratory Astrophysics, University of Colorado, Boulder, Colorado). The JILA data compilations and bibliographies are listed in the bibliographies by McDaniel et al. cited below.

J. Phys. Chem. Ref. Data. Exhaustive compilations of data on selected topics in atomic and molecular structure and collisions.

E. W. McDaniel, M. R. Flannery, E. W. Thomas, and S. T. Manson, *ADNDT* **33**, 1–148 (1985). A selected bibliography on atomic collisions: data collections, bibliographies, review articles, books, and papers of particular tutorial value.

E. W. McDaniel and E. J. Mansky, "Guide to Bibliographies, Reviews, and Compendia of Data on Atomic Collisions," *AAMOP*, **31** (1993). An update of McDaniel et al. (1985).

AUTHOR INDEX

639

SUBJECT INDEX

655